Green Extraction and Valorization of By-Products from Food Processing

Green Extraction and Valorization of By-Products from Food Processing

Edited by

Francisco J. Barba
Elena Roselló-Soto
Mladen Brnčić
Jose M. Lorenzo

CRC Press
Taylor & Francis Group
Boca Raton London New York

CRC Press is an imprint of the
Taylor & Francis Group, an **informa** business

CRC Press
Taylor & Francis Group
6000 Broken Sound Parkway NW, Suite 300
Boca Raton, FL 33487-2742

CRC Press is an imprint of Taylor & Francis Group, an Informa business

Printed on acid-free paper

International Standard Book Number-13: 978-1-138-54404-8 (Hardback)

Visit the Taylor & Francis Web site at
http://www.taylorandfrancis.com

and the CRC Press Web site at
http://www.crcpress.com

Contents

Preface

Over the last decades, an increased public interest has been shown for the big challenge that constitutes food wastage in Europe. Generation of Agricultural Wastes and By-products (AWB) during the production, processing, and consumption of agricultural commodities is unavoidable. Either referring to nonedible parts of each commodity (e.g., straws, pruning, manure, and bones) or to edible parts and quantities that are removed or discarded so as to reassure quality and health appropriateness of the final product/food (e.g., peels, rotten fruits and vegetables, spoiled meat and milk), AWB are estimated to more than 700 million tons per year in Europe (Pavwelczyk, 2005). In this line, different normatives and resolutions have been adopted by the European Parliament 19 January 2012 (2011/2175(INI). Apart from its significant quantities, the physicochemical characteristics of the various AWB denote that there is immense potential for their reuse/recycle/valorization through various different processes.

The valorization or reuse of AWB includes many different processes and ways. For instance, straw can be used as fodder (with or without processing), bedding material, for energy exploitation through direct burning, or preferably through various processes such as pyrolysis, anaerobic digestion, and hydrothermal carbonization, as organic fertilizer/soil conditioner to replenish soil organic matter, or for the production of insulation panels or substrate for growing mushrooms. Waste disposal and by-product management in food processing industry pose problems in the areas of environmental protection and sustainability. However, they could be a great source of valuable nutraceuticals, which can be used to deal with the prospects of feeding fast growing population in the 21st century. Typical categories of bioactive compounds found in AWB include: carotenoids, anthocyanins, betalains, phenolic acids, monoterpenes, isoflavones, flavonols, stilbenes, procyanidins, isothiocyanates, glucosinolates, glycosides, peptides, fatty acids, steroids, alkaloids, terpenes, waxes, etc. Perspectives originated from the enormous amounts of food-related materials that are discharged worldwide and the existing technologies, which promise the recovery, recycling, and sustainability of high-added value ingredients inside food chain will be discussed in this book.

This possibility arises from the existing minimal technologies, which provide gentle extraction and recovery of healthy ingredients. Moreover, nonthermal technologies allow preserving the physicochemical and nutritional properties of by-products to be used in the formulation of new and traditional food products.

About the Editors

Dr. Francisco J. Barba is Doctor and Professor at the University of Valencia, Valencia, Spain. He holds a European PhD in Pharmacy (with distinction) from the University of Valencia and holds degrees in Pharmacy and in Food Science and Technology. He has more than 235 publications (SCOPUS; h-index: 38). He is an associate editor of the journals *Food Research International, Journal of Food Composition and Analysis, Journal of Food Processing and Preservation, Molecules*, and others.

Dr. Elena Roselló-Soto is a pharmacist. She obtained her MSc degree in Food Science and Food Safety and PhD in Pharmacy from the University of Valencia. She holds a degree in Pharmacy. Her research focuses on valorization of food wastes and by-products.

Dr. Mladen Brnčić is full-time Professor and Head of the Department of Process Engineering and Head of Laboratory for Thermodynamics at Faculty of Food Technology and Biotechnology, University of Zagreb, Croatia. He obtained his BS in Biochemical Engineering and MSc and PhD in Food Engineering from the University of Zagreb. His research output summarizes in over 100 peer-review publications and various international meeting and conference contributions as plenary and invited lectures.

Dr. Jose Manuel Lorenzo is Head of Research at the Meat Technology Centre of Galicia, Ourense, Spain. His research focuses on development of new product, processes, and food packaging. He has more than 260 publications (SCOPUS; h-index: 31). He is an associate editor of the *Food Research International, Journal of the Science Food and Agriculture, Food Analytical Methods*, and *Canadian Journal of Animal Science.*

Contributors

João Carlos Martins Barreira
Centro de Investigação de Montanha
Instituto Politécnico de Bragança, Campus de Santa Apolónia
Bragança, Portugal

Sonia Barba-Orellana
Centro Sanitario Integrado de Xirivella
Consorci Hospital General Universitari València
Xirivella, Spain

Paula Borrajo
Centro Tecnológico de la Carne de Galicia
Ourense, Spain

Erasmo Cadena
Vertech Group SAS
Nice, France
Université Nice Sophia Antipolis, IMREDD
Nice, France

Javier Carballo
Área de Tecnología de Alimentos, Facultad de Ciencias
Universidad de Vigo
Ourense, Spain

Tomás Landete-Castillejos
Animal Science Techniques Applied to Wildlife Management Research Group
Instituto de Investigación en Recursos Cinegéticos
Albacete Section of CSIC-UCLM-JCCM
Universidad de Castilla-La Mancha
Albacete, Spain

Sección de Recursos Cinegéticos y Ganaderos
Instituto de Desarrollo Regional of Universidad de Castilla-La Mancha
Albacete, Spain
Departamento de Ciencia y Tecnología Agroforestal y Genética
Escuela Técnica Superior de Ingenieros Agrónomos y Montes of Universidad de Castilla-La Mancha
Albacete, Spain

Juan A. Centeno
Área de Tecnología de Alimentos, Facultad de Ciencias
Universidad de Vigo
Ourense, Spain

Paulo Cezar Bastianello Campagnol
Universidade Federal de Santa Maria
Santa Maria, Rio Grande do Sul
Brazil

Maria G. Daskalaki
School of Medicine
University of Crete
Heraklion, Greece

Gabriela I. Denoya
Instituto Tecnología de Alimentos
Instituto Nacional de Tecnología Agropecuaria (INTA)
Hurlingham, Buenos Aires, Argentina
Consejo Nacional de Investigaciones Científicas y Técnicas (CONICET)
Argentina

Rubén Domínguez
Centro Tecnológico de la Carne de Galicia
Ourense, Spain

Cátia Dourado
QOPNA & LAQV-REQUIMTE, Department of Chemistry
University of Aveiro
Aveiro, Portugal

Ana I. Durán
Grupo de Reciclado y Valorización de Materiales Residuales (REVAL)\
Instituto de Investigacións Mariñas (IIM-CSIC)
Galicia, Spain

Noemí Echegaray
Área de Tecnología de Alimentos, Facultad de Ciencias
Universidad de Vigo
Ourense, Spain

Patricia Esquivel
Escuela de Tecnología de Alimentos
Universidad de Costa Rica
San Pedro, Costa Rica

Isabel Cristina Fernandes Rodrigues Ferreira
Centro de Investigação de Montanha
Instituto Politécnico de Bragança, Campus de Santa Apolónia
Bragança, Portugal

Juana Fernández-López
Grupo de Industrialización de Productos de Origen Animal (IPOA)
Universidad Miguel Hernández
Alicante, Spain

Mathilde Fiorletta
Vertech Group SAS
Nice, France

Javier Fraguas
Grupo de Reciclado y Valorización de Materiales Residuales (REVAL)\
Instituto de Investigacións Mariñas (IIM-CSIC)
Galicia, Spain

Cyrielle Garcia
QualiSud, Université de La Réunion, CIRAD
Université Montpellier, Montpellier SupAgro, Université d'Avignon
Sainte Clotilde, France

Kashif Ghafoor
Department of Food Science and Nutrition
King Saud University
Riyadh, Saudi Arabia

Belén Gómez
Centro Tecnológico de la Carne de Galicia
Ourense, Spain

Víctor M. Jiménez
CIGRAS/IIA
Universidad de Costa Rica
San Pedro, Costa Rica
Food Security Center
University of Hohenheim
Stuttgart, Germany

Lars Leonhardt
Chair of Food Engineering
Technische Universität Dresden
Dresden, Germany
Elea Vertriebs- und Vermarktungsgesellschaft mbH
Quakenbrück, Germany

Francisco J. Marti-Quijal
Nutrition and Food Science Area
Faculty of Pharmacy, Universitat de València, Avda.
València, Spain

Araceli Menduíña
Grupo de Reciclado y Valorización de Materiales Residuales (REVAL)\
Instituto de Investigacións Mariñas (IIM-CSIC)
Galicia, Spain

Jesús Mirón
Grupo de Reciclado y Valorización de Materiales Residuales (REVAL)\
Instituto de Investigacións Mariñas (IIM-CSIC)
Galicia, Spain

Andrés Moreno
University of Castilla-La Mancha
Faculty of Chemical Sciences and Technologies (San Alberto Magno Building)
Department of Organic Chemistry
Ciudad Real, Spain

Paulo E.S. Munekata
Centro Tecnológico de la Carne de Galicia
Ourense, Spain

Margarita Nogueira
Grupo de Reciclado y Valorización de Materiales Residuales (REVAL)\
Instituto de Investigacións Mariñas (IIM-CSIC)
Galicia, Spain

Oleksii Parniakov
Elea Vertriebs- und Vermarktungsgesellschaft mbH
Quakenbrück, Germany

Mirian Pateiro
Centro Tecnológico de la Carne de Galicia
Ourense, Spain

José Angel Pérez-Álvarez
Grupo de Industrialización de Productos de Origen Animal (IPOA)
Universidad Miguel Hernández
Alicante, Spain

Martina Pérez-Serrano
Animal Science Techniques Applied to Wildlife Management Research Group
Instituto de Investigación en Recursos Cinegéticos
Albacete Section of CSIC-UCLM-JCCM
Universidad de Castilla-La Mancha
Albacete, Spain
Sección de Recursos Cinegéticos y Ganaderos
Instituto de Desarrollo Regional of Universidad de Castilla-La Mancha
Albacete, Spain
Departamento de Ciencia y Tecnología Agroforestal y Genética
Escuela Técnica Superior de Ingenieros Agrónomos y Montes of Universidad de Castilla-La Mancha
Albacete, Spain

Fabienne Remize
QualiSud, Université de La Réunion, CIRAD
Université Montpellier, Montpellier SupAgro, Université d'Avignon
Sainte Clotilde, France

Suzana Rimac-Brnčić
Faculty of Food Technology and Biotechnology
University of Zagreb
Zagreb, Croatia

Luis Roca
Compostajes y Abonos Desco S.L.
València, Spain

Harald Rohm
Elea Vertriebs- und Vermarktungsgesellschaft mbH
Quakenbrück, Germany

Jorge A. Saraiva
QOPNA & LAQV-REQUIMTE, Department of Chemistry
University of Aveiro
Aveiro, Portugal

Stefan Toepfl
Elea Vertriebs- und Vermarktungsgesellschaft mbH
Quakenbrück, Germany

Igor Tomasevic
Department of Animal Source Food Technology
University of Belgrade, Faculty of Agriculture
Belgrade, Serbia

Jesús Valcárcel
Grupo de Reciclado y Valorización de Materiales Residuales (REVAL)\
Instituto de Investigacións Mariñas (IIM-CSIC)
Galicia, Spain

José Antonio Vázquez
Grupo de Reciclado y Valorización de Materiales Residuales (REVAL)\
Instituto de Investigacións Mariñas (IIM-CSIC)
Galicia, Spain

Manuel Viuda-Martos
Grupo de Industrialización de Productos de Origen Animal (IPOA)
Universidad Miguel Hernández
Alicante, Spain

Fang Wang
College of Food Science and Engineering
Wuhan Polytechnic University
Wuhan, China

Julian Witt
Elea Vertriebs- und Vermarktungsgesellschaft mbH
Quakenbrück, Germany

Zhenzhou Zhu
College of Food Science and Engineering
Wuhan Polytechnic University
Wuhan, China

1

Green Extraction

MARTINA PÉREZ-SERRANO AND TOMÁS LANDETE-CASTILLEJOS

Animal Science Techniques Applied to Wildlife Management Research Group, Instituto de Investigación en Recursos Cinegéticos, Albacete Section of CSIC-UCLM-JCCM, Universidad de Castilla-La Mancha, Campus Universitario sn, 02071, Albacete, Spain
Sección de Recursos Cinegéticos y Ganaderos, Instituto de Desarrollo Regional of Universidad de Castilla-La Mancha, Campus Universitario sn, 02071, Albacete, Spain
Departamento de Ciencia y Tecnología Agroforestal y Genética, Escuela Técnica Superior de Ingenieros Agrónomos y Montes of Universidad de Castilla-La Mancha, Campus Universitario sn, 02071, Albacete, Spain

LUIS ROCA-PÉREZ

Department of Vegetal Biology, Faculty of Pharmacy, Universitat de València, Avda. Vicent Andrés Estellés, s/n 46100 Burjassot, València, Spain

MLADEN BRNČIĆ AND SUZANA RIMAC-BRNČIĆ

Faculty of Food Technology and Biotechnology; University of Zagreb; Pierottijeva 6; Zagreb; Croatia

JOSE M. LORENZO

Centro Tecnológico de la Carne de Galicia, Rúa Galicia Nº 4, ParqueTecnológico de Galicia, San Cibrán das Viñas, 32900 Ourense, Spain

FRANCISCO J. MARTI-QUIJAL AND FRANCISCO J. BARBA

Nutrition and Food Science Area, Preventive Medicine and Public Health, Food Science, Toxicology and Forensic Medicine Department, Faculty of Pharmacy, Universitat de València, Avda. Vicent Andrés Estellés, s/n, 46100 Burjassot, València, Spain

Contents

1 Introduction

The legislative framework developed in several countries around the world prioritizes recycling and recovery of waste and by-products to elimination. For instance, the European Directive 2006/12/EEC (Anonymous 2006) sets out the importance of encouraging the recovery of waste/by-products and the use of recycled materials in order to preserve natural resources. In this framework, Directive 2008/98/EC (Anonymous 2008) on waste lays down measures to protect the environment and human health by preventing or reducing overall impacts of resource use and improving the efficiency of such use.

The cultivation and processing of the products obtained entail the generation of waste that can be valued as by-products. In this sense, Lal (2005) defined the residue of a crop "as the part of the inedible plant that remains in the field after harvest." Some researchers also include as such the organic remains that are produced in the crop treatment plants, or those that are discarded after the management of the crop. Recently, Alexander et al. (2017) quantified the losses, inefficiencies, and waste in the global food system. These authors concluded that, due to cumulative losses, the proportion of global agricultural dry biomass consumed as food is just 6% (9.0% for energy and 7.6% for protein) and of harvest biomass is 24.8% (31.9% for energy and 27.8% for protein). The highest rates of loss are associated with livestock production, although the largest absolute losses of biomass occur prior to harvest. Losses of harvested crops were also found to be substantial, with 44.0% of crop dry matter (36.9% of energy and 50.1% of protein) lost prior to human consumption. On the other hand, about 38% of food waste is produced during food processing (Helkar, Sahoo, and Patil 2016).

According to the US Environmental Protection Agency (EPA), food waste represents the single largest type of waste entering landfills (Nishida 2014). A large amount of food waste and by-products, as well as their microbial decomposition, creates huge environmental, human health, economic, and social problems (Helkar, Sahoo, and Patil 2016; Koubaa et al. 2015; Mourad 2016; Parniakov et al. 2015; Putnik et al. 2017b). In addition, the large cost for waste treatment is an additional charge for the food manufacturers. However, wastes and by-products derived from agri-food industry contain large amounts of high-added-value compounds such as nutrients and bioactive compounds (e.g., vitamins, polyphenols, and glucosinolates/isothiocyanates), among others that have beneficial effects attributed, mainly, to their antioxidant power (Barba, Esteve, and Frígola 2014; Barba et al. 2016b; Deng et al. 2015; Granato, Nunes, and Barba 2017; Putnik et al. 2017a; Roselló-Soto et al. 2016; Vinceković et al. 2017). Food by-products have numerous applications in several industries such as pharmaceutical, biomedicine, cosmetic, paper industry, as well as human and animal nutrition, among others. In fact, a healthy trend in the food industry consists of converting the food wastes or by-products to functional food ingredients (Helkar, Sahoo, and Patil 2016).

In order to valorize the waste and by-products from different agri-food industries, it is important to evaluate first the amount produced, as well as the microbial load they present and the added value they can present from an economic point of view, especially due to their content in high-added-value compounds (Barba et al. 2015a, 2015b, 2016a; Barba and Orlien 2017; Putnik et al. 2017b).

Most of the wastes and by-products that have been selected in this book to extract high-added-value compounds, among other criteria, have been chosen for their remarkable production worldwide. **Table 1.1** shows the global production data of some of the crops obtained from FAO in 2015. The waste generated by the crops can be estimated by multiplying the annual production by the values of residue-production-ratio (RPR) for each crop. The proportion between the harvested product and the waste can vary depending on a set of factors such as the variety of crop, the supply of water and nutrients during cultivation, the water content of the waste, the use of chemical growth regulators, environmental factors of each zone, harvested method, and so on (Abou Hussein and Sawan 2010; Eisentraut 2010; Singh 2015). The RPR depends on the crop and the approximate amount of waste generated worldwide for some of the waste generated.

Table 1.1 World production of crops in 2015

MATRIX	WORLD 2015 PRODUCTION (MILLIONS TONS)	RPR[1]	ESTIMATED RESIDUE PRODUCTION (MILLIONS TONS)
Cereals total	2,795	1.5 Rice and wheat (Lal 2005) 2.0 Maize (Iye and Bilsborrow 2013)	4,751
Sweet potato	103	0.25 (Iye and Bilsborrow 2013)	26
Olive oil	3	1.5 (Eisentraut 2010)	–
Nuts with nutshell	3	–	–
Wine	29	0.25 (OIV 2015)	7.25
Oilseed (soya+rapeseed +sunflower+peanut+corn +sesame)	98	1.5 (Eisentraut 2010)	147
Coffee green	8.8	2.1 (Duku, Gu, and Hagan 2011)	18.5
Leaf vegetable	287	1.25 (Eisentraut 2010)	359

[1] Residue-production-ratio (RPR)

Most of the waste from crops is produced by cereals. Other important food by-products are pulses, fruits, vegetables, meat, dairy, marine, sugarcane, winery, and plantation waste. In particular, production and applications of waste and by-products originated from fruit and vegetable, grain proccesing, coffee, winery, marine (including fish and algae), meat, and dairy industries are reviewed in this book. Moreover, other by-products have been selected due to their importance at the local level such as tiger nuts (*Cyperus esculentus*). Given their high global annual production, to deliver insight into the current state of art and innovative extraction technologies to extract valuable products from food processing by-products is necessary (Koubaa et al. 2015; Roselló-Soto et al. 2018a, 2018b).

Conventional methods of by-products' disposal, animal feed, soil fertilizers, and/or biofuels do not provide the industry with the sufficient economic incentive to increase their competitiveness and address environmental problems, which is aggravated by legal restrictions. Traditionally, different techniques have been used for the recovery of bioactive compounds from waste and by-products of the agri-food industry, such as milling, extraction with solvents, etc. (Misra et al. 2015, 2017a; Roselló-Soto et al. 2015b). However, this type of techniques, generally, uses a large amount of solvents, which in many cases can be toxic. Moreover, they also need long extraction times to obtain maximum performance, which supposes an additional operating cost. That is why research is necessary to find new alternative technologies to eliminate toxic solvents, reduce operating costs with the maximization of yields, and use techniques that are ecological, thus generating a great benefit for both industries and the society. Innovative extraction technologies are, in general, more "green" technologies than conventional ones for the extraction of bioactive compounds. As global waste production rates are expected to continue rising during the following decennium, the development and selection of sustainable waste management solutions become more and more interesting (Six et al. 2016).

Different classifications for waste and by-products from agri-food industries are available. Food by-products can be classified into the following six categories: (a) crop waste and residues; (b) fruit and vegetables by-products; (c) sugar, starch, and confectionary industry by-products; (d) oil industry by-products; (e) grain and legume by-products; and (f) distilleries and breweries by-products (Ajila et al. 2012). In the current book, waste and by-products from food industries will be categorized, based on Helkar et al. (2016) classification, as follows: (a) fruit and vegetable (including olive oil), (b) grain processing, (c) winery (includes also brewery), (d) marine, (e) meat, and (f) dairy industry.

2 Main Wastes and By-Products from Food Industries

2.1 Fruit and Vegetable By-Products

Vegetable and fruit processing industries produce millions of tons of by-products such as peels, seeds, stones, residual pulp, and discarded pieces that generate important environmental management problems for the industry (Barba et al. 2017b; Parniakov

Table 1.2 Percentage of fruit and vegetable wastes and by-products according to the origin (AWARENET 2004; De Las Fuentes et al. 2004).

PRODUCTION PROCESS	WASTE AND BY-PRODUCTS (%)
White wine production	20–30
Red wine production	20–30
Fruit and vegetable juice production	30–50
Fruit and vegetable processing and preservation	5–30
Vegetable oil production	40–70
Sugar production from sugar beet	85

et al. 2014, 2015, 2016a, 2016b). As shown in Table 1.2, the percentage of fruit and vegetable wastes and by-products varies from 5% to 85% depending on the origin (AWARENET 2004; De Las Fuentes et al. 2004). As average, the production of likely waste is estimated to be approximately 30% of the processed material for most fruits and vegetables. Tropical and subtropical fruits processing has considerably higher ratios of by-products than the temperate fruits (Barba et al. 2017b).

The reuse of vegetable and fruit by-products to obtain biologically active phytochemicals (phenolic compounds, carotenoids, minerals, dietary fiber, etc.), which can be used as natural functional ingredients that include antitumoral, antibacterial, cardioprotective, and antimutagenic activities, could be a solution to the economic and environmental problems involved in removing vegetable and fruit processing waste. Moreover, the use of vegetable and fruit processing by-products as a source for obtaining phytochemicals, which can be used in the design of new functional foods, is a topic of great interest today (Barba, Boussetta and Vorobiev 2015; Barba et al. 2015; Koubaa et al. 2016; Roselló-Soto et al. 2015a).

The waste obtained from fruits and processing industry is extremely diverse due to the use of a wide variety of fruits and vegetables, the broad range of processes, and the multiplicity of the product (William 2005). In general, fruit and vegetable by-products are low in fat and rich in fibers (gastrointestinal system health), minerals, and antioxidants (prevention of degenerative diseases). In fact, in some cases, antioxidants present a higher activity in the residues than in the pulp (Gorinstein et al. 2001). Fiber derived from fruit wastes (95% skin or pulp tissues) is used as food ingredients (enriched food products), for functional foods, and also for animal nutrition. Moreover, pectins are used in pharmaceutical industries as drug carriers and excipients. The main functional ingredients from fruit and vegetable by-products of higher consumption (apple skin/pomace, grape pomace/skin/seeds, carrot peel, potato peel, red beet, and olive pomace) are polyphenols that are antioxidant compounds with antihypertensive, anticancer, antidiabetic, and hypolipidemic activities (Rabetafika et al. 2014). Polyphenols are commonly used as additives in foods to prevent peroxidation and also for protecting against oxidative damage in living systems by scavenging oxygen-free radicals as well as nutritional supplements (Barba, Esteve and Frigola 2013, 2014; Horita et al. 2018; Lorenzo et al. 2018, 2018; Makris, Boskou, and Andrikopoulos 2007). In addition, hesperidin from orange peel; carotenoids from orange pulp, carrot peel, and tomato skin (lycopene); unsaturated fatty acids from tomato seeds; and phytosterols from sunflower seed and soybean seed can also be obtained (Galanakis 2012; Oreopoulou and Tzia 2007; Putnik et al. 2017c; Roselló-Soto et al. 2015a). Fruit and vegetable by-products in industrial applications have been recently extensively reviewed (Cheok et al. 2018; Galanakis et al. 2015; Gómez and Martinez 2017; Kasapidou, Sossidou, and Mitlianga 2015; Kodagoda and Marapana 2017; Okino Delgado and Fleuri 2016; O'Shea, Arendt, and Gallagher 2012). **Table 1.3** shows the main components and their industrial applications for fruit and vegetable by-products.

Recently, many research studies have focused on innovative technologies that promise to overcome the issues of convential techniques such as time-consuming, inefficiency, and low yield, among others (Barba et al. 2017a; Ferrentino, Asaduzzaman, and Scampicchio 2018; Misra et al. 2015, 2017a, 2017b; Ran et al. 2018; Ribas-Agustí et al. 2017), such as sub/supercritical, ultrasound-, and enzyme-assisted extractions (Ben Rahal et al. 2015; Koubaa et al. 2017b; Poojary et al. 2016; Roselló-Soto et al. 2016; Zhu et al. 2017). The main advantages of ultrasonic-assisted extraction (UAE) and (microwave-assisted extraction (MAE) consist in large reduction in extraction time, higher yield of oils and active principles, improved selectivity, and higher stability and organoleptic quality of the extracts (Bouras et al. 2015; Carbonell-Capella et al. 2017; Koubaa et al. 2016; Roselló-Soto et al. 2015; Zhu et al. 2016a, 2016b). Each matrix, however, needs to have its operating conditions carefully optimized to achieve high quality standards and yields.

By-products from coffee processing can also be considered as potential functional ingredients for the food industry. The coffee husks, peel, and pulp, which comprises nearly 45% of the cherry, are one of the main by-products of coffee agro-industry and might be a valuable material for several purposes, including extraction of caffeine and polyphenols (Bondesson 2015). Other by-products originating from coffee processing, which have been less studied, are the mucilage and the parchment. These compounds also might have a high potential as a source of important ingredients as well (Esquivel and Jiménez 2012; Pérez-Sariñana and Saldaña-Trinidad 2017) (**Figure 1.1**). Moreover, the spent coffee grounds composting seems to be a practical approach that offers an alternative and direct reuse for this by-product, extendable to other crops, providing value-added vegetable products (Cruz 2014).

Furthermore, other by-products have been selected due to their importance at Mediterranean or at the local level as, for example, olive oil and tiger nuts (*Cyperus esculentus*). Olive tree culture is especially important in the Mediterranean area, Spain being first about total culture surface and number of productive trees. The use of residues and by-products from the olive oil production chain has been extensively studied (Anastopoulos, Massas, and Ehaliotis 2015; Koubaa et al. 2017a;

Table 1.3 Main components and industrial applications of fruit and vegetable by-products (Source: own ellaboration based on Helkar et al. (2016) and Kodagoda and Marapana (2017))

FRUIT OR VEGETABLE SUBPRODUCT	PART	COMPOUND	APPLICATIONS
Apple	Pomace[1]	Phlorizin (phenolic compound)	Component of an oral antidiabetic drug class. Substitution of manufactured azocolors for foods, some of which have been associated with health risks
		Polyphenolic extracts (antioxidant)	Prevent colon cancer in vitro[2]
		Pectin (10–15% w/w dry basis)	Food ingredient. Drug carrier and excipient for medicines
		Non starch polysaccharides: 36.5% insoluble fiber and 14.6% soluble fiber	Enriched bakery products replacing wheat flour
		Others: chlorogenic acid, phloretin glycosides, quercetin glycosides, catechins, and procyanidisms	Functional foods
	Skin	Polyphenolic extracts (antioxidant)	Human health. Food industry
Pomegranate	Skin and seed	Polyphenols, including primarily hydrolyzable ellagitannins, anthocyanins, and other polyphenols	Animal nutrition. Hypocholesterolemic. Supplemental foods as antioxidants. Nutraceutical properties
Grape	Pomace (includes stems, seeds, and skins)	Anthocyanin-based pigments	Substitution of manufactured azocolors for foods, some of which have been associated with health risks
		Phenolic extracts linked to fiber	Antioxidant. Prevent cancer and cardiovascular diseases
		Fiber	Functional ingredient in bakery products. To reduce rancidity on ice storage of seafood. Alternative fining agents for red wines. To remove red wine tannings. To increase the dietary fiber and total phenolic content in dairy products. To delay lipid oxidation in yogurt and salad dressings
		Oil (mainly for seed): oleic and linoleic acids, phenolic extracts	Antioxidant, hepatoprotective, neuroprotective, liver cholesterol reduction ability, meat industry
Citrus	Skin	Flavanones, polymethoxylated flavones, flavonols, and phenolic acid	Antioxidants in pharmaceutical, biotechnological, and food industries
		Oil (lime and lemon)	Aroma flavor for beverages. Food (e.g., preserving mozzarella cheese by the inhibitory effect on the microorganisms responsible for spoilage phenomena without affecting the functional microbiota of the product). Pharmaceutical (to mask unpleasant tastes of drugs)

(*Continued*)

Table 1.3 (Cont.)

FRUIT OR VEGETABLE SUBPRODUCT	PART	COMPOUND	APPLICATIONS
Banana	Skin	Phenolic, anthocyanin, carotenoids, catecholamines, sterols, and triterpenes	Low calorie food products (fiber-enriched food). Heavy metals sorption removing chromium (III) and (IV)
		Polymers as lignin, hemicellulose, and pectin	Synthesis of nanoparticles displayed antimicrobial activity toward pathogenic fungi and most of the tested bacterial cultures
Granadilla	Pomace	Pectins	Anticytotoxic compounds
Mango	Skin	Carotenoids	Antioxidants for human health. Colorants for food industry
		Fiber	Decrease intestinal absorption of cholesterol and its level in blood
	Kernel (powder)	Oil (oleic and linolenic fatty acids)	Health benefits
		Flavonoids, terpenes, tannins, and coumarins	Antimicrobial and antifungal activity
Papaya	Skin and seed	Papain (enzyme with strong proteolytic activity)	Meat tenderizers
	Petioles		Synthesis of oleic acid esters
Pineapple	Skin	Sugars	Nutrient in fermentation processes: substrate for methane, ethanol, hydrogen generation
		Carbohydrates and proteins	Biogas generation
		Ferulic acid	Precursor for vanillic acid
	Core	Fiber	Production of pineapple juice concentrates, beverages, and vinegar Fiber-enriched food products Production of lactic and citric acids
	Stem	Bromelain (enzyme with strong proteolytic activity)	Meat tenderizer. Bread dough improver. Fruit anti-browning agent. Beer clarifier. Tooth whitening agent. Animal feed
Artichoke	Leaves, outside bracts, and stems	Bioactive phenolic compounds	Hepatoprotective Anti-HIV, anticarcinogenic, antibacterial, bile-expelling, antioxidative and urinative activities Ability to inhibit cholesterol biosynthesis and LDL oxidation Nutraceuticals
		Inulin (low calorie fiber)	Production of fat reduced foods
		Cynarin (1,3-O-dicaffeoylquinic acid)	Liver protestations
Tomato	Pulp and skin	Polysaccharides (similar to xyloglucan biopolymer)	Antioxidant activity
Onion	Tunic and the first two layers	Thiol compounds	Polyphenol oxidase enzyme inhibitory effect. Antioxidant and antibrowning properties. Nutraceuticals. Antibacterial and anti-inflammatory properties

(Continued)

Table 1.3 (Cont.)

FRUIT OR VEGETABLE SUBPRODUCT	PART	COMPOUND	APPLICATIONS
Sunflower[3]	Hull and seeds without hull	Protein	Alternative to soy and egg protein as nourishment protein being devoid of toxic substances and low in antinutrients Oil, hull, and flour for animal nutrition
Coconut pulp[4]			Free milk, no lactose, low fat, and no cholesterol food

[1] 25% of the original fruit mass.
[2] McCann, Gill, Brien, Rao, Mcroberts, Hughes, Mcentee, and Rowland (2007).
[3] De-oiled press cake.
[4] By-product from water coco industry.

Moubarik, Barba, and Grimi 2015; Putnik et al. 2017b; Roselló-Soto et al. 2015b; Şahin et al. 2017, 2018). One important alternative from a quantitative point of view is their utilization as a source of nutrients for animal nutrition, in particular for small ruminants' nutrition (Molina Alcaide and Nefzaoui 1996). Recently, Guermazi et al. (2017) have proposed a new facility composed of a destoner and a two-phase extraction system to improve the working capacity and contribute to decreasing the environmental impact during all the processes of olive oil production. The proposed new system produces, on the one hand, an olive pulp that gives an extra virgin olive oil rich in natural antioxidants and an olive paste with a smoky taste for food consumption, and on the other hand, stones that give an

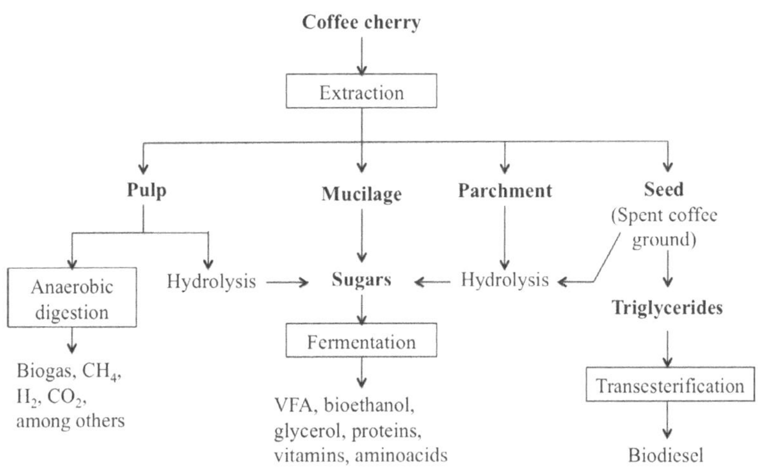

Figure 1.1 Applications for pulp, mucilage, parchment, and seed of coffee (Source: (Pérez-Sariñana and Saldaña-Trinidad 2017)).

olive oil for cosmetic application and a raw material to produce energy. The tiger nut's health aspects, composition, properties, and food applications were revised by Sánchez-Zapata, Fernández-López, and Angel Pérez-Alvarez (2012). Tiger nut is a weed plant (yellow nut sedge) of tropical and Mediterranean regions. Its sweet almond-like tubers are highly appreciated for their health benefits and nutritive value: high content of fiber, proteins, and sugars. They are rich in oleic acid and glucose, as well as in P, K, and vitamins C and E. In Spain, these tuberous "nuts" are mainly used to manufacture a milky beverage called "horchata de chufa." Tiger nut has attracted very little scientific and technological interest, except for the production of "horchata de chufa" and some studies on its oil (Roselló-Soto et al. 2018a). The development of new products from the tubers could enhance more interest in this crop. In this respect, various opportunities are offered: source of dietary fiber, use of its oil in cooking or salad preparation, and production of caramel to be used as a food additive. Recently, Roselló-Soto et al. (2018b) have studied the tiger nut and its by-products valorization from extraction of oil and valuable compounds to the development of new healthy products. This review provided an overview of some of the most relevant innovative processing technologies to allow the industrial sustainability and green recovery of oil from tiger nuts and their by-products, and tries to expose potential avenues for future research in commercial exploitation of tiger nuts and its by-products as source of ingredients to be incorporated in new food matrices to improve their technological and functional aspects.

Novel scientific and alternative technologies should be used to extract the optimum levels of bioactive compounds as well as other compounds of economic importance from the fruit wastes (Kodagoda and Marapana 2017). The combined effort of waste minimization and sustainable utilization of the by-products would substantially reduce the large quantities of fruit wastes accumulated globally.

2.2 Grain Processing Industry By-Products

Cereal by-products represent abundant and low-cost resources of phytochemicals with potential nutraceutical and pharmaceutical applications as it was recently reviewed by Dapćević-Hadnadev, Hadnadev, and Pojić (2018). γ-Oryzanols found in rice bran as well as in rice bran oil have 10 times higher antioxidant properties than tocopherols. Corn bran is one of the best sources of the potent antioxidant ferulic acid, compared to other cereals, fruits, and vegetables. Sorghum bran is a unique dietary source of 3-deoxyanthocyanidin, a rare type of flavonoid that has shown strong cytotoxic activities. Wheat and rye bran contain arabinoxylans, which contribute toward reduction in blood glucose, while oat bran is a starting raw material for the extraction of dietary fibers—β-glucans (Dapćević-Hadnadev, Hadnadev, and Pojić 2018). **Figure 1.2** shows a schematic diagram of the main industrial processing of wheat, corn, rice, and barley products and their by-products (Elmekawy et al. 2013). Due to the high amount produced, these wastes

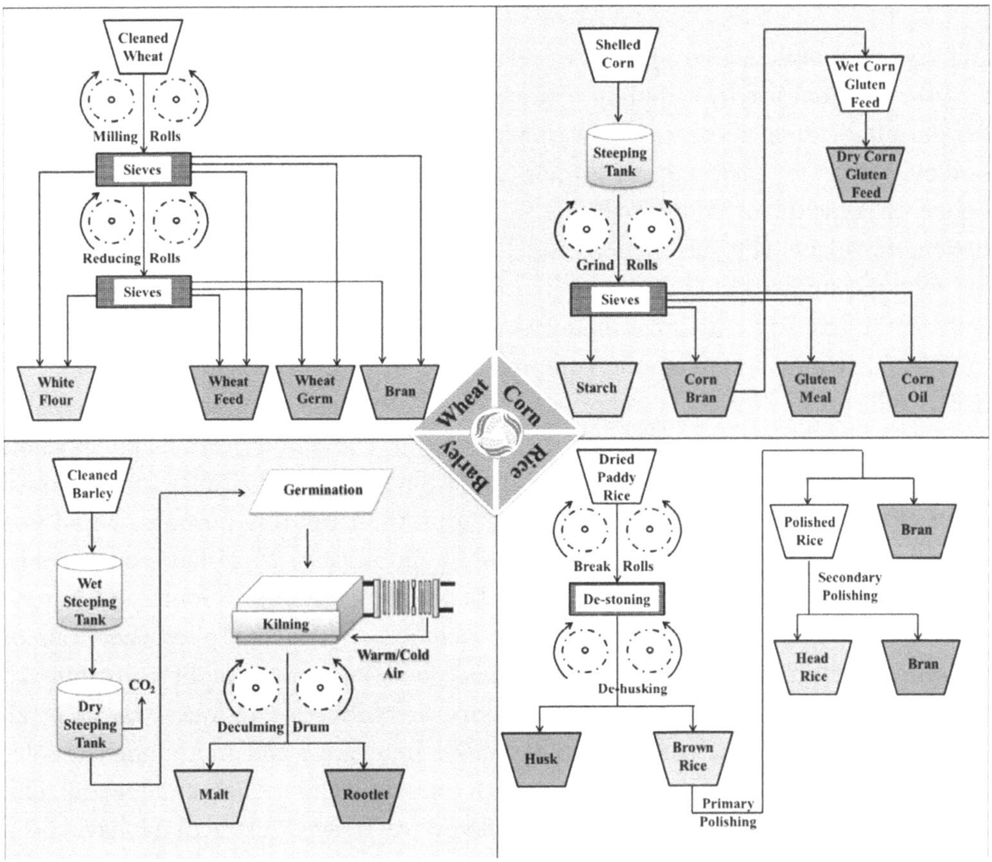

Figure 1.2 Schematic diagram of the main industrial processing of cereals illustrating products (green shade) and by-products (red shade) (Source: Elmekawy et al. (2013)).

represent serious economic and environmental challenges. However, they contain extensive amounts of potentially reusable materials and energy.

Cereal valorization approaches are divided into biochemical (organic acids, enzymes, vanillin, and antibiotics) and bioenergetic (biohydrogen, bioethanol, biobutanol, biogas, and bioelectricity) (Helkar, Sahoo, and Patil 2016). Some of these by-products have a special importance in pharmaceutical and medicinal industries such as in the case of rice by-products whose anticancer effects have been widely demonstrated (Esa, Ling, and Peng 2013). In addition, the reutilization of cereal processing by-products in bread making has been recently evaluated (Duţă, Culeţu, and Mohan 2018). Conventionally, various techniques have been used for the valorization of cereals by-products, and the available information in this regard is very extensive. Moreover, the extrussion processing of some raw food materials and by-products (Nikmaram et al. 2017), including cereals, is an alternative to conventional methods as it was recently reviewed by Offiah et al. (2018).

2.3 Winery Industry By-Products

This socioeconomic activity generates a large amount of solid waste (up to 30% w/w of the material used), including vine prunings, grape stalks, pomace, seeds, yeast lees, tartrate, carbon dioxide, and wastewater, which become fertilizer, animal feed, or fuel. However, their most common application is in food industry based on their high antioxidant content (Barba et al. 2016b) that are much safer than synthetic antioxidants (Helkar, Sahoo, and Patil 2016). Teixeira et al. (2014) reviewed the potential numerous health benefits of winery wastes polyphenols with antioxidant, antimicrobial, anti-inflammatory, anticancer, and cardiovascular protection activities, in support of phenolic compound use in pharmaceutical, food, and cosmetic industries. However, there are several constraints in currently available options for reusing these unprofitable materials. For example, certain polyphenols found in winery by-products are known to be phytotoxic and display antimicrobial effects during composting, impairing their utilization for this purpose. Regarding their use in livestock feed, some animals show intolerance to certain components, such as condensed tannins, which negatively affect digestibility (González-Centeno et al. 2014). Hence, their valorization as a source of bioactive phytochemicals of application in pharmaceutical, cosmetic, and food industries might constitute an efficient, profitable, and environment-friendly alternative for residues (Barba et al. 2016b; Makris, Boskou, Andrikopoulos 2007). Innovative and more efficient solvents and extraction methods such as high pressure and temperature extraction, supercritical fluids, or ultrasound and microwave-assisted extractions have been reported in an attempt to enhance the efficiency of the extraction of phytochemicals from vinification residues (Delgado-Torre et al. 2012; Ju and Howard 2003). However, the use of alternative/nonconventional processes/technologies to the extraction and purification of high-added-value compounds from by-products of the winemaking chain is possible (Barba et al. 2015, 2016b).

2.4 Marine Industry By-Products

Marine industry includes fish (oils and proteins), macro- and microalgae, among other products. By-products from seafood processing may account for up to 80% of the weight of the harvest depending on the species (Helkar, Sahoo, and Patil 2016), and include a variety of constituents with important added value as nutraceuticals, bioactive compounds, and functional food ingredients (Gormley 2013). These include ω-3 polyunsaturated fatty acids (PUFAs) from the livers of white lean fish, waste flesh parts of fatty fish, blubber of marine animals, hydrolysates from fish guts/cleanings, peptides, and products from crustaceans such as chitosan, chitosan oligomers, and glucosamines. In addition, shells from crustaceans, fish bones, and shark cartilages are extensively used for obtaining Ca for supplements in human (Kadam and Prabhasankar 2010) and animal nutrition (Safaa et al.

2008). On the other hand, marine algae are well-known natural sources of gums, such as alginate, agar, and carrageenan. Once their usefulness were in the food, pharmaceutical, and other industries because of their high content of fiber, antioxidants, and carotenoids.

Based on the evidence of potential health benefits, bioactive peptides derived from fish by-products have promising applications as natural nutraceuticals. Until now, however, a limited number of cell-based as well as in vivo studies on antiproliferative and antioxidant activity of peptides from fish by-products have been performed. Further research on the utilization of fish by-products for treatment and management of cancer is essential to improve our understanding of its mechanism and application (Nurdiani et al. 2017).

2.5 Meat Industry By-Products

Meat by-products (including carcasses, skins, bones, meat trimmings, blood, fatty tissues, horns, feet, hoofs, or internal organs) are rich in lipids, carbohydrates, and proteins. Bioactive peptides can be obtained from meat proteins utilizing hydrolysis, cooking, or fermentation. These bioactive peptides are known to have numerous activities (Lafarga and Hayes 2014) such as antimicrobial, antioxidative, antithrombotic, antihypertensive, anticarcinogenic, satiety regulating and immunomodulatory activities that may affect the cardiovascular, immune, nervous, and digestive systems. Peptides may also be effective in the treatment of mental health diseases, cancer, diabetes, and obesity (Lafarga and Hayes 2014). Recently, Chernukha et al. (2017) have confirmed that meat products based on porcine hearts and aortas ameliorates serum lipid profile and inflammation in hyperlipidemic rats.

2.6 Dairy Industry By-Products

Whey is the main by-product generated by the dairy industry. It contains several profitable constituents, particularly proteins, for example, β-lactoglobulin, α-lactalbumin, immunoglobulin, bovine serum albumin, lactoferrin, and lactoperoxidase (Ramos et al. 2015) with numerous applications including animal nutrition (Shershenkov and Suchkova 2015). Whey is widely accepted as a food ingredient in a few sustenance details (e.g., confectionery, bakery, health, and sports supplements), normally in dry form. Whey items have pertinent nutritious (e.g., high content of essential amino acids), functional (e.g., gelation, foaming, and emulsifying agent), and biological (e.g., antimicrobial, anticarcinogenic, and immunomodulatory activities) properties for well-being. The enriched whey can be used for the production of functional foods for meeting consumers' demands. Curd is another type of by-product generated by the milk industry. It acts as a probiotic functional food. Moreover, potential uses of either crude or purified caseins include the production of non-food biomaterials such as plastics, adhesives, gels, composites, and films (Ryder et al. 2017)

Scientific information about the applications of waste and by-products from the food industry are very extensive. The objective of this book is to collect the most recent information about the new innovative technologies that allow to improve the extraction yields of these by-products, even improving their stability and the food security.

References

Abou Hussein, S D, and O M Sawan 2010. "The Utilization of Agricultural Waste as One of the Environmental Issues in Egypt (a Case Study)". *Journal of Applied Sciences Research* 6 (8): 1116–24.

Ajila, C M, S K Brar, M Verma, and U P Rao 2012. *Environmental Protection Strategies for Sustainable Development*. In edited by A Malik and E Grohmann, 65–109. Springer Science/Business Media: Dordrecht, The Netherlands.

Alexander, P, C Brown, A Arneth, J Finnigan, D Moran, and M D A Rounsevell 2017. "Losses, Inefficiencies and Waste in the Global Food System". *Agricultural Systems* 153: 190–200.

Anastopoulos, I, I Massas, and C Ehaliotis 2015. "Use of Residues and by-Products of the Olive-Oil Production Chain for the Removal of Pollutants from Environmental Media: A Review of Batch Biosorption Approaches". *Journal of Environmental Science and Health - Part A Toxic/Hazardous Substances and Environmental Engineering* 50 (7): 677–718.

Anonymous. 2006. "Directive 2006/12/EC of the European Parliament and of the Council of 5 April 2006 on Waste". *Official Journal of the European Union* 114: 9–21.

Anonymous. 2008. "Directive 2008/98/EC of the European Parliament and of the Council of 19 November 2008 on Waste and Repealing Certain Directives". *Official Journal of the European Union* 312: 3–30.

Anonymous. 2017. *FAOSTAT Agricultural Production*. www.fao.org/faostat/es) (Consulta: 20 december 2017).

AWARENET 2004. *Handbook for the Prevention and Minimisation of Waste and Valorisation of by-Products in European Agro-Food Industries*. Edited by AWARENET. Grahamstown, South Africa. Gaiker Centro Technologico: Vizcaya, Spain.

Barba, F J, N Boussetta, and E Vorobiev 2015. "Emerging Technologies for the Recovery of Isothiocyanates, Protein and Phenolic Compounds from Rapeseed and Rapeseed Press-Cake: Effect of High Voltage Electrical Discharges". *Innovative Food Science and Emerging Technologies*: 31. 10.1016/j.ifset.2015.06.008.

Barba, F J, S Brianceau, M Turk, N Boussetta, and E Vorobiev 2015. "Effect of alternative physical treatments (Ultrasounds, Pulsed Electric Fields, and High-Voltage Electrical Discharges) on selective recovery of bio-compounds from fermented grape pomace". *Food and Bioprocess Technology*. 10.1007/s11947-015-1482-3.

Barba, F J, M J Esteve, and A Frigola 2013. *Gallic acid as a useful indicator of the antioxidant capacity of liquid foods?. Handbook on gallic acid: Natural occurrences, antioxidant properties and health implications*. Nova Science Publishers, Inc.: Hauppauge, New York, USA.

Barba, F J, M J Esteve, and A Frigola 2014. "Bioactive Components from Leaf Vegetable Products". *Studies in Natural Products Chemistry* 41: 321–46.

Barba, F J, C M Galanakis, M J Esteve, A Frigola, and E Vorobiev 2015. "Potential Use of Pulsed Electric Technologies and Ultrasounds to Improve the Recovery of High-Added Value Compounds from Blackberries". *Journal of Food Engineering* 30 (4): 455–64.

Barba, F J, L R B Mariutti, N Bragagnolo, A Z Mercadante, G V Barbosa-Cánovas, and V Orlien 2017a. "Bioaccessibility of Bioactive Compounds from Fruits and Vegetables after Thermal and Nonthermal Processing". *Trends in Food Science & Technology* 67: 195–206. 10.1016/j.tifs.2017.07.006.

Barba, F J, and V Orlien 2017. "Processing, Bioaccessibility and Bioavailability of Bioactive Sulfur Compounds: Facts and Gaps". *Journal of Food Composition and Analysis* 61: 1–3. 10.1016/j.jfca.2017.05.013.

Barba, F J, V Orlien, M J Mota, R P Lopes, S A Pereira, and J A Saraiva 2016a. "Implementation of Emerging Technologies". *Innovation Strategies in the Food Industry: Tools for Implementation*. 10.1016/B978-0-12-803751-5.00007-6.

Barba, F J, O Parniakov, M Koubaa, and N Lebovka 2017b. *Pulsed Electric Fields-Assisted Extraction from Exotic Fruit Residues. Handbook of Electroporation*. Vol. 4. Springer, Cham: Switzerland.

Barba, F J, N S Terefe, R Buckow, D Knorr, and V Orlien 2015a. "New Opportunities and Perspectives of High Pressure Treatment to Improve Health and Safety Attributes of Foods. A Review". *Food Research International* 77 (Part 4 (November)): 725–42.

Barba, F J, O Parniakov, S A Pereira, A Wiktor, N Grimi, N Boussetta, J A Saraiva et al. 2015b. "Current Applications and New Opportunities for the Use of Pulsed Electric Fields in Food Science and Industry". *Food Research International*: 77. 10.1016/j.foodres.2015.09.015.

Barba, F J, Z Zhu, M Koubaa, A S Sant'Ana, and V Orlien 2016b. "Green Alternative Methods for the Extraction of Antioxidant Bioactive Compounds from Winery Wastes and by-Products: A Review". *Trends in Food Science & Technology* 49: 96–109.

Ben Rahal, N, F J Barba, D Barth, and I Chevalot 2015. "Supercritical CO_2 Extraction of Oil, Fatty Acids and Flavonolignans from Milk Thistle Seeds: Evaluation of Their Antioxidant and Cytotoxic Activities in Caco-2 Cells". *Food and Chemical Toxicology: An International Journal Published for the British Industrial Biological Research Association* 83 (September): 275–82.

Bondesson, E 2015. *A Nutritional Analysis on the by-Product Coffee Husk and Its Potential Utilization in Food Production - A Literature Study*. http://stud.epsilon.slu.se.

Bouras, M, M Chadni, F J Barba, N Grimi, O Bals, and E Vorobiev 2015. "Optimization of Microwave-Assisted Extraction of Polyphenols from Quercus Bark". *Industrial Crops and Products* 77: 590–601.

Carbonell-Capella, J M, J Šic Žlabur, S Rimac Brnčić, F J Barba, N Grimi, M Koubaa, M Brnčić, and E Vorobiev 2017. "Electrotechnologies, Microwaves, and Ultrasounds Combined with Binary Mixtures of Ethanol and Water to Extract Steviol Glycosides and Antioxidant Compounds from Stevia Rebaudiana Leaves". *Journal of Food Processing and Preservation* e13179: n/a.

Cheok, C Y, N Mohd Adzahan, R Abdul Rahman, N H Zainal Abedin, N Hussain, R Sulaiman, and G H Chong 2018. "Current Trends of Tropical Fruit Waste Utilization". *Critical Reviews in Food Science and Nutrition* 58 (3): 335–61.

Chernukha, I M, E A Kotenkova, and L V Fedulova. 2017. "Meat Product Based on Porcine Hearts and Aortas Ameliorates Serum Lipid Profile and Inflammation in Hyperlipidemic Rats." In *IOP Conference Series: Earth and Environmental Science*. Vol. 85.

Cruz, R. 2014. "Coffee By-Products Sustainable Agro-Industrial Recovery and Impact on Vegetables Quality. Dissertation of 2th Cycle of Studies Leading to the Master's Degree in Quality Control Specialization in Food and Water." Faculdade de Farmacia. Universidade do Porto, Portugal. https://repositorio-aberto.up.pt/bitstream/10216/76762/2/32890.pdf.

Dapćević-Hadnadev, T, M Hadnadev, and M Pojić 2018. "The Healthy Components of Cereal by-Products and Their Functional Properties". *Sustainable Recovery and Reutilization of Cereal Processing By-Products.* 10.1016/B978-0-08-102162-0.00002-2.

De Las Fuentes, L, B Sanders, A Lorenzo, and S Alber 2004. "Total Food Exploiting Co-Products-Minimizing Waste-Workshop Report". In: eds. A C Smith Waldron, K W C B Faulds, 233–44. Institute of Food Research: Norwich, United Kingdom.

Delgado-Torre, M P, C Ferreiro-Vera, F Priego-Capote, P M Perez-Juan, and M D Luque De Castro 2012. "Comparison of Accelerated Methods for the Extraction of Phenolic Compounds from Different Vine-Shoot Cultivars". *Journal of Agricultural and Food Chemistry* 60 (12): 3051–60.

Deng, Q, K G Zinoviadou, C M Galanakis, V Orlien, N Grimi, E Vorobiev, N Lebovka, and F J Barba 2015. "The Effects of Conventional and Non-Conventional Processing on Glucosinolates and Its Derived Forms, Isothiocyanates: Extraction, Degradation, and Applications". *Food Engineering Reviews* 7 (Springer US): 357–81.

Duku, M H, S Gu, and E B Hagan 2011. "A Comprehensive Review of Biomass Resources and Biofuels Potential in Ghana". *Renewable and Sustainable Energy Reviews* 15 (1): 404–15.

Duţă, D E, A Culeţu, and G Mohan 2018. *Reutilization of Cereal Processing by-Products in Bread Making. Sustainable Recovery and Reutilization of Cereal Processing By-Products.* Elsevier: Amsterdam, Netherlands.

Eisentraut, A. 2010. *Sustainable Production of Second-Generation Biofuels. Potential and Perspectives in Major Economies and Developing Countries.* In Edited by A Eisentraut. OECD/IEA. www.oecd.org/berlin/44567743.pdf Consultado el 30-6-2013).

Elmekawy, A, L Diels, H De Wever, and D Pant 2013. "Valorization of Cereal Based Biorefinery Byproducts: Reality and Expectations". *Environmental Science and Technology* 47 (16): 9014–27.

Esa, N M, T B Ling, and L S Peng 2013. "By-Products of Rice Processing: An Overview of Health Benefits and Applications". *Journal of Rice Research* 1 (1): 1–11.

Esquivel, P, and V M Jiménez 2012. "Functional Properties of Coffee and Coffee by-Products". *Food Research International* 46 (2): 488–95.

Ferrentino, G, M Asaduzzaman, and M M Scampicchio 2018. "Current Technologies and New Insights for the Recovery of High Valuable Compounds from Fruits by-Products". *Critical Reviews in Food Science and Nutrition* 58 (3): 386–404.

Galanakis, C M. 2012. "Recovery of High Added-Value Components from Food Wastes: Conventional, Emerging Technologies and Commercialized Applications". *Trends in Food Science & Technology* 26 (2): 68–87.

Galanakis, C M, N Martinez-Saez, M D Del Castillo, F J Barba, and V S Mitropoulou 2015. *Patented and Commercialized Applications. Food Waste Recovery: Processing Technologies and Industrial Techniques.* Elsevier: Amsterdam, Netherlands.

Gómez, M, and MM Martinez 2017. "Fruit and Vegetable by-Products as Novel Ingredients to Improve the Nutritional Quality of Baked Goods". *Critical Reviews in Food Science and Nutrition* 0 (0): 1–17.

González-Centeno, M R, K Knoerzer, H Sabarez, S Simal, C Rosselló, and A Femenia 2014. "Effect of Acoustic Frequency and Power Density on the Aqueous Ultrasonic-Assisted Extraction of Grape Pomace (Vitis Vinifera L.) - A Response Surface Approach". *Ultrasonics Sonochemistry* 21 (6): 2176–84.

Gorinstein, S, Z Zachwieja, M Folta, H Barton, J Piotrowicz, M Zemser, M Weisz, S Trakhtenberg, and O Màrtín-Belloso 2001. "Comparative Contents of Dietary Fiber, Total Phenolics, and Minerals in Persimmons and Apples". *Journal of Agricultural and Food Chemistry* 49 (2): 952–57.

Gormley, R 2013. *Fish as a Functional Food: Some Issues and Outcomes. SeaHealth-UCD 9A.* www.ucd.ie/t4cms/Supplement%209A%20Fish%20as%20a%20functional%20food.pdf.

Granato, D, D S Nunes, and F J Barba 2017. "An Integrated Strategy between Food Chemistry, Biology, Nutrition, Pharmacology, and Statistics in the Development of Functional Foods: A Proposal". *Trends in Food Science & Technology* 62: 13–22.

Guermazi, Z, M Gharsallaoui, E Perri, S Gabsi, and C Benincasa 2017. "Integrated Approach for the Eco Design of a New Process through the Life Cycle Analysis of Olive Oil: Total Use of Olive by-Products". *European Journal of Lipid Science and Technology* 119 (9): 1700009.

Helkar, P B, A K Sahoo, and N J Patil 2016. "Review: Food Industry by-Products Used as a Functional Food Ingredients". *International Journal of Waste Resources* 6 (3): 1000248.

Horita, C N, R C Baptista, M Y R Caturla, J M Lorenzo, F J Barba, and A S Sant'Ana 2018. "Combining Reformulation, Active Packaging and Non-Thermal Post-Packaging Decontamination Technologies to Increase the Microbiological Quality and Safety of Cooked Ready-to-Eat Meat Products". *Trends in Food Science & Technology* 72: 45–61.

Iye, E L, and P E Bilsborrow 2013. "Assessment of the Availability of Agricultural Residues on a Zonal Basis for Medium- to Large-Scale Bioenergy Production in Nigeria". *Biomass and Bioenergy* 48: 66–74.

Ju, Z Y, and L R Howard 2003. "Effects of Solvent and Temperature on Pressurized Liquid Extraction of Anthocyanins and Total Phenolics from Dried Red Grape Skin". *Journal of Agricultural and Food Chemistry* 51 (18): 5207–13.

Kadam, S U, and P Prabhasankar 2010. "Marine Foods as Functional Ingredients in Bakery and Pasta Products". *Food Research International* 43 (8): 1975–80.

Kasapidou, E, E Sossidou, and P Mitlianga 2015. "Fruit and Vegetable Co-Products as Functional Feed Ingredients in Farm Animal Nutrition for Improved Product Quality". *Agriculture* 5: 1020–34.

Kodagoda, K H G K, and R A U J Marapana 2017. "Utilization of Fruit Processing by-Products for Industrial Applications: A Review". *International Journal of Food Science and Nutrition* 2 (6): 24–30.

Koubaa, M, F J Barba, N Grimi, H Mhemdi, W Koubaa, N Boussetta, and E Vorobiev 2016. "Recovery of Colorants from Red Prickly Pear Peels and Pulps Enhanced by Pulsed Electric Field and Ultrasound". *Innovative Food Science and Emerging Technologies* 37 (Part C): 336–44.

Koubaa, M, F J Barba, H Mhemdi, N Grimi, W Koubaa, and E Vorobiev 2015. "Gas Assisted Mechanical Expression (GAME) as a Promising Technology for Oil and Phenolic Compound Recovery from Tiger Nuts". *Innovative Food Science and Emerging Technologies* 32: 172–80.

Koubaa, M, L Lepreux, F J Barba, H Mhemdi, and E Vorobiev 2017a. "Gas Assisted Mechanical Expression (GAME) for the Selective Recovery of Lipophilic and Hydrophilic Compounds from Olive Kernel". *Journal of Cleaner Production* 166: 387–94.

Koubaa, M, H Mhemdi, F J Barba, A Angelotti, F Bouaziz, S E Chaabouni, and E Vorobiev 2017b. "Seed Oil Extraction from Red Prickly Pear Using Hexane and Supercritical CO2: Assessment of Phenolic Compound Composition, Antioxidant and Antibacterial Activities". *Journal of the Science of Food and Agriculture* 97 (2): 613–20.

Koubaa, M, H Mhemdi, F J Barba, S Roohinejad, R Greiner, and E Vorobiev 2016. "Oilseed Treatment by Ultrasounds and Microwaves to Improve Oil Yield and Quality: An Overview". *Food Research International* 85: 59–66.

Koubaa, M, E. Rosello-Soto, J. Šic Žlabur, A. Režek Jambrak, M. Brnčić, N. Grimi, N. Boussetta, and F J Barba 2015. "Current and New Insights in the Sustainable and

Green Recovery of Nutritionally Valuable Compounds from Stevia Rebaudiana Bertoni". *Journal of Agricultural and Food Chemistry* 63: 6835–46.

Lafarga, T, and M Hayes 2014. "Bioactive Peptides from Meat Muscle and by-Products: Generation, Functionality and Application as Functional Ingredients". *Meat Science* 98 (2): 227–39.

Lal, R. 2005. "World Crop Residues Production and Implications of Its Use as a Biofuel". *Environment International* 31 (4): 575–84.

Lorenzo, J M, P E S Munekata, A S Sant'Ana, R B Carvalho, F J Barba, F Toldrá, L Mora, and M A Trindade 2018. "Main Characteristics of Peanut Skin and Its Role for the Preservation of Meat Products". *Trends in Food Science & Technology* 77: 1–10.

Lorenzo, J M, M Pateiro, R Domínguez, F J Barba, P Putnik, D B Kovacević, A Shpigelman, D Granato, and D Franco 2018. "Berries Extracts as Natural Antioxidants in Meat Products: A Review". *Food Research International* 106: 1095–104.

Makris, D P, G Boskou, and N K Andrikopoulos 2007. "Polyphenolic Content and in Vitro Antioxidant Characteristics of Wine Industry and Other Agri-Food Solid Waste Extracts". *Journal of Food Composition and Analysis* 20 (2): 125–32.

McCann, M. J., C. I. R. Gill, G. O'Brien, J. R. Rao, W. C. McRoberts, P. Hughes, R. McEntee, and I. R. Rowland. 2007. "Anti-Cancer Properties of Phenolics from Apple Waste on Colon Carcinogenesis in Vitro." *Food and Chemical Toxicology* 45 (7): 1224–1230.

Misra, N N, P J Cullen, F J Barba, C Lik Hii, H Jaeger, J Schmidt, A Kovacs, and H Yoshida 2015. *Emerging Macroscopic Pretreatment. Food Waste Recovery: Processing Technologies and Industrial Techniques*. Elsevier: Amsterdam, Netherlands.

Misra, N N, M Koubaa, S Roohinejad, P Juliano, H Alpas, R S Inacio, J A Saraiva, and F J Barba 2017a. "Landmarks in the Historical Development of Twenty First Century Food Processing Technologies". *Food Research International* 97: 318–39.

Misra, N N, A Martynenko, F Chemat, L Paniwnyk, F J Barba, and A Rezek Jambrak 2017b. "Thermodynamics, Transport Phenomena and Electrochemistry of External Field Assisted Non-Thermal Food Technologies". *Critical Reviews in Food Science and Nutrition* 0 (ja): 0.

Molina Alcaide, E, and A Nefzaoui 1996. "Recycling of Olive Oil by-Products: Possibilities of Utilization in Animal Nutrition". *International Biodeterioration and Biodegradation* 38 (3–4): 227–35.

Moubarik, A., F J Barba, and N. Grimi 2015. "Understanding the Physicochemical Properties of Olive Kernel to Be Used as a Potential Tool in the Development of Phenol-Formaldehyde Wood Adhesive". *International Journal of Adhesion and Adhesives* 61: 122–26.

Mourad, M. 2016. "Recycling, Recovering and Preventing 'Food Waste': Competing Solutions for Food Systems Sustainability in the United States and France". *Journal of Cleaner Production* 126: 461–77.

Nikmaram, N, S Y Leong, M Koubaa, Z Zhu, F J Barba, R Greiner, I Oey, and S Roohinejad 2017. "Effect of Extrusion on the Anti-Nutritional Factors of Food Products: An Overview". *Food Control* 79: 62–73.

Nishida, J. 2014. *Reducing Food Waste and Promoting Food Recovery Globally, EPA Connect. The Official Blog of the EPA Leadership*. https://blog.epa.gov/blog/2014/10/reducing-food-waste-and-promoting-food-recovery-globally/.

Nurdiani, R, T Vasiljevic, T K Singh, and O N Donkor 2017. "Bioactive Peptides from Fish by-Products with Anticarcinogenic Potential". *International Food Research Journal* 24 (5): 1840–49.

O'Shea, N, E K Arendt, and E Gallagher 2012. "Dietary Fibre and Phytochemical Characteristics of Fruit and Vegetable by-Products and Their Recent Applications as Novel Ingredients in Food Products". *Innovative Food Science and Emerging Technologies* 16: 1–10.

Offiah, V, V Kontogiorgos, and K O Falade 2018. "Extrusion Processing of Raw Food Materials and by-Products: A Review". *Critical Reviews in Food Science and Nutrition* 22: 1–20.

OIV. 2015. *Www.ecoticias.com/residuos-reciclaje/104842/Residuos-Elaboracion-Desarrollar-Productos-Alimenticios-Cosmeticos-Farmaceuticos.*

Okino Delgado, C H, and L F Fleuri 2016. "Orange and Mango by-Products: Agro-Industrial Waste as Source of Bioactive Compounds and Botanical versus Commercial description—A Review". *Food Reviews International* 32 (1): 1–14.

Oreopoulou, V, and C Tzia 2007. *Utilization of Plant by-Products for the Recovery of Proteins, Dietary Fibers, Antioxidants, and Colorants. Utilization of By-Products and Treatment of Waste in the Food Industry.* Springer: Boston, MA, USA.

Parniakov, O, F J Barba, N Grimi, N Lebovka, and E Vorobiev 2014. "Impact of Pulsed Electric Fields and High Voltage Electrical Discharges on Extraction of High-Added Value Compounds from Papaya Peels". *Food Research International* Recovery and Utilization of Valuable Compounds from Food Processing By-products 65 (Part C (November)): 337–43.

Parniakov, O, F J Barba, N Grimi, N Lebovka, and E Vorobiev 2016a. "Extraction Assisted by Pulsed Electric Energy as a Potential Tool for Green and Sustainable Recovery of Nutritionally Valuable Compounds from Mango Peels". *Food Chemistry* 192: 842–48.

Parniakov, O, F J Barba, N Grimi, N Lebovka, and E Vorobiev 2016b. "Electro-Biorefinery as a Potential Tool for Valorization of Mango and Papaya by-Products". *IFMBE Proceedings* 53: 418–21.

Parniakov, O, E Rosello-Soto, F J Barba, N Grimi, N Lebovka, and E Vorobiev 2015. "New Approaches for the Effective Valorization of Papaya Seeds: Extraction of Proteins, Phenolic Compounds, Carbohydrates, and Isothiocyanates Assisted by Pulsed Electric Energy". *Food Research International* 77 (4): 711–717.

Pérez-Sariñana, B Y, and S Saldaña-Trinidad 2017. *The Question of Caffeine. World's Largest Science, Technology & Medicine.* Edited by B Y Pérez-Sariñana and S Saldaña-Trinidad. Intechopen: London, United Kingdom.

Poojary, M M, F J Barba, B Aliakbarian, F Donsì, G Pataro, D A Dias, and P Juliano 2016. "Innovative Alternative Technologies to Extract Carotenoids from Microalgae and Seaweeds". *Marine Drugs* 14: 11.

Putnik, P, F J Barba, J M Lorenzo, D Gabrić, A Shpigelman, G Cravotto, and D Bursać Kovačević 2017a. "An Integrated Approach to Mandarin Processing: Food Safety and Nutritional Quality, Consumer Preference, and Nutrient Bioaccessibility". *Comprehensive Reviews in Food Science and Food Safety* 16 (6): 1345–58.

Putnik, P, F J Barba, I Španić, Z Zorić, V Dragović-Uzelac, and D Bursać Kovačević 2017b. "Green Extraction Approach for the Recovery of Polyphenols from Croatian Olive Leaves (Olea Europea)". *Food and Bioproducts Processing* 106 (Supplement C): 19–28.

Putnik, P, D Bursać Kovacević, A Režek Jambrak, F J Barba, G Cravotto, A Binello, J M Lorenzo, and A Shpigelman 2017c. "Innovative 'Green' and Novel Strategies for the Extraction of Bioactive Added Value Compounds from Citruswastes - A Review". *Molecules* 22: 5.

Rabetafika, H N, B Bchir, C Blecker, and A Richel 2014. "Fractionation of Apple by-Products as Source of New Ingredients: Current Situation and Perspectives". *Trends in Food Science & Technology* 40 (1): 99–114.

Ramos, O L, R N Pereira, R M Rodrigues, J A Teixeira, A A Vicente, and F X Malcata 2015. *Whey and Whey Powders: Production and Uses. Encyclopedia of Food and Health.* Elsevier: Amsterdam, Netherlands.

Ran, X-l, M Zhang, Y Wang, and B Adhikari 2018. "Novel Technologies Applied for Recovery and Value Addition of High Value Compounds from Plant Byproducts: A Review". *Critical Reviews in Food Science and Nutrition* 59 (3): 450–461.

Ribas-Agustí, A, O Martín-Belloso, R Soliva-Fortuny, and Elez-Martínez 2017. "Food Processing Strategies to Enhance Phenolic Compounds Bioaccessibility and Bioavailability in Plant-Based Foods". *Critical Reviews in Food Science and Nutrition* 58 (15): 2531–2548.

Roselló-Soto, E, C M Galanakis, M Brnčić, V Orlien, F J Trujillo, R Mawson, K Knoerzer, B K Tiwari, and F J Barba 2015a. "Clean Recovery of Antioxidant Compounds from Plant Foods, by-Products and Algae Assisted by Ultrasounds Processing. Modeling Approaches to Optimize Processing Conditions". *Trends in Food Science & Technology* 42 (2): 134–49.

Roselló-Soto, E, M. Koubaa, A. Moubarik, R. P. Lopes, J. A. Saraiva, N. Boussetta, N. Grimi, and F J Barba 2015b. "Emerging Opportunities for the Effective Valorization of Wastes and by-Products Generated during Olive Oil Production Process: Non-Conventional Methods for the Recovery of High-Added Value Compounds". *Trends in Food Science & Technology* 45 (2): 296–310.

Roselló-Soto, E, O Parniakov, Q Deng, A Patras, M Koubaa, N Grimi, N Boussetta et al. 2016. "Application of Non-Conventional Extraction Methods: Toward a Sustainable and Green Production of Valuable Compounds from Mushrooms". *Food Engineering Reviews* 8 (2): 214–34.

Roselló-Soto, E, M M Poojary, F J Barba, M Koubaa, J M Lorenzo, J Mañes, and J C Moltó 2018a. "Thermal and Non-Thermal Preservation Techniques of Tiger Nuts' Beverage 'Horchata de Chufa'. Implications for Food Safety, Nutritional and Quality Properties". *Food Research International* 105: 945–51.

Roselló-Soto, E, M M Poojary, F J Barba, J M Lorenzo, J Mañes, and J C Moltó 2018b. "Tiger Nut and Its by-Products Valorization: From Extraction of Oil and Valuable Compounds to Development of New Healthy Products". *Innovative Food Science and Emerging Technologies* 45: 306–12.

Ryder, K, M A Ali, A Carne, and J Billakanti 2017. "The Potential Use of Dairy by-Products for the Production of Nonfood Biomaterials". *Critical Reviews in Environmental Science and Technology* 47 (8): 621–42.

Safaa, H M, M P Serrano, D G Valencia, M Frikha, E Jiménez-Moreno, and G G Mateos 2008. "Productive Performance and Egg Quality of Brown Egg-Laying Hens in the Late Phase of Production as Influenced by Level and Source of Calcium in the Diet". *Poultry Science* 87 (10): 2043–51.

Şahin, S, E Elhussein, M Bilgin, J M Lorenzo, F J Barba, and S Roohinejad 2018. "Effect of Drying Method on Oleuropein, Total Phenolic Content, Flavonoid Content, and Antioxidant Activity of Olive (Olea Europaea) Leaf". *Journal of Food Processing and Preservation* 42 (5): e13604.

Şahin, S, R Samli, A S Birteks, Z Tan, F J Barba, F Chemat, G Cravotto, and J M Lorenzo 2017. "Solvent-Free Microwave-Assisted Extraction of Polyphenols from Olive Tree Leaves: Antioxidant and Antimicrobial Properties". *Molecules* 22 (7): e1056.

Sánchez-Zapata, E, J Fernández-López, and J Angel Pérez-Alvarez 2012. "Tiger Nut (*Cyperus Esculentus*) Commercialization: Health Aspects, Composition, Properties, and Food Applications". *Comprehensive Reviews in Food Science and Food Safety* 11 (4): 366–77.

Shershenkov, B, and E Suchkova 2015. "The Direct Microbial Synthesis of Complex Bioactive Compounds as Perspective Way of Milk Whey Utilization". *Energy Procedia* 72: 317–21.

Singh, J. 2015. "Overview of Electric Power Potential of Surplus Agricultural Biomass from Economic, Social, Environmental and Technical Perspective - A Case Study of Punjab". *Renewable and Sustainable Energy Reviews* 42: 286–97.

Six, L, F Velghe, S Verstichel, and S De Meester 2016. Sustainability considerations on the valorization of organic waste. In edited by P Poltronieri and O F D'Urso, 287–307. *The Food, Feed, Fibre, Fuel (4F) Economy*. Elsevier: Amsterdam, Netherlands.

Teixeira, A, N Baenas, R Dominguez-Perles, A Barros, E Rosa, D A Moreno, and C Garcia-Viguera 2014. "Natural Bioactive Compounds from Winery by-Products as Health Promoters: A Review". *International Journal of Molecular Sciences* 15 (9): 15638–78.

Vinceković, M, M Viskić, S Jurić, J Giacometti, D Bursać Kovacević, P Putnik, F Donsì, F J Barba, and A Režek Jambrak 2017. "Innovative Technologies for Encapsulation of Mediterranean Plants Extracts". *Trends in Food Science & Technology* 69: 1–12.

William, P T. 2005. "Water Treatment and Disposal". In: ed. J Wally, Vol. 9. www.iwapublishing.com/sites/default/files/ebooks/9781780402086.pdf: Department of Sanitary and Environmental Engineering Federal University of Minas Gerais, Brasil.

Zhu, Z, Q Guan, Y Guo, J He, G Liu, S Li, F J Barba, and M Y Jaffrin 2016a. "Green Ultrasound-Assisted Extraction of Anthocyanin and Phenolic Compounds from Purple Sweet Potato Using Response Surface Methodology". *International Agrophysics* 30 (1): 113–22.

Zhu, Z, T Jiang, J He, F J Barba, G Cravotto, and M Koubaa 2016b. "Ultrasound-Assisted Extraction, Centrifugation and Ultrafiltration: Multistage Process for Polyphenol Recovery from Purple Sweet Potatoes". *Molecules* 21: 11.

Zhu, Z, R Zhang, S Zhan, J He, F J Barba, G Cravotto, W Wu, and S Li 2017. "Recovery of Oil with Unsaturated Fatty Acids and Polyphenols from Chaenomelessinensis (Thouin) Koehne: Process Optimization of Pilot-Scale Subcritical Fluid Assisted Extraction". *Molecules* 22: 10.

2

Dairy By-Products as Source of High Added Value Compounds

Conventional and Innovative Extraction Methods

NOEMÍ ECHEGARAY, JUAN A. CENTENO, AND JAVIER CARBALLO

Área de Tecnología de Alimentos, Facultad de Ciencias, Universidad de Vigo. Campus Universitario, s/n, 32004 Ourense, Spain

Contents

1 Introduction

In 2016, total milk produced in the world reached 845 million tons. Cow's (82%) and buffalo's (14%) milk accounts for 96% of the total milk produced, and 62% of the total cow's and buffalo's milk was delivered to dairies where it was transformed into dairy products (IFCN [International Farm Corporation Network] 2017).

Dairy industry generates different by-products of different composition and characteristics during the manufacturing of dairy products. Recovering and treatment of these by-products was initially an environmental need due to the contaminating nature of many of them. However, some of the molecules that make up these by-products have interesting functional, biological, or medical properties, which make them high-added value compounds whose recovery can be an important source of income for the dairy industry. The extraction and obtaining of these compounds from the by-products in their native state with all their chemical and physical properties intact has been a challenge for the transformation industries. In this chapter, after the description of the main by-products and their nature and composition, the extraction methods used for recovering the molecules from the dairy by-products are reviewed and described.

2 Dairy By-Products: Importance and Characteristics

In the manufacturing processes of the dairy industry, several by-products of different nature are generated, but only three of them have qualitative and quantitative importance, and therefore specific technologies were developed for the separation of their components and conditioning for further applications. These are the cheese whey, buttermilk, and ghee residue (GR). All of them contain valuable compounds for specific uses and purposes.

2.1 Cheese Whey

During cheese manufacture, after milk coagulation (under the action of the acid and/or proteolytic enzymes), coagulated milk is cut after having reached the right consistency and two phases are separated: the curd that is later molded and pressed forming the cheese, and the whey, a yellow-green liquid that is drained and separated. Whey represents about 82–90% of the total volume of milk transformed into cheese (Carvalho et al. 2013); it contains on average 55% of the nutrients originally present in the milk (Varnam and Sutherland 1994), and its amount supposes on average nine-fold the amount of cheese manufactured. Worldwide production of cheese whey is estimated at around 190–200 million tons per year, with a 2% increase every year (Agyei et al. 2015; Ryan and Walsh 2018).

Based on the procedure followed for milk coagulation in cheese manufacturing, two different types of cheese whey can be obtained. Milk can be coagulated by lactic acid action (when achieving a pH value of 4.6 corresponding to the isoelectric point of the caseins), in which case an acid whey (pH value ≤ 5) is obtained. In other cases, casein destabilization and subsequent milk coagulation are achieved by the addition of proteolytic enzymes (rennet, cynarases, etc.) and a sweet whey (pH 6–7) is obtained. In industrial practice, most of the cheeses are produced through a mixed acid/enzymatic coagulation and the coagulant enzyme is added after variable acidification resulting from the development of a lactic acid bacterial culture. In any case, acid and sweet whey differ in their chemical composition (see for guidance Table 2.1 taken from Yadav et al. 2015), and the whey produced has a composition more similar to that of sweet whey or to that of an acid whey depending on whether the cheese has been made predominantly by enzymatic or acid coagulation. The main compositional differences between the two types are related to fat, lactose, and lactate contents and to the specific composition of the protein fraction. Of course, the cheese whey composition is highly influenced by the animal species (cow, goat, ewe, buffalo, or camel) from which the milk used to make the cheese comes from. For example, bovine cheese whey contains 70–80% of lactose, 9% of proteins, 8–20% of

Table 2.1 Compositional characteristics of sweet whey and acid whey (Yadav et al. 2015)

CONSTITUENTS	SWEET WHEY (g/L)	ACID WHEY (g/L)
Total solids	63.0–70.0	63.0–70.0
Lactose	46.0–52.0	44.0–46.0
Protein	6.0–10.0	6.0–8.0
Fat	5.0	0.4
Lactate	2.0	6.4
Ash	5.0	8.0
Calcium	0.4–0.6	1.2–1.6
Phosphate	1.0–3.0	2.0–4.5
Chloride	1.1	1.1

minerals (all the components expressed as percentage of the total solids) and other less abundant components such as some peptides from the κ -casein hydrolysis, and lipids (Daufin et al. 1998). Sheep cheese whey has a total nitrogen, when expressed as percentage of the dry matter, much higher than the cow cheese whey, being double the soluble protein content (Assenat 1985).

2.2 Buttermilk

Butter is the most popular and widely manufactured milk fat product. It is obtained from cream (around 40% of fat) through a churning process after which butter is separated from the non-fat liquid phase called "buttermilk." The overall production of liquid buttermilk could be estimated at around 4.2 million tons worldwide, based on the data provided by the Food and Agricultural Organization of the United Nations (FAOSTAT) for 2014 (AHDB [Agriculture and Horticulture Development Board] 2017; FAO [Food and Agriculture Organization] 2018).

Most of the buttermilk produced is sweet buttermilk obtained from churning of sweet cream, but sour buttermilk from cultured cream and whey buttermilk from whey cream are also manufactured. Because of their nature, composition characteristics of sweet and cultured buttermilk are close to that of the skimmed milk, whereas that of the whey buttermilk are similar to that of the cheese whey with the exception of fat content, which is higher in buttermilk compared to skimmed milk or cheese whey (Ali 2018). Buttermilk contains all the water-soluble compounds of cream, such as lactose, milk proteins, and minerals, and it also includes components from the milk fat globule membrane (such as phospholipids) that breaks in the course of cream churning operations and migrates to the buttermilk fraction (Corredig and Dalgleish 1997). Polar lipids in buttermilk account for 1.2–2.1% of the total solids (Vanderghem et al. 2010). Composition of the dehydrated buttermilk from different creams is shown in Table 2.2.

Table 2.2 Gross composition (%) on a dry matter basis and pH values of buttermilk powders from different creams (Sodini et al. 2006)

SAMPLES	CRUDE PROTEIN (%) N × 6.38	FAT (%)	PHOSPHOLIPIDS (%)	ASH (%)	LACTOSE (%)	pH
Buttermilk powders						
Buttermilk from sweet cream	31.5–32.9	5.7–13.1	1.27–1.34	6.7–7.6	48.7–53.8	6.46–6.61
Buttermilk from cultured cream	27.8	22.3	1.15	6.2	43.7	5.39
Buttermilk from whey cream	14.1	15.5	1.87	7.0	63.4	5.98

2.3 *Ghee Residue*

Between 2014 and 2015, milk production in India reached 146 million tons per year (Munirathnamma et al. 2017). About 30–35% of the total milk production in India is converted into ghee (Gandhi et al. 2013). Ghee is clarified butter fat made by fermenting whole milk until it is transformed into curd and then churning it until butter is produced, followed by clarification by heating it to 105–145 °C. During clarification of ghee, the solid non-fat (SNF) compounds settle down in the form of small particles and constitute a moist brownish sediment known as "ghee residue" (GR) (Janghu et al. 2014). GR is one of the largest by-products of the dairy industry. The amount of GR generated during manufacture and its proximate composition depend on the method of preparation of ghee. GR contains variable quantities of milk constituents such as fats, proteins, lactose, and minerals (Munirathnamma et al. 2017). Table 2.3 depicts the proximate composition of GR from different origins.

2.4 *Main Components of the Dairy By-Products*

According to the manufacturing processes in which these by-products are generated, they contain the components of milk other than casein and fat that are the milk components retained in the manufactured products (cheese, butter, and ghee). Therefore, the by-products contain mainly the following compounds.

2.4.1 Lactose Lactose, the milk carbohydrate accounts approximately for two-third of the total solids present in cheese whey (Yadav et al. 2015) and around 50% of the total solids present in buttermilk (Sodini et al. 2006). Lactose can be incorporated in several foodstuffs with different functions (flavor, aroma, and color enhancer, improving texture, reducing sweetness, anticaking agent, shelf-life extender, etc.) (Yang and Silva 1995). Lactose is also the basis of microbial fermentation processes of the whey to produce several interesting products such as single-cell protein, long-chain polysaccharides, ethanol, amino acids, enzymes, organic acids, bacteriocins, oils, vitamins, bioplastics, and insecticides (Yang and Silva 1995; Mollea et al. 2013).

Table 2.3 Proximate composition of ghee residue

GHEE RESIDUE	MOISTURE (%)	FAT (%)	ASH (%)	PROTEIN (%)	LACTOSE (%)
Direct cream method*	26.64±0.43	33.13±0.58	3.27±0.12	30.91±0.98	6.05***
Creamery butter method*	17.71±0.22	41.83±0.47	2.56±0.14	31.69±0.54	6.21***
Undefined manufacture method**	21.81±0.22	35.99±0.39	3.81±0.31	19.97±0.27	17.88±0.10

Data are presented as mean ± SEM (n=3).
* Janghu et al. (2014).
** Munirathnamma et al. (2017).
*** Determined by difference.

2.4.2 Beta-Lactoglobulin Belonging to the family of the lipocalins, β-lactoglobulin is the main protein in whey and it accounts for 50–55% of the total whey proteins. From the nutritional point of view, β-lactoglobulin is rich in essential amino acids (for in-depth description and characteristics, see Sawyer 2003) and binds long-chain fatty acids, retinoids, and steroids (Kontopidis et al. 2004), increasing the bioavailability of nutrients that have this type of nature, including fat-soluble vitamins.

2.4.3 Alpha-Lactalbumin Alpha-lactalbumin (alpha-LA) is quantitatively the second protein in whey, accounting for 20–25% of the total whey proteins. Alpha-lactalbumin is a Ca-binding protein having the ability to bind metal cations (Permyakov and Berliner 2000) and may facilitate the absorption of essential minerals (Stănciuc and Râpeanu 2010). From the nutritional point of view, it is a unique protein, being rich in essential amino acids, mainly tryptophan. Tryptophan is the precursor in the synthesis of serotonin (a neurotransmitter implicated in the control of the appetite and sleep and that improves the mood, cognitive performance, and the control of stress) and melatonin (a hormone involved in the regulation of the sleep–wake cycle). Relative recent studies show that some folding variants of alpha-LA have bactericidal activity (López Expósito and Recio 2006) and some others cause apoptosis of tumor cells (Gustafsson et al. 2005).

2.4.4 Glycomacropeptide Glycomacropeptide is a C-terminal peptide (f 106–169) of the κ-casein released by the coagulant enzyme (chymosin) in the first stage of the enzymatic coagulation of milk during the cheese manufacture; the peptide is glycosylated in the $Thr_{131,\ 133,\ 135,\ 142}$ and Ser_{141} positions of the backbone (Eigel et al. 1984). Being released from the casein micelles, it passes to the serum and is eliminated with it during the whey drainage. Glycomacropeptide accounts for 0–15% of the whey proteins, being its content maximum in sweet whey from an enzymatic coagulation of the milk. It is a very interesting molecule showing functional properties, such as emulsification, foaming and gel formation ability, and also biological properties including reduction of gastric secretion, growth factor for *Bifidobacterium*, anti-carcinogenic effect, inhibition of cholera toxin, prevention of intestinal infection, hemagglutinin inhibition, modulation of immune response, stimulation of cholecystokinin release, and nutritional management of phenylketonuria (Neelima et al. 2013). Being its sole origin the hydrolysis of the κ-casein by chymosin, its presence has been used for detection of cheese whey adulteration in milk (Neelima et al. 2013).

2.4.5 Immunoglobulins Immunoglobulins account for 10–15% of the total whey proteins. Immunoglobulins provide passive immunity (Schroeder and Cavacini 2010) and their intake may be of importance to people who are immunocompromised (Bell 2000). Three different immunoglobulins (IgG, IgM, and IgA) were detected in cheese whey in this order of quantitative importance (El-Loly 2007).

2.4.6 Bovine Serum Albumin Bovine serum albumin accounts for 5–10% of the total whey proteins in the whey obtained from cow's milk. It is a globular protein with a good essential amino acid profile; it is involved in maintaining homeostatic colloidal pressure and binds lipids, acting as a carrier for fatty acids, catecholamines, and other hormones. Recent studies have shown that the inclusion of the bovine serum albumin in the diet has anti-obesity effects (McManus et al. 2015).

2.4.7 Lactoferrin

Lactoferrin accounts for 1–2% of the total whey proteins. It is an iron-binding glycoprotein characterized by its ability to bind and transport iron ions, having immunomodulatory, anti-inflammatory, antimicrobial, and anti-tumoral activities (Brock 2002). Due to its iron-binding ability, bovine lactoferrin is also recognized as a valuable functional food ingredient (Stănciuc et al. 2013) as a natural solubilizer of iron in foods.

2.4.8 Lactoperoxidase Lactoperoxidase is a glycoprotein with both bactericidal and bacteriostatic effects (Dajanta et al. 2008; Tayefi-Nasrabadi and Asadpour 2008) that represents 0.5% of the total whey proteins, varying its content with season and diet (Kussendrager and van Hooijdonk 2000). Because of the absence of caseins, whey is the preferential source for isolation and obtaining of this protein (Borzouee et al. 2016).

2.4.9 Lysozyme Lysozyme (*N*-acetylmuramide glycan-hydrolase) is a hydrolytic enzyme widely distributed in the animal kingdom (Callewaert and Michiels 2010) that hydrolyzes peptidoglycan of the bacterial cell wall and exhibits antibacterial activity against Gram-positive bacteria. Although it has higher concentrations in tears and egg white fraction, cheese whey can be a suitable source for obtaining it. This enzyme is widely used in the industry as a food additive for the preservation of several foods, and in the treatment of bacterial infections, leukemia, and cancer (Proctor et al. 1988).

2.4.10 Minerals All the minerals in milk are present in whey, but in quantities that differ according to the nature of the whey. Among the main minerals, K is the most abundant followed by Ca, P, Na, and Mg. Among the trace elements, Zn, Fe, Cu, and Mn are the most abundant in this quantitative order. It was observed (Wong et al. 1978) that the Ca concentration was three times higher, the Mn concentration four times higher, and the Zn concentration twenty times higher in acid whey than in sweet whey. The other minerals, with the exception of Na, were also more abundant in acid whey than in sweet whey, but with lesser differences.

2.4.11 Vitamins As milk, whey contains appreciable amounts of A, C, D, E, and B-complex vitamins. Important vitamins such as B12 are more available in milk than in supplements from other origins (Matte et al. 2012).

2.4.12 Other Milk Compounds Occasionally Present In addition to the components already described, and depending on the type of dairy by-products considered and the manufacturing processes through which they are obtained, other important compounds can be present. Cheese whey contains milk fat whose quantities depend on the acidity of the whey (see Table 2.1). Buttermilk, particularly sweet and sour buttermilk, contain high quantities of caseins (>77% of the total nitrogen content) (Sodini et al. 2006). Also, as previously indicated, buttermilk contains phospholipids from milk fat membrane globule. Several authors reported that buttermilk contains between 4 and 7 times more phospholipids than whole milk (Christie et al. 1987; Elling et al. 1996).

3 Conventional Transformation and Extraction Methods Applied to Dairy By-Products

The primitive technique for whey fractionation consisted of fat removal, acidification till a pH value of 4.6, and subsequent flocculation of whey proteins through heating by boiling or by direct steam injection. After protein decantation, lactose was recovered from the supernatant by a classical crystallization process consisting of heat concentration until supersaturation and crystallization was attained. Heating at 90 °C for 30 minutes only allowed the flocculation of 80% of soluble proteins and it was necessary to reach 96–97 °C to obtain total flocculation. The proteins flocculated were completely denatured and their functional properties could not be exploited in later food uses. Moreover, this denaturation made at high temperatures and in the presence of lactose resulted in a significant loss of essential amino acids such as lysine, so that the nutritional value of the proteins was also reduced. To prepare non-heated proteins having intact functional properties and nutritive value, diverse flocculating agents (i.e., sodium polyphosphate, sodium hexametaphosphate, and ferric salts) were assayed, but these attempts had no future in the food industry and were quickly abandoned. The emerging of the membrane techniques meant definitive abandonment of these primitive techniques of fractionation.

3.1 Membrane Techniques

3.1.1 Fundamentals and General Concepts From the late 60s of the last century, the membrane techniques have been used massively and successfully in the dairy industry for the fractionation of the components of liquid by-products (Pouliot 2008).

Milk compounds integrating the dairy by-products that were briefly described in Section 2.4 have different chemical nature and properties but also very different molecular masses and sizes (see Table 2.4 made from data adapted from Walstra et al. 1999). This later characteristic is the fundamental basis of the membrane separation techniques.

If a solution containing chemical molecules having diverse sizes is driven toward a semipermeable porous membrane with a sole and defined pore size, the membrane will allow the passing of compounds having molecular sizes smaller to the size of its pores (these compounds are called "permeate"), but it will retain those having molecular sizes larger than its pore size (compounds called "retentate"). To force the passage of the components of the solution through the membrane, a gradient, either pressure or electrostatic, is used and the molecules are separated on the basis of their size, and even their molecular charge. The pressure required to force the passage of the compounds through the membrane is usually proportional to the size of the pores, so it is necessary to increase the pressure as the pore size decreases (Brans et al. 2004). The pressure gradient established between both sides of the membrane is called "transmembrane pressure" and this is an important parameter associated with the performance of separation processes that use this technique (Gésan-Guiziou et al. 2000). The permeability of a membrane is adequate when there is a linear relationship between the flow of low hardness water and the transmembrane pressure that is recorded in the said membrane (Carneiro et al. 2002). The membrane performance depends not only

Table 2.4 Size distribution of the components of milk (Walstra et al. 1999)

	SIZE RANGE AND AVERAGE (AT WEIGHT AVERAGE)
Fat globules	0.1–15 μm, average 3.4 μm
Casein (in micelles)	20–300 nm, average 110 nm
Serum proteins	3–6 nm
α-Lactalbumin	14 kDa
β-Lactoglobulin	18 kDa
BSA	66 kDa
Proteose-peptone	4–40 kDa
Immunoglobulins	150–900 kDa
Lactoferrin	86 kDa
Transferrin	76 kDa
Lactose	0.35 kDa

on the transmembrane pressure, but also on the feed flow rate, temperature, concentration of solution that is being separated, and physicochemical characteristics of the membrane (Gésan-Guiziou et al. 2000).

Membranes used are diverse in materials and configurations. Regarding the materials, a suitable material for membrane manufacture must meet the following requirements: high porosity, narrow pore distribution or sharp molecular weight cutoff, high polymer strength (elongation capacity, high burst, and collapse pressure), good polymer flexibility, permanent hydrophilic character, stability in a wide range of pH values, good chlorine tolerance for cleaning operations, and low cost. The most usual materials are cellulose acetate, organic polymers (polyamides, polypropylene, polysulfone, polysulfur, polytetrafluoroethylene, propylene, etc.) or ceramic (inorganic materials) (Al_2O_3, TiO_2, ZrO_2, and SiO_2, or a combination of these materials) (Li 2007).

Cellulose acetate membranes tend to be less resistant to temperature (work at maximum temperature 50 °C) and to the chemical (pH 3–8) and microbiological deterioration. For these reasons, the use of these membranes is nowadays limited.

The polymeric membranes are variable in composition depending on the technical characteristics and working conditions desired. They have a wide range of working conditions (pH 2–12, and temperatures until 80 °C), but they can not stand too well to compaction and chlorine agents. Cleaning of these membranes is complicated and therefore their lifetime is short (GEA Process Engineering 2012). However, these membranes are the most used in the fractionation of whey. They are arranged in different configurations (spiral bound, hollow fibers, flat sheet—plate and frame) and provide a high membrane area per element that leads to smaller and cheaper plant designs. Polymeric membranes come in a wide range of pore sizes, and they can be used for diverse dairy filtration applications.

The ceramic membranes are most resistant to temperature (>300 °C), chemicals (pH 0–14, chlorine resistant), and pressure (>1 MPa) (Mahaut 1992) and they are easy to clean. The lifetime of the ceramic membranes is longer, but ceramic membranes come in a limited range of pore sizes and due to the limited membrane area per element they are relatively expensive.

According to the relationship between the direction of the flux of the solution and the direction of the pressure applied, the process may be tangential or frontal. In the dairy industry, processes are generally of the tangential type. The permeate is collected and the retentate is subjected to several cycles of recirculation according to the characteristics of the desired final retentate (James et al. 2003). Figure 2.1 shows the general scheme of a tangential membrane installation usually used in the food industry (Chacón-Villalobos 2006).

3.1.2 Membrane Processes in the Dairy Industry Considering the wide variety of sizes, shapes, and distributions of milk molecules, milk and by-products of the dairy

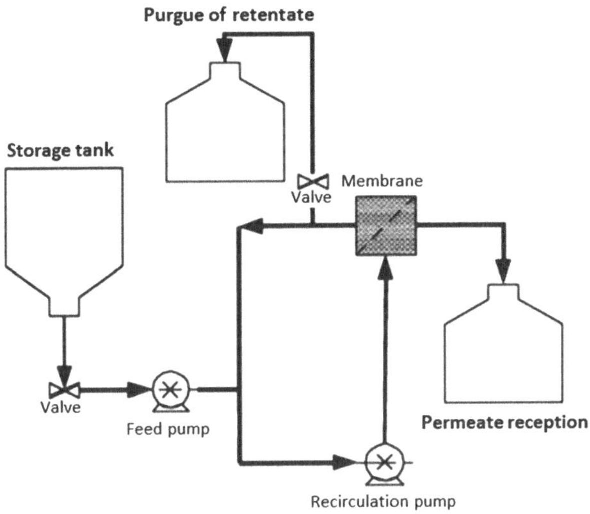

Figure 2.1 Principle of the tangential industrial process of membrane separation (Chacón-Villalobos 2006).

industry constitute a truly challenging product for processing through membrane technologies (Brans et al. 2004).

The membrane processes that use pressure as a driving force basically differ in the size of the pores of the membranes used and therefore on the nature of compounds that are retained or that are allowed to pass through the process. The most common of these processes are reverse osmosis (RO) or hyperfiltration, nanofiltration (NF), ultrafiltration (UF), and microfiltration (MF). Figure 2.2 taken from Brans et al. (2004) itemizes all these processes in relation to the milk components that allow permeating according to the pore size of the membranes used. The behavior of all these processes in relation to the different components present in milk or in dairy by-products is also schematized in Figure 2.3.

In addition to the processes that use pressure as a driving force, there are someother processes that separate particles according to their electric charge and notexclusively according to their molecular size. For this purpose, an electric field is usedas a driving force, and electrostatically charged membranes serve as a separationinstrument (Bazinet 2005). These membrane processes are generally called as electrodialysis(ED). Because of the nature of the driving force and the membranesused, ED extracts ions through selective ion-exchange membranes. Membranesused are of two types: cation-exchange membranes that only allow cations to transport,and anion-exchange membranes that only allow anions to go through. These membranesare impermeable to liquids.

In the following paragraphs, we will briefly describe all these processes beginning with those that use membranes with a smaller pore size.

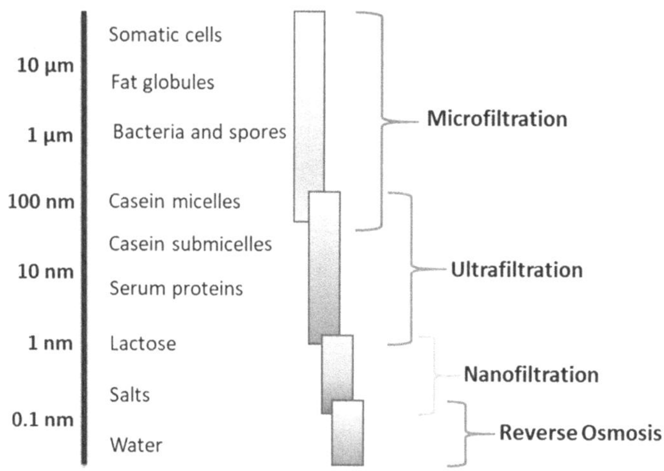

Figure 2.2 Components of milk: size indication and membrane processes (Brans et al. 2004).

Reverse osmosis or hyperfiltration is a high pressure-driven filtration process that uses very dense membranes (pore size from 0.5 to 1.5 nm). In this process, only the water (molecular mass lower than 150 Da) passes through the membrane layer. The working pressure is 700 psi (around 5,000 kPa) and the temperature from 40 to 90 °C depending on the nature of the membrane used.

Reverse osmosis is normally used for concentration or volume reduction of whey and unlike the classic processes of concentration does not use heat and allows obtaining retentates with a maximum solid content of 30% (Twiford 2004).

Nanofiltration is a medium-to-high pressure-driven filtration process. In general, NF can be considered as a type of RO in which the membrane has a slightly more open structure (pore size of ≤ 2 nm). In practice, the membrane allows to pass water and monovalent ions, retaining the di- and polyvalent ions (Mucchetti et al. 2000). In the dairy industry, this process is used for concentration and partial demineralization of whey.

Ultrafiltration is a medium pressure-driven filtration process. Membranes have medium-open structure (pore size from 2 to 100 nm), retaining molecules that have from 450 to 300,000 Da. In such conditions, they allow to pass water, minerals, and lactose (see Figure 2.3), retaining the whey proteins. The process uses transmembrane pressures around 1,500 kPa, temperatures of 50–60 °C, and polysulfone membranes (Akoum et al. 2003; Baldasso et al. 2011).

The permeate is generally composed of lactose (76–80% of total solids in this fraction), minerals, and other low molecular mass compounds such as non-protein nitrogen molecules. The retentate generally has a protein content of 60% of the total solids.

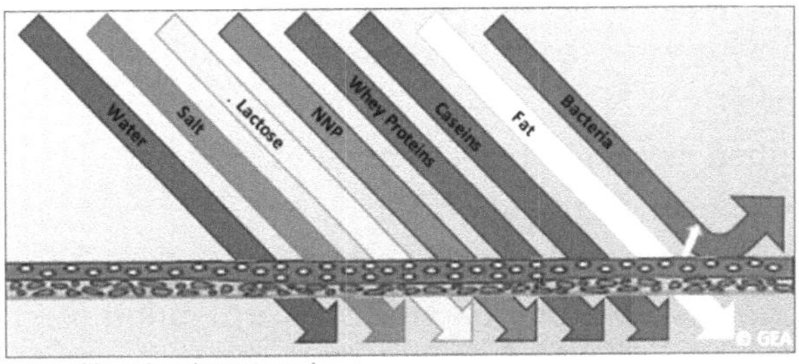

Microfiltration (1400 nm)

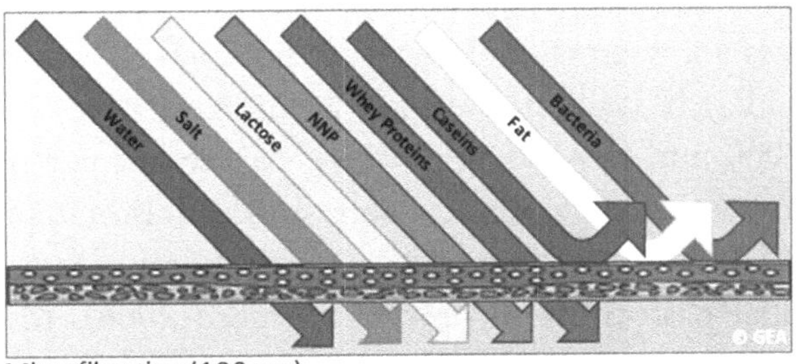

Microfiltration (100 nm)

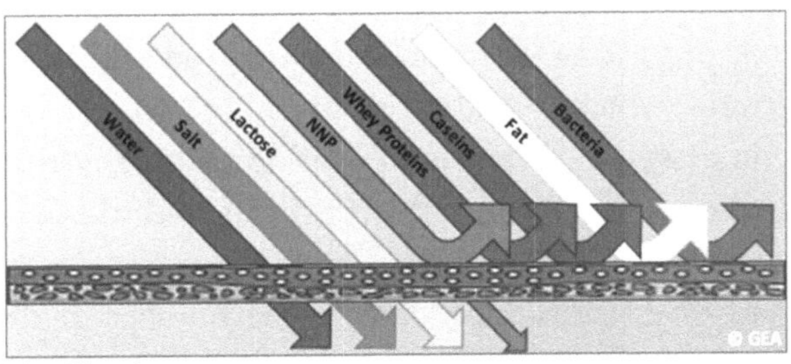

Ultrafiltration

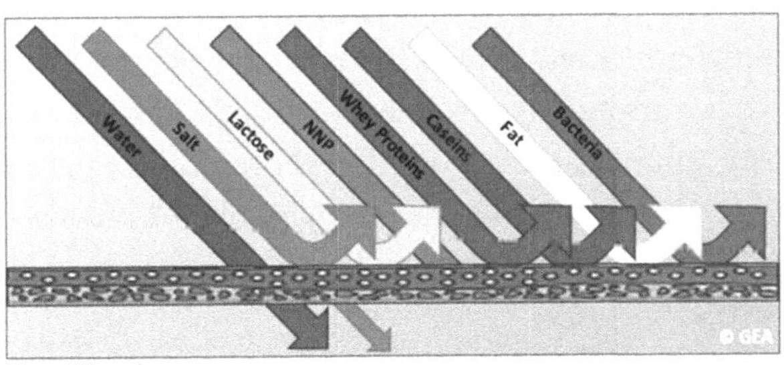

Nanofiltration

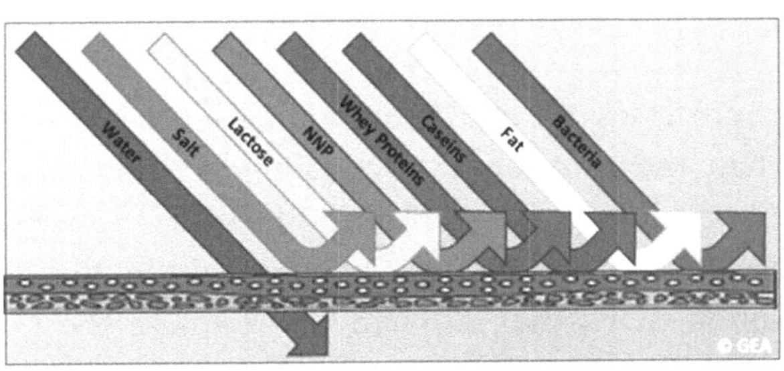

Reverse Osmosis

Figure 2.3 Behavior of the membrane filtration processes against the milk components (GEA Process Engineering 2012).

A particular UF process is the diafiltration (DF). In the DF processes, the retentate is diluted with water and recirculated, thus achieving a protein content of 80% of the total solids.

Retentate obtained by UF or DF processes is spray dried and is generally known as whey protein concentrate (WPC).

By combining MF (the process which is described below) for fat removal and UF followed by DF for later stage, and optimizing the filtration parameters, a retentate containing 90–95% of proteins in the total solids can be obtained (Synder filtration 2019). This retentate after spray-drying is known as whey protein isolate (WPI).

Microfiltration is a low pressure-driven filtration process (pressure values generally lower than 70 kPa) (Cavalli et al. 2002). Membranes have open structures with pore size of 100–2000 nm. Depending on the pore size, membranes retain bacteria, spores, molds, and yeasts (for the highest pore sizes) or also fat and caseins (for the lowest ones) (Figure 2.3). By retaining bacteria, molds, and yeasts, MF can be considered a nonthermal pasteurization process (Twiford 2004). MF membranes are usually made of ceramic (inorganic) materials. Although more expensive, ceramic membranes are very resistant to temperature and chemicals, easy to clean, and with long lifetimes. In the processing of liquid dairy by-products, the MF processes are usually used for bacteria reduction and fat removal.

Electrodialysis processes deserve special consideration. ED is an electrochemical separation process where, under the influence of an electric field, the ions travel through selective membranes. This movement is directly proportional to the specific conductivity and inversely proportional to the number of dissolved molecules (Gardais 1990). Charged ions can be removed from the dairy by-products (i.e., whey) using membranes containing ion-exchangeable groups and subjected to potentials of 1–2 V. Membranes, generally made from styrene divinylbenzene, contain sulfonic, carboxylic, arsenic, or phosphoric groups in the case of cationic membranes, or quaternary ammonium groups for the anionic membranes. To achieve efficient operation, an anionic membrane must be placed after a cationic one, generally separated by a distance of 1 mm. ED is used for whey desalting processes (Diblíková et al. 2010).

3.2 Ion-Exchange Techniques

The ion-exchange processes for purifying proteins are based on the ability of some resins to retain the proteins at a certain pH value and leave them free, thus being able to be eluted, at a different pH value. Ion-exchange resins are polymers that are able to exchange specific ions within the polymer with ions contained in a solution that passes through the resin. In short, it is about changing the surface charge attraction between the desired protein molecules and the ion-exchange resin to retain and isolate such proteins. Although it is possible to use membrane filtration for the fractionation of proteins with similar molecular weight (Zydney 1998), none of these techniques has been effectively implemented at a commercial scale. In relation to the classical membrane techniques, the ion-exchange processes allow selective retention and purification of the different whey proteins and separation of proteins that have little or no differences in molecular mass and size. This is an important aspect because purified individual milk proteins show better functionality than in their native protein mixtures (Imafidon et al. 1997).

To purify the protein whey fraction by using this technique, the pH value of the pretreated whey is adjusted to 3–3.5, generally using phosphoric acid. Next, the whey is introduced into a resin tank under stirring conditions, and the whey proteins are adsorbed to the resins. After removal the deproteinized whey from the tank, the proteins adsorbed are eluted (removed from the resin) by increasing and adjusting the pH value with a sodium hydroxide solution until reaching the isoelectric points of the retained proteins. Finally, the pH value of the eluted fraction is restored to the desired pH value. This process allows obtaining of WPI with a protein content of 90–96% of total solids.

WPI obtained using ion-exchange techniques have high levels of α-lactalbumin, β-lactoglobulin, bovine serum albumin, and immunoglobulins, but contrary to the WPI obtained by membrane filtration lacks important protein fractions such as lactoferrin, lactoperoxidase, and glycomacropetide. Although lactoferrin and lactoperoxidase are bounded by the ion-exchange resin, their isoelectric points (7.9 and 9.0–10.0, respectively) demands to achieve very high pH values in the eluate, and the process is not viable in economic terms. These retained proteins are allowed to remain on the resin until the cleaning-in-place operation. Also, the WPI obtained from membrane processes have peptide fragments coming from the enzymatic activity upon the other proteins present. These peptide fragments, which are absent in the ion-exchange WPI are easily digested and absorbed and add nutritional value to the WPI. Moreover, ion-exchange resins do not act as filters and therefore they do not remove bacteria or other microorganisms.

Finally, another inconvenience has been described. The pH adjustments can denature proteins and therefore reduce their biological and functional activities. Also, due to protein denaturation, ion-exchange WPI has noticeable flavor differences, and it tastes "processed," salty, and with fewer dairy notes.

Ion-exchange techniques are also a real alternative or complementary step for whey demineralization. For this purpose, whey is first circulated through a synthetic resin that is negatively charged (cation exchanger). In this exchanger, the ions H^+ from the resin are exchanged and replaced with Na^+, K^+, Ca^{2+}, Mg^{+2}, etc. ions from the whey that are trapped in the column. Next, whey passes through an anion exchanger, which is a positively charged resin. In this second resin, OH^- ions from the resin are replaced with anions from the whey (PO_4^{3-}, Cl^-, etc.) that were trapped in the resin. Cation resins typically contain sulfonic acid groups (e.g., sodium polystyrene sulfonate or poly (2-acrylamido-2-methyl-1-propanesulfonic acid®) or carboxylic groups (weakly acidic resins), while anion resins contain quaternary amino groups (e.g., trimethylammonium groups - poly(acrylamido-*N*-propyltrimethylammonium chloride -) or primary, secondary, and/or tertiary amino groups (weakly basic resins). These techniques require a small initial economic investment, but the costs of operation are high. The most complex and expensive step is the resin regeneration using acid and alkaline solutions. For the rest, the automation of installation process does not present any difficulty, and the energetic requirements are minimal.

After a pretreatment for fat removal, the use of a NF step, followed by an ED and an ion-exchange demineralization allow elimination of 90% of the minerals initially present in whey (Novasep Inc. 2014). Demineralized whey is destined for the manufacture of infant formula and of dietetic foods (e.g., for chronic kidney disease) that require the presence of lactose and nutritional quality of whey proteins, but do not tolerate appreciable mineral contents.

3.3 Protein Hydrolysis and Recovery of Bioactive Peptides

3.3.1 General Aspects Whey proteins whose extraction procedures we just described have the highest nutritional quality, with protein digestibility-corrected amino acid score (PDCCAS) = 1. They are rich in human-essential amino acids, including branched-chain amino acids and sulfur-containing amino acids with significant roles as metabolic regulators, and precursors of molecules such as glutathione with important roles in live cells (Smithers 2015).

Whey proteins have also proved to have antimicrobial, anti-cancerous, and anti-inflammatory properties, also promoting several health and therapeutic benefits on the cardiovascular, gastrointestinal, endocrine, immune, locomotor and nervous human systems, and delayed ageing process (Gupta and Prakash 2017).

Because of these biological functions, whey proteins are more suitable for use in functional foods. However, most of the bioactivities of whey proteins are due to peptides (Brandelli et al. 2015) that are encrypted in their native polypeptidic sequences and that must be liberated so that these bioactivities can manifest themselves. Liberation of these peptides is preferentially carried out by proteolytic processes, generally enzyme catalyzed (Pihlanto-Leppälä 2000). The natural liberation of these bioactive peptides only occurs during the digestion of the milk proteins under the action of digestive enzymes or during some food processes such as fermentation (fermented milks) or ripening (ripened cheeses) under the enzymatic activities of the lactic acid bacteria (Korhonen 2009). However, natural digestion of the whey proteins from the dairy products consumed, or supply made by fermented/ripened dairy products in the diet could not be sufficient to achieve a positive health effect. In some other cases, specific enzymatic activities operating in the gastrointestinal tract or displayed by the lactic acid bacteria in the dairy products may lack the specificity needed to release some specific peptides with defined and concrete activities. In such a scenario, to make use of their potential, the peptides responsible for the beneficial health effects must be produced via industrial processes of hydrolysis of the whey proteins (Gauthier et al. 2006).

3.3.2 Strategies and Processes for Obtaining Bioactive Peptides The common experimental procedure for recovery and study of bioactive peptides follows a protocol with several steps: (i) isolation of individual proteins or a protein mixture; (ii) hydrolysis of proteins by means of proteases—these can include gastrointestinal enzymes such as pepsin, trypsin, and chymotrypsin, proteases of plant or microbial origin such as alcalase, bromelain, and ficin, or enzymes from starter lactic acid bacteria (LAB), non-starter LAB, or from yeasts associated with fermentation/ripening of dairy products; (iii) preliminary screening of the activities of protein hydrolysates; (iv) separation and purification of the bioactive peptides, a task combined with intermediate bioactivity tests; (v) amino acid sequencing and identification of peptides having interesting activities; (vi) confirmation of the activity in such peptides synthesized de novo; and (vii) performance of tests in vitro and in vivo conditions (Li-Chan 2015; Agyei et al. 2016; Capriotti et al. 2016).

The production at industrial level of whey protein hydrolysates (WPH) with high content of bioactive peptides includes the following steps: (i) isolation of whey proteins using some of the fractionation techniques already described in this chapter; (ii) pretreatment steps before protein hydrolysis—heat or high pressure; (iii) enzymatic hydrolysis of whey proteins—using commercial enzymes such as pancreatin, pepsin, or trypsin, bacterial or fungal proteases from microbial fermentations, or plant enzymes such as bromelain or papain; (iv) post enzymatic treatments that can include diverse processes—separation by filtration techniques, centrifugation, chromatographic steps, etc. ; and (v) spray-drying and packaging of the manufactured hydrolysates (Pasupuleti and Braun 2008). This whole industrial process includes crucial operations that should be carefully carried out for optimal results; otherwise, the bioactivity of peptides could be altered or even destroyed. These operations are the pretreatment steps, the steps of protein hydrolysis, the purification of the bioactive peptides, and the final operation of spray-drying (Dullius et al. 2018).

The usual protocols of protein hydrolysis at large-scale level make use of food-grade enzymes (Lafarga and Hayes 2017). The use of proteases from selected LAB strains seem to be the best alternative (Agyei and Danquah 2011) because of the cheap culture of the LAB and the less laborious enzyme purification. Moreover, the proteolytic systems of LAB include a wide diversity of proteases that offers a variety of enzymatic specificities to produce concrete desired bioactive peptides (Liu et al. 2010; Hati et al. 2018). The simplest method to perform the protein hydrolysis is a continuous process using free enzymes coupled to an UF membrane module

(Korhonen and Pihlanto 2006) that retains the enzyme and at the same time acts as the first purification step of the bioactive peptides (Dullius et al. 2018). The use of immobilized enzymes, however, offers several advantages in terms of control of the process and recycling of the immobilized enzymes. To purify the bioactive peptides for the manufacture of peptide-based food ingredients the membrane separation techniques (NF and UF) using appropriate pore size in the membranes are the only methods with the possibility of industrial application in economically viable processes (Agyei et al. 2016). Of course, more effective purification processes such as chromatography techniques would offer more effective and specific enrichment; however, these techniques are very expensive and difficult to apply for large-scale production. Moreover, these purification processes could affect the structure of peptides reducing or even eliminating their bioactive properties (Agyei et al. 2016).

The processes for obtaining bioactive peptides-enriched ingredients from whey proteins for the manufacture of functional foods are still laborious, complex, and expensive, and it is necessary to travel a long and hard road of research and development to scale this level until putting such products in the market. A proof of these difficulties is the fact that, despite the efforts made in this field in the past decades, only seven products of this type from whey proteins are available nowadays in the market (Dullius et al. 2018). Such are *Biozate® product line* (from Davisco, USA, that reduces the blood pressure), *BioPureGMP™* (from Davisco, USA, with its activity in preventing dental caries, blood clotting, antibacterial, and antivirus), *Vivinal® ALPHA* (from Borculo Domo Ingredients (BDI), the Netherlands, that promotes relaxation and sleep), *Praventin™* (from DMV International, the Netherlands, with positive effects in acne reduction), *Dermylex™* (from Advitec, Inc., Canada, that reduces the symptoms of psoriasis), *Hilmar™8390* (from Hilmar Ingredients, USA, with antidiabetic activity), and *NOP-47™* (from Glanbia Nutritionals, USA, with anti-inflammatory properties).

4 Innovative Extraction Methods Applied to Dairy By-Products

4.1 Affinity Chromatography

Affinity chromatography is a promising method for the isolation of concrete proteins from whey. Usually, this method is based on the use of specific antibodies against the milk protein that will be isolated, which are first immobilized to a bed of an appropriate nature (e.g., resins) packed in columns. By circulating the whey or the WPI through the column, proteins are adsorbed by binding with their specific antibodies and they get trapped in the column. After adsorption, usually, the column is rinsed with several volumes of washing buffer (e.g., phosphate buffer). Next, the proteins are desorbed by rinsing the column with a buffer with an appropriate pH value. Following this technique, several whey proteins such as lactoferrin (Tu et al. 2002; Chen et al. 2007), β-lactoglobulin

(Puerta et al. 2002), and some minor proteins (Ben Ounis et al. 2008) or bovine serum albumin (Besselink et al. 2015) have been isolated. However, this procedure is very expensive, and its suitability is limited to analytical purposes (Besselink et al. 2015).

4.2 Magnetic Separation

The magnetic nanotechnology emerges as a valuable alternative to purify the whey proteins. The existing information in the literature regarding this purification procedure was very recently reviewed (Nicolás et al. 2019). Magnetic nanoparticles (MNPs) can be formulated to bind selectively the whey proteins. Next, magnetic decantation allows protein isolation by means of a magnetic field. Finally, the proteins are eluted from the MNPs. Therefore, MNPs can be considered as a particular stationary phase in liquid affinity chromatography, but in contrary to the column-filling materials, MNPs offer the possibility of solid-liquid extraction in batch. MNPs have shown a high efficiency in the protein purification from their native media (Gädke et al. 2017). One of the advantages of this procedure is that, after appropriate elution of bound proteins, the regenerated MNPs can be used in successive purification cycles (Zhou et al. 2018). Moreover, their production and final discarding are harmless to the environment.

The MNPs consist of an inorganic magnetic core (usually magnetite/maghemite) coated with a shell of diverse nature (both small and polymeric molecules can be used for this purpose). Their preparation can be made through several methods (Nicolás et al. 2019). The magnetic core can be chemically modified to provide them a selective affinity toward the target protein (Nicolás et al. 2019).

4.3 Supercritical Fluid Extraction

As indicated in previous sections, buttermilk contains not negligible quantities of components of the milk fat globule membrane (MFGM), mainly phospholipids. Apart from the health beneficial effects of these components (Noh and Koo 2004; Spitsberg 2005; Fong et al. 2007), phospholipids have emulsifying properties, and they are largely used on the basis of this characteristic to improve the sensory attributes of several foodstuffs (Dewettinck et al. 2008).

The outstanding properties of the MGFM components increased in the past years the interest to effective methods for their isolation and concentration. The microfiltration has been applied to buttermilk (Morin et al. 2007) and to whey-cream buttermilk (Morin et al. 2006) to this purpose, but the presence of casein micelles in buttermilk restricts the concentration of MGFM components, since microfiltration concentrates together the casein micelles and the MGFM fragments. Ultrafiltration followed by diafiltration and by supercritical fluid extraction applied to whey buttermilk (Costa et al. 2010) and microfiltration or microfiltration followed by diafiltration, in both cases followed by supercritical fluid extraction, applied to sweet-cream buttermilk and to whey buttermilk (Olabi et al. 2015) allowed the obtaining of products enriched in MGFM phospholipids and proteins.

The supercritical fluid extraction is carried out using CO_2 as a solvent because it can easily reach supercritical conditions, and has low toxicity and cost, wide availability, and environmental friendliness (Herrero et al. 2010). In addition to all these advantages, CO_2 is a good solvent for milk triglycerides that does not dissolve polar lipids of milk (Olabi et al. 2015). Supercritical fluid extraction conditions applied to filtration products were conveniently described by Astaire et al. (2003).

4.4 Use of Ultrasound in Processing Steps

Ultrasounds are defined as sound waves of frequency higher than 20 kHz, which when they pass through a medium generate acoustic cavitation and induce vibrations on the surfaces. The acoustic cavitation is a result of the pressure fluctuations generated within the liquids as a consequence of the passage of ultrasound wave. The cavitation generates pulses, turbulence, liquid circulation, and shock waves (Zisu et al. 2010; Chemat et al. 2011). The surface vibrations are useful in avoiding the caking formation in the filtration processes and also in fouling disruption for cleaning. All these physical effects promote intensification of the processes of mass transport.

The available information on the use of ultrasound for intensification of the recovery of products from cheese whey was reviewed by Gajendragadkar and Gogate (2016). Briefly, ultrasound is applied in cheese whey fractionation in four different steps of whey processing: pretreatment, UF, spray drying, and crystallization of lactose.

For many decades, thickening or gelling of protein-based dairy products and ingredients upon heating has been a recurrent problem in dairy processing. This phenomenon limits options for the manufacture of dairy products and reduces manufacturing efficiencies by limiting the type and extent of heat treatment applicable. To induce whey protein aggregate formation in whey products, preheating treatments are used. However, the application of ultrasound for a short time after such a heating step breaks down these aggregates and prevents their reformation on subsequent heating, thereby reducing the viscosity increase that is usually associated with this process (Ashokkumar et al. 2009b). This treatment therefore increases the possibilities of using whey proteins in different foodstuffs in food industry. Such ultrasound treatment also increases the heat stability of the whey proteins that retain their quality even after freeze- or spray-drying and subsequent reconstitution (Ashokkumar et al. 2009b).

Although sonication results in small changes in the thermal behavior of whey proteins increasing their heat stability (Ashokkumar et al. 2009b), these treatments do not produce significant changes in protein structure (Chandrapala et al. 2011) nor in the functional properties, and therefore the nutritional value and the functional performances of the whey proteins (the two main purposes of their use) are maintained (Ashokkumar et al. 2009b; Zisu et al. 2010).

The fouling of UF membranes is a major problem in the dairy industry. Fouling is produced by the deposition and adhesion of particles on the surface of the filtration membranes, which reduces the permeate flux. Ultrasound application during filtration enhances permeate flux and reduces the need for cleaning cycles in the whey UF process. Physical effects of ultrasound cause removal of the fouling layer and the disruption of the aggregates of whey proteins, thus enhancing the flux by a factor between 1.2 and 1.7 (Muthukumaran et al. 2005).

As previously indicated, drying or spray-drying is the final step in the manufacture of WPC or WPI powders. In industrial practice, centrifugal and pressure atomizers are the most commonly used devices for the atomization of the concentrate during spray dehydration. The use of the ultrasound atomization provides an alternative to the centrifugal and pressure atomizers in spray-drying with several important advantages (Gajendragadkar and Gogate 2016). Ultrasonic nozzles are more reliable, precise, and controllable. Besides, they can handle very low velocities without obstruction. Ultrasonic spray dryers with either nozzles or nebulizers produce a spray with a very low velocity and therefore the size of the chamber for drying can be reduced when compared to conventional spray-drying that reduces the cost of the process (Ashokkumar et al. 2009a).

The crystallization of solutes involves three different steps: the supersaturation of the solute, the nucleation or formation of nuclei, and the crystal growth. The use of ultrasound during the crystallization process can enhance the rate of nuclei formation and crystal growth and this alternative is called "sonocrystallization". Also, sonocrystallization allows obtaining crystals of uniform and desired size through the choice of the appropriate sonication conditions (i.e., frequency, intensity, power, and pulsing) (Ruecroft et al. 2005).

The crystallization of lactose is the final step in the lactose recovery from whey for further use. The improvement of the classical crystallization procedure (consisting of concentration by evaporation until supersaturation and crystal formation under stirring conditions) would be an undoubted advantage for industries in this sector. The sonocrystallization of lactose has been studied in different types of lactose solutions, using diverse compounds such as anti-solvents, and using different ultrasound variables (Gajendragadkar and Gogate 2016). When sonocrystallization of lactose is performed in concentrated whey solution, ultrasound exposure provoked a rapid nucleation, and the rate of crystallization was higher compared to the mechanical stirring. The yield of lactose cannot be improved in a significant way by sonocrystallization, but a narrow crystal size distribution and smaller lactose crystal sizes were obtained (Zisu et al. 2014).

4.5 Improved Separation of Proteins Using Charged Ultrafiltration Membranes

As indicated in Section 4.1, affinity chromatography is a very powerful technique available for purification of biomolecules. However, membrane separation processes are simplest, more efficient in terms of energy consumption, and more easily scalable from laboratory to industry level. Despite the efforts carried out in developing new membrane materials and more efficient conformations and configurations, the membrane technologies failed when trying to separate proteins with a similar size (Valiño et al. 2014). A recent report, however, demonstrates that the use of charged UF membranes in combination with optimum process conditions allows the separation of a mixture of bovine serum albumin and lactoferrin (Valiño et al. 2014), two proteins with comparable molecular weights and with functional properties that make its separation very difficult.

5 Properties and Applications of Dairy By-Products and Their Molecules

Because of the wide nature and properties of their components, the dairy by-products (mainly cheese whey) and their isolated compounds have wide applications in the food industry to such an extent that this particular topic could be per se the subject of a chapter or even of a whole book. Some of the properties and applications of whey were already commented or discussed in the previous sections of this chapter.

The whey is part of a large number of processed foods. The regulations in different countries allow using whey powder in bread, ice creams, melted cheese, and in manufactured meat, fish, and poultry products. It is also used in many non-standardized foods such as candies and cakes. One possibility of using whey powder is by mixing with other ingredients (e.g., proteins from animal or plant origin) to prepare mixes for specific purposes and with concrete functional properties adapted to the intended application. There is a great variety of these mixes that are obtained by adjusting the protein content and by choosing the most appropriate protein and the most adequate flavor. Examples of typical mixes are the following: (i) whey powder plus caseinates for ice cream manufacture; (ii) whey powder plus soy proteins for cakes; (iii) whey powder plus skim milk powder for ice creams; (iv) whey powder plus proteins of animal or plant origin for chocolate products, etc.

The delactosed permeate, that is, the liquid remaining after obtaining the WPC or WPI and after crystallization of the lactose, is currently still underutilized, but it has a promising potential for future applications in the food industry (Oliveira et al. 2018).

The whey and whey permeate can be biotransformed (through yeast fermentation) into proteinaceous feed, food-grade bioprotein/single-cell protein, and some other interesting products (Mollea et al. 2013; Yadav et al. 2015).

Whey proteins are, of course, the most valuable fraction with the wider and most important applications. In addition to its already commented nutritional value, whey proteins have important functional properties such as emulsifying (Moussier et al. 2019), gelling (Lorenzen and Schrader 2006), foaming (Pein et al. 2018), and thickening agent (Wang et al. 2017). These properties, which can be modified or improved through several processing operations (Uluko et al. 2016), enhance positive rheological and sensory attributes of foods and drinks (Kelleher et al. 2018). In this sense, yogurt and the positive effects of WPI adding on this dairy product is the most recurrent topic (Mahomud et al. 2017).

The gelling properties of whey proteins are the basis of the formation of films and coatings that are flexible, transparent, colorless, tasteless, and edible (if the addition of any compound that compromises this property is not done), and that are promissing in the field of food packaging and preservation (Jooyandeh 2011; Murrieta-Martínez et al. 2018). WPI were used as the basis for producing films with good gas barrier and wetting properties (Galus and Kadzinska 2019). Mechanically strength and oxygen barrier performance of these films can be improved by incorporation of nanofillers such as vermiculite and montmorillonite clays (Müller et al. 2017), so that these films could become a real alternative to plastic polymers in the manufacture of food packaging. WPC- or WPI-based films can even behave like active packages by releasing lysozyme (antimicrobial) (Ozer et al. 2016) or essential oils (antioxidant) (Ribeiro-Santos et al. 2018) previously incorporated into the coatings.

Due to their specific properties, whey proteins are also suitable compounds for serving as carriers for encapsulation and release of bioactive compounds such as proteins (Raei et al. 2017), phenolic compounds (Akdeniz et al. 2018; Farias-Cervantes et al. 2018; Murthy et al. 2018; Mohammadian et al. 2019), essential oils (Fernandes et al. 2018), and plant extracts (Hu et al. 2019).

Finally, the health promoting effects of whey proteins deserve a special mention. Apart from the therapeutic properties described for the whey proteins and for their encrypted peptides already mentioned in Sections 2.4 and 3.3 of this chapter, WPI has demonstrated bone regeneration effects (Douglas et al. 2017). Whey protein dietary supplements showed an anti-tumor effect in some urogenital cancer (Bounous 2000). Also, diet supplementation with WPI increased the glutathione levels and improved nutritional status and immunity in cancer patients undergoing chemotherapy (Bumrungpert et al. 2018). Some of the beneficial effects of whey proteins at a cellular level are mediated by glutathione (Bounous and Gold 1991; Bounous 2000; Kent et al. 2003), a compound synthesized from Cys amino acid, which is largely supplied by the whey proteins. Glutathione (γ-glutamylcysteinylglycine) is the most abundant nonprotein thiol in mammalian cells. It acts as an antioxidant compound that limits oxidation-induced damage to lipids, proteins, and genetic material (Anderson 1998; Griffith 1999).

6 Conclusion

Dairy by-products are not a waste and neither one more among the valorizable residues from the food industry. They are valuable ingredients in the manufacture of several foods and are source of high-added value compounds with a promising future in food and health fields. Currently, the membrane filtration techniques are the most performing procedures for isolation or enrichment of interesting compounds from these by-products at the industrial level. These techniques still have room for improvement if they incorporate developments that allow the separation of molecules not only attending to their molecular size. However, more research and studies are needed to improve the extraction and treatment processes with the aim of obtaining more purified compounds and preserving or even enhancing the beneficial properties of these molecules.

References

Agyei, D., and Danquah, M. K. 2011. Industrial-scale manufacturing of pharmaceutical grade bioactive peptides. *Biotechnology Advances* 29:272–277.

Agyei, D., Ongkudon, C.M., Wei, C.Y., Chan, A.S., and Danquah, M.K. 2016. Bioprocess challenges to the isolation and purification of bioactive peptides. *Food and Bioproducts Processing* 98:244–256.

Agyei, D., Shanbhag, B.K., and He, L. 2015. Enzymes for food waste remediation and valorization. In: *Improving and Tailoring Enzymes for Food Quality and Functionality* (Yada, R.Y., ed.). pp. 123–145. Woodhead Publishing Series in Food Science, Technology and Nutrition, Woodhead Publishing, Cambridge, UK.

AHDB (Agriculture and Horticulture Development Board). 2017. https://dairy.ahdb.org.uk/market-information/processing-trade/dairy-product-production/world-dairy-product-production/#.XFIwUqCCGpo Accessed on 15 December 2018.

Akdeniz, B., Sumnu, G., and Sahin, S. 2018. Micorencapsulation of phenolic compounds extracted from onion (*Allium cepa*) skin. *Journal of Food Processing and Preservation* 42: e13648.

Akoum, O., Chotard-Ghodsnia, O., Ding, L.H., and Jaffrin, M.Y. 2003. Ultrafiltration of low-heat and UHT skim milks with shear-enhanced vibrating filtration system. *Separation Science and Technology* 38:571–589.

Ali, A.H. 2018. Current knowledge of buttermilk: Composition, applications in the food industry, nutritional and beneficial health characteristics. *International Journal of Dairy Technology* 70:1–14.

Anderson, M.E. 1998. Glutathione: An overview of biosynthesis and modulation. *Chemico-Biological Interactions* 111–112:1–14.

Ashokkumar, M., Bhaskaracharya, R., Kentish, S., Lee, J., Palmer, M., and Zisu, B. 2009a. The ultrasonic processing of dairy products – an overview. *Dairy Science and Technology* 90:147–168.

Ashokkumar, M., Lee, J., Zisu, B., Bhaskaracharya, R., Palmer, M., and Kentish, S. 2009b. Sonication increases the heat stability of whey proteins. *Journal of Dairy Science* 92:5353–5356.

Assenat, L. 1985. Le lait de brebis: composition et propriétés. In: *Laits et Produits Laitiers: vache, brebis, chèvre. Tome 3, Qualité, énergie et tables de composiction* (Luquet, F.M., ed.). pp. 281–319. Tec & Doc Lavoisier, Paris, France.

Astaire, J.C., Ward, R., German, J.B., and Jiménez-Flores, R. 2003. Concentration of polar MFGM lipids from buttermilk by microfiltration and supercritical fluid extraction. *Journal of Dairy Science* 86:2297–2307.

Baldasso, C., Barros, T.C., and Tessaro, I.C. 2011. Concentration and purification of whey proteins by ultrafiltration. *Desalination* 278:381–386.

Bazinet, L. 2005. Electrodialytic phenomena and their applications in the dairy industry: A review. *Critical Reviews in Food Science and Nutrition* 45:307–326.

Bell, S.J. 2000. Whey protein concentrates with and without immunoglobulins: A review. *Journal of Medicinal Food* 3:1–13.

Ben Ounis, W., Gauthier, S.F., Turgeon, S.L., Roufik, S., and Pouliot, Y. 2008. Separation of minor protein components from whey protein isolates by heparin affinity chromatography. *International Dairy Journal* 18:1043–1050.

Besselink, T., Janssen, A.E.M., and Boom, R.M. 2015. Isolation of bovine serum albumin from whey using affinity chromatography. *International Dairy Journal* 41:32–37.

Borzouee, F., Reza-Modif, M., Varshosaz, J., and Shariat, S.Z.A.S. 2016. Purification of lacto-peroxidase from bovine whey and investigation of kinetic parameters. *Advanced Biomedical Research* 5:189.

Bounous, G. 2000. Whey protein concentrate (WPC) and glutathione modulation in cancer treatment. *Anticancer Research* 20:4785–4792.

Bounous, G., and Gold, P. 1991. The biological activity of undenatured dietary whey proteins: Role of glutathione. *Clinical and Investigative Medicine* 14:296–309.

Brandelli, A., Daroit, D.J., and Corrêa, A.P.F. 2015. Whey as a source of peptides with remarkable biological activities. *Food Research International* 73:149–161.

Brans, G., Schroën, C.G.P.H., van der Sman, R.G.M., and Boom, R.M. 2004. Membrane fractionation of milk: State of the art and challenges. *Journal of Membrane Science* 243:263–272.

Brock, J.H. 2002. The physiology of lactoferrin. *Biochemistry and Cell Biology* 80:1–6.

Bumrungpert, A., Pavadhgul, P., Nunthanawanich, P., Sirikanchanarod, A., and Adulbhan, A. 2018. Whey protein supplementation improves nutritional status, glutathione levels, and immune function in cancer patients: A randomized, double-blind controlled trial. *Journal of Medicinal Food* 21:612–616.

Callewaert, L., and Michiels, C.W. 2010. Lysozymes in the animal kingdom. *Journal of Biosciences* 35:127–160.

Capriotti, A. L., Cavaliere, C., Piovesana, S., Samperi, R., and Laganà, A. 2016. Recent trends in the analysis of bioactive peptides in milk and dairy products. *Analytical and Bioanalytical Chemistry* 408:2677–2685.

Carneiro, L., Dos Santos Sa, I., Dos Santos Gomes, F., Martins Matta, V., and Corrêa Cabral, L.M. 2002. Cold sterilization and clarification of pineapple juice by tangential microfiltration. *Desalination* 148:93–98.

Carvalho, F., Prazeres, A.R., and Rivas, J. 2013. Cheese whey wastewater: Characterization and treatment. *Science of the Total Environment* 445–446:385–396.

Cavalli, A., Cortellino, G., Gervasoni, E., Monti, L., Carminati, D., and Rampili, M. 2002. Prove sperimentali di microfiltrazione del latte di capra: Aspetti compositivi e microbiologici. *Scienza e Tecnica Lattiero-Casearia* 53:107–120.

Chacón-Villalobos, A. 2006. Tecnologías de membranas en la agoindustria láctea. *Agronomía Mesoamericana* 17:243–264.

Chandrapala, J., Zisu, B., Palmer, M., Kentish, S., and Ashokkumar, M. 2011. Effect of ultrasound on the thermal and structural characteristics of proteins in reconstituted whey protein concentrate. *Ultrasonics Sonochemistry* 18:951–957.

Chemat, F., Zill-e-Huma, and Khan, M.K. 2011. Applications of ultrasound in food technology: Processing, preservation and extraction. *Ultrasonics Sonochemistry* 18:813–835.

Chen, L., Guo, C., Guan, Y., and Liu, H. 2007. Isolation of lactoferrin from acid whey by magnetic affinity separation. *Separation and Purification Technology* 56:168–174.

Christie, W.W., Noble, R.C., and Davies, G. 1987. Phospholipids in milk and dairy products. *International Journal of Dairy Technology* 40:10–12.

Corredig, M., and Dalgleish, D.G. 1997. Isolates from industrial buttermilk: Emulsifying properties of materials derived from the milk fat globule membrane. *Journal of Agricultural and Food Chemistry* 45:4595–4600.

Costa, M.R., Elias-Argote, X.E., Jiménez-Flores, R., and Gigante, M.L. 2010. Use of ultrafiltration and supercritical fluid extraction to obtain a whey buttermilk poder enriched in milk fat globule membrane phospholipids. *International Dairy Journal* 20:598–602.

Dajanta, K., Chukeatirote, E., and Apichartsrangkoon, A. 2008. Effect of lactoperoxidase system on keeping quality of raw cow's milk in Thailand. *International Journal of Dairy Science* 3:112–116.

Daufin, G., René, F., and Aimar, P. 1998. *Les separations par membrane dans les procédés de l'industrie alimentaire*. Tec & Doc Lavoisier, Paris, France.

Dewettinck, K., Rombaut, R., Thienpont, N., Le, T.T., Messens, K., and Van Camp, J. 2008. Nutritional and technological aspects of milk fat globule membrane material. *International Dairy Journal* 18:436–457.

Diblíková, L., Čurda, L., and Homolová, K. 2010. Electrodialysis in whey desalting process. *Desalination and Water Treatment* 14:208–213.

Douglas, T.E.L., Vandrovková, M., Krocilova, N., Keppler, J.K., Zárubová, J., Skirtach, A.G., and Bacáková, L. 2017. Application of whey protein isolate in bone regeneration: Effects on growth and osteogenic differentiation of bone-forming cells. *Journal of Dairy Science* 101:28–36.

Dullius, A., Goettert, M.I., and Volken de Souza, C.F. 2018. Whey protein hydrolysates as a source of bioactive peptides for functional foods – Biotechnological facilitation of industrial scale-up. *Journal of Functional Foods* 42:58–74.

Eigel, W.N., Butler, J.E., Ernstrom, C.A., Farrell, H.M., Harwalkar, V.R., Jenness, R., and Whitney, R.M. 1984. Nomenclature of proteins of cow's milk: Fifth revision. *Journal of Dairy Science* 67:1599–1631.

Elling, J.L., Duncan, S.E., Keenan, T.W., Eigel, W.N., and Boling, J. 1996. Composition and microscopy of reformulated creams form reduced-cholesterol butteroil. *Journal of Food Science* 61:48–53.

El-Loly, M.M. 2007. Identification and quantification of whey Immunoglobulins by reversed phase chromatography. *International Journal of Dairy Science* 2:268–274.

FAO (Food and Agriculture Organization). 2018. www.fao.org/faostat/en/#data. Accessed on 15 December 2018.

Farias-Cervantes, V.S., Chávez-Rodríguez, A., García –Salcedo, P.A., García-López, P.M., Casas-Solís, J., and Andrade-González, I. 2018. Antimicrobial effect and in vitro release of anthocyanins from berries and Roselle obtained via microencapsulation by spray drying. *Journal of Food Processing and Preservation* 42:e13713.

Fernandes, R.V.B., Brotel, D.A., Monteiro, P.S., Borges, S.V., Souza, A.U., and Mendes, L.E. S. 2018. Microencapsulated oregano essential oil in grated Parmesan cheese conservation. *International Food Research Journal* 25:661–669.

Fong, B.Y., Norris, C.S., and Macgibbon, A.K.H. 2007. Protein and lipid composition of bovine milk-fat-globule membrane. *International Dairy Journal* 17:275–288.

Gädke, J., Kleinfeldt, L., Schubert, C., Rohde, M., Biedendieck, R., Garnweitner, G., and Krull, R. 2017. In situ affinity purification of his-tagged protein A from *Bacillus megaterium* cultivation using recyclable superparamagnetic iron oxide nanoparticles. *Journal of Biotechnology* 242:55–63.

Gajendragadkar, C.N., and Gogate, P.R. 2016. Intensified recovery of valuable products from whey by use of ultrasound in processing steps – A review. *Ultrasonics Sonochemistry* 32:102–118.

Galus, S., and Kadzinska, J. 2019. Gas barrier and wetting properties of whey protein isolate-based emulsion films. *Polymer Engineering and Science* 59:E375–E383.

Gandhi, K., Pawar, N., Kumar, N., and Arora, S. 2013. Effect of vidarikand (extracts) on oxidative stability of ghee: A comparative study. *Research and Reviews: Journal of Dairy Science and Technology* 2:1–11.

Gardais, D. 1990. *Environnement et electricité. Procédés electriques de traitement des rejets industriels.* Dopée Editeur, Paris.

Gauthier, S.F., Pouliot, Y., and Saint-Sauveur, D. 2006. Immunomodulatory peptides obtained by the enzymatic hydrolysis of whey proteins. *International Dairy Journal* 16:1315–1323.

GEA Process Engineering. 2012. *Membrane Filtration in the Dairy Industry.* GEA Filtration, Hudson, WI.

Gésan-Guiziou, G., Daufin, G., and Boyaval, E. 2000. Critical stability conditions in skimmed milk crossflow microfiltration: Impact on operating modes. *Lait* 80:129–140.

Griffith, O.W. 1999. Biologic and pharmacologic regulation of mammalian glutathione synthesis. *Free Radical Biology and Medicine* 27:922–935.

Gupta, C., and Prakash, D. 2017. Therapeutic potential of milk whey. *Beverages* 3:31, 14 pages.

Gustafsson, L., Hallgren, O., Mossberg, A.K., Pettersson, J., Fischer, W., Aronsson, A., and Svanborg, C. 2005. HAMLET kills tumour cells by apoptosis: Structure, cellular mechanisms and therapy. *Journal of Nutrition* 135:1299–1303.

Hati, S., Patel, N., Sakure, A., and Mandal, S. 2018. Influence of whey protein concentrate on the production of antibacterial peptides derived from fermented milk by lactic acid bacteria. *International Journal of Peptide Research and Therapeutics* 24:87–98.

Herrero, M., Mendiola, J.A., Cifuentes, A., and Ibáñez, A. 2010. Supercritical fluid extraction: Recent advances and applications. *Journal of Chromatography A* 1217:2495–2511.

Hu, Y., Kou, G., Chen, Q., Li, Y., and Zhou, Z. 2019. Protection and delivery of mandarin (*Citrus reticulata* Blanco) peel extracts by encapsulation of whey protein concentrate nanoparticles. *LWT-Food Science and Technology* 99:24–33.

IFCN (International Farm Corporation Network). 2017. The Global Dairy World 2016/2017. International Farm Corporation Network, Kiel, Germany.

Imafidon, G.I., Farkye, N.Y., and Spanier, A. 1997. Isolation, purification, and alteration of some functional groups of major milk proteins: A review. *Critical Reviews in Food Science and Nutrition* 37:663–689.

James, B.J., Jing, Y., and Chen, X.D. 2003. Membrane fouling during filtration of milk - a microstructural study. *Journal of Food Engineering* 60:431–437.

Janghu, S., Kaushik, R., Bansal, V., Sharma, P., and Dhindwal, S. 2014. Physico-chemical analysis of ghee residue and conversion into confectionary food products. *Indian Journal of Dairy Science* 67:1–6.

Jooyandeh, H. 2011. Whey protein films and coatings: Review. *Pakistan Journal of Nutrition* 10:296–301.

Kelleher, C.M., O'Mahony, J.M., Kelly, A.L., O'Callaghan, D.J., Kilcawley, K.N., and McCarthy, N.A. 2018. The effect of direct and indirect heat treatment on the attributes of whey protein beverages. *International Dairy Journal* 85:144–152.

Kent, K.D., Harper, W.J., and Bomser, J.A 2003. Effect of whey protein isolate on intracellular glutathione and oxidant-induced cell death in human prostate epithelial cells. *Toxicology in Vitro* 17:27–33.

Kontopidis, G., Holt, C., and Sawyer, L. 2004. Invited Review: β-Lactoglobulin: Binding properties, structure, and function. *Journal of Dairy Science* 87:785–796.

Korhonen, H. 2009. Milk-derived bioactive peptides: From science to applications. *Journal of Functional Foods* 1:177–187.

Korhonen, H., and Pihlanto, A. 2006. Bioactive peptides: Production and functionality. *International Dairy Journal* 16:945–960.

Kussendrager, K.D., and van Hooijdonk, A.C. 2000. Lactoperoxidase: Physico-chemical properties, occurrence, mechanism of action and applications. *British Journal of Nutrition* 84: S19–S25.

Lafarga, T., and Hayes, M. 2017. Bioactive protein hydrolysates in the functional food ingredient industry: Overcoming current challenges. *Food Reviews International* 33:217–246.

Li, K. 2007. Ceramic membranes and membrane processes. In: *Ceramic Membranes for Separation and Reaction* (Li, K., ed.). pp. 1–20. John Wiley & Sons Ltd., Chichester, UK.

Li-Chan, E.C. 2015. Bioactive peptides and protein hydrolysates: Research trends and challenges for application as nutraceuticals and functional food ingredients. *Current Opinion in Food Science* 1:28–37.

Liu, M., Bayjanov, J.R., Renckens, B., Nauta, A., and Siezen, R.J. 2010. The proteolytic system of lactic acid bacteria revisited: A genomic comparison. *BMC Genomics* 11:36.

López Expósito, I., and Recio, I. 2006. Antibacterial activity of peptides and folding variants from milk proteins. *International Dairy Journal* 16:1294–1305.

Lorenzen, P.C., and Schrader, K. 2006. A comparative study of the gelation properties of whey protein concentrate and whey protein isolate. *Lait* 86:259–271.

Mahaut, M. 1992. Aplicaciones de la ósmosis inversa, ultrafiltración y microfiltración en la industria láctea. *Revista Argentina de Lactología* 6:19–44.

Mahomud, Md.S., Katsuno, N., Zhang, L., and Nishizu, T. 2017. Physical, rheological and microstructural properties of whey protein enriched yogurt influenced by heating the milk at different pH values. *Journal of Food Processing and Preservation* 41:e13236.

Matte, J.J., Guay, F., and Girard, C.L. 2012. Bioavailability of vitamin B12 in cows' milk. *British Journal of Nutrition* 107:61–66.

McManus, B.L., Korpela, R., Speakman, J.R., Cryan, J.F., and Cotter, P.D. 2015. Nilaweera KN. Bovine serum albumin as the dominant form of dietary protein reduces subcutaneous fat mass, plasma leptin and plasma corticosterone in high fat-fed C57/BL6J mice. *British Journal of Nutrition* 114:654–662.

Mohammadian, M., Salami, M., Momen, S., Alavi, F., and Emam-Djomeh, Z. 2019. Fabrication of curcumin-loaded whey protein microgels: Structural properties, antioxidant activity and in vitro release behaviour. *LWT-Food Science and Technology* 103:94–100.

Mollea, C., Marmo, M., and Bosco, L. 2013. Chapter 24. Valorisation of cheese whey, a by-product from the dairy industry. In: *Food Industry* (Mazzalupo, I., ed.). Intech Open Ltd., London, UK. http://dx.doi.org/10.5772/53159.

Morin, P., Britten, M., Jiménez-Flores, R., and Pouliot, Y. 2007. Microfiltration of buttermilk and washed cream buttermilk for concentration of milk fat globule membrane components. *Journal of Dairy Science* 90:2132–2140.

Morin, P., Pouliot, Y., and Jiménez-Flores, R. 2006. A comparative study of the fractionation of regular buttermilk and whey buttermilk by microfiltration. *Journal of Food Engineering* 77:521–528.

Moussier, M., Bosc, V., Michon, C., Pistre, V., and Chaudemarche, C. 2019. Multi-scale understanding of the effects of the solvent and process on whey protein emulsifying properties: Application to dairy emulsion. *Food Hydrocolloids* 87:869–879.

Mucchetti, G., Zardi, G., Orlandini, F., and Gostoli, C. 2000. The pre-concentration of milk by nanofiltration in the production of Quarg-type fresh cheeses. *Lait* 80:43–50.

Müller, K., Jesdinski, M., and Schmid, M. 2017. Modification of functional properties of whey protein isolate nanocomposite films and coatings with nanoclays. *Journal of Nanomaterials* Article ID 6039192:10 pages.

Munirathnamma, V., Gupta, V.K., and Meena, G.S. 2017. Effect of different extraction processes on the recovery of ghee residue proteins. *Indian Journal of Animal Science* 87:366–372.

Murrieta-Martínez, C.L., Soto-Valdez, H., Pacheco-Aguilar, R., Torres-Arreola, V., Rodríguez-Félix, F., and Márquez-Ríos, E. 2018. Edible protein films: Sources and behavior. *Packaging Technology and Science* 31:113–122.

Murthy, K.N.C., Monika, P., Jayaprakasha, G.K., and Patil, B.S 2018. Nanoencapsulation: An advanced nanotechnological approach to enhance the biological efficacy of curcumin. *ACS Symposium Series* 1286:383–405.

Muthukumaran, S., Kentish, S.E., Ashokkumar, M., and Stevens, G.W. 2005. Mechanisms for the ultrasonic enhancement of dairy whey ultrafiltration. *Journal of Membrane Science* 258:106–114.

Neelima, Sharma R., Rajput, Y.S., and Mann, B. 2013. Chemical and functional properties of glycomacropeptide (GMP) and its role in detection of cheese whey adulteration in milk: A review. *Dairy Science and Technology* 93:21–43.

Nicolás, P., Ferreira, M.L., and Lassalle, V. 2019. A review of magnetic separation of whey proteins and potential application to whey proteins recovery, isolation and utilization. *Journal of Food Engineering* 246:7–15.

Noh, S.K., and Koo, S.L. 2004. Milk sphingomyelin is more effective than egg sphingomyelin in inhibiting intestinal absorption of cholesterol and fat in rats. *Journal of Nutrition* 134:2611–2616.

Novasep Inc. 2014. *Processes for Demineralized Whey. Novasep and MEGA High-Efficiency Solutions*. Novasep LLC, Boothwyn, PA.

Olabi, A., Jinjarak, S., Jiménez-Flores, R., Walker, J.H., and Daroub, H. 2015. Compositional and sensory differences of products of sweet-cream and whey buttermilk produced by microfiltration, diafiltration, and supercritical CO_2. *Journal of Dairy Science* 98:3590–3598.

Oliveira, D., Puri, R., Fenelon, M.A., and O'Mahory, J.A. 2018. Delactosed permeate as a dairy processing co-product with major potential value: A review. *International Journal of Food Science and Technology* https://doi.org/10.1111/ijfs.14064.

Ozer, B.B.P., Uz, M., Oymaci P., and Altinkaya, S.A. 2016. Development of a novel strategy for controlled release of lysozyme from whey protein isolate based active food packaging films. *Food Hydrocolloids* 61:877–886.

Pasupuleti, V.K., and Braun, S. 2008. State of the art manufacturing of protein hydrolysates. In: *Protein Hydrolysates in Biotechnology* (Pasupuleti, V.K., and Demain, A.L., eds.). pp. 11–32. Springer, The Netherlands.

Pein, D., Clawin-Rädecker, I., and Lorenzen, P.C. 2018. Peptic treatment of beta-lactoglobulin improves foaming properties substantially. *Journal of Food Processing and Preservation* 42:e13543.

Permyakov, E.A., and Berliner, L.J. 2000. Alpha-lactalbumin: Structure and function. *FEBS Letters* 473:269–274.

Pihlanto-Leppälä, A. 2000. Bioactive peptides derived from bovine whey proteins: Opioid and ace-inhibitory peptides. *Trends in Food Science and Technology* 11:347–356.

Pouliot, Y. 2008. Membrane processes in dairy technology-From a simple idea to worldwide panacea. *International Dairy Journal* 18:735–740.

Proctor, V.A., Cunningham, F.E., and Fung, D.Y.C. 1988. The chemistry of lysozyme and its use as a food preservative and a pharmaceutical. *Critical Reviews in Food Science and Nutrition* 26:359–395.

Puerta, A., Jaulmes, A., De Frutos, M., Diez-Masa, J.C., and Vidal-Madjar, C. 2002. Adsorption kinetics of β-lactoglobulin on a polyclonal immunochromatographic support. *Journal of Chromatography A* 953:17–30.

Raei, M., Shahidi, F., Farhoodi, M., Jafari, S.M., and Rafe, A. 2017. Application of whey protein-pectin nano-complex carriers for loading of lactoferrin. *International Journal of Biological Macromolecules* 105:281–291.

Ribeiro-Santos, R., Ramos de Melo, N., Andrade, M., Azevedo, G., Machado, A.V., Carvalho-Costa, D., and Sanches-Silva, A. 2018. Whey protein active films incorporated with a blend of essential oils: Characterization and effectiveness. *Packaging Technology Science* 31:27–40.

Ruecroft, G., Hipkiss, D., Ly, T., Maxted, N., and Cains, P.W. 2005. Sonocrystallization: The use of ultrasound for improved industrial crystallization. *Organic Process Research and Development* 9:923–932.

Ryan, M.P., and Walsh, G. 2018. The biotechnological potential of whey. *Reviews in Environmental Science and Biotechnology* 15:479–498.

Sawyer, L. 2003. β-Lactoglobulin. In: *Advanced Dairy Chemistry 3rd ed.* (Fox, P.F., and McSweeney, P., eds.). pp. 319–386. Kluwer, Amsterdam, The Netherlands.

Schroeder, H.V., and Cavacini, L. 2010. Structure and function of immunoglobulins. *Journal of Allergy and Clinical Immunology* 125:S41–S52.

Smithers, G.W. 2015. Whey-ing up the options - Yesterday, today and tomorrow. *International Dairy Journal* 48:2–14.

Sodini, I., Morin, P., Olabi, A., and Jiménez-Flores, R. 2006. Compositional and functional properties of buttermilk: A comparison between sweet, sour, and whey buttermilk. *Journal of Dairy Science* 89:525–536.

Spitsberg, V.L. 2005. Bovine milk fat globule membrane as a potential nutraceutical. *Journal of Dairy Science* 88:2289–2294.

Stănciuc, N., Aprodu, I., Râpeanu, G., van der Plancken, I., Bahrim, G., and Hendricks, M. 2013. Analysis of the thermally induced structural changes of bovine lactoferrin. *Journal of Agricultural and Food Chemistry* 61:2234–2243.

Stănciuc, N., and Râpeanu, G. 2010. An overview of bovine α-lactalbumin structure and functionality. *Annals of the University* Dunarea de Jos *of Galati, Fascicle 6 – Food Technology* 34:82–93.

Synder filtration. 2019. http://synderfiltration.com/applications/industries/dairy-industry/whey-protein-concentrationisolate-production/. Accessed on 31 January 2019.

Tayefi-Nasrabadi, H., and Asadpour, R. 2008. Effect of heat treatment on buffalo (*Bubalus bubalis*) lactoperoxidase activity in raw milk. *Journal of Biological Sciences* 8:310–315.

Tu, Y.Y., Chen, C.C., Chang, J.H., and Chang, H.M. 2002. Characterization of lactoferrin (LF) from colostral whey using anti-LF antibody immunoaffinity chromatography. *Journal of Food Science* 67:996–1001.

Twiford, R. 2004. Membrane filtration offers a variety of applications for dairy. *Dairy Foods* 105:76–78.

Uluko, H., Liu, L., Lv, J.-P., and Zhang, S.-W. 2016. Functional characteristics of milk protein concentrates and their modification. *Critical Reviews in Food Science and Nutrition* 56:1193–1208.

Valiño, V., San Román, M.F., Ibáñez, R., and Ortiz, I. 2014. Improved separation of bovine serum albumin and lactoferrin mixtures using charged ultrafiltration membranes. *Separation and Purification Technology* 125:163–169.

Vanderghem, C., Bodson, P., Danthine, S., Paquot, M., Deroanne, C., and Blecker, C. 2010. Milk fat globule membrane and buttermilks: From composition to valorization. *Biotechnologie, Agronomie, Société et Environnement* 14:485–500.

Varnam, A.H., and Sutherland, J.P. 1994. Dairy protein products. In: *Milk and Milk Products. Technology, Chemistry and Microbiology* (Varnam, A.H., and Sutherland, J.P., eds.). pp. 159–182. Springer, New York.

Walstra, P., Geurts, T.J., Noomen, A., Jellema, A., and van Boekel, M.A.J.S. 1999. *Dairy Technology: Principles of Milk Properties and Processes*. Marcel Dekker, New York.

Wang, H., Wang, C., Wang, M., and Guo, M. 2017. Chemical, physiochemical, and microstructural properties, and probiotic survivability of fermented goat milk using polymerized whey protein and starter culture kefir mild 01. *Journal of Food Science* 82:2650–2658.

Wong, N.P., LaCroix, D.E., and McDonough, F.E. 1978. Minerals in whey and whey fractions. *Journal of Dairy Science* 61:1700–1703.

Yadav, J.S.S., Yan, S., Pilli, S., Kumar, L., Tyagi, R.D., and Surampalli, R.Y. 2015. Cheese whey: A potential source to transform into bioprotein, functional/nutritional proteins and bioactive peptides. *Biotechnology Advances* 33:756–774.

Yang, S.T., and Silva, E.M. 1995. Novel products and new technologies for use of a familiar carbohydrate, milk lactose. *Journal of Dairy Science* 78:2541–2562.

Zhou, Y., Yan, D., Yuan, S., Chen, Y., Fletcher, E.E., Shi, H., and Han, B. 2018. Selective binding, magnetic separation and purification of histidine-tagged protein using bio-polymer magnetic core-shell nanoparticles. *Protein Expression and Purification* 144:5–11.

Zisu, B., Bhaskaracharya, R., Kentish, S., and Ashokkumar, M. 2010. Ultrasonic processing of dairy systems in large scale reactors. *Ultrasonics Sonochemistry* 17:1075–1081.

Zisu, B., Sciberras, M., Jayasena, V., Weeks, M., Palmer, M., and Dincer, T.D. 2014. Sonocrystallisation of lactose in concentrated whey. *Ultrasonics Sonochemistry* 21:2117–2121.

Zydney, A.L. 1998. Protein separations using membrane filtration: New opportunities for whey fractionation. *International Dairy Journal* 8:243–250.

3

Extraction of Valuable Compounds from Meat By-Products

MIRIAN PATEIRO, PAULA BORRAJO, RUBÉN DOMÍNGUEZ, PAULO E.S. MUNEKATA, AND JOSE M. LORENZO

Centro Tecnológico de la Carne de Galicia, Rúa Galicia Nº 4, Parque Tecnológico de Galicia, San Cibrán das Viñas, 32900 Ourense, Spain

PAULO CEZAR BASTIANELLO CAMPAGNOL

Universidade Federal de Santa Maria, CEP 97105-900, Santa Maria, Rio Grande do Sul, Brazil

IGOR TOMASEVIC

Department of Animal Source Food Technology, University of Belgrade, Faculty of Agriculture, Nemanjina 6, 11080 Belgrade, Serbia

FRANCISCO J. BARBA

Nutrition and Food Science Area, Preventive Medicine and Public Health, Food Sciences, Toxicology and Forensic Medicine Department, Faculty of Pharmacy, Universitat de València, Avda. Vicent Andrés Estellés, s/n, 46100 Burjassot, València, Spain

Contents

1 Introduction

Meat is one of the most consumed foods of the world, since it is very nutritious due to the large amounts of vitamins B, amino acids, proteins, and minerals (Lombardi-Boccia et al., 2005). Its main highlight is its essential amino acids (Franco et al., 2010), Zn, and Fe content (Lombardi-Boccia et al., 2005). The amount of meat produced worldwide represents around 330 million tons, of which 63 million tons are produced in Europe, and Germany, Spain, France, and Italy being the countries with the largest population of primary livestock (FAOSTAT, 2016). Germany and Spain have the highest population of pigs (59.5 and 47.3 million heads, respectively), France of bovines (4.7 million heads), United Kingdom of sheep (14.6 million heads), and Spain and France of goats (1.2 and 1.1 million heads, respectively).

However, not all the body parts are suitable and safe for human consumption (Figure 3.1). It is estimated that the total production of these by-products is directly related with the number of animals slaughtered and the dressing percentages (Ockerman and Hansen, 2000). More than half of by-products generated from the live weight were not appropriate for human consumption (Irshad and Sharma, 2015). These products represent 66, 52, and 68% of live weight of cattle, pigs, and lambs (Jayathilakan et al., 2012). Animal by-products can be defined as the entire body of an animal, the parts of its body or products derived from animals that are not intended for human consumption (EC, 2009). To protect the population from potential hazards of these products, these are classified into three categories (EC, 2009). Category 1 includes those by-products that can only be used for disposal. Bone marrow, brains and spinal cords from cattle, sheep, goats, and from sick and dead animals are included in this category. This risk material is especially related with bovine spongiform encephalopathy (BSE). Category 2 includes those by-products that are not suitable for animal consumption, and category 3 those that are not intended for human consumption but for animal feed. Blood, feather, feet, heads, hides, hoof, horn, and skin were included in the last category (Lasekan et al., 2013).

Europe is the second major producer of these products (37.1% of world production), and together with Asia are the two largest consumers of meat by-products, especially those that have beef and lamb as source (Liu and Ockerman, 2001). Despite this classification, discussion about edible and nonedible by-products depends on the customs of each country and of the animal species used (Swan, 1999). The edible parts include brains, feet, head, heart, liver, lungs, spleen, stomach, throat, tongue, and tripes. These products can be used for human consumption, while nonedible parts include hides,

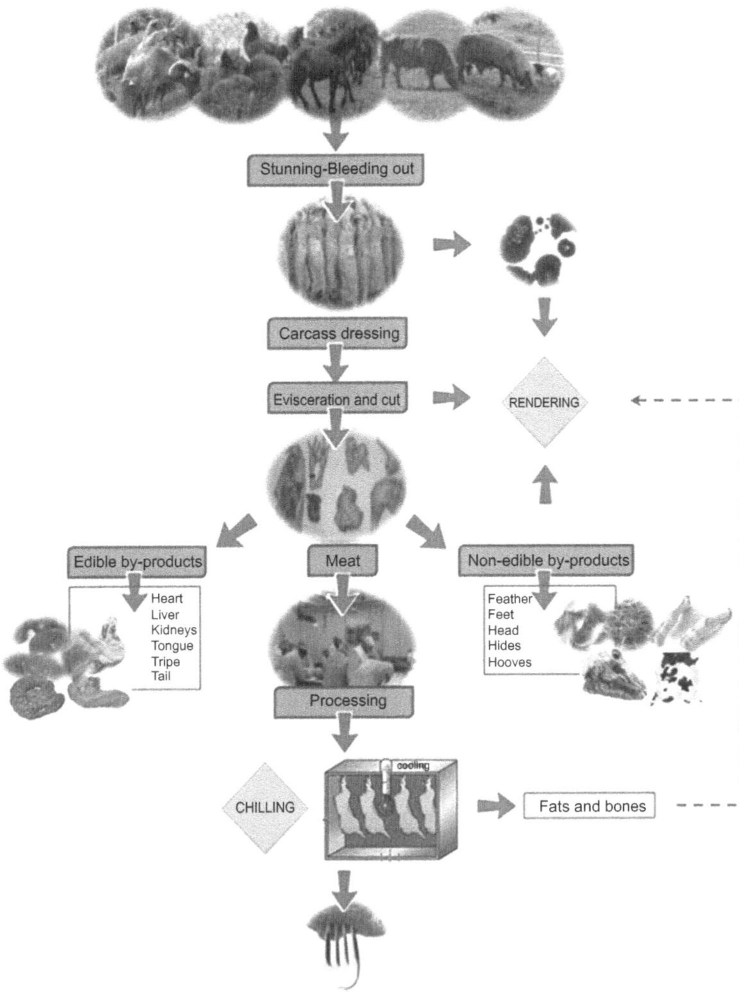

Figure 3.1 Generation of edible and nonedible meat by-products during the slaughtering process.

hoofs, skins, and stomach contents that can be transformed into high-added value products (Toldrá and Reig, 2011; Ockerman and Basu, 2014). These by-products have a high nutritional content, since proteins, lipids, minerals, and trace elements, as well as B vitamins and some liposoluble vitamins are part of their composition (Ockerman and Hansen, 2000; Pearl, 2004).

Processes such as fusion, composting, and other chemical, microbial, and thermal treatments applied to animal waste have been widely described and used. However, following the outbreak of BSE in the United Kingdom, the United States and some European countries have decreased the conversion of meat by-products into animal feed. This attached to the possible relationship between the consumption of meat infected with BSE and the development of the Creutzfeldt-Jakob disease has led to numerous investigations that are underway, trying to find alternative methods for the exploitation of animal by-products, seeking new sources of animal by-products that replace beef by chicken and pork.

Meat by-products have been the subject of many studies due to the large number of bioactive compounds that can be obtained from them, being blood and collagen among the most evaluated (Ryder et al., 2016). Some of the studies carried out in recent years about the use of viscera are focused from a descriptive point of view, such as the case of Florek et al. (2012) in which they determined the chemical composition, the mineral content, and the fatty acid profile of tongues, hearts, kidneys, and beef livers. On thc other hand, there are other works that show procedures to carry out this use, such as Salminen and Rintala (2002), which describes one of the most common methods to convert solid meat residues into products with added value, or "rendering." Another method frequently used today to enhance animal by-products is described by Gbogouri et al. (2004), consisting of the extraction of the protein by hydrolysis and obtaining a hydrolysate that will contain a mixture of peptides of different sizes and different amino acid composition. Hydrolyzed protein products have interesting functional properties such as emulsifying, foaming, gelling, and solubility, what makes them suitable for the production of derivatives (Klompong et al., 2007; Selmane et al., 2008). The future applications of the aforementioned by-products include the extraction of bioactive compounds with high-value added, which can be used in human nutrition, and with potential medicine and pharmaceutical uses (Toldrá et al., 2016). Collagens, enzymes, protein hydrolysates, and polyunsaturated fatty acids are present in these biomolecules (Ravindran and Jaiswal, 2016; Mullen et al., 2017).

The objective of the present chapter is to study the potential use of the residues generated by the meat industry through the use of blood, cuttings, organs and/or viscera to obtain products of high-added value, leading to the reduction of waste generated by the meat industry and the increase in the economic performance of this activity.

2 Nutritional Value of Meat By-Products

Meat by-products are rich in carbohydrates, lipids, proteins, micronutrients, and minerals (Tables 3.1 and 3.2), which makes them acceptable by consumers who become part of the diet (Aristoy and Toldrá, 2011; García-Llatas et al., 2011; Kim, 2011; Nollet and Toldrá, 2011).

In the case of proteins, by-products offer a good source of protein, inasmuch as today, new sources of protein are being sought, which would increase their value (Alao et al., 2017). The contents ranged between 9 and 22% in bovine, ovine, porcine, and poultry offal (Fornias, 1996; Seong et al., 2014a, 2014b, 2015; Mullen et al., 2017). Amino acid content determines the biological activity of the protein (Aristoy and Toldrá, 2011; Mullen et al., 2017). Essential amino acids are included in the composition of meat by-products; however, the important contents of connective tissue in by-products make amino acids such as glycine, hydroxyproline, and proline constitute an important part of their composition (Aristoy and Toldrá,

2011). These nutrients are present in significant amounts as by-products in ears, feet, lungs, stomach, and tripe. Low amounts of tryptophan and tyrosine are also present.

The fat contents are below 5% in organs such as heart, kidney, liver, lung, spleen, and thymus, while higher contents were found in brain, chitterlings, pancreas, skin, stomach, tongue, and rendering fats like lard, suet, and tallow. Carbohydrate content was as low as 1%, except for liver that showed percentages between 3.5 and 15%, glycogen being the main compound isolated (Honikel, 2011).

Vitamin content is usually found in higher amounts than those found in muscle tissues. In this sense, the liver is a complete organ because of the amount of ascorbic and folic acids, niacin, riboflavin, and vitamins A, B_6, and B_{12} present. The kidney can also be considered an important source of vitamins (folic acid, riboflavin, vitamins B_6, and B_{12}).

Important contents of macro-minerals (calcium, magnesium, phosphorous, potassium, and sodium) and micro-minerals (iron and copper) are isolated from kidney, liver, lungs, spleen, sweetbreads, and thymus (García-Llatas et al., 2011). Higher amounts of iron, copper, and manganese were found in liver, while phosphorous and potassium were mainly present in sweetbreads and thymus. Sodium is detected in low amounts in by-products; in fact, kidney, lungs, and spleen have minute amounts of sodium. Mechanically deboned meat stands out for its calcium content (Liu, 2002; Irshad and Sharma, 2015).

Furthermore, the presence of unsaturated fatty acids with contents usually higher than 50% confirms the nutritional value of meat by-products (Honikel, 2011). Some health organizations recommend increasing the consumption of these fatty acids, especially polyunsaturated fatty acids, as they are essential for a healthy diet (Gálvez et al., 2018). Notable contents were found in glands such as the brain, heart, kidney, liver, and lungs.

Regarding nutritional indices, the ratios PUFA/SFA and n-6/n-3 give an idea of the healthiness of these products. The recommendations for PUFA/SFA advise values around 0.85 in the human diet (FAO, 2010), whereas typical values of the Mediterranean diet are between 0.5 and 0.7 (Ulbricht and Southgate, 1991). Most by-products meet these specifications (Tables 3.1 and 3.2). The n-6/n-3 ratio is also very important in the human diet since a balanced ratio could prevent cardiovascular and other chronic diseases (Sanders, 1988; Williams, 2000; Simopoulos, 2008). Nutritional recommendations advised an increase in the consumption of n-3 to decrease the n-6/n-3 ratio (Wood et al., 2004; FAO, 2010). The values obtained in meat by-products (Tables 3.1 and 3.2) exceed the FAO nutritional recommendations for human diet (n-6/n-3 <4) (FAO, 2010).

Besides, the brain has beneficial effects on diet due to its phospholipid content (Küllenberg et al., 2012). In contrast, the high contents of cholesterol in the aforementioned organs limited their use, inasmuch as it is associated with cardiovascular diseases, one of the major causes of death in developed countries (WHO, 2003). The average contents are approximately around 150 mg/100 g; however, in organs

Table 3.1 Proximate composition and nutritional composition of main animal by-products

	HEART			LIVER			LUNG			SMALL INTESTINE		
	BOVINE	POULTRY	PORCINE	BOVINE	POULTRY	PORCINE	BOVINE	POULTRY	PORCINE	BOVINE	POULTRY	PORCINE
Proximate composition (%)												
Moisture	78.52	77.36	75.87	70.73	76.68	71.59	77.35	83.23	78.71	82.91	82.61	82.48
Protein	18.74	13.83	17.62	22.11	17.70	22.05	17.64	10.96	16.60	10.19	11.78	11.99
Fat	1.51	4.53	4.55	3.14	2.89	2.94	2.68	1.90	1.79	4.94	1.82	1.16
Ash	1.23	n/a	0.81	1.52	n/a	1.34	0.44	n/a	0.76	0.53	n/a	0.30
Cholesterol (mg/100 g)	150	175	150	300	550	300	290	n/a	320	160	n/a	n/a
Fatty acids (%)												
C16:0	15.50	24.26	16.50	10.1	22.79	21.50	34.52	33.37	33.65	26.68	25.22	26.64
C16:1*n*-7	0.94	4.86	1.58	0.78	1.69	1.60	1.45	4.43	0.91	1.86	4.64	0.90
C18:0	17.50	10.96	13.00	34.00	20.84	21.00	16.92	9.33	14.02	19.35	8.73	16.62
C18:1*n*-9	20.60	31.40	19.50	13.40	21.76	25.50	31.91	32.95	27.26	36.26	32.09	27.38
C18:2*n*-6	25.50	20.59	29.00	16.00	18.12	14.00	4.84	12.84	14.46	7.01	17.79	18.14
C18:3*n*-3	0.72	0.59	1.39	0.85	0.37	0.76	0.12	0.57	0.09	0.16	0.67	0.08
C20:4*n*-6	8.95	5.40	14.10	9.00	10.72	10.05	7.01	3.85	6.56	7.27	7.39	8.66
SFA	36.50	36.01	32.00	49.00	43.96	43.50	54.33	43.85	49.49	49.47	34.97	44.55
MUFA	22.50	36.82	22.00	15.00	23.98	29.50	33.77	37.89	28.65	38.68	37.26	28.63
PUFA	38.00	27.17	49.00	34.00	32.05	28.00	11.90	18.27	21.86	11.85	27.78	26.83
$\sum$ *n*-6	35.34	26.36	45.18	28.48	29.64	25.43	11.77	17.33	21.36	11.69	26.61	26.24
$\sum$ *n*-3	2.24	0.82	3.69	5.63	2.41	3.15	0.12	0.94	0.51	0.16	1.17	0.59
n-6/*n*-3	15.78	32.17	12.24	5.05	12.35	8.07	98.08	18.51	42.72	73.06	22.75	52.68
PUFA/SFA	1.04	0.76	1.53	0.69	0.73	0.64	0.22	0.42	0.44	0.27	0.80	0.62

Amino acids (%)												
Essential												
Lysine	1.53	1.15	1.05	1.42	1.34	1.33	1.14	0.79	1.17	0.80	0.89	0.67
Methionine	0.38	0.34	0.21	0.32	0.38	0.36	0.25	0.18	0.20	0.17	0.26	0.15
Threonine	0.85	0.63	0.39	0.84	0.79	0.45	0.65	0.46	0.22	0.42	0.53	0.21
Leucine	1.66	1.14	0.71	1.78	1.41	1.05	1.37	0.82	0.72	0.70	0.90	0.50
Isoleucine	0.69	0.53	0.76	0.66	0.65	1.02	0.46	0.35	0.61	0.31	0.44	0.46
Phenylalanine	0.76	0.60	0.94	1.01	0.76	1.29	0.80	0.41	1.61	0.41	0.48	1.01
Valine	0.79	0.62	1.52	0.84	0.83	2.01	0.82	0.50	1.31	0.40	0.52	0.88
Histidine	0.50	0.2	0.72	0.65	0.43	1.10	0.60	0.28	0.79	0.41	0.26	0.43
Arginine	1.10	0.85	1.00	1.02	1.04	1.22	0.96	0.65	0.91	0.66	0.80	0.78
Nonessential												
Serine	0.79	0.60	0.61	0.86	0.80	0.87	0.74	0.49	0.40	0.46	0.55	0.33
Cysteine	0.22	0.18	0.47	0.29	0.27	0.64	0.23	0.14	0.47	0.13	0.15	0.32
Tyrosine	0.59	0.48	2.53	0.66	0.59	2.93	0.50	0.29	1.97	0.29	0.41	1.47
Alanine	1.21	0.77	0.74	1.20	0.98	1.11	1.26	0.65	0.73	0.71	0.68	0.52
Aspartic acid	1.67	1.24	1.43	1.68	1.49	1.69	1.31	0.87	1.09	0.80	1.03	0.89
Glutamic acid	2.93	2.03	1.49	2.24	2.23	2.11	1.85	1.32	1.30	1.24	0.15	0.81
Glycine	0.88	0.69	0.55	0.93	0.81	0.77	1.57	0.72	0.48	0.87	0.78	0.35
Proline	0.80	0.61	0.78	0.85	0.71	1.11	0.97	0.55	1.08	0.63	0.58	0.69
Minerals (mg/kg)												
Macro-minerals (g/kg)												
Ca	0.05	0.09	0.06	0.06	0.09	0.07	0.09	0.15	0.13	0.09	0.14	0.18
Mg	0.20	0.20	0.20	0.17	0.22	0.21	0.11	0.14	0.14	0.08	0.16	0.13
P	2.17	1.94	1.87	2.90	2.93	2.48	2.15	1.94	1.89	1.20	1.47	1.19
Na	0.91	0.95	1.33	0.67	1.03	0.95	1.60	0.75	1.93	0.50	0.41	0.54
K	2.87	2.14	2.35	3.07	2.91	3.03	2.10	1.75	2.03	0.93	1.18	1.13

(*Continued*)

Table 3.1 (Cont.)

	HEART			LIVER			LUNG			SMALL INTESTINE		
	BOVINE	POULTRY	PORCINE	BOVINE	POULTRY	PORCINE	BOVINE	POULTRY	PORCINE	BOVINE	POULTRY	PORCINE
Micro-minerals (mg/kg)												
Fe	58.88	32.99	34.58	66.71	79.29	148.40	100.77	59.59	63.01	18.42	23.35	22.56
Zn	20.44	23.22	1.83	36.64	29.91	7.04	17.55	14.26	1.53	11.68	20.09	1.92
Cu	2.87	2.23	3.06	122.36	3.67	5.50	1.6	1.75	1.17	1.12	1.32	1.41
Mn	0.92	0.67	0.65	3.05	2.51	2.98	0.29	0.88	0.36	0.67	1.00	1.22

Data from Aristoy and Toldrá (2011); Honikel (2011); Prates et al. (2011); Florek et al. (2012); Seong et al. (2014a, 2014b); Seong et al. (2015); Tomovic et al. (2015); Mora et al. (2019); n/a: data not available

Table 3.2 Proximate composition and nutritional composition of secondary animal by-products

	BRAIN		KIDNEY		LARGE INTESTINE		PANCREAS		SPLEEN		TONGUE	
	BOVINE	PORCINE	BOVINE	PORCINE	BOVINE	PORCINE	BOVINE	PORCINE	BOVINE	PORCINE	BOVINE	PORCINE
Proximate composition (%)												
Moisture	n/a	n/a	79.03	n/a	82.98	69.73	56.21	72.20	77.76	78.58	75.14	n/a
Protein	11.88	12.30	17.53	20.00	13.28	8.45	13.38	20.98	18.21	17.79	17.22	16.00
Fat	10.80	9.72	1.89	3.10	3.34	19.54	26.29	7.18	1.40	0.97	5.32	9.00
Ash	n/a	n/a	1.35	n/a	0.48	0.15	0.87	1.26	1.15	1.12	1.02	n/a
Cholesterol (mg/100 g)	3,162	2,200	350	360	n/a	n/a	200	195	260	360	100	100
Fatty acids (%)												
C16:0	16.50	19.00	17.50	25.50	22.15	27.80	22.50	25.50	27.46	27.97	22.50	25.50
C16:1*n-7*	0.39	1.56	1.00	2.14	2.91	0.83	3.44	3.56	1.01	0.53	2.77	2.16
C18:0	20.50	21.00	15.50	15.50	13.02	21.46	19.00	19.50	22.40	20.00	13.50	12.00
C18:1*n-9*	24.00	24.50	21.00	29.00	45.21	33.15	31.50	28.50	21.69	17.74	38.50	42.50
C18:2*n-6*	0.53	1.15	19.00	12.10	7.44	13.61	11.45	12.30	8.55	12.99	6.90	9.45
C18:3*n-3*	0.13	1.16	0.91	0.30	0.17	0.09	0.77	0.57	0.10	0.21	0.57	0.42
C20:4*n-6*	6.40	10.05	13.50	11.20	5.71	0.58	3.45	3.35	16.66	20.27	2.85	1.65
SFA	42.00	41.50	36.50	42.00	38.14	50.48	45.00	47.00	51.30	48.85	41.00	40.00
MUFA	30.00	30.00	23.00	32.00	48.45	34.60	35.50	33.50	23.11	20.99	43.50	47.50
PUFA	26.00	27.50	37.50	27.00	13.41	14.92	16.50	17.00	25.59	30.16	11.95	13.00
Σ *n-6*	14.54	18.82	33.95	25.55	13.23	14.29	15.25	16.21	25.48	29.36	10.35	12.24
Σ *n-3*	13.48	11.18	3.05	2.08	0.17	0.63	1.68	1.04	0.70	0.80	1.36	0.75
n-6/n-3	1.08	1.68	11.13	12.28	76.87	23.06	9.08	15.59	36.40	36.70	7.61	16.32
PUFA/SFA	0.62	0.66	1.02	0.64	0.37	0.30	0.37	0.36	0.50	0.62	0.29	0.33

(Continued)

Table 3.2 (Cont.)

	BRAIN		KIDNEY		LARGE INTESTINE		PANCREAS		SPLEEN		TONGUE	
	BOVINE	PORCINE	BOVINE	PORCINE	BOVINE	PORCINE	BOVINE	PORCINE	BOVINE	PORCINE	BOVINE	PORCINE
Amino acids (%)												
Essential												
Lysine	0.72	0.96	1.15	1.20	0.94	0.47	0.96	1.11	1.38	1.05	1.05	2.04
Methionine	0.24	0.24	0.26	0.34	0.22	0.10	0.26	0.42	0.34	0.23	0.35	0.60
Threonine	0.60	0.60	0.72	0.69	0.53	0.14	0.61	0.32	0.76	0.32	0.70	1.08
Leucine	0.96	1.08	1.39	1.46	0.87	0.36	1.02	1.11	1.44	0.80	1.05	1.92
Isoleucine	0.48	0.60	0.54	0.86	0.39	0.33	0.47	0.95	0.51	0.75	0.70	1.08
Phenylalanine	0.60	0.60	0.78	0.77	0.51	0.77	0.54	1.23	0.88	1.07	0.70	0.96
Valine	0.60	0.72	0.72	0.94	0.49	0.64	0.64	2.09	0.88	1.54	0.70	1.32
Histidine	0.36	0.36	0.62	0.39	0.44	0.30	0.32	1.06	0.72	0.79	0.35	0.60
Arginine	0.72	0.60	0.89	1.03	0.91	0.62	0.62	1.07	1.10	1.00	1.05	1.56
Nonessential												
Serine	0.72	0.60	0.76	0.86	0.63	0.24	0.70	0.81	0.81	0.53	0.70	-
Cysteine	0.24	0.24	0.26	0.34	0.16	0.24	0.72	0.53	0.24	0.53	0.35	-
Tyrosine	0.48	0.48	0.55	0.60	0.36	1.06	0.45	2.34	0.56	2.39	0.35	-
Alanine	0.72	0.72	0.97	1.03	0.84	0.41	0.74	0.97	1.12	0.80	0.70	-
Aspartic acid	1.20	1.20	1.37	1.55	1.02	0.75	1.18	1.36	1.53	1.39	1.40	2.28
Glutamic acid	1.56	1.44	1.86	1.97	1.99	0.57	1.42	1.54	2.16	1.41	2.10	3.12
Glycine	0.60	0.60	0.96	1.03	1.22	0.25	0.83	0.83	1.06	0.57	1.05	1.80
Proline	0.48	0.72	0.79	1.03	0.86	0.46	0.62	0.94	0.79	0.82	0.70	-
Minerals (mg/kg)												
Macro-minerals (g/kg)												
Ca	0.05	0.10	0.09	0.10	0.10	0.20	0.13	0.22	0.06	0.07	0.07	0.11
Mg	0.15	0.14	0.14	0.18	0.09	0.10	0.16	0.23	0.16	0.18	0.18	0.18

P	1.48	2.96	2.15	2.37	1.29	0.61	3.04	3.15	2.52	2.57	1.75	1.90
Na	1.10	1.22	1.89	1.02	0.74	0.23	0.73	0.85	0.82	0.88	0.75	1.15
K	2.49	2.40	2.02	2.35	1.47	0.41	2.53	3.08	3.43	3.40	2.20	2.55
Micro-minerals (mg/kg)												
Fe	26.40	28.00	68.23	59.00	12.25	11.69	45.17	37.97	440.0	178.54	25.00	45.00
Zn	42.00	14.00	23.40	26.00	14.37	1.24	44.03	3.83	26.32	2.63	30.00	26.00
Cu	1.20	2.50	3.72	8.00	2.54	0.88	1.65	1.55	2.01	1.18	1.00	2.50
Mn	–	0.10	1.24	1.50	0.58	0.63	1.69	1.98	0.33	0.53	0.50	5.00

Data from Aristoy and Toldrá (2011); Honikel (2011); Prates et al. (2011); Florek et al. (2012); Mora et al. (2019); Seong et al. (2014a, 2014b); n/a: data not available.

such as kidney, liver, and spleen values of 550 mg/100 g are reached. The organ that presents higher amounts of cholesterol is the brain, with values that exceed 1,000 mg/100 g (Honikel, 2011).

3 Treatment of Meat By-Products

Once the meat by-products are located and extracted, they must be submitted to several processes prior to their use. Washing, trimming, chilling, packaging, and cooling are included in these processes (Liu and Ockerman, 2001).

Some meat by-products are subjected to a process that allows the transformation of edible and nonedible raw animal tissues into fat, mineral, and protein products (Meeker and Hamilton, 2006; Jayathilakan et al., 2012). This process is known as rendering. It is estimated that approximately 40% of the animal body weight is treated in the rendering plants (Ockerman and Basu, 2006). Blood, bones, fat trim, feathers, and viscera are the edible meat by-products produced during the process, while dried blood, feather meal, hydrolyzed hair, and poultry by-product meals are among the primary products that are the result of this process. The main use of these products is as ingredients in the preparation of animal feed. In contrast, gelatin extraction and edible tissue separated from bone could be a food application (Ockerman and Basu, 2006). Besides, other animal by-products processed by rendering and included in human food are lard and tallow, which are obtained from the muscle and bone tissues and contribute to energy intake (Awan et al., 2015).

The operations of drying and fat separation (Meeker and Hamilton, 2006) included in rendering process depend on the raw materials processed and the content of protein of the solids (meal) (Gooding and Meeker, 2016).. In the rendering plant, the material is ground and then thermally treated (115–145 °C) in a cooker for moisture removal, dissolving fat, and inhibiting pathogenic microorganisms (Anderson, 2006). The resulting liquid passes through a drainer conveyor into a settling tank to be centrifuged to obtain fine solids, which contain proteins as sludge. Furthermore, solids that remain in the drainer conveyor are taken into a screw press, resulting in a cake rich in protein, which is destined to meal processing (Gooding and Meeker, 2016).

4 Bioactive Compounds Extracted from Meat By-Products

4.1 Proteins and Peptides

Proteins are found in the nutrients present in meat by-products that can have functional properties (Liu, 2002; Matak et al., 2015; Toldrá et al., 2016). Bovine (heart) and porcine (lungs) (DeWitt et al., 2002; Selmane et al., 2008; Darine et al., 2010) and chicken-meat processing by-products (Hrynets et al., 2011; Tahergorabi et al., 2011) are used for the extraction of proteins. Although there is not much information about the extraction methods of these compounds in meat by-products, usually mild conditions are used to avoid the alteration of the functional properties of these proteins (Lynch et al., 2018). For this reason, emerging nonthermal technologies are increasingly used since their conditions allow maintaining the properties of proteins. These techniques include isoelectric solubilization/precipitation (ISP) that recovers the protein in several steps through changes of pH. The process involves solubilization of the protein, an isoelectric precipitation to remove the lipids and the recuperation of the initial pH (Tahergorabi et al., 2011; Tahergorabi et al., 2012). Selmane et al. (2008) have studied different extraction conditions (pH, temperature, and extraction time) to optimize the method of protein extraction. These conditions will depend on the material from which the extraction is made. In the same way as in the extraction of hormones, prior to the extraction, the glands are chopped, vacuum-packaged, and frozen. Once defrosted, the samples are homogenized with water in a proportion of 20% (w/v) in an alkaline medium at 20 °C for 60 min. The extract obtained after the centrifugation of the previous suspensions is purified and concentrated using the membrane technology and constant conditions of pressure and flow. The concentration continues with a pH adjustment and centrifugation for isoelectric precipitation to occur. Finally, purification consists of the removal of lipid fraction by stirring the precipitate with an organic solvent (hexane and isopropanol, 3:2, v/v) (1:5, w/v) at 20 °C for 1 h. The final product is obtained once the solvent is removed by evaporation.

Bioactive peptides originated from proteins of meat and meat by-products using extraction procedures based on enzymatic hydrolysis and microbial fermentation (Lafarga and Hayes, 2014; Mora et al., 2014). These compounds can also be released in gastrointestinal digestion and food processing (Arihara and Ohata, 2010; Escudero et al., 2012).

As previously mentioned, amino acid content determines the biological activity of the proteins (Aristoy and Toldrá, 2011; Mullen et al., 2017). Thus, anticancerogenic, antihypertensive (ACE-inhibitory), antimicrobial, antioxidative, antithrombotic, and immunomodulatory regulating activities that may affect the cardiovascular, immune, nervous, and digestive systems are between the activities associated with bioactive peptides isolated from meat by-products (Lafarga and Hayes, 2014; Helkar et al., 2016). Besides, the peptide chain has an influence on its properties, so that the greater its length, the lower is its activity (Vermeirssen et al., 2004). Moreover, although the peptides isolated from meat by-products have lower activity than those from synthetic products, several researches have proved the option of using them as a potential substitute for synthetic products. Thus, the potential toxicity and carcinogenesis of artificial products would be avoided (Lee et al., 2012a, 2012b). These beneficial properties suggest that these bioactive compounds could be used as potential nutraceuticals.

To avoid hypertension, ACE-I-inhibitory and antihypertensive peptides can be introduced in the diet as functional foods. For this effect, it is necessary that this activity must be tested both in vitro and in vivo, since the action of the proteolytic enzymes during digestion can hydrolyze some of them, which would result in inactive peptides (Mora et al., 2014). Regarding antimicrobial and antioxidant peptides, they would allow to extend the shelf-life of foods, reducing the formation of reactive oxygen species and inhibiting the action of strains of pathogenic microorganisms (Di Bernardini et al., 2011a; Liu et al., 2016; Lorenzo et al., 2018). In this way, Przybylski et al. (2016) have evaluated the potential application of a peptide isolated from bovine hemoglobin as antioxidant in minced meat. The obtained results showed that its hydrolysate would be able to inhibit the lipid oxidation at the same level as that of known synthetic antioxidants, which would confirm the above-mentioned study.

As regards the general extraction method, this is carried out by hydrolysis, which allows increasing the yield of valuable compounds (Aspevik et al., 2017). A chemical or enzymatic protein hydrolysis of the meat by-product can be used. The first one uses an acid or an alkali solution to split the peptide bonds, while the enzymatic hydrolysis uses endogenous or exogenous enzymes. The second one gives rise to quality peptides, with lower molecular weights, higher ionizable groups and solubility, and better nutritional quality than those obtained with chemical hydrolysis (Panyam and Kilara, 1996). The enzymatic procedure is performed in a batch-fed or in an ultrafiltration membrane reactor in combination with proteases (Figure 3.2). The characteristics of the hydrolyzed product depends on the source from which they are extracted and the enzymes involved in the procedure. The following steps depend on the use to which the resulted hydrolyzed product is destined. When it is destined to biological activities, the hydrolyzed product is purified using chromatography, membrane filtration (ultra or nanofiltration, the selection of one or another is function of the molecular weight of the bioactive peptides), or ion-exchange technologies. This step could enable obtaining only those peptides that are bioactive, which means that they can offer health benefits (Arihara and Ohata, 2010). Finally, the hydrolyzed product is concentrated by evaporation and dried to obtain bioactive peptides. Recently, improvements found in the literature propose previous isolation of the protein to optimize the hydrolysis extraction (Mullen et al., 2017). As mentioned previously, other possible source of proteins could be the products that result from rendering treatment (Toldrá et al., 2012). The hydrolyzed proteins can be used in food applications as flavor enhancers and for improving physicochemical properties (Maehashi et al., 1999).

Blood is an edible by-product that is used since ancient times in Asia and Europe, and which are recognized as a ingredient in meat products (Ghost, 2001; Jayathilakan et al., 2012). However, blood is still rejected by some cultures or for health reasons (Hsieh and Ofori, 2011). Moreover, this product is considered as a source of bioactive compounds due to its important protein constituents (Wan et al., 2002; Bah et al., 2013).

The collection of sample is a key step since it is easily contaminated. The use of a closed draining system avoids the contact of blood with air, thus guaranteeing

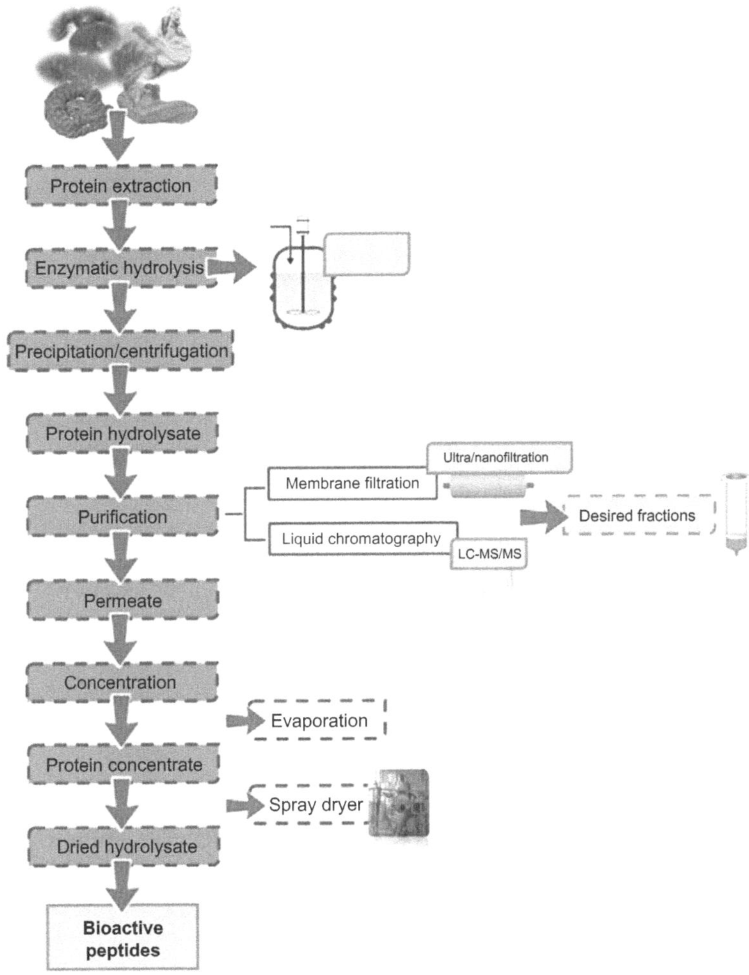

Figure 3.2 Isolation of bioactive peptides from meat by-products using an enzymatic hydrolysis.

its safety (Bah et al., 2013). An anticoagulant is added to the collected blood to separate its fractions. After centrifugation, red blood cells and plasma are obtained, which once subjected to different methods of extraction lead to proteins and bioactive peptides. Blood is composed of two fractions: cellular and plasma. Hemoglobin is the most important protein isolated from the cellular fraction, while immunoglobulins, fibrinogen, and serum albumin are found in plasma. The last fraction highlights for its functional and therapeutic properties that enable its use as food ingredient and as dietary supplement (Del Hoyo et al., 2008; Jayathilakan et al., 2012). Several plasma proteins have been subject to patents, being commercialized under different trademarks (Mullen et al., 2017): Fibrimex® (purified fibrinogen), Immunolin® (concentrated immunoglobulins), Plasma Powder FG (plasma enriched in fibrinogen), and Vepro 95 HV (purified globin).

The cellular fraction that includes red and white blood cells and platelets is usually used as a color enhancer in meat products (Hsieh and Ofori, 2011). New utilities have been found when hemoglobin is extracted from this fraction, obtaining a heme-iron enriched peptide that prevents iron deficiencies (Toldrá et al., 2012). This heme-rich enriched peptide is isolated prior to an enzymatic digestion of the dried hemoglobin using a combination of endo- and exopeptidases (In et al., 2002). The yield of this procedure increases with the use of a dialysis membrane with ultrafiltration and diafiltration (Vanhoute et al., 2010) or by using a protease-modified alumina complex (Ticu et al., 2005). The iron contents obtained in the resulting peptide are greater than those found in hemoglobin. Emerging technologies are recently applied in the extraction of hemoglobin hydrolysates. The combination of enzymatic hydrolysis with high hydrostatic pressure treatments enhance the action of enzymes and therefore resulting in protein denaturation and yield of the hydrolysis (Toldrá et al., 2011).

The hydrolysates isolated from hemoglobin and plasma have antihypertensive (ACE-inhibitory), antimicrobial, antioxidant, and opioid activities (Liu et al., 2010a, 2010b; Adje et al., 2011; Nyberg et al., 2013). Table 3.3 shows some peptides obtained by enzymatic hydrolysis of meat by-products.

4.2 Collagen and Its Derivatives

Collagen is one of the most important products obtained from animal by-products. This structural protein is the predominant protein present in animals since it can represent up to 30% of total animal proteins (Ferraro et al., 2016). It is found in the connective tissues of bovine and porcine bones, skins, and tendons (Boire et al., 2017). Several functional and bioactive properties are attributed to collagen and its hydrolysates (Saiga et al., 2008; Gómez-Guillén et al., 2011; Herregods et al., 2011; Ferraro et al., 2013), which has increased their food and pharmaceutical applications. Besides, several studies attribute to its peptides ACE-inhibitory, antimicrobial, and antioxidant activities (Ohba et al., 2003; Li et al., 2007; Saiga et al., 2008; Di Bernardini et al., 2011a; Liu et al., 2015). The bioactive peptides obtained from collagen take part in regeneration processes (cartilage and nerve regeneration, cicatrization, vascularization), and anticancer, antihypertensive, and antithrombotic properties are attributed to them (Alovskaya et al., 2007; Gómez-Guillén et al., 2011).

It is to be noted that the properties of collagen (bioactivity and functionality) depend on the extraction method (Ferraro et al., 2016). The selection of one or the other is also influenced by the age of the animal from which it is extracted. Usually, organic acids (acetic acid), salts, or enzymes are involved in its extraction without modifying its structure (Ferraro et al., 2016). Organic acids and cold neutral salt solutions are used in young collagen or when it is not cross-linked, while enzymes are used when it is strongly cross-linked (Liu et al., 2015). Alcalase,

Table 3.3 Bioactive peptides isolated from meat by-products after enzyme hydrolysis

BY-PRODUCT	SOURCE	PARENTAL PROTEIN	ENZYME	IC_{50}	MAIN EFFECTS	REFERENCE
Blood		α-hemoglobin β-hemoglobin	Papain, Pepsin, Trypsin	4.92 μM 6.02 μM	ACE	Yu et al. (2006)
	Porcine hemoglobin	Hemoglobin	Pepsin	0.25 mg/mL 0.02 mg/mL	ACE	Ren et al. (2011) Deng et al. (2014)
		Hemoglobin	Alcalase–Flavourzyme	n/a	AOX	Chang et al. (2007)
	Bovine hemoglobin	Hemoglobin	Pepsin	2.9, 4.3, 4.6, 45.2, 46.3, 80.5 μM	OPI	Zhao et al. (1994)
		Hemoglobin	Pepsin	34.3 μM, 78.2 μM	ACE	Zhao et al. (1997)
		α-hemoglobin	Pepsin	–	AMB	Froidevaux et al. (2001); Daoud et al. (2005); Nedjar-Arroume et al. (2008)
		α-hemoglobin	Pepsin	42.5, 366, 518.3, 1095.5 μM	ACE, AMB	Adje et al. (2011)
	Porcine plasma	Albumin and globin	Alcalase	–	AOX	Wang et al. (2008)
		Plasma protein	Alcalase	236.4 μM	AOX	Liu et al. (2010a)
		Plasma protein	Pepsin/Papain	n/a	AOX	Xu et al. (2009)
	Red deer plasma	Plasma protein	Trypsin	3.1 μM	ACE	Liu et al. (2010b)
	Silkie fowl plasma	Plasma protein	Alcalase	–	AOX	Cheng et al. (2016)
	Chicken blood meal	–	Alcalase	0.34 mg/mL	ACE	Huang and Liu (2010)
Bones	Duck processing by-products	–	Protease	75 μg/mL	AOX	Lee et al. (2010)
Legs	Chicken legs collagen	Collagen	*Aspergillus* species derived enzymes	29.4–60.8 μg/mL	ACE	Saiga et al. (2008)
Ligament	Bovine ligamentum nuchae	Elastin	Pepsin	n/a	AOX	Hattori et al. (1998)
Liver	Bovine	Sarcoplasmic protein	Thermolysin	n/a	AOX	Di Bernardini et al. (2011b)

(Continued)

Table 3.3 (Cont.)

BY-PRODUCT	SOURCE	PARENTAL PROTEIN	ENZYME	IC_{50}	MAIN EFFECTS	REFERENCE
	Chicken	–	Pepsin	n/a	AOX	Chou et al. (2014)
	Porcine skin collagen	Collagen	Cocktail proteases	n/a	AOX	Li et al. (2007)
	Bovine skin gelatin	Collagen	Alcalase, collagenase, pronase E	n/a	AOX	Kim et al. (2001)
Skin	Yak	Gelatin	Alcalase, proteinase K	35 μg/mL	ACE, AOX	Tian et al. (2017)
	Duck	Gelatin	Collagenase, pepsin	22.7 μg/mL	AOX	Lee et al. (2012a)
			Alcalase Collagenase Flavozyme Neutrase Papain Pepsin Protamex Trypsin α-chymotrypsin	0.632 mg/mL	ACE, AOX	Lee et al. (2012b)
Tendon	Bovine Achilles tendon collagen	Collagen	Bacterial collagenase	51.1 μM, 79.85 μM	ACE	Banerjee and Shanthi (2012)
			Alcalase Collagenase Thermolysin Proteinase K Pepsin Trypsin	n/a	ACE	Zhang et al. (2013)
			Protease	n/a	ACE	Ryder et al. (2016)
Viscera	Sheep stomach, large and small intestines	–	Fungal protease	–	FPP	Bhaskar et al. (2007)

ACE=Inhibition of angiotensin I-converting enzyme; AMB=antimicrobial; AOX=Antioxidant; FPP=Food properties; OPI=Opioid; n/a: data not available bromelain, chymotrypsin, flavourzyme, neutrase, papain, pepsin, and trypsin are the enzymes with which greater activity is obtained (Saiga et al., 2008; Gómez-Guillén et al., 2011; Di Bernardini et al., 2011a).

Although there are several studies that evaluate different collagen extraction methods (Langmaier et al., 2001; Li et al., 2009), it is difficult to find an appropriate method that does not have interferences in the process (Mokrejs et al., 2009). The most common method used is ultrasonic irradiation, but the use of long treatment sometimes can also have disadvantages in the extraction process, thus reducing the percentage of the extracted collagen (Kim et al., 2012). The use of ultrasonic irradiation combined with the proteolytic enzyme pepsin can be a good option (Ran and Wang, 2014). It is convenient to start the extraction with a degreasing step to remove non-collagen proteins and lipoid compounds. The first ones are eliminated by treating the sample with an alkaline solution, while an organic solvent is used to degrease once the samples have been neutralized. Previous to the ultrasonic and enzymatic hydrolysis treatment, the sample was soaked in an acetic acid solution that in addition to obtaining the acid soluble collagen, promotes the action of the enzyme. Pepsin is very selective since it allows removal of telopeptides (non-helical ends) of the collagen. Besides, to avoid temperature increases that could harm the extraction, the tandem treatment is carried out in a water bath. Thereafter, the collagen solution is precipitated and centrifuged, and the sediment obtained was dialyzed using acetic acid and distilled water to finally lyophilize it and obtain the final product. This procedure is more suitable than those that use only the enzymatic hydrolysis (Langmaier et al., 2001; Mokrejs et al., 2009).

On the other hand, the extraction of collagen hydrolysate peptides occurs through hydrolysis with bacterial enzymes and separated by membrane ultrafiltration, and results in functional peptides with ACE-inhibitory activity. Their capacity has been tested through in vitro and in vivo trials. The results obtained prove that they can be included in the diet. This is the case of the bioactive peptides released from chicken legs that have the beneficial effects of stabilization of blood pressure, avoiding cardiovascular diseases (Saiga et al., 2008). Even though there are several studies that evaluate the activity of peptides extracted from other by-products, only in vitro tests are performed to understand their effects (Banerjee and Shanthi, 2012). This is because it would be convenient to make more studies to ensure the safety of these products.

Anther peptide that has a relevant importance in cosmetic, food, and pharmaceutical fields for its emulsifying, foaming, and gelling properties is gelatin (Zhou et al., 2006; Sarbon et al., 2013). This polypeptide that results from the thermal denaturation of insoluble collagen is commonly isolated from mammalian skin (porcine), bones, and connective tissues (bovine) (Morrison et al., 1999). However, the controversial use of these gelatin sources has made new alternatives (Jun et al., 2000; Lim et al., 2001; Karim and Bhat, 2009; Sarbon et al., 2013).

The properties of gelatin are functional and rheological and depend on the extraction method, as well as the species and the tissues from which it originates from (Sarbon et al., 2013). Several studies confirm the use of poultry by-products (bones, feet, and skin) as alternatives of bovine and porcine gelatin (Rahman et al., 2012; Sarbon et al., 2013; Bichukale et al., 2018). The investigations carried out with skin conclude that chicken gelatin has similar proximate composition than bovine gelatin, with more contents of the amino acids hydroxyproline and proline, and better thermostable properties than those obtained from bovine and porcine sources (Sarbon et al., 2013). In contrast, gelatin extracted from feet and bones showed inferior physicochemical and sensory quality to bovine gelatin (Rahman and Jamalulail, 2012). Moreover, its chemical composition is also different, with protein and hydroxyproline contents lower than those found in other bovine and chicken sources (Rahman and Jamalulail, 2012; Bichukale et al., 2018).

Gelatin extraction usually takes place using a combination of acid, alkali, or high-pressure steam treatments. The procedure should include three steps: non-collagenous removal, collagen hydrolysis, and gelatin recovery and drying. To a large extent, the samples from which gelatin is going to be obtained are freeze-dried and defatted. Then, the samples were extracted three times with alkaline solution for removing pigments and non-collagenous proteins. The procedure continues with an acid treatment of the obtained pellets after rinsing with distilled water. The acid extraction has two steps to denature the collagen. The first step is treatment with sulfuric acid and the second one with citric acid. This procedure is repeated three times, and rinsing with distilled water is carried out again. Finally, the pellets, previously centrifuged, are extracted with distilled water kept overnight at 45 °C. The resulting mixture is filtered and deionized with a resin after adjusting the pH and evaporated under vacuum at 45 °C. The obtained gelatin is allowed to stabilize overnight for lyophilization (Sarbon et al., 2013; Bichukale et al., 2018).

When the extraction is carried out with acid treatment, as in the case of pig bones and skins, the proceeding begins with treatment with an inorganic acid such as hydrochloric; phosphoric, or sulfuric acid that favors collagen solubilization to continue with a similar alkaline procedure mentioned above. The gelatin obtained has a different quality than those resulting from alkaline treatment. Both products are soluble and allow forming gels, but the first one has better gel strength and translucent characteristics. These properties make the gelatin useful in food and pharmaceutical applications as capsules and tablets are manufactured from it.

A similar extraction procedure is used for extracting gelatin from chicken bones and feet (Rahman and Jamalulail, 2012; Bichukale et al., 2018). The difference is, in the pre-treatment of the sample before being dried, bones are cleaned in boiling water to remove cuticles, fat, and skin. Besides, the defatted sample is subjected to an acid treatment with hydrochloric acid to remove minerals and non-collagenous compounds, and in this way obtain osein on which it will be made the alkaline extraction.

Bovine tendons, and more specifically, short tendons, are also an important source of collagen, and therefore of gelatin (Mokrejs et al., 2012). Unlike the aforementioned method, the extraction includes an enzymatic treatment with a proteolytic enzyme, which is a key factor in the procedure since it allows the rupture of collagen that favors the extraction process.

Other relevant fibrous structural proteins are elastin and keratin (Ferraro et al., 2016). These proteins are part of the bovine (ligaments), porcine (hairs, ligaments, and skins), ovine (wool), and poultry (feathers, ligaments, and skins) connective tissues. The recovered products obtained from these by-products have biomedical applications, and they can even be considered as a potential substitute for synthetic products (Daamen et al., 2007; Vasconcelos and Cavaco-Paulo, 2013).

The alkaline treatment and the high temperatures used on by-products of ligaments result in the release of non-collagenous compounds, such as elastin. In contrast, the extraction of keratins needs an oxidative and enzymatic procedure (Vasconcelos and Cavaco-Paulo, 2013; Ferraro et al., 2016). A lot of applications are known about keratin peptides obtained from wool and feathers, which include skin hydration and elasticity, and antihypertensive and antioxidant activities (Di Bernardini et al., 2011a; Fontoura et al., 2014). Elastin can also be extracted using enzymatic procedures and mild hydrolysis. The peptides obtained have beneficial effects on thrombosis, regeneration activities, and can be used as food additives with a protective effect against lipid oxidation (Hattori et al., 1998; Daamen et al., 2007; Di Bernardini et al., 2011a).

4.3 Glycosaminoglycans

Heparin is an important anticoagulating agent included in the List of the Essential Medicines of the World Health Organization (WHO, 2015), which is able to prevent blood clotting. This hormone is usually extracted from the lungs of cattle, and liver and lining of the small intestine of pigs. However, the appearance of mast cells together with the problems associated with the BSE in by-products of these species has made it look for new sources of extraction of heparin. Turkey and mouse can also be used to obtain glycosaminoglycan (Bland et al., 1982; Warda et al., 2003). To exercise its function as an anticoagulant, it is necessary to inject it with a dose of 10 to 20 μg/mL of plasma, which together with antithrombin helps in preventing thrombin action, and therefore the conversion of fibrinogen into an insoluble fibrin clot.

The extraction from the lining of the small intestine of pigs needs a previously described step, which aims to reduce the bacterial load of this by-product using sodium metabisulfite (1%). The extraction of heparin begins with alkaline hydrolysis of the mucosa of the intestine and continues with digestion using proteolytic enzymes (Figure 3.3). The resulting extract is treated with an ion-exchange resin that will adsorb the heparin, and then desorbed with a concentrated salt solution. Before precipitation with alcohol, the crude heparin is subjected to several

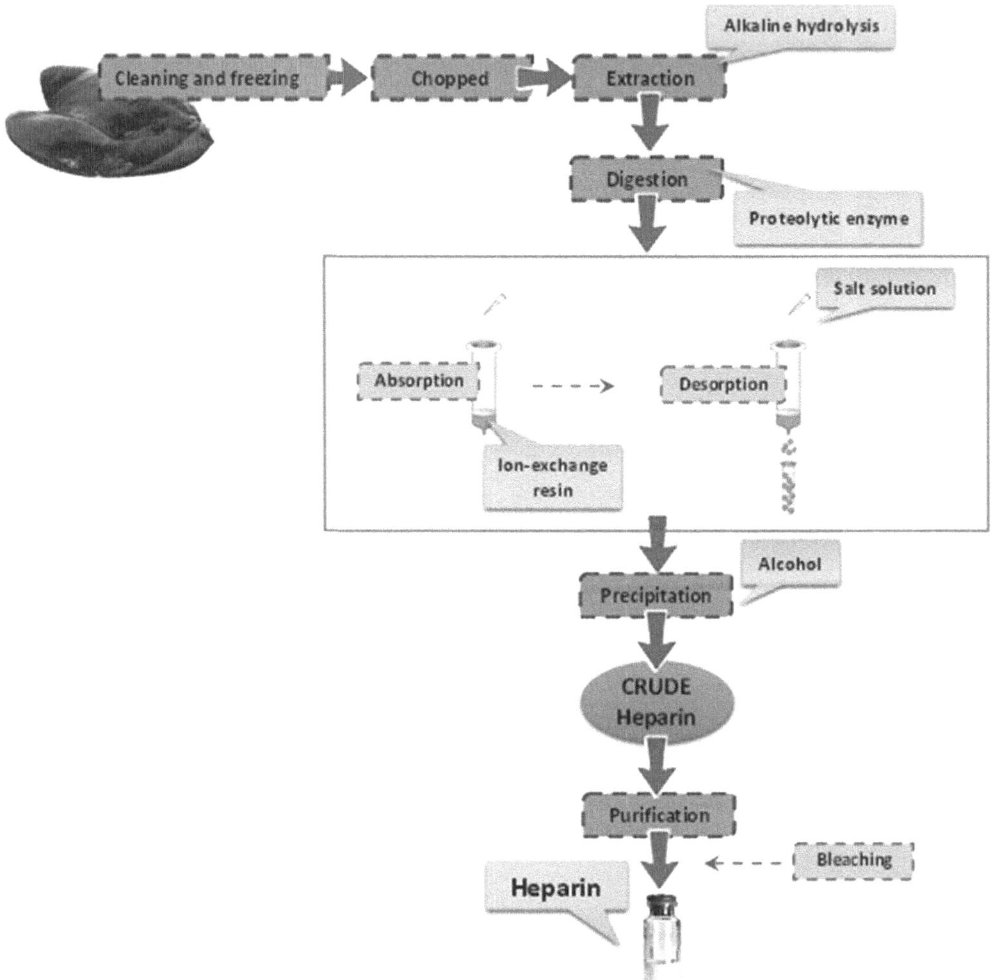

Figure 3.3 Extraction procedure for the recovery of heparin from meat by-products.

purifications to obtain the final product. It is important to note that the pharmaceutical use of heparin (as calcium or sodium salt) requires bleaching (using permanganate or peroxide) to get a white powder (Khouw et al., 1989).

Similar extraction method was carried out by Warda et al. (2003) in porcine and turkey intestines. Prior to proteolytic digestion, the pH of the homogenized sample is adjusted to 9.0 using sodium carbonate. Alcalase is the enzyme selected to digest the sample at 50 °C for 12 h. After this, its protease is inactivated in a water bath at 100 °C. The extract obtained before centrifugation is concentrated using a resin that was activated using methanol, water, and NaCl (2 M), and water. The desorption process from the resin is performed in two steps. In the first step, the resin is washed with water and sodium chloride (3%), and only residual proteins are eluted. In the second step, glycosaminoglycans are desorbed after washing the resin with sodium chloride (16%). The eluted extract is precipitated with methanol 80% (v/v), centrifuged and dialyzed overnight in cooling conditions. The obtained product is lyophilized and solubilized in a buffer (Tris-HCl buffer, pH=8). The process is repeated with a new digestion procedure, this time with an endonuclease at 37 °C for 12 h, eluted with sodium chloride (16%), precipitated with methanol 80% (v/v), and dialyzed overnight using demineralized water at cooling conditions. The freeze-drying of the obtained glycosaminoglycan results in the final product.

4.4 Hormones

The presence of enzymes, hormones, pigments, and vitamins in the composition of several meat by-products makes them susceptible to be used in the medical field (Irshad and Sharma, 2015). Antigens, antitoxins, serums, and vaccines can be made from them (Pearl, 2004).

Hormones are signaling molecules that are synthesized and secreted in specialized hormone-synthesizing animal glands. They are transferred to specific target sites in another part of the body through the blood stream. These organic substances are obtained from only healthy animals for reasons of food safety and quality. The importance and the functions of each one depend on the species, age, and sex of the animal.

Adrenal, corpus luteum, follicle, kidney, liver, lungs, ovary, pancreas, parathyroid, pituitary, stomach, thymus, and thyroid are the endocrine glands responsible for secreting hormones that control the normal functioning of the metabolism. This is the reason why these compounds are an important source in the development of medicines (Awan et al., 2015). China, India, and Japan have traditionally used these by-products in medicine. Adrenocorticotropic hormone (ACTH), epinephrine, estrogen, insulin, nor-epinephrine, parathyroid hormone, progesterone, testosterone, thymosin, thyroid stimulating hormone, thyroxin, and trypsin are the medicines that are obtained from the slaughtered animal glands (Awan et al., 2015).

Once glands are excised from the animal, they are cleaned by removing the fat and the connective tissue to immediately freeze them, avoiding the possible microbial growth. The process continues in the pharmaceutical industry, where these by-products are checked prior to the extraction procedure. These consist of mincing the sample and then mixing with different solvents to remove the fat. The sample after defatting was dried in a vacuum drier and ground to obtain a powder that is introduced in a capsule. Finally, it is necessary to test the resulting product in vitro and in vivo to check its safety.

The pancreas is an endocrine and exocrine gland that produces several hormones. Glucagon and insulin are two of these hormones; both hormones are related to the regulation of blood glucose. The insulin extraction must be carried out to get a purity of 99.9%, reducing the presence of proteins as far as possible that can give rise to immunologic reactions. In this way, after preparing the pancreas for the extraction procedure (trimmed and frozen), the gland is grounded with alcohol in acid conditions. Once clarified, the alcoholic fraction is evaporated under room temperature to avoid insulin losses. In the next step, the protein is precipitated and meticulously purified (Khouw et al., 1989). The human use of bovine and porcine insulin implies one last step, the change of the amino acid alanine by threonine (present in human insulin). Although there are other models of insulin production in molecular biology that allow coding the chains of human insulin. These models are rarely used because they are very expensive if compared to its extraction from animal origin.

The pineal gland is a small endocrine gland located in the epithalamus near the center of the brain that produces the hormone melatonin. This hormone is used in the treatment of headaches (Peres et al., 2006), epigastric pain (Kandil et al., 2010), insomnia (especially in people with mental disorders such as autism, hyperactivity, narcolepsy, etc.; Braam et al., 2009), and schizophrenia. The pituitary gland located in the base of the brain secretes several hormones; however, it secretes adrenocorticotropic hormone (ATCH). This hormone is effective in the treatment of arthritis, eye inflammation, multiple myeloma, and rheumatism.

Steroids extracted from adrenal glands of cattle, pigs, and sheep are important anti-inflammatory agents in the disease management of asthma. Epinephrine and norepinephrine belonging to the catecholamine family are released from the medulla of the aforementioned animal species and are used as anti-hemorrhaging agents and stimulate heart agents. Prednisone and cortisone are the two hormones obtained from bile. They can be used in the medical field in the treatment of arthritis, allergies, breathing problems, cancer, blood/hormone/immune system disorders, and eye and skin problems (Liu, 2002).

4.5 Enzymes

Kallikrein is an enzyme secreted by pancreas, which is used in Europe and Japan since 1940. This enzyme is responsible for regulation of blood pressure (Khouw et al., 1989). Porcine pancreas is usually used for the extraction of this enzyme, and as in the case of insulin, the extraction begins once the gland is frozen and chopped. The aqueous extract obtained before the extraction with a buffer is fractionated using ammonium sulfate, and then it is purified with an ion-exchange resin. The kallikrein obtained is lyophilized (Khouw and Kesler, 1974) and it is used for commercial purposes such as in pills for health benefits in countries such as Japan. The recommendations are 1–2 capsules a day. Each blister contains 10–50 μg of pure kallikrein.

Other enzymes isolated from the pancreas are proteolytic enzymes, trypsin, and chymotrypsin, which can be used in the treatment of cancer, since they are considered to be anti-inflammatory agents that inhibit the induction of protease. Furthermore, they can be used for early recovery after a surgical operation.

Chymosin and catalase are two important enzymes in the dairy industry that are isolated from meat by-products. Chymosin is a proteolytic enzyme present in the stomach of bovines. It is extracted with a salt solution and purified using an anion-exchange resin that allows the adsorption of chymosin and removal of mucin, which is the main contaminant associated with this protease. The eluate is obtained through a salt solution, which is precipitated with the salt to obtain the final product (Elagamy, 2000; Mohanty et al., 2003). Catalase is isolated from the liver of mammals, mainly beef. Its extraction first implies an aqueous extraction of liver followed by an organic fractionation. The procedure continues with a purification that utilizes anion-exchange chromatography, and a stabilization with a buffer until the final product is obtained (Nadeem et al., 2015). This enzyme is mainly commercialized as Catalase L.

Moreover, the intestine of chicken is an important source of proteases (Raj and Mahendrakar, 2010). The activity of these enzymes is similar to those found in other tissues that are isolated such as liver or spleen (Jamdar and Harikumar, 2005).

4.6 Polyunsaturated Fatty Acids

Viscera are considered an important source of compounds with functional and bioactive properties (Lasekan et al., 2013). In addition to protein and its hydrolysates, polyunsaturated fatty acids could be isolated from these meat by-products (Jayathilakan et al., 2012).

Lipids are essential for adequate functioning of the human metabolism. As mentioned earlier, polyunsaturated fatty acids have beneficial effects on human health, highlighting the long chain polyunsaturated *n*-3 fatty acids (eicosapentanoic acid—EPA, docosahexanoic acid—DHA), although the presence of linoleic acid is also necessary to avoid health disorders (Simopoulos, 2008).

Patil and Nag (2011) evaluated the potential use of chicken viscera for the production of PUFA concentrates. To this regard, three steps are necessary for obtaining these bioactive concentrates: extraction of free fatty acids, acetone treatment, and urea fractionation. First, chicken viscera are subjected to an alkaline hydrolysis that includes saponification and neutralization of viscera for the extraction of free fatty acids. The organic phase (containing free fatty acids) was extracted with a mixture of hexane:diethyl ether. Then another organic extraction with acetone allows removal of saturated fatty acids. Mono- and polyunsaturated fatty acids are subjected to urea fractionation that leads to the formation of complex urea-fatty acid crystals that are washed with isooctane to remove impurities of the filtrate. This filtrate is extracted again with the mixture of hexane:diethyl ether to finally obtain free fatty acids enriched in PUFA. The results confirm the possibility of use of these by-products in producing PUFA concentrates. The obtained percentages show that these can be isolated from 84% of PUFA concentrates from chicken viscera that can contain 82% of linoleic acid.

The temperatures used in this extraction procedure can cause losses of these valuable compounds during the extraction, reducing the recovery percentages. Moreover, the toxicity associated with the use of organic solvents makes it necessary to search alternative extraction methods (Roselló-Soto et al., 2018). Supercritical carbon dioxide (SC-CO_2) can include these techniques since it is non-destructive, and it allows obtaining an extract free of solvent and without residues. Although there are studies that use SC-CO_2 with fish by-products, it is necessary to develop more researches that evaluate it for the extraction of PUFA from meat by-products (Kang et al., 2005; Lasekan et al., 2013).

4.7 Minerals and Vitamins

The liver is a very nutritious gland because in addition to proteins, it is also an important source of vitamins and minerals. Its iron and zinc contents are also notable (Devatkal et al., 2004). In addition, this iron is in the form of iron heme compared to those present in other organs since it is present as a non-heme iron. This is the reason why the iron present in the liver is more easily absorbed than those present in other meat by-products (Hedrick et al., 1994).

It also contains vitamins such as vitamin B12, which can be used as a dietary supplement for the treatment of anemia (Devatkal et al., 2004; Kim, 2011). For this, bovine or porcine liver was chopped and treated with hot water in mild acid media. The obtained extract is evaporated in vacuum conditions and room temperature until a precipitate is obtained, which is used in the pharmaceutical field (Jayathilakan et al., 2012).

5 Conclusions

Meat by-products are an important source of bioactive compounds, which convert them into products with a great added value and its continuous positive economic impact in the meat sector. Since ancient times the main use of these products was as ingredients in the preparation of animal feed. However, the nutritional value associated with these products (carbohydrates, lipids, proteins, micronutrients, and minerals), the increased demand of alternative protein sources, and the concern over health have led to new technological applications and a more efficient use of these products.

The antimicrobial and antioxidant properties associated with the hydrolyzed proteins allow extending the shelf-life of meat products, thus balancing the results obtained with that of synthetic products. On the other hand, the ACE inhibitory and opioid activities of the recovered products, as well as the presence of enzymes and hormones have their applications in the biomedical field. To consider the safety of these bioactive compounds, it is necessary to complement the existing information with in vivo studies.

Finally, the application of emerging technologies in the extraction of these bioactive compounds will increase their quality, since they hardly affect the composition and the structure of these compounds.

Acknowledgments

This research was supported by Grant RTA 2017-00024-CO4-04 from INIA (Spain). Acknowledgments to INIA for granting Paula Borrajo with a pre-doctoral scholarship (grant number CPD2016-0030). Paulo E. Munekata acknowledges postdoctoral fellowship support from the Ministry of Economy and Competitiveness (MINECO, Spain) "Juan de la Cierva" program (FJCI-2016-29486).

References

Adje, E.Y., R. Balti, M. Kouach, P. Dhulster, D. Guillochon and N. Nedjar-Arroume. 2011. Obtaining antimicrobial peptides by controlled peptic hydrolysis of bovine hemoglobin. *International Journal of Biological Macromolecules* 49(2): 143–153.

Alao, B.O., A. B. Falowo, A. Chulayo and V. Muchenje. 2017. The potential of animal by-products in food systems: Production, prospects and challenges. *Sustainability* 9(7): 1089.

Alovskaya, A., T. Alekseeva, J. B. Phillips, V. King and R. Brown. 2007. Fibronectin, collagen, fibrin - components of extracellular matrix for nerve regeneration. In: N. Ashammakhi, R. Reis and E. Chiellini (eds.), *Topics in tissue engineering*. University of Oulu Publisher, Oulu, Finland, pp. 1–27.

Anderson, D. 2006. Rendering operations. In: D. L. Meeker (ed.), *Essential rendering. All about the animal by-products industry*. National Renderers Association, Washington, DC, pp. 31–52.

Arihara, K., Ohata, M. 2010. Functional meat products. In: F. Toldrá (ed.), *Handbook of meat processing*. Wiley-Blackwell, Ames, Iowa, pp. 423–439.

Aristoy, M.C., Toldrá, F. 2011. Essential amino acids. In: L. M. L. Nollet and F. Toldrá (eds.), *Handbook of analysis of edible animal by-products*. CRC Press, Boca Raton, FL, pp. 123–136.

Aspevik, T., Å. Oterhals, S.B. Rønning, T. Altintzoglou, S. G. Wubshet, A. Gildberg, ... D. Lindberg. 2017. Valorization of proteins from co-and by-products from the fish and meat industry. *Topics in Current Chemistry* 375(3): 53.

Awan, Z.A., M. Tariq, M. M. Awan, N. W. Satti, T. Mukhtar, W. Akram and M. Faizan Yasin. 2015. Edible by-products of meat. *Veterinaria* 3(1): 33–36.

Bah, C.S., A.E.D.A. Bekhit, A. Carne, and M. A. McConnell. 2013. Slaughterhouse blood: An emerging source of bioactive compounds. *Comprehensive Reviews in Food Science and Food Safety* 12(3): 314–331.

Banerjee, P., and C. Shanthi. 2012. Isolation of novel bioactive regions from bovine Achilles tendon collagen having angiotensin I-converting enzyme-inhibitory properties. *Process biochemistry* 47(12): 2335–2346.

Bhaskar, N., V. K. Modi, K. Govindaraju, C. Radha, and R. G. Lalitha. 2007. Utilization of meat industry by products: Protein hydrolysate from sheep visceral mass. *Bioresource technology* 98(2): 388–394.

Bichukale, A.D., J. M. Koli, A. E. Sonavane, V. V. Vishwasrao, K. H. Pujari, P. E. Shingare. 2018. Functional properties of gelatin extracted from poultry skin and bone waste. *International Journal of Pure and Applied Bioscience* 6(4): 87–101.

Bland, C.E., H. Ginsburg, J. E. Silbert, D. D. Metcalfe. 1982. Mouse heparin proteoglycan. Synthesis by mast cell-fibroblast monolayers during lymphocyte-dependent mast cell proliferation. *Journal of Biological Chemistry* 257(15): 8661–8666.

Boire, A., A. Bouchoux, S. Bouhallab, A. L. Chapeau, T. Croguennec, V. Ferraro, ... V. Santé-Lhoutellier. 2017. Proteins for the future: A soft matter approach to link basic knowledge and innovative applications. *Innovative Food Science and Emerging Technologies* 46: 18–28.

Braam, W., M. G. Smits, R. Didden, H. Korzilius, I. M. Van Geijlswijk, L. M. Curfs. 2009. Exogenous melatonin for sleep problems in individuals with intellectual disability: A meta-analysis. *Developmental Medicine and Child Neurology* 51(5): 340–349.

Chang, C.Y., K. C. Wu, S. H. Chiang. 2007. Antioxidant properties and protein compositions of porcine haemoglobin hydrolysates. *Food Chemistry* 100(4): 1537–1543.

Cheng, F.Y., I. C. Lai, L. C. Lin, R. Sakata. 2016. The in vitro antioxidant properties of alcalase hydrolysate prepared from silkie fowl (*Gallus gallus*) blood protein. *Animal Science Journal* 87(7): 921–928.

Chou, C.H., S. Y. Wang, Y. T. Lin, Y. C. Chen. 2014. Antioxidant activities of chicken liver hydrolysates by pepsin treatment. *International Journal of Food Science and Technology* 49(7): 1654–1662.

Daamen, W.F., J. H. Veerkamp, J. C. M. van Hest, T. H. van Kuppevelt. 2007. Elastin as a biomaterial for tissue engineering. *Biomaterials* 28: 4378–4398.

Daoud, R., V. Dubois, L. Bors-Dodita, N. Nedjar-Arroume, F. Krier, N. E. Chihib, P. Marya, M. Kouach, G. Briand, D. Guillochon. 2005. New antibacterial peptide derived from bovine hemoglobin. *Peptides* 26, 713-719.

Darine, S., V. Christophe and D. Gholamreza. 2010. Production and functional properties of beef lung protein concentrates. *Meat Science* 84(3): 315–322.

Del Hoyo, P., M. Rendueles and M. Díaz. 2008. Effect of processing on functional properties of animal blood plasma. *Meat Science* 78(4): 522–528.

Deng, H., J. Zheng, F. Zhang, Y. Wang, J. Kan. 2014. Isolation of angiotensin I-converting enzyme inhibitor from pepsin hydrolysate of porcine hemoglobin. *European Food Research and Technology* 239: 933–940.

Devatkal, S., S. K. Mendiratta, N. Kondaiah, M. C. Sharma, A. S. R. Anjaneyulu. 2004. Physicochemical, functional and microbiological quality of buffalo liver. *Meat Science* 68(5): 79–86.

DeWitt, C.M., G. Gomez and J. M. James. 2002. Protein extraction from beef heart using acid solubilization. *Journal of Food Science* 67(9): 3335–3341.

Di Bernardini, R., P. Harnedy, D. Bolton, J. Kerry, E. O'Neill, A. M. Mullen, M. Hayes. 2011a. Antioxidant and antimicrobial peptidic hydrolysates from muscle protein sources and by-products. *Food Chemistry* 124: 1296–1307.

Di Bernardini, R., D. K. Rai, D. Bolton, J. Kerry, E. O'neill, A. M. Mullen, P. Harnedy, M. Hayes. 2011b. Isolation, purification and characterization of antioxidant peptidic fractions from a bovine liver sarcoplasmic protein thermolysin hydrolyzate. *Peptides* 32(2): 388–400.

EC. 2009. Regulation (EC) No 1069/2009 of the European Parliament and of the Council of 21 October 2009 Laying Down Health Rules as Regards Animal By-Products and Derived Products not Intended for Human Consumption and Repealing Regulation (EC) No. 1774/2002 (Animal By-Products Regulation), pp. L301–L333.

Elagamy, E.I. 2000. Physicochemical, molecular and immunological characterization of camel calf rennet: A comparison with buffalo rennet. *Journal of Dairy Research* 67(1): 73–81.

Escudero, E., F. Toldrá, M. A. Sentandreu, H. Nishimura, K. Arihara. 2012. Antihypertensive activity of peptides derived from the in vitro gastrointestinal digestion of pork meat. *Meat Science* 91: 306–311.

FAO. 2010. *Fat and fatty acid requirements for adults. Fats and fatty acids in human nutrition.* Food and Agriculture Organization of the United Nations, Rome, Italy, pp. 55–62.

FAOSTAT. 2016. www.fao.org/faostat/en/#data/QL/metadata.

Ferraro, V., M. Anton, V. Santé-Lhoutellier. 2016. The "sisters" α-helices of collagen, elastin and keratin recovered from animal by-products: Functionality, bioactivity and trends of application. *Trends in Food Science and Technology* 51: 65–75.

Ferraro, V., A. P. Carvalho, C. Piccirillo, M. Santos, M. P. L. Castro, E. M. Pintado. 2013. Extraction of high added value biological compounds from sardine, sardine-type fish and mackerel canning residues - a review. *Material Science and Engineering C* 33(6): 3111–3120.

Florek, M., Z. Litwińczuk, P. Skałecki, M. Kędzierska-Matysek, T. Grodzicki. 2012. Chemical composition and inherent properties of offal from calves maintained under two production systems. *Meat Science* 90: 402–409.

Fontoura, R., D. J. Daroit, A. P. F. Correa, S. M. M. Meira, M. Mosquera, A. Brandelli. 2014. Production of feathers hydrolysates with antioxidant, angiotensin-I converting enzyme- and dipeptidyl peptidase-IV- inhibitory activities. *New Biotechnology* 31: 506–513.

Fornias, O.V. 1996. Edible by-products of slaughter animals. *FAO Animal Production and Health Paper* 123: 9.

Franco, D., L. González, E. Bispo, P. Rodríguez, J. I. Garabal, T. Moreno. 2010. Study of hydrolyzed protein composition, free amino acid, and taurine content in different muscles of Galician blonde beef. *Journal of Muscle Foods* 21: 769–784.

Froidevaux, R., F. Krier, N. Nedjar-Arroume, D. Vercaigne-Marko, E. Kosciarz, C. Ruckebusch, ... D. Guillochon. 2001. Antibacterial activity of a pepsin-derived bovine hemoglobin fragment. *Febs Letters* 491(1–2): 159–163.

Gálvez, F., A. Maggiolino, R. Domínguez, M. Pateiro, S. Gil, P. De Palo, ... J. M. Lorenzo. 2018. Nutritional and meat quality characteristics of seven primal cuts from 9-month old female veal calves: A preliminary study. *Journal of the Science of Food and Agriculture 99*(6): 2947-2956.

García-Llatas, G., A. Alegría, R. Barberá, R. Farré. 2011. Minerals and trace elements. In: L. M. L. Nollet and F. Toldrá (eds.), *Handbook of analysis of edible animal by-products.* CRC Press, Boca Raton, FL, pp. 183–206.

Gbogouri, G.A., M. Linder, J. Fanni, M. Parmentier. 2004. Influence of hydrolysis degree on the functional properties of salmon by-products hydrolysates. *Journal of Food Science* 69: C615–C622.

Ghost, R. 2001. Fractionating of biological macromolecules using carrier phase ultrafiltration. *Biotechnology and Bioengineering* 74: 1–11.

Gómez-Guillén, M.C., B. Giménez, M. A López-Caballero, M. P. Montero. 2011. Functional and bioactive properties of collagen and gelatin from alternative sources: A review. *Food hydrocolloids* 25(8): 1813–1827.

Gooding, C.H., D. L. Meeker. 2016. Comparison of 3 alternatives for large-scale processing of animal carcasses and meat by-products. *The Professional Animal Scientist* 32(3): 259–270.

Hattori, M., K. Yamaji-Tsumamoto, H. Kumagai, Y. Feng and K. Takahashi. 1998. - Antioxidative activity of soluble elastin peptides. *Journal of Agriculture and Food Chemistry* 46: 2167–2170.

Hedrick, H.B., E. D. Aberle, J. C. Forrest, M. D. Judge, R. A. Merkel. 1994. Nutritive value of meat. In: E.D. Aberle, J. C. Forrest, D. E. Gerrard, E. W. Mills (eds.), *Principles of Meat Science.* Kendall Hunt Publishing Company, Dubuque, Iowa, pp. 289–298.

Helkar, P.B., Sahoo, A.K., Patil, N.J. 2016. Review: Food industry by-products used as a functional food ingredients. *International Journal of Waste Resources* 6(3): 1–6.

Herregods, G., Van Camp, J., Morel, N., Ghesquière, B., Gevaert, K., Vercruysse, L., ... Smagghe, G. 2011. Angiotensin I-converting enzyme inhibitory activity of gelatin hydrolysates and identification of bioactive peptides. *Journal of Agricultural and Food Chemistry* 59: 552–558.

Honikel, K.O. 2011. Composition and calories. In: Nollet, L.M.L., Toldrá, F. (eds.), *Handbook of analysis of edible animal by-products.* CRC Press, Boca Raton, FL, pp. 105–123.

Hrynets, Y., Omana, D.A., Xu, Y., Betti, M. 2011. Comparative study on the effect of acid-and alkaline-aided extractions on mechanically separated turkey meat (MSTM): Chemical characteristics of recovered proteins. *Process biochemistry* 46(1): 335–343.

Hsieh, Y.H.P., Ofori, J.A. 2011. Blood-derived products for human consumption. *Revelation and Science* 1(01): 14–21.

Huang, S.C., Liu, P.J. 2010. Inhibition of angiotensin I-converting enzymes by enzymatic hydrolysates from chicken blood. *Journal of Food and Drug Analysis* 18(6): 458–463.

In, M.J., Chae, H.J., Oh, N.S. 2002. Process development for heme-enriched peptide by enzymatic hydrolysis of hemoglobin. *Bioresource technology* 84(1): 63–68.

Irshad, A., Sharma, B.D. 2015. Abattoir by-product utilization for sustainable meat industry: A review. *Journal of Animal Production Advances* 6: 681–696.

Jamdar, S.N., Harikumar, P. 2005. Autolytic degradation of chicken intestinal proteins. *Bioresource Technology* 96(11): 1276–1284.

Jayathilakan, K., Sultana, K., Radhakrishna, K., Bawa, A.S. 2012. Utilization of byproducts and waste materials from meat, poultry and fish processing industries: A review. *Journal of Food Science and Technology* 49(3): 278–293.

Jun, M., Lee, J.M., Lee, K.S., Kim, K.O. 2000. The effect of preparation conditions on the properties of Jokpyun (traditional Korean gel type food) model system. *Food Science and Biotechnology* 9(27): 27–31.

Kandil, T.S., Mousa, A.A., El-Gendy, A.A., Abbas, A.M. 2010. The potential therapeutic effect of melatonin in gastro-esophageal reflux disease. *BMC Gastroenterology* 10: 7.

Kang, K.Y., Ahn, D.H., Jung, S.M., Kim, D.H., Chun, B.S. 2005. Separation of protein and fatty acids from tuna viscera using supercritical carbon dioxide. *Biotechnology and Bioprocess Engineering* 10(4): 315–321.

Karim, A.A., Bhat, R. 2009. Fish gelatin: Properties, challenges, and prospects as an alternative to mammalian gelatins. *Food Hydrocolloids* 23: 563–576.

Khouw, B., Kesler, J. 1974. *U.S. Patent No. 3,809,748*. Patent and Trademark Office, Washington, DC.

Khouw, B.T., Rubin, L.J., Berry, B. 1989. Meat animal by-products of pharmaceutical and food interest. In: *Proceedings-Annual Reciprocal Meat Conference of the American Meat Science Association (USA)*. American Meat Science Association (AMSA), Champaign, IL USA.

Kim, H.K., Kim, Y.H., Kim, Y.J., Park, H.J., Lee, N.H. 2012. Effects of ultrasonic treatment on collagen extraction from skins of the sea bass *Lateolabrax japonicus*. *Fisheries Science* 78(2): 485–490.

Kim, S.K., Kim, Y.T., Byun, H.G., Park, P.J., Ito, H. 2001. Purification and characterization of antioxidative peptides from bovine skin. *BMB Reports* 34(3): 219–224.

Kim, Y.N. 2011. Vitamins. In: Nollet, L.M.L., Toldrá, F. (eds.), *Handbook of analysis of edible animal by-products*. CRC Press, Boca Raton, FL, pp. 161–182.

Klompong, V., Benjakul, S., Kantachote, D., Shahidi, F. 2007. Antioxidative activity and functional properties of protein hydrolysate of yellow stripe trevally (*Selaroides leptolepis*) as influenced by the degree of hydrolysis and enzyme type. *Food Chemistry* 102: 1317–1327.

Küllenberg, D., Taylor, L.A., Schneider, M., Massing, U. 2012. Health effects of dietary phospholipids. *Lipids in Health and Disease* 11: 3.

Lafarga, T., Hayes, M. 2014. Bioactive peptides from meat muscle and by-products: Generation, functionality and application as functional ingredients. *Meat Science* 98(2): 227–239.

Langmaier, F., Mládek, M., Kolomazník, K., Sukop, S. 2001. Collagenous hydrolysates from untraditional sources of proteins. Reaction condition and the yield of enzymatic hydrolysis of short cattle tendons. *International Journal of Cosmetic Science* 23(4): 201–206.

Lasekan, A., Bakar, F.A., Hashim, D. 2013. Potential of chicken by-products as sources of useful biological resources. *Waste Management* 33(3): 552–565.

Lee, S.J., Kim, E.K., Hwang, J.W., Oh, H.J., Cheong, S.H., Moon, S.H., … Park, P.J. 2010. Purification and characterisation of an antioxidative peptide from enzymatic hydrolysates of duck processing by-products. *Food Chemistry* 123(2): 216–220.

Lee, S.J., Kim, K.H., Kim, Y.S., Kim, E.K., Hwang, J.W., Lim, B.O., ... Park, P.J. 2012b. Biological activity from the gelatin hydrolysates of duck skin by-products. *Process Biochemistry* 47(7): 1150–1154.

Lee, S.J., Kim, Y.S., Hwang, J.W., Kim, E.K., Moon, S.H., Jeon, B.T., ... Park, P.J. 2012a. Purification and characterization of a novel antioxidative peptide from duck skin by-products that protects liver against oxidative damage. *Food Research International* 49(1): 285–295.

Li, B., Chen, F., Xang, X., Ji, B., Wu, Y. 2007. Isolation and identification of antioxidative peptides from porcine collagen hydrolysate by consecutive chromatography and electrospray ionization-mass spectrometry. *Food Chemistry* 102: 1135–1143.

Li, D., Mu, C., Cai, S., Lin, W. 2009. Ultrasonic irradiation in the enzymatic extraction of collagen. *Ultrasonics Sonochemistry* 16(5): 605–609.

Lim, J.Y., Oh, S.S., Kim, K.O. 2001. The effects of processing conditions on the properties of chicken feet gelatin. *Food Science and Biotechnology* 10(6): 638–645.

Liu, D., Nikoo, M., Boran, G., Zhou, P., Regenstein, J.M. 2015. Collagen and gelatin. *Annual Reviews of Food Science and Technology* 6: 527–557.

Liu, D.C., 2002. Better utilization of by-products from the meat industry. www.agnet.org /library/eb/515/.

Liu, D.C., Ockerman, H.W. 2001. Meat co-products. In: Hui, Y.H., Nip, W.K., Rogers, R.W., Young, O.A. (eds.), *Meat science and applications*. CRC Press, Boca Raton, FL, pp. 583–606.

Liu, Q., Kong, B., Xiong, Y.L., Xia, X. 2010a. Antioxidant activity and functional properties of porcine plasma protein hydrolysate as influenced by the degree of hydrolysis. *Food Chemistry* 118(2): 403–410.

Liu, R., Xing, L., Fu, Q., Zhou, G.H., Zhang, W.G. 2016. A review of antioxidant peptides derived from meat muscle and by-products. *Antioxidants* 5(3): 32.

Liu, X., Song, C., Chen, R., Jiang, X., Jin, Y., Zou, H. 2010b. Identification of angiotensin I-converting enzyme inhibitors in peptides mixture of hydrolyzed red deer plasma with proteomic approach. *Chinese Journal of Chemistry* 28: 1665–1672.

Lombardi-Boccia, G., Lanzi, S., Aguzzi, A. 2005. Aspects of meat quality: Trace elements and B vitamins in raw and cooked meats. *Journal of Food Composition and Analysis* 18: 39–46.

Lorenzo, J.M., Munekata, P.E., Gómez, B., Barba, F.J., Mora, L., Pérez-Santaescolástica, C., Toldrá, F. 2018. Bioactive peptides as natural antioxidants in food products - A review. *Trends in Food Science and Technology* 79: 136–147.

Lynch, S.A., Mullen, A.M., O'Neill, E., Drummond, L., Álvarez, C. 2018. Opportunities and perspectives for utilisation of co-products in the meat industry. *Meat Science* 144: 62–73.

Maehashi, K., Matsuzaki, M., Yamamoto, Y., Udaka, S. 1999. Isolation of peptides from an enzymatic hydrolysate of food proteins and characterization of their taste properties. *Bioscience, Biotechnology, and Biochemistry* 63(3): 555–559.

Matak, K.E., Tahergorabi, R., Jaczynski, J. 2015. A review: Protein isolates recovered by isoelectric solubilization/precipitation processing from muscle food by-products as a component of nutraceutical foods. *Food Research International* 77: 697–703.

Meeker, D.L., Hamilton, C.R. 2006. An Overview of the Rendering Industry. In: Meeker, D. L. (ed.), *Essential rendering. All about the animal by-products industry*. National Renderers Association, Washington, DC, pp. 1–16.

Mohanty, A.K., Mukhopadhyay, U.K., Kaushik, J.K., Grover, S., Batish, V.K. 2003. Isolation, purification and characterization of chymosin from riverine buffalo (*Bubalos bubalis*). *Journal of Dairy Research* 70(1): 37–43.

Mokrejs, P., Janacova, D., Svoboda, P. 2012. Three-stage extraction of gelatines from tendons of abattoir cattle: 1 - Reaction Conditions. *Applied biochemistry and biotechnology* 168(4): 917–927.

Mokrejs, P., Langmaier, F., Mládek, M., Janacova, D., Kolomazník, K., Vasek, V. 2009. Extraction of collagen and gelatine from meat industry by-products for food and non food uses. *Waste Management and Research* 27(1): 31–37.

Mora, L., Reig, M., Toldrá, F. 2014. Bioactive peptides generated from meat industry by-products. *Food Research International* 65: 344–349.

Mora, L., Toldrá-Reig, F., Reig, M., Toldrá, F. 2019. Possible uses of processed slaughter byproducts. In: Galanakis, C.M. (ed.), *Sustainable meat production and processing*. Elsevier, London, pp. 145–160.

Morrison, N.A., Clark, R.C., Chen, Y.L., Talashek, T., Sworn, G. 1999. Gelatin alternatives for the food industry. *Progress in Colloid Polymer Science* 114: 127–131.

Mullen, A.M., Álvarez, C., Zeugolis, D.I., Henchion, M., O'Neill, E., Drummond, L. 2017. Alternative uses for co-products: Harnessing the potential of valuable compounds from meat processing chains. *Meat Science* 132: 90–98.

Nadeem, S.M.S., Khan, J.A., Murtaza, B.N., Muhammad, K., Rauf, A. 2015. Purification and properties of liver catalase from water buffalo (*Bubalus bubalis*). *South Asian Journal of Life Sciences* 3(2): 51–55.

Nedjar-Arroume, N., Dubois-Delval, V., Adje, E.Y., Traisnel, J., Krier, F., Mary, P., ... Guillochon, D. 2008. Bovine hemoglobin: An attractive source of antibacterial peptides. *Peptides* 29(6): 969–977.

Nollet, L.M.L., Toldrá, F. 2011. Introduction. Offal meat: Definitions, regions, cultures, generalities. In: Nollet, L.M.L., Toldrá, F. (eds.), *Handbook of analysis of edible animal by-products*. CRC Press, Boca Raton, FL, pp. 3–11.

Nyberg, F., Carlsson, A., Hallberg, M. 2013. Casomorphins/hemorphins. In: Kastin, A.J. (ed.), *Handbook of biologically active peptides*. Elsevier, San Diego, CA, pp. 1550–1555.

Ockerman, H.W., Basu, L. 2006. Edible Rendering-Rendered Products for Human Use. In: Meeker, D.L. (ed.), *Essential rendering. All about the animal by-products industry*. National Renderers Association, Washington, DC, pp. 95–110.

Ockerman, H.W., Basu, L. 2014. Hides and Skins. In: Jensen, W., Devine, C., Dikemann, M. (eds.), *Encyclopaedia of meat science*. Elsevier Science Ltd., London, pp. 125–138.

Ockerman, H.W., Hansen, C.L. 2000. Introduction and history of processing animal by-products. In: Ockerman, H.W., Hansen, C.L. (eds.), *Animal By-Product Processing and Utilization*. CRC Press, Boca Raton, FL, pp. 1–21.

Ohba, R., Deguchi, T., Kishikawa, M., Arsyad, F., Morimura, S., Kida, K. 2003. Physiological functions of enzymatic hydrolysates of collagen or keratin contained in livestock and fish waste. *Food Science and Technology Research* 9(1): 91–93.

Panyam, D., Kilara, A. 1996. Enhancing the functionality of food proteins by enzymatic modification. *Trends in Food Science and Technology* 7(4): 120–125.

Patil, D., Nag, A. 2011. Production of PUFA concentrates from poultry and fish processing waste. *Journal of the American Oil Chemists' Society* 88(4): 589–593.

Pearl, G.G. 2004. Inedible. In: Jensen, W., Devine, C., Dikemann, M. (eds.), *Encyclopedia of Meat Sciences*. Elsevier Science Ltd., London, pp. 112–125.

Peres, M.F., Masruha, M.R., Zukerman, E., Moreira-Filho, C.A., Cavalheiro, E.A. 2006. Potential therapeutic use of melatonin in migraine and other headache disorders. *Expert Opinion on Investigational Drugs* 15(4): 367–375.

Prates, J.A.M., Alfaia, C.M., Alves, S.P., Bessa, R.J.B. 2011. Fatty Acids. In: Nollet, L.M.L., Toldrá, F. (eds.), *Handbook of analysis of edible animal by-products*. CRC Press, Boca Raton, FL, pp. 137–159.

Przybylski, R., Firdaous, L., Châtaigné, G., Dhulster, P., Nedjar, N. 2016. Production of an antimicrobial peptide derived from slaughterhouse by-product and its potential application on meat as preservative. *Food Chemistry* 211: 306–313.

Rahman, M.N.A., Jamalulail, S.A.S.K. 2012. Extraction, physicochemical characterizations and sensory quality of chicken feet gelatin. *Borneo Science* 30: 1–13.

Raj, K.R., Mahendrakar, N.S. 2010. Effect of ensiling and organic solvents treatment on proteolytic enzymes of layer chicken intestine. *Journal of Food Science and Technology* 47(3): 320–324.

Ran, X.G., Wang, L.Y. 2014. Use of ultrasonic and pepsin treatment in tandem for collagen extraction from meat industry by-products. *Journal of the Science of Food and Agriculture* 94 (3): 585–590.

Ravindran, R., Jaiswal, A.K. 2016. Exploitation of food industry waste for high-value products. *Trends in Biotechnology* 34(1): 58–69.

Ren, Y., Wan, D.-G., Lu, X., Chen, L., Zhang, T., Guo, J. 2011. Isolation and characterization of angiotensin I converting enzyme inhibitor peptides derived from porcine hemoglobin. *Scientific Research and Essays* 6: 6262–6269.

Roselló-Soto, E., Barba, F.J., Lorenzo, J.M., Dominguez, R., Pateiro, M., Mañes, J., Moltó, J. C. 2018. Evaluating the impact of supercritical-CO_2 pressure on the recovery and quality of oil from "horchata" by-products: Fatty acid profile, α-tocopherol, phenolic compounds, and lipid oxidation parameters. *Food Research International* https://doi.org/10.1016/j.foodres.2018.11.054.

Ryder, K., Bekhit, A.E.D., McConnell, M., Carne, A. 2016. Towards generation of bioactive peptides from meat industry waste proteins: Generation of peptides using commercial microbial proteases. *Food Chemistry* 208: 42–50.

Saiga, A., Iwai, K., Hayakawa, T., Takahata, Y., Kitamura, S., Nishimura, T., Morimatsu, F. 2008. Angiotensin I-converting enzyme-inhibitory peptides obtained from chicken collagen hydrolysate. *Journal of Agricultural and Food Chemistry* 56: 9586–9591.

Salminen, E., Rintala, J. 2002. Anaerobic digestion of organic solid poultry slaughterhouse waste – A review. *Bioresource Technology* 83: 13–26.

Sanders, T.B.A. 1988. Essential and trans fatty acids in nutrition. *Nutrition Research Reviews* 1: 57–78.

Sarbon, N.M., Badii, F., Howell, N.K. 2013. Preparation and characterisation of chicken skin gelatin as an alternative to mammalian gelatin. *Food Hydrocolloids* 30(1): 143–151.

Selmane, D., Christophe, V., Gholamreza, D. 2008. Extraction of proteins from slaughterhouse by-products: Influence of operating conditions on functional properties. *Meat Science* 79(4): 640–647.

Seong, P.N., Cho, S.H., Park, K.M., Kang, G.H., Park, B.Y., Moon, S.S., Van Ba, H. 2015. Characterization of chicken by-products by mean of proximate and nutritional compositions. *Korean Journal for Food Science of Animal Resources* 35(2): 179–188.

Seong, P.N., Kang, G.H., Park, K.M., Cho, S.H., Kang, S.M., Park, B.Y., ... Van Ba, H. 2014b. Characterization of Hanwoo bovine by-products by means of yield, physicochemical and nutritional compositions. *Korean Journal for Food Science of Animal Resources* 34(4): 434–447.

Seong, P.N., Park, K.M., Cho, S.H., Kang, S.M., Kang, G.H., Park, B.Y., ... Van Ba, H. 2014a. Characterization of edible pork by-products by means of yield and nutritional composition. *Korean Journal for Food Science of Animal Resources* 34(3): 297–306.

Simopoulos, A. 2008. Minireview: The importance of the omega-6/omega-3 fatty acid ratio in cardiovascular disease and other chronic diseases. *Experimental Biology and Medicine* 233: 674–688.

Swan, J.E. 1999. Animal by-product processing. In: Francis, F.J. (ed.), *Wiley Encyclopedia of Food Science and Technology*. John Wiley & Sons Inc., New York, NY, pp. 35–43.

Tahergorabi, R., Beamer, S.K., Matak, K.E., Jaczynski, J. 2011. Effect of isoelectric solubilization/precipitation and titanium dioxide on whitening and texture of proteins recovered from dark chicken-meat processing by-products. *LWT-Food Science and Technology* 44(4): 896–903.

Tahergorabi, R., Sivanandan, L., Jaczynski, J. 2012. Dynamic rheology and endothermic transitions of proteins recovered from chicken-meat processing by-products using isoelectric solubilization/precipitation and addition of TiO2. *LWT-Food Science and Technology* 46(1): 148–155.

Tian, L., Liu, J., Ma, L., Zhang, L., Wang, S., Yan, E., Zhu, H. 2017. Isolation and Purification of Antioxidant and ACE-Inhibitory Peptides from Yak (*Bos grunniens*) Skin. *Journal of Food Processing and Preservation* 41(5): e13123.

Ticu, E.L., Vercaigne-Marko, D., Froidevaux, R., Huma, A., Artenie, V., Guillochon, D. 2005. Use of a protease-modified-alumina complex to design a continuous stirred tank reactor for producing bioactive hydrolysates. *Process Biochemistry* 40(8): 2841–2848.

Toldrá, F., Aristoy, M.C., Mora, L., Reig, M. 2012. Innovations in value-addition of edible meat by-products. *Meat Science* 92(3): 290–296.

Toldrá, F., Mora, L., Reig, M. 2016. New insights into meat by-product utilization. *Meat Science* 120: 54–59.

Toldrá, F., Reig, M. 2011. Innovations for healthier processed meats. *Trends in Food Science and Technology* 22(9): 517–522.

Toldrá, M., Parés, D., Saguer, E., Carretero, C. 2011. Hemoglobin hydrolysates from porcine blood obtained through enzymatic hydrolysis assisted by high hydrostatic pressure processing. *Innovative Food Science and Emerging Technologies* 12(4): 435–442.

Tomovic, V., Jokanovic, M., Sojic, B., Skaljac, S., Tasic, T., Ikonic, P. 2015. Minerals in pork meat and edible offal. *Procedia Food Science* 5: 293–295.

Ulbricht, T.L.V., Southgate, D.A.T. 1991. Coronary heart disease: Seven dietary factors. *Lancet* 338: 985–992.

Vanhoute, M., Firdaous, L., Bazinet, L., Froidevaux, R., Lecouturier, D., Guillochon, D., Dhulster, P. 2010. Effect of haem on the fractionation of bovine haemoglobin peptic hydrolysate by electrodialysis with ultrafiltration membranes. *Journal of Membrane Science* 365: 16–24.

Vasconcelos, A., Cavaco-Paulo, A. 2013. The use of keratin in biomedical applications. *Current Drug Targets* 14(5): 612–619.

Vermeirssen, V., Van Camp, J., Verstraete, W. 2004. Bioavailability of angiotensin I-converting enzyme inhibitory peptides. *British Journal of Nutrition* 92(3): 357–366.

Wan, Y., Ghost, R., Cui, Z. 2002. High resolution plasma protein fractionation using ultrafiltration. *Desalination* 144: 301–306.

Wang, J.Z., Zhang, H.A.O., Zhang, M., Yao, W.T., Mao, X.Y., Ren, F.Z. 2008. Antioxidant activity of hydrolysates and peptide fractions of porcine plasma albumin and globulin. *Journal of Food Biochemistry* 32(6): 693–707.

Warda, M., Mao, W., Toida, T., Linhardt, R.J. 2003. Turkey intestine as a commercial source of heparin? Comparative structural studies of intestinal avian and mammalian glycosaminoglycans. Comparative Biochemistry and Physiology. *Part B: Biochemistry and Molecular Biology* 134(1): 189–197.

WHO. 2003. *Diet, nutrition and the prevention of chronic diseases.* WHO technical report series 916. World Health Organization, Geneva, Switzerland, ISBN: 92-4-1209-16X.

WHO. 2015. *WHO Model List of Essential Medicines (19th List).* World Health Organization, Geneva, Switzerland.

Williams, C. 2000. Dietary fatty acids and human health. *Annales de Zootechnie* 49: 165–180.

Wood, J., Richardson, R., Nute, G., Fisher, A., Campo, M., Kasapidou, E., Sheard, P., Enser, M. 2004. Effects of fatty acids on meat quality: A review. *Meat Science* 66: 21–32.

Xu, X., Cao, R., He, L., Yang, N. 2009. Antioxidant activity of hydrolysates derived from porcine plasma. *Journal of the Science of Food and Agriculture* 89(11): 1897–1903.

Yu, Y., Hu, J., Miyaguchi, Y., Bai, X., Du, Y., Lin, B. 2006. Isolation and characterization of angiotensin I-converting enzyme inhibitory peptides derived from porcine hemoglobin. *Peptide* 27(11): 2950–2956.

Zhang, Y., K. Olsen, A. Grossi, J. Otte. 2013. Effect of pretreatment on enzymatic hydrolysis of bovine collagen and formation of ACE-inhibitory peptides. *Food Chemistry* 141(3): 2343-2354.

Zhao, Q.Y., Garreau, I., Sannier, F., Piot, J.M. 1997. Opioid peptides derived from hemoglobin: Hemorphins. *Peptide Science* 43(2): 75–98.

Zhao, Q.Y., Sannier, F., Garreau, I., Guillochon, D., Piot, J.M. 1994. Inhibition and inhibition kinetics of angiotensin converting enzyme activity by hemorphins, isolated from a peptic bovine hemoglobin hydrolysate. *Biochemical and Biophysical Research Communications* 204(1): 216–223.

Zhou, P., Mulvaney, S.J., Regenstein, J.M. 2006. Properties of Alaska pollock skin gelatin: A comparison with Tilapia and pork skin gelatins. *Journal of Food Science* 71: C313–C321.

4

Tailor-Made Process to Recover High Added Value Compounds from Fishery By-Products

JOSÉ ANTONIO VÁZQUEZ, ANA I. DURÁN, ARACELI MENDUÍÑA, MARGARITA NOGUEIRA, JAVIER FRAGUAS, JESÚS MIRÓN, AND JESÚS VALCÁRCEL

Grupo de Reciclado y Valorización de Materiales Residuales (REVAL), Instituto de Investigacións Mariñas (IIM-CSIC). r/Eduardo Cabello, 6. Vigo-36208, Galicia, Spain

Contents

1 Introduction

The catches of fish and other marine organisms worldwide, which include aquaculture, amount to 240 million tons (FAO, 2016). More than 25% of these catches are considered as by-products because these include animal fractions (heads, viscera, skeletons, bones, etc.) that are separated in the initial steps of its manufacture for human feeding, without commercial value. For example, in Galicia (Spain), the largest fishing region in Spain and Europe, considering both the size of the fleet and the number of jobs involved and the economic value derived from the fishing activity, more than 120,000 tons of fresh fish (from several species such as hake, megrim, shark, blue whiting, ray, pouting, and mackerel) was landed to different ports of the region in 2017 (IGE, 2018; Mapama, 2018; Port of Vigo, 2018), which resulted in the generation of around 35,000 tons of by-products. Such huge amounts of wastes need to be well managed to avoid causing serious environment-related health problems. To these volumes, we must also add the frozen fish that is transformed in the processing and canning industries (some of the biggest European companies in both sectors have a presence in Galicia).

In addition, millions of tons of several species are returned to sea after capture (as discards and by-catch) because of not meeting legal size requirements, lack of fishing quota, low quality, or individuals not having any commercial interest. In this context, the new common fishery EU policy (*toward zero discard*) implemented from 2019 obliges to adopt a set of measures toward the complete elimination of discards (EU, 2018). This regulation led to a new scenario and presents a challenge to develop a process aimed at managing and valorizing new biomass.

The traditional way of valorization, which has been the only process for decades, of wastes from fishing is determined by the production of fish meal (mainly protein material) and oils (from livers or decanted from fish meal process), which are in huge demand in the production of animal feed, nutraceutical supplements, and paints (Bowyer et al., 2013; Ferraro et al., 2013). However, the search of biotechnological alternatives to recover and result in added-value compounds from by-products as well as discards is peremptory, and in this context an extensive range of biomolecules, products, and bioactivities has been reported in the last years (Blanco et al., 2007; Olsen et al., 2014; Ordoñez et al., 2014; Silva et al., 2012). For instance, skins, bones, cartilages, eyes, viscera, and skeletons of crustaceans and cephalopods are raw materials for obtaining collagen, gelatins, apatites, fish protein hydrolysates (FPH), glycosaminoglycans (GAGs), hyaluronic acid (HA), enzymes, biosilage, and chitosan using chemical and enzymatic hydrolysis operations, selective precipitation by solvents, membrane purification, and so on (Beaney et al., 2005; Ghaly et al., 2013; Gómez-Guillén et al., 2002; Halim et al., 2016; Maccari et al., 2015; Safari et al., 2012; Silva et al., 2014a; Vázquez et al., 2013). The application of these products is very broad which ranges from animal feed, the food and cosmetic sector, to the pharmaceutical and medical field (Carvalho et al., 2018; Kerton et al., 2013; Shahidi and Kamil, 2002; Stephen et al., 2010; Teves and Ragaza, 2016; Yuan et al., 2012).

Further, the most valuable biocompounds isolated and products obtained from fishery wastes are described along with some of the most important fields of application. In this chapter, we have not explored the different alternatives of valorization to formulate products for direct human consumption, such as fish mince, surimi-like materials, and restructured fish foods. The valorization strategies for marine algae is not included in this review either, since they are not a by-product generated from fishing activities.

2 Fish Meal and Fish Oils

Fish meal is defined as a powder with humidity of less than 12% and variable proportions of minerals (ashes), fats, and proteins (main component) obtained by cooking, pressing, drying, and milling of whole fish, by-products, and/or discards. These processes are perfectly established and optimized at an industrial scale, and the composition of the final fish meal is dependent only on the kind of raw material and fish utilized. When fish meal contains more than 65% of protein, and the digestibility is higher than 80%, this product is considered to be of high quality and can double the commercial value of the low protein content and low quality (Olsen et al., 2014). Its application is basically in the formulation of feedstuffs with high protein concentration especially demanded by aquaculture. Fish meal is an essential ingredient in the formulation of aquaculture feed (Bowyer et al., 2013; Kousoulaki et al., 2015; Teves and Ragaza, 2016) since its total or almost total substitution by meal from vegetables is nonviable in all cultivated species due to inflammatory problems that provoke certain proteins and plant peptides in the digestive system of target fish (Bowyer et al., 2013; Kokou and Fountoulaki, 2018; Zhou et al., 2018). Nevertheless, the industrial fishing of low-price species destined for fish meal production for feeding important aquatic species (e.g., salmon and turbot) is reasonably low, environmentally unsustainable, and constitutes an important factor in skimming the oceans.

In the production of fish meal, especially when fatty fish are processed, it is very important to separate the oils by means of (1) a pressing step among cooking and drying stages and (2) subsequent centrifugation of liquid phase to recover differentially oil and tailwater. The latter after concentration by evaporation or membrane treatment (Afonso and Borquez, 2002; Afonso et al., 2004) can be reutilized in fish meal or can be applied as a nutrient source in organic fertilizers (Arvanitoyannis and Kassaveti, 2008). Besides the interesting commercial value of such fish oil, the presence of a remarkable amount of oil in the dry powder enhances the risk of spontaneous combustion of fish meal.

On the other hand, beginning with livers, skins, and heads of fatty fish, the procurement of crude oil is simple, requiring grinding, pressing at temperature 50–70 °C, homogenization, centrifugation, or decantation and stabilization with commercial antioxidants to avoid/reduce oxidation. This procedure can be assisted or without assisted by enzymatic proteolysis to enhance the recovery of a lipid fraction (Batista et al., 2010; Kechaou et al., 2009; Linder et al., 2005). The set of techniques to refine the oils, in terms of polyunsaturated fatty acids, include adsorption chromatography, fractional distillation, isoelectric solubilization/precipitation, and complexation with urea or extraction in supercritical conditions (Chen et al., 2007; Dave and Routray, 2018; Haq et al., 2018; Rubio-Rodríguez et al., 2012; Sahena et al., 2010). The uses and commercial prices of fish oils vary depending on the content of polyunsaturated fatty acids (PUFAs), mainly omega-3 as docosahexaenoic acid (DHA); thus, levels higher than 60% of the fatty acid drive the oil to the formulation of nutraceuticals and antiarrhythmic, anti-inflammatory, and antiplatelet preparations (Fiori et al., 2012; Kralovec et al., 2012; Sears et al., 2010; Shahidi and Miraliakbari, 2006). Although the best yields of fish oil recovery are obtained from fatty species and mainly in tissues as skin, viscera, and heads, the composition of fatty acids from almost all commercial fish species and tissues have been extensively reported (Dave and Routray, 2018; Nettleton, 1995; Parrish, 1999; Pike and Jackson, 2010; Rubio-Rodríguez et al., 2010; Sahena et al., 2009; Shahidi, 2008). Fish oils without nutraceutical applications have been applied as a supplement for aquaculture feed (Bendiksen et al., 2011; Goosen et al., 2014; Watanabe, 2002) and as substrate for the production of biodiesel, after chemical processing (transesterification), describing satisfactory potentialities, in some cases finding better outcomes than vegetable oils (Behçet, 2011; Enascuta et al., 2018; Fadhil et al., 2017; Gharehghani et al., 2017; Jayasinghe and Hawboldt, 2012; Lin and Li, 2009a, 2009b).

Although squalene is a hydrocarbon intermediary in the biosynthesis of cholesterol and precursor of vitamin D, it is a fundamental component of the unsaponified fraction of the oil from livers of some deep-sea shark species. The content of squalene ranges from 10–90% (w/w) in the shark livers, being nurse shark (*Ginglymostoma cirratum*) with more than 80% (Remme et al., 2005). The steps of squalene refining are based on sequential steps of distillation and supercritical fluid extraction (Pietsch and Jaeger, 2007). Squalene applications have high commercial values, including lubricants for the aeronautical industry, sun filters for sun creams, an ingredient of perfumes, and formulations with antibacterial and anticancer activities (Gopakumar, 2012; Naziri et al., 2011). However, little knowledge in several ecological aspects of deep-sea shark species (e.g., population dynamics, reproduction, fish-stock, and recruitment) together with the increasing fishery effort on such species and the presence of persistent contaminants in shark livers (PCB, PAH, etc.) put in difficulty the viability of this bioproduction in the future (Popa et al., 2015).

From an upgrading point of view, the by-products and discards that can be valorized to obtain biocompounds of higher added value (e.g., skins or cartilages for collagen and GAGs productions) should not be directed to fish meal. In those cases, fish meal could be the ultimate end of the wastes that such processes generate in agreement with the circular economy principles. Furthermore, fish meal equipment are commonly dimensioned to large scale to treat tens or hundreds of tons per day of raw materials to justify the investments and running costs (Naylor et al., 2009). For small ports with amounts of wastes inferior to fish meal requirements, alternative as FPH or silage production are the best options to stabilize and process that biomass.

3 Fish Protein Hydrolysates and Bioactive Peptides

FPH refers to those protein materials that are obtained by enzymatic hydrolysis of whole fish or fish by-products using proteases, as a catalyst, under controlled experimental conditions. In some conditions of hydrolysis, this product is rich in proteins, peptides, and free amino acids, including peptides with biological activities (antioxidants, antihypertensive, antimicrobial, etc.). Figure 4.1 shows the basic stages to generate FPH together with fish oil (when fatty fish are processed), mineral materials, marine peptones, and bioactive peptides.

Initially, the substrates must be crushed to facilitate the homogenization of raw materials previous to their enzymatic proteolysis. The hydrolysates obtained are then filtrated and/or centrifuged (for instance, using a tricanter) to separate the solid fraction (fish bones) and two liquid phases: fish oil and pre-FPH. The latter, after thermal enzyme inactivation, is the final FPH that can be used as nitrogen source for the formulation of microbial culture media (Aspmo et al., 2005a; Safari et al., 2011) or dried, previously concentrated by evaporation, to be incorporated as ingredient in aquaculture feed (Martínez-Álvarez et al., 2015; Zheng et al., 2012, 2013). In cases in which the biological activities of FPH present remarkable values, additional steps of purification should be performed, including membranes of ultrafiltration and/or nanofiltration (depending on the molecular weights of bioactive peptides) and/or chromatography techniques (Amado et al., 2013; Bourseau et al., 2009; Ferreira et al., 2007; Ishak and Sarbon, 2018; Je et al., 2007; Jeon et al., 2000; Najafian and Babji, 2012; Suwal et al., 2014).

Many enzymes, basically proteases, have been reported to be interesting in digestion of fish tissues (Halim et al., 2016; He et al., 2013) changing the optimal experimental conditions of catalysis (pH, temperature, enzyme concentration, enzyme/substrate ratio, solid/liquid ratio, etc.) to obtain different degrees of

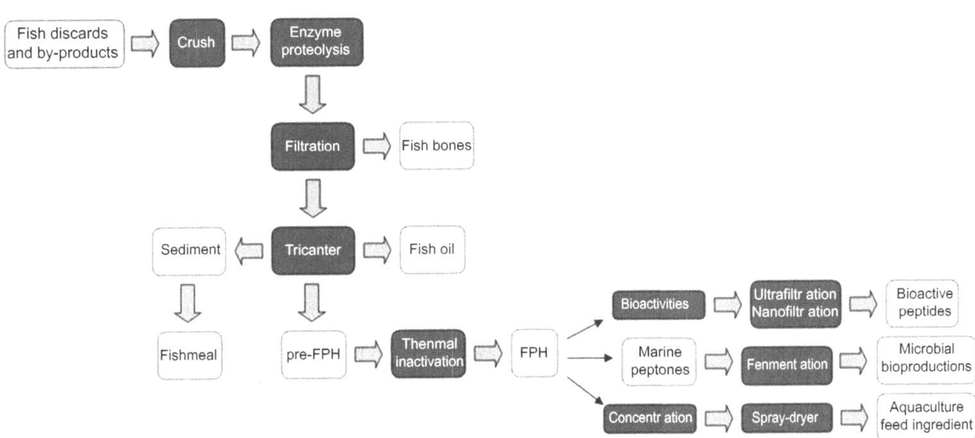

Figure 4.1 Flowchart describing the basic steps to produce FPH and bioactive peptides from fish wastes.

hydrolysis and yields in the function of the type of enzyme and the substrate to hydrolyze. The list of commercial enzymes and endogenous proteolytic activities reported to produce FPH is very extensive, including acidic, neutral, and alkaline enzymes such as alcalase, neutrase, papain, pepsin, trypsin, protamex, flavourzyme, and bromelain (Chalamaiah et al., 2012; Villamil et al., 2017). In addition, several bioactivities and chemical properties from peptides generated in FPH have been determined and isolated, including antioxidant, antihypertensive, antimicrobial, anti-inflammatory, antiproliferative, immunomodulatory, anticancer, and antihyperglycemic functionalities (Harnedy and FitzGerald, 2012; García-Moreno et al., 2014; Ishak and Sarbon, 2018; Zamora-Sillero et al., 2018), as well as solubility, fat-binding, and water-holding capacities and emulsifying and foaming properties (Halim et al., 2016; He et al., 2013).

The number of fish species and by-products explored to produce FPH along with the conditions, enzymes utilized, and bioactivities and chemical properties evaluated is extensive. A summarized description of the most recent FPH reported is shown in Table 4.1. As it can be observed, tuna and tilapia are two of the most species studied whose antioxidant and antihypertensive biological activities are more commonly determined. In some cases, the peptides that showed the highest biological activities were isolated by membranous and chromatographic methods or combining both and the sequence of amino acids and molecular weights were elucidated. The corresponding activities in terms of inhibition % and IC_{50} values were generally lower than synthetic controls (antioxidants as butylated hydroxytoluene-BHT and propyl gallate and antihypertensives as captotril and alacepril) (Chi et al., 2015a; Himaya et al., 2012; Ngo et al., 2014; Zhang et al., 2012); however, various authors suggest the possibility of nutraceutical uses of these peptides avoiding the side effects described for the commercial ones (e.g., inflammatory response and skin eruptions) (Intarasirisawat et al., 2014; Zamora-Sillero et al., 2018). The applications of FPH as an ingredient in aquaculture feeds (AF) induced, in all studies reported (Table 4.1), had excellent fish responses in terms of fish growth (increasing weight and length), feed efficiency, innate immunity, survival rate, protein digestibility, etc.

4 Enzymes

In the pancreas of chondrichthyes fish and in the pyloric of osteichthyes, digestive enzymes are present catalyzing the processes hydrolysis of foods and nutrients. Mainly, these enzymes are proteases (both endo- and exoproteases) that hydrolyze, with different steric restrictions and peptide bonds. A special group of proteases are collagenases, capable of hydrolyzing collagen, a protein that is an important part of skin, bones, and tendons in mammals. In comparison with enzymes from terrestrial and microbial origins, unique properties of marine organisms in terms of high specific activity has been proposed, at low temperature and stability in the presence of

Table 4.1 Recent FPH produced from fish by-products along with its bioactivities and chemical properties determined or in vivo applications evaluated.

FISH	TISSUES	ENZYME	BIOACTIVITIES/ PROPERTIES/ APPLICATION	REFERENCES
Bluefin leather jacket	Heads	Papain	AO	Chi et al. (2015b)
Nile tilapia	Scales	Alcalase	AO	Ngo et al. (2010)
Salmon	Pectoral fin	Alcalase, pepsin, neutrase, protamex, flavourzyme	AO, ACE, AI	Ahn et al. (2014)
Skate	Cartilage	Trypsin, alcalase	AO	Pan et al. (2016)
Smooth hound	Viscera	Neutrase, esperase	AO, AM (gram+ and gram-)	Abdelhedi et al. (2016)
Tuna	Dark muscle	Alcalase	AO	Saidi et al. (2014)
Atlantic mackerel	Viscera, liver, heart, intestine	Protamex	AM (gram+ and gram-)	Ennaas et al. (2015)
Rainbow trout	Viscera	Pepsin	AM (gram positive and gram negative)	Wald et al. (2016)
Tilapia	heads, frames, víscera	Protamex	AM (gram positive and gram negative)	Robert et al. (2015)
Anchovy	Cooking wastewaters	Protamex	AM (gram positive and gram negative)	Tang et al. (2015)
Smooth hound	Heads, viscera	Protease, gastric extract from smooth hound intestine	ACE, AO	Sayari et al. (2016)
Tilapia	Frames	Flavourzyme	ACE, AO	Chuesiang and Sanguandeekul (2015)
Cod	Backbones	Protease P	AO	Halldorsdottir et al. (2014)
Persian sturgeon	Viscera	Alcalase		Ovissipour et al. (2009)
Hoki	Frames	Pepsin, trypsin, papain, alcalase, neutrase, α-chymotrypsin	AO	Kim et al. (2007)
Alaska pollack	Frames	Crude proteinase from mackerel intestine	AO, ACE	Je et al. (2005)
Sole	Frames	Alcalase	ACE	Jung et al. (2006)
Tuna	Frames	Pepsin, trypsin, papain, α-chymotrypsin	ACE	Lee et al. (2010)
Shrimp	Whole shrimp	Protease from *Bacillus* sp.	ACE	Hai-Lun et al. (2006)
Shrimp	Cephalothorax	Flavourzyme	ACE	Benjakul et al. (2009)
Skipjack tuna	Roe	Alcalase	ACE, AO	Intarasirisawat et al. (2014)

(*Continued*)

Table 4.1 (Cont.)

FISH	TISSUES	ENZYME	BIOACTIVITIES/ PROPERTIES/ APPLICATION	REFERENCES
Tuna	Dark muscle by-product	Alcalase	AO	Kim et al. (2011)
Hoki	Frames	Alcalase	AO	Je et al. (2007); Mendis et al. (2005)
Ornate threadfin bream	Muscle	Trypsin from skipjack tuna	AO	Nalinanon et al. (2011)
Herring	Flesh	Endogenous enzyme	–	Prabha et al. (2013)
Sardinelle	Viscera, heads	Alcalase, α-chymotrypsin	ACE	Bougatef et al. (2008)
Bluewing searobin	Muscle	Alcalase	S, EA, ES, FBC, WHC	Dos Santos et al. (2011)
Catla (Indian carp)	Muscle	Alcalase, protamex, bromelain, flavourzyme	AO, EA, FC, FBC	Elavarasan et al. (2014)
Striped catfish	Muscle, frames	Papain, bromelain	S, EA, FC, FBC	Tanuja et al. (2012)
Rainbow trout	Viscera	Alcalase	S, EA, ES, FC,FS, WHC, FBC	Taheri et al. (2013)
Dagaa	Muscle	–	FBC, S, AO	Betty et al. (2014)
Striped snakehead Rohu	Roe	Alcalase	S, EA, FC, FS, FBC, AO	Galla et al. (2012)
Tilapia	Muscle	Alcalase	S, EA, FC, WHC, FBC	Foh et al. (2011)
Sardinelle	Muscle	Fermentation	AO, AM, EA, ES, FC, FS, WHC, FBC	Jemil et al. (2014)
Zebra blenny, cobia	Muscle	Alcalase	AO, S, EA, FC, FS, FBC, WHC	Razali et al. (2015)
Ornate threadfin bream	Muscle	Pepsin	S, EA, ES, FC, FS, AO	Nalinanon et al. (2011)
Blue whiting, cod	Heads, gut	Alcalase	ACE, S, FBC, WHC	Geirsdottir et al. (2011)
Yellowfin sole	Frames	α-chymotrypsin	ACE	Jung et al. (2006)
Tuna	Cark muscle	Papain, protease XXIII	AP	Hsu et al. (2011)
Tuna	Cooking juice	Protease XXIII	AP	Hung et al. (2014)
Atlantic salmon	Frames	Papain, trypsin, α-chymotrypsin	AD	Roblet et al. (2016)
Sardinelle	Heads, viscera	Alcalase	AO	Souissi et al. (2007)
Hake	Mince	Validase, flavourzyme	AO	Samaranayaka and Li-Chan (2008)
Dojo loach	Meat	Papain, pepsin, pancreatin	AO	You et al. (2010a, 2010b)
Chinese carp	Muscle	Papain, pancreatin, neutrase, alcalase, bromelain	AO	Ren et al. (2008)
Salmon	Fish milt	Pancreatin	AO	Wang et al. (2008)

Smooth hound	Muscle	Pepsin, alkaline protease	AO	Bougatef et al. (2008)
Sardinelle	Heads, viscera	Alcalase, extracts from sardinelle and *Aspergillus*	AO	Barkia et al. (2010); Bougatef et al. (2010)
Black scabbardfish	By-products	Protamex	AO	Batista et al. (2010)
Silver carp	Processing by-products	Alcalase, pepsin, neutrase, papain, protamex, trypsin, flavourzyme	AO	Zhong et al. (2011)
Black pomfret	Viscera	Pepsin, trypsin, α-chymotrypsin	AO	Samparth Kumar et al. **(2011)**
Tuna	Viscera	Alcalase	AF	Chotikachinda et al. (2013)
Pacific hake	Whole fish	Alcalase	AF	Ho et al. (2014)
Pollock	Heads	Alcalase, flavourzyme	AO, AF	Zheng et al. (2013)
Tuna	Viscera	–	AF	Ovissipour et al. (2014)
Shrimp	Heads	–	AF	Bui et al. (2014)
Tilapia	Carcass	–	AF	Lewandowski et al. (2013)
Pollock	Heads, gut	Acid protease	AF	Liang et al. (2006)
Herring	Whole fish	Alcalase	AF	Hevrøy et al. (2005)
Tuna	Heads	Protamex	AF	Nguyen et al. (2012)
Salmon	Heads, frames	Papain	AF	Grey et al. (2009)
Pollock	Fillets	Alcalase, protamex	AF	Hermannsdottir et al. (2009)
Round scad	Muscle	Alcalase, flavourzyme	AO	Thiansilakul et al. (2007)
Blue whiting	Whole fish	Subtilisin, trypsin	AO, EA, FC, FBC, AM	García-Moreno et al. (2017)
Yellow stripe trevally	Whole fish	Alcalase, flavourzyme	AO	Klompong et al. (2007)
Boardfish	Whole fish	Trypsin, alcalase, extract from catshark pancreas	S, FC, EA, AO	Blanco et al. (2015b)
Small-spotted catshark	Muscle	Alcalase, esperase, protamex	AO, ACE	Vázquez et al. (2017a)
Channel catfish Tilapia	Fillets	Protamex	AO	Theodore et al. (2008)

AO: antioxidant activities, ACE: angiotensin-I converting enzyme-inhibitory method (antihypertensive activity), AI: anti-inflammatory activity, AM: antimicrobial (against gram positive and/or gram negative), AP: antiproliferative, IM: immunomodulatory activitiy, S: solubility properties, EA: emulsifying activity, ES: emulsifying stability, FC: foaming capacity, FS: foaming stability, WHC: water holding capacity, FBC: fat binding capacity, AF: aquaculture feed application, AD: antidiabetic.

surfactants, heavy metals, and oxidizing agents, potency of action, and cold thermostability (Jesús-de la Cruz et al., 2018; Joo et al., 2001; Ketnawa et al., 2013; Simpson, 2000; Sriket, 2014). Other interesting enzymes have also been isolated from fish by-products such as uracil-DNA glycosylase from cod (Lanes et al., 2000, 2002), alkaline phosphatase from shrimp processing wastewaters (Olsen et al., 1990), and lipase from sardine (Castro-Ceseña et al., 2012) but, as indicated, the majority are proteases (Khandagale et al., 2017; Stefansson et al., 2017).

The range of optimal values of temperature and pH for proteases isolated from marine by-products is 30–70 °C and 6.5–10, respectively (Jesús de la Cruz et al., 2018; Murado et al., 2009). Their applications are mainly directed to food processing (Balti et al., 2009; Freitas et al., 2012), as an ingredient in the formulation of detergents (Bkhairia et al., 2016; Nasri et al., 2012; Younes et al., 2015), production of FPH (Ketnawa et al., 2016; Tavano, 2013; Villamil et al., 2017), biotechnological developments (Bougatef, 2013), and valorization of marine by-products (Poonsin et al., 2017; Younes et al., 2015).

The economical viability of enzyme production from fishery wastes to industrial scale is generally limited due to the following reasons (Olsen et al., 2014): (a) the concentration of the enzymes presents in the corresponding fish organs, as well as the volume and quality of the required by-products are too small in many cases; (b) the costs of extraction, recovery, and complete purification of enzymes are expensive; (c) the production of enzymes with high activities from biotechnological origin is the most profitable, stable, reproducible, and scalable option. In the search for the exhaustive purification of enzymes, a combination of initial homogenization, chemical fractionation/precipitation (e.g., ammonium sulfate is one of the most commonly used solvents), and subsequent affinity, gel-filtration, or ion-exchange chromatography has been routinely performed (Bkhairia et al., 2016; Bougatef et al., 2007; Khangembam and Chakrabarti, 2015; Marcuschi et al., 2010; Silva et al., 2011). Based on this protocol, trypsin of 27.5 kDa from crovalle jack pyloric caeca (Costas et al., 2013), serine proteinase from pancreas of *Scyliorhinus canicula* (Blanco et al., 2014), chymotryptic, elastolytic, tryptic, and collagenolytic activities from snow crab (Souchet and Laplante, 2011), trypsin from spleen of skipjack tuna (*Katsuwonus pelamis*) (Klomklao et al., 2007), and endogenous proteinases from sardine (*Sardinops melanostictus*) (Klomklao et al., 2008) were purified. With an additional step of ultrafiltration, trypsin from sardinelle viscera was also recovered (Khaled et al., 2008). Recent methods of purification of marine enzyme based on the partition in aqueous two-phase system, extraction with supercritical CO_2, and on the application of ferromagnetic particles are being developed with remarkable results (Alves et al., 2017; Lamas, 2015; Poonsin et al., 2017).

However, in some cases, it is most important to reduce the cost of purification and to reach partial purifications with mixed enzymatic activities, to be more effective than the sum of activities separately. For example, a simple process based on homogenization, centrifugation, and selective precipitation using overcooled acetone (easily rectifiable solvent for reuse) is enough to isolate proteolytic and collagenolytic activities from ray pancreas (Murado et al., 2009), which are highly effective for the digestion of ray cartilages (Murado et al., 2010). In concordance, various proteolytic (trypsin, chymotrypsin, aminopeptidase, and pepsin) and lipase activities were also extracted from Monterey sardine viscera using acetone (Castro-Ceseña et al., 2012). These extracts showed threefold higher activity than alcalase in protein hydrolysis applications. Lamas and coworkers (Lamas et al., 2017) recovered alkaline trypsin of 25 kDa from viscera and head of *Engraulis anchoita* precipitated with acetone. Nevertheless, crude extracts obtained by simple homogenization were also a source of remarkable activities: caseinolytic from zebra blenny viscera with extreme stability in presence of nonionic surfactants and oxidizing agents (Ktari et al., 2012), diverse proteases from fillets, heads, livers, and viscera of pink Alaska salmon (Bower et al., 2011a) and acidic-aspartic proteases from head and viscera of shrimp *Pleoticus muelleri* (Pereira and Fernánez-Giménez, 2017). In the first case, the crude extract was applied in shrimp waste deproteinization at 40 °C for 3 h and in liquid detergents, and the last acidic protease extract was successfully employed as milk clotting agents for cheese making.

5 Glycosaminoglycans

There are two main sources of GAGs in fish by-products and by-catch: humor vitreous from eyes and, especially, cartilage present in chondrichthyan fish. Skin and bones have been also explored as a source of GAGs (Cheng et al., 2013; Maccari et al., 2015), but the levels and yields are much lower and not comparable to the previous one. Cartilage is a tissue constituted by a complex matrix of chondrocytes, HA, collagen, and proteoglycans (different sulfated GAGs united with proteins). GAGs are linear heteropolysaccharides, without branched chains, constituted by a repeating disaccharide unit of an uronic acid (commonly, glucuronic acid) and an amino monosaccharide (*N*-acetylgalactosamine in chondroitin sulfate [CS] or *N*-acetylglucosamine in HA) united by glycosidic bonds (generally β1–4 between disaccharide unit and β1–3 between an uronic acid and amino sugar). The presence of GAGs in cartilaginous tissues, cell surfaces, extracellular matrix, and joint fluids are due to their physicochemical properties (hygroscopicity, viscoelasticity, and pseudoplasticity) and that they act in the given structure and bind and regulate different protein signals (cytokines, growth factors) among cells. The most abundant and important GAGs both commercially and for their physicochemical properties are CS and HA. The types of CS are chemically classified in terms of sulfation patterns (Valcarcel et al., 2017): CS-A and CS-B or dermatan sulfate are sulfated in C4 of *N*-acetylgalactosamine, CS-E in both C4 and C6, CS-C in C6, and CS-D in C6 of *N*-acetylgalactosamine and C2 of glucuronic acid. Keratan sulfate (KS) is an analog with dimers of galactose and *N*-acetylglucosamine sulfated in C6. In squaliformes (sharks), CS is the main GAG with remarkable concentrations of HA and KS present, but some rajiformes (rays) do not contain KS. CS from fish is mostly sulfated in C6 than C4 in contrast with CS from mammalian, which is C4 predominant (Duan et al., 2013; Novoa-Carballal et al., 2017).

GAGs have varied and growing uses in tissue engineering where their high biocompatibility permit them for repair of bones, cartilages, and skin lesions, dental implants, arthritis and arthrosis treatments, anti-inflammatory and anticoagulant solutions, the formulation of hydrogels and nanodevices for drug delivery, injectables for synovial fluid replacement and dressing, and topical applications for wound healing (Agnello et al., 2016; Cardoso et al., 2016; Carvalho et al., 2018; Cerqueira et al., 2014; Da Silva et al., 2016; Mourão, 2015; Napavichayanun and Aramwit, 2017; Pan et al., 2018; Piai et al., 2017; Volpi, 2006). The therapeutic potential of GAGs: antiproliferative cells activity, neurite outgrowth promoting and anti-cancer activity has been also reported (Hashiguchi et al., 2011; Kaczmarek et al., 2017; Krichen et al., 2017; Sadhasivam et al., 2013; Valcarcel et al., 2017). In the recent years, HA is the star ingredient in all antiaging cosmetic creams and one of most applied compounds in anti-wrinkle treatments and dermo-aesthetic surgery (Narurkar et al., 2016; Papakonstantinou et al., 2012; Salwowska et al., 2016).

Although marine invertebrates are not strictly species generated in fisheries activities due to their lack of commercial interest and uncontrolled stocks, the isolation of GAGs from them has been extensively investigated (Glauser et al., 2013; Pavão, 2014; Pomin, 2014; Pomin and Mourão, 2014; Saha et al., 2014): CS fucosylated from sea cucumber, dermatan sulfate (DS) from ascidians, GAGs like heparin, sulfated glucans, and galactans from sponges.

The protocols developed to isolate GAGs from fish by-products require preparing, homogenizing, and hydrolyzing the organic tissues for the removal of proteins and for purifying selective polysaccharides. A mixture of HA and CS was obtained from lumpsucker fish using alcalase for the hydrolysis of the dorsal hump, separation with anion-exchange resin, precipitation with MetOH, and purification by dialysis at 8 kDa membrane cutoff (Panagos et al., 2014). CS from bones of cod and monkfish, dogfish, salmon, and tuna was recovered after defatted wastes with acetone, solubilization with sodium acetate (AcNa), papain hydrolysis, precipitation with EtOH-AcNa, dissolution in NaCl, QAE sephadex anion-exchange resin purification, and precipitation again with EtOH (Maccari et al., 2015). Zhao and co-workers isolated GAGs from Chinese sturgeon cartilage using alkaline hydrolysis and maintaining chromatography separation using DEAE cellulose and Sephadex G-100 columns (Zhao et al., 2013).

Dorsal fin of yellowfin sole and skin, head, fins, scales, and internal organs of bluefin tuna, nile tilapia, bream, Japanese jack mackerel, and Atlantic mackerel were explored in search of DS and CS with a scheme in which two enzyme steps were sequentially applied (extended for 1 week); instead, only one enzyme hydrolysis, trifluoroacetic (TFA) was added and anion-exchange chromatography was substituted by dialyzation (Arima et al., 2013). CS, DS, and heparin sulfate of low molecular weights (15–40 kDa) and biological activities were very recently recovered from different tissues and organisms, shellfish, mollusc cockle, skin of corn, and swim bladder using similar protocols (Aldairi et al., 2018; Bougatef et al., 2018; Okamoto et al., 2018; Pan et al., 2018).

In this context, a more prolix protocol including papain plus trypsin proteolysis for 24 h, ethanolic precipitation at 4°C for 12 h, saline redissolution combined with alkaline hydrolysis, protein removal by organic solvents and final dialysis for 3 days has been performed for the CS and DS production from skin and bones of salmon, snakehead, monkfish, and skipjack tuna (Lin et al., 2017a). The use of trypsin and papain in the first step of proteolysis and TFA for removing proteins were checked for GAGs from viscera of Chinese carp, beltfish, and yellow croaker (Liu et al., 2017). Another specific precipitating reagent for GAGs is the cetylpyridinium chloride that along with several EtOH treatments was applied to fish skins for CS purification (Takeda et al., 2016).

However, all these processes are far from the optimal operating conditions, reagents are very expensive, time-consuming (e.g., enzyme hydrolysis or dialysis steps of more than 4 days), and never validated at the pilot plant scale. A most efficient and cheaper flowchart has been developed and optimized (Figure 4.2) for the production of high purity CS from cartilages of ray, blue shark, and small-spotted catshark (Blanco et al., 2015a; Murado et al., 2010; Vázquez et al., 2016a). The sequential proposed strategy is based on enzymatic proteolysis, generating bioapatites and protein substrate for FPH production as solid waste and a liquid hydrolysate that was precipitated with an alkaline-hydroalcoholic-saline solution and CS was finally purified by ultrafiltration-diafiltration processing using the membrane of 30 kDa cutoff. A huge reduction of volume and concentration of reagents and time of operation was reached with the mentioned alternative. A similar procedure was also employed for the isolation of high purity and high molecular weight HA from the vitreous humor of blue shark, swordfish, and tuna (Murado et al., 2012). Novel techniques based on the application of high-intensity pulsed electric fields and the combination of ultrasonic and microwave have been

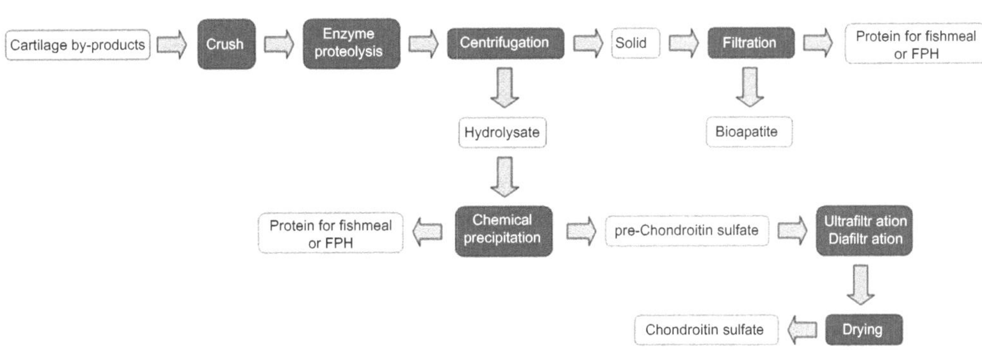

Figure 4.2 Scheme showing the different operations to produce chondroitin sulfate from fish cartilage wastes.

reported for the extraction of CS from tilapia by-products and fish bones (Cheng et al., 2013; Cheng et al., 2011; He et al., 2014).

6 Chitin and Chitosan and Chitooligosaccharides

After cellulose, chitin is the most extended polysaccharide in nature since it is present in the exoskeleton of insects and crustaceans (shells), cell wall of microorganisms, among others. Chemically it is a long linear chain constituted of successive units of *N*-acetylglucosamine (β1–4 glycosidic bonds). When more than 50% of chitin is deactylated, the macromolecule is called "chitosan." The main marine sources of chitin are the shells of crustaceans (composed of α-chitin that presents antiparallel microfibril orientation with strong intra- and intermolecular hydrogen bonds) and the endoskeleton (pen) of cephalopods (composed of β-chitin that presents parallel chains held by weak intrachain hydrogen bonds) (Minke and Blackwell, 1978; Rinaudo, 2006).

Because chitin, chitosan, and chitooligosaccharides have distinctive biological and physicochemical properties (Kurita, 2006; Malerba and Cerana, 2018; Mourya and Inamdar, 2008; Salehi et al., 2016), they have been intensively studied in the past in a wide range of applications (Aam et al., 2010; Fu et al., 2016; Ganiari et al., 2017; Ramrakhiani et al., 2016; Sánchez et al., 2011; Shuai et al., 2017). They are used as a filling matrix in adsorption columns for water and soil depuration, ingredient in cosmetic formulations, support for the immobilization of enzymes, flocculant, immunostimulant, antitumor, antimicrobial and antioxidant agent, material for tissue regeneration devices and structural biopolymer in the preparation of hydrogels, and bioactive films for the preservation of food as well as nanodevices for drug delivery (Bernocchi et al., 2017; Corinaldesi et al., 2017; Huang et al., 2012; Karaki et al., 2016; Liu and Chen, 2016; Logithkumar et al., 2016; López-Moreno et al., 2018; Wang et al., 2016; Yuan et al., 2016; Zou et al., 2016).

Chitin in crustacean shell by-products is linked to proteins and associated with mineral salts (basically $CaCO_3$) forming a complex matrix with the following composition (dry weight basis): 30–50% mineral salts, 13–42% chitin, and 30–40% protein, variable ranges depending on the species and season (Kurita, 2006; Percot et al., 2003). The content in chitin is higher the more flexible the shells are (e.g., shrimp and prawns in comparison with spider crab). The conventional procedures to purify chitin from exoskeletons, including (Synowiecki and Al-Khateeb, 2003; Tolamaite et al., 2003; Valdez-Peña et al., 2010; Vázquez et al., 2017b): (a) preparatory stage of crushing, drying, and sieving to obtain a particulate material with a diameter lower than 1 mm; (b) step of deproteinization by chemical (with alkalis) and/or enzymatic (using proteases) hydrolysis; (c) acid demineralization of deproteinized shells; (d) additional hot alkaline hydrolysis (optional); and (e) bleaching step (optional) to remove remaining pigments (astaxanthin). Astaxanthin is a carotenoid with antioxidant properties and is useful as a colorant in aquaculture feed (Barros et al., 2011; Mata-Gómez et al., 2014). Although the production of astaxanthin by microorganisms (microalgae as *Haematococcus pluvialis* or yeast as *Phaffia rhodozyma*) leads to higher production yields than those obtained from crustacean shells, this simple alternative can contribute to the economic interest of the integral chitin valorization step (Amado et al., 2016; Amado and Vázquez, 2015; Amado et al., 2015; Ambati et al., 2014; Del Campo and García-González, 2007; García-López et al., 2016; Lopes et al., 2018).On the other hand, some authors suggested that it is more appropriate to eliminate mineral salts before deproteinization (Tolaimate et al., 2003). In addition, the effluent from (b), especially when enzymes are used for proteolysis, could be exploited as sources of marine peptones or FPH (Vázquez et al., 2016b).

In the case of cephalopods endoskeleton (pen), the content in minerals is lower than 10%, the protein material is higher than 40%, and no pigments are present in pens. This is why the extraction of chitin from pens does not need the steps (c–e) described for shells (Cuong et al., 2016; Youn et al., 2013). Figure 4.3 shows a double way of purifying chitin by enzymatic or chemical proteolysis and then to continue with its deacetylation to produce chitosan using alkaline hydrolysis at temperatures higher than 70 °C (Vázquez et al., 2017b). The purity, degree of deacetylation, and molecular weight of chitosan obtained depend on temperature,

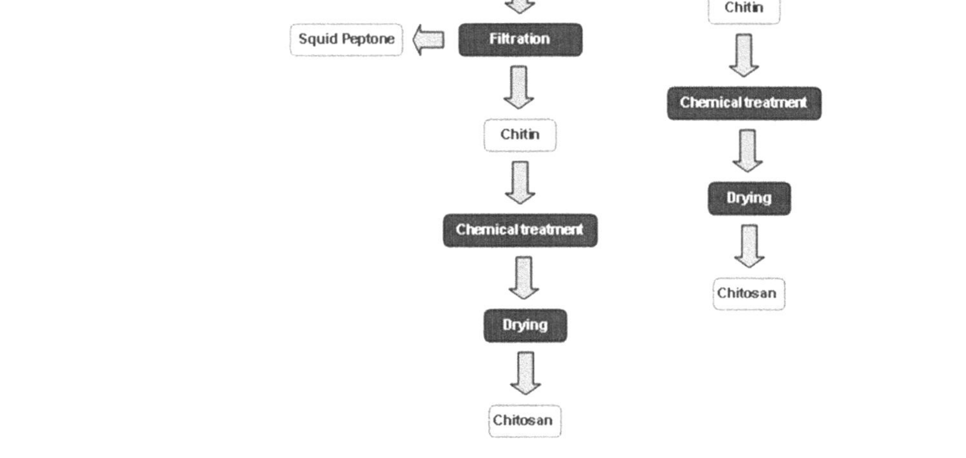

Figure 4.3 Production of chitin and chitosan from squid pens by enzymatic and/or chemical processing.

time, and alkali concentration (Cuong et al., 2016; Kurita, 2006; Vázquez et al., 2017c). To reduce and mitigate the management and pollution of effluents generated in bioproductions, the protein hydrolysates from pen deproteinization were successfully applied as marine peptones in biotechnological developments for the growth of several bacteria (Vázquez et al., 2016b).

Although this scheme is simple, and the application of commercial proteases (e.g., alcalase and papain) together with the partial valorization of their effluents reduces the potential environmental problems caused, the use of large volumes of concentrated alkaline and acid solutions at high temperatures in many cases convert the production of chitosan in hazardous and energy consuming procedure. Some environmental friendly alternatives have been evaluated: (1) fermentation of shells and pens with lactic acid bacteria (LAB) and/or protease-producing bacteria for the extraction of chitin (Bajaj et al., 2015; Mao et al., 2017; Rao and Stevens, 2005) and (2) chitinase and chitin deacetylase-producing fungi for the production of chitosan (Chatterjee et al., 2005; Martinou et al., 1998). In the first case, the lactic and acetic acids produced by sugar metabolic pathway of LAB contribute to demineralization and the proteases released from LAB and bacteria from *Bacillus* genus help in removing peptides and proteins (Chen et al., 2017; Ghorbel-Bellaaj et al., 2013; Liu et al., 2014). In the second case, extracts rich in chitin deacetylase activity from fungi such as *Mucor rouxii* and *Rhizopus nigricans* or solid-state fermentation of such fungi on the chitinous substrate were studied (Chatterjee et al., 2010; Jeraj et al., 2006; Trutnau et al., 2009). However, the activity of such enzyme and deacetylation yields achieved in this option was lower than conventional alkaline treatment (Chatterjee et al., 2010; Win and Stevens, 2001).

7 Collagen, Gelatins, and Derivatives

Collagen is the most important structural protein present in the connective tissues of animals; specifically, it is the most abundant protein in mammals (Lodish et al., 2000). It is a heterotrophic protein rich in glycine, proline, and hydroxyproline with triple helical conformation (tropocollagen) forming fibrillar structures with a molecular weight in its native form, higher than 300 kDa and insoluble in water but soluble in organic acids such as acetic acid. It is mainly present in the skin, tendons, and ligaments of mammals and skin, cartilage, and bones of fish (Nagai and Suzuki, 2000). Several types of collagen have been defined (from type I to more than type XX) depending on its structural properties being type-I and type-II the most common in fish by-products such as skin and cartilage, respectively (Khan et al., 2009; Kittiphattanabawon et al., 2010a; Silva et al., 2014a). The most habitual and commercialized collagens are obtained from porcine and bovine skins and tendons but due to bovine spongiform encephalopathy (BSE) disease and because of religious restrictions derived from the consumption of porcine and bovine products (in Muslim and Jewish communities), have revealed collagen from marine organisms as the most important alternative source.

From the point of view of the isolation procedure, two types of collagens are produced from fishing by-products: acid soluble collagen (ASC) and pepsin soluble collagen (PSC) (Silva et al., 2014a). In both cases, the initial step is to soak the skin in alkalis (NaOH) solution for several hours to remove the non-collagenous proteins; subsequently, skins are neutralized and defatted when fatty species are processed. The defatted skins must be cut into small pieces for improving extraction. For ASC, skins are treated with acid (being acetic acid the most effective) for at least 1 day. For PSC, fish skins are suspended in acetic acid and a protease (commonly pepsin) is added for collagen extraction. ASC and PSC extracts are filtered, the filtrates are dialyzed to neutrality with cold distilled water and the purified collagens are freeze-dried (Muralidharan et al., 2013; Nagai et al., 2002; Tang et al., 2015). The commercial value of this biopolymer is high but the expenses of reagents (solid:liquid ratio higher than 1:10, up to 1:55 in some steps), energy demand (all operation conditions are performed at 4 °C to avoid collagen denaturation), and the problems associated with the management of effluents generated hinder its production at industrial scale.

Following these procedures, different types of collagen have been recovered from several species and by-products: ASC-PSC-brownbanded bamboo shark skin (Kittiphattanabawon et al., 2010a), ASC-PSC-blacktip shark and ASC-PSC-brownbanded bamboo shark cartilages (Kittiphattanabawon et al., 2010b), ASC-tilapia and ASC-silver sarp-skins (Tang et al., 2015), ASC-PSC-ocellate puffer fish skin (Nagai et al., 2002), ASC-PSC-redtoothed triggerfish (Muralidharan et al., 2013), ASC-PSC-brownstripe red snapper-skin (Jonglareonrak et al., 2005, 2006), ASC-PSC-bigeye snapper-skin-bone (Kittiphattanabawon et al., 2005), ASC-horse mackerel-scale, ASC-gray mullet-scale, ASC-flying fish-scale and ASC-yellowback seabream-scale (Thuy et al., 2014), ASC-PSC-cuttlefish-skin (Shanmugam et al., 2012), ASC-PSC-largefin longbarbel catfish-skin (Zhang et al., 2009), ASC-PSC-seaweed pipefish (Khan et al., 2009), PSC-brown backed toadfish-skin (Senaratne et al., 2006), and ASC-bigeye tuna-bone (Jeong et al., 2013).

The partial hydrolysis of collagen by chemical, thermal, and/or enzymatic processing led to a water soluble material (reduction of protein fibrils into smaller peptides) called "gelatin," when such protein solutions form films (gelling) under specific conditions having broad molecular weight ranges (50–300 kDa). The applications of gelatins are wider, including thickeners and food clarifiers, precursors of industrial dyes and glues, materials to build micro- and nanodevices for drug and bioactive delivery and substrate to formulate films and hydrogels for food and tissue regenerative applications (An et al., 2010; Coelho et al., 2017; Fernandes-Silva et al., 2013; Gómez-Guillén et al., 2011; Gómez-Guillén et al., 2007).

The protocols reported to isolate gelatin are mainly based on pretreatments of skins with different alkalis and mineral and organic acids (Gómez-Guillén and

Montero, 2001; Jongjareonrak et al., 2006; Kolodziejska et al., 2008). The alkaline step hydrolyzes and removes noncollagenous proteins without degradation and loss of collagen, and the acid wash prepares the skin for minimizing the loss of collagen for the subsequent thermal extraction of gelatin (Zhou and Regenstein, 2005). The ratio (S:L, skin:reagent liquid) applied to the soaking step are higher than 1:6 and the reagents commonly employed are NaOH and H_2SO_4 but $Ca(OH)_2$, NaCl, H_3PO_4, butyl alcohol and acetic, lactic, and citric acids are also reported (Karayannakidis and Zotos, 2016). The final extraction of gelatins is almost always performed with hot water at 40–80 °C, in one or two steps, for several hours (among 2–16 h) (Arnesen and Gildberg, 2007; Cheow et al., 2007; Duan et al., 2011; Nikoo et al., 2011). Enzyme-assisted treatment or high-pressure application for gelatin extraction from the skin is moreover indicated for improving the chemical properties of gelatin obtained (Gómez-Guillén et al., 2005; Kawahara and Tanihata, 2005; Lin et al., 2017b; Nalinanon et al., 2008)

Figure 4.4 shows a process developed to maximize gelatin yield isolated from yellowfin tuna, blue shark, and Greenland halibut skin by-products and it was also validated to pilot plant scale (Sousa et al., 2017). This procedure was optimized in terms of reducing water, reagents, and time-consumption, improving the protocols previously reported (Grossman and Bergman, 1992; Gudmundsson and Hafsteinsson, 1997; Limpisophon et al., 2009).

Briefly, the skins were initially cut in portions of 2 × 2 cm and washed in water and then were treated with various alkaline and acid solutions (solid:liquid ratio of 1:4, at room temperature) for a short time, and the gelatin is thermically extracted in water solution at 45 °C/16 h. The gelatin solutions were finally deodorized with an adsorbent (active charcoal) or deodorized and desalinated by ultrafiltration-diafiltration techniques and dried by oven or spray-drying. The purified solid gelatins generated gels with strengths higher than 150 blooms for tuna and blue shark origins. These values were in agreement with gelatins purified from cobia, croaker, skipjack tuna, and hoki skins (Mohtar et al., 2010; Shyni et al.,

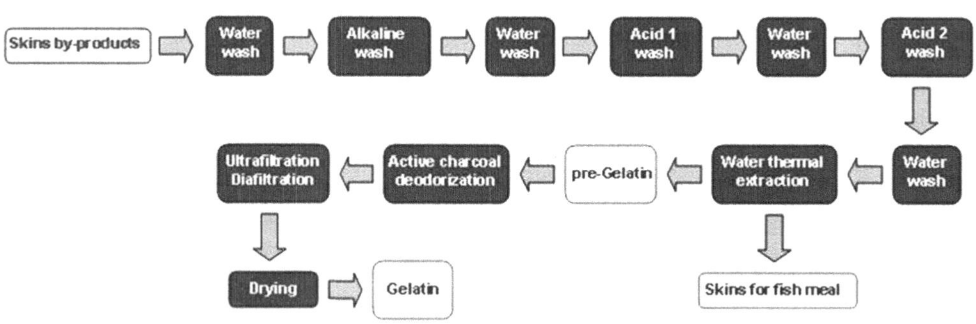

Figure 4.4 Flowchart describing the steps to produce gelatins from fish skins.

2014; Silva et al., 2014b). The production yields were 12.5% (w/w of fresh skin), 8.7% (w/w), and 1.3% (w/w) for tuna, blue shark, and halibut. Similar values were found for gelatin extraction from giant catfish, red tilapia, walking catfish, and striped catfish (Eysturskarð et al., 2009; Jamilah and Harvinder, 2002; Sai-Ut et al., 2012).

However, skins from species such as Nile tilapia, pangas catfish, and Asian red-tail catfish have released yields around 20% (Ratnasari et al., 2013). Bones, head, and scales of species such as blue withing, Atlantic mackerel, red snapper, Nile perch, king weakfish, and carp were also explored as source of gelatin, but the production yields in comparison to skin are much lower, and the procedures of extraction are more expensive and time longer (da Trindade Alfaro et al., 2009; Ferraro et al., 2010; Herpandi and Adzitey, 2011; Khiari et al., 2013; Muyonga et al., 2004; Shakila et al., 2012; Zhang et al., 2011).

The most important derivative product from marine collagen and gelatin is the enzymatic production of collagen/gelatin hydrolysates (releasing peptides ranging from 200–10,000 Da), profusely studied and marketed in the last two decades (Himaya et al., 2012; Lassoued et al., 2015; Liu et al., 2015). The main applications of these hydrolysates are focused on its incorporation as an ingredient of nutraceutical and cosmetic formulations claiming health benefits for joint diseases, antiaging, anti-wrinkle, and antioxidant properties, collagen synthesis in fibroblast cultures, and devices for regenerative tissues (Karayannakidis and Zotos, 2016; Kim, 2014; Sánchez et al., 2018; Silva et al., 2014a). A brief representation of the collagen/gelatin hydrolysates and the bioactivities reported in recent scientific literature is documented in Table 4.2. As in the case of FPH, a great variety of commercial proteases has been used to produce collagen/gelatin hydrolysates from several fish species. Antioxidant and antihypertensive were the most habitual activities determined but also antidiabetic, antiproliferative, and immunomodulatory activities along with chemical properties have been studied.

8 Mineral Fraction and Bioapatites

Mineral fraction recovered from fish bones, scales, and heads has been commonly seen as waste without utility. This mineral fraction, clean of muscle rests, is mainly obtained when proteolytic enzymes are applied to the fish wastes to produce FPH. A subsequent thermal process of calcination (T>550 °C) led to a mineral material without any organic matter. The degree of crystallization and type of porous material obtained depends on the initial particle size, the previous treatment performed, the final temperature of calcination, and the gradient of temperatura to reach it (Mondal et al., 2012; Piccirillo et al., 2014a).

For instance, boiled salmon frames, saithe, and cod bones treated with enzymes generated a mineral fraction after calcination that showed better adsorption of

Table 4.2 Recent collagen hydrolysates produced from fish skins along with its bioactivities and chemical properties determined.

FISH	TISSUE	ENZYME	BIOACTIVITIES/ PROPERTIES/ APPLICATIONS	REFERENCES
Atlantic salmon	Skin collagen	Alcalase, papain	ACE	Gu et al. (2011)
Pacific cod	Skin gelatin	Pepsin, trypsin, α-chymotrypsin	ACE, AO	Himaya et al. (2012)
Thornback Ray	Skin gelatin	Alcalase, neutrase	AO	Lassoued et al. (2015)
Bluefin leatherjacket	Skin	Trypsin, neutrase, alcalase, pepsin, flavourzyme	AO	Chi et al. (2015b)
Giant catfish	Skin gelatin	Trypsin	AO	Ketnawa et al. (2016)
Blue shark	Skin gelatin	Protamex	AO	Weng et al. (2014)
Amur sturgeon	Skin gelatin	Alcalase, flavourzyme	AO	Nikoo et al. (2011)
Grass carp	Skin	Alcalase	AO	Cai et al. (2015)
Tilapia	Skin gelatin	Alcalase, bromelain, neutrase, papain, trypsin	ACE, AO	Choonpicharn et al. (2015)
Unicorn leatherjacket	Skin gelatin	Autolysis	AP, AO, IM	Karnjanapratum et al. (2016)
Grass carp	Skin	Alcalase	EC, WHC, FBC	Wasswa et al. (2007)
Snapper bigeye	Skin gelatin hydrolysate	Extract from pyloric caeca	AO	Phanturat et al. (2010)
Snapper	Skin gelatin hydrolysate	Protease from pyloric caeca	AO	Khantaphant and Benjakul (2008)
Skate	Skin	α-chymotrypsin	ACE	Lee et al. (2011)
Hoki	Skin gelatin	Trypsin	AO	Mendis et al. (2005)
Tilapia	Skin gelatin	Neutrase, properase F	AO	Zhang et al. (2012)
Horse mackerel	Skin	Pepsin, trypsin, α-chymotrypsin	AO	Sampath Kumar et al. (2011)
Seabass	Skin	Alcalase	AO	Senphan and Benjakul (2014)
Skate	Skin gelatin	Alcalase, protease	ACE, AO	Ngo et al. (2014)
Catfish	Skin	Alcalase	ACE	Mahmoodani et al. (2014)
Spanish mackerel	Skin	Pepsin	S, EA, FC, AO	Chi et al. (2014)
Zebra blenny, cobia	Skin gelatin	Alcalase	AO, S, EA, FC, FS, FBC, WHC	Razali et al. (2015)
Tilapia	Skin	Flavourzyme	AD	Wang et al. (2015)
Atlantic salmon	Skin gelatin	Alcalase, bromelain, flavourzyme	AD	Li-Chan et al. (2012)
Asian seabrass	Skin gelatin	Alcalase	IM, AP	Sae-Leaw et al. (2016)
Cobia	Skin	Papain, pancreatin, bromelain, trypsin	AO	Yang et al. (2008)
Alaska pollock	Skin collagen	Trypsin, flavourzyme	AO	Zhuang et al. (2009)
Sole	Skin gelatin	Alcalase	AO	Giménez et al. (2009)
Red bigeye	Skin gelatin	Alcalase, neutrase	AO	Phanturat et al. (2010)
Tuna, halibut	Skin gelatin	Alcalase, collagenase, trypsin, pepsin	AO	Aleman et al. (2011)

AO: antioxidant activities, ACE: antihypertensive activity (angiotensin-converting enzyme inhibitor method), AI: anti-inflammatory activity, AM: antimicrobial (against gram positive and/or gram negative), AP: antiproliferative, IM: immunomodulatory activity, S: solubility properties, EA: emulsifying activity, ES: emulsifying stability, FC: foaming capacity, FS: foaming stability, WHC: water holding capacity, FBC: fat binding capacity, AD: antidiabetic.

calcium and phosphorous by growing pigs when enzyme hydrolysis was previously performed (Malde et al., 2010). Codfish bones processed thermically at 900–-1200 °C produced two types of inorganic materials (bioapatites), hydroxyapatite (HAp: $Ca_{10}(PO_4)_6(OH)_2$) and β-tri-calcium phosphate (β-TCP: β-$Ca(PO_4)_3$) in ratio 75:25. Additionally, before thermal treatment using washes with $CaCl_2$ and NaF, a new bioapatite with enhanced biological properties was found: chlorapatite (ClAp: $Ca_{10}(PO_4)_6Cl_2$) and fluorapatite (FAp: $Ca_{10}(PO_4)_6F_2$), respectively (Piccirillo et al., 2013a). Similar biphasic materials HAp and β-TCP were observed in bones and scales of European sardine treated at 600–1000 °C (Piccinillo et al., 2014b). This bioceramic material was highlighted by the good biocompatibility of HAp and the good resorbability of β-TCP. These authors also observed that higher T increases β-TCP proportion.

Fish bones of swordfish and tuna processed at dual temperatures, 600 °C and 950 °C, produced two types of inorganic materials β-type HAp at 600 °C and HAp+β-TCP (87:13 ratio) at 950 °C with remarkable concentrations of Sr and Mg, which is biologically much beneficial than synthetic apatites (Boutinguiza et al., 2012). When pyrolysis was applied to codfish bone, carbonate apatite (CAp: $Ca_{10}(PO_4)_6(CO_3)_2$) and a mixture of oxyapatite (OAp: $Ca_{10}(PO_4)_6O$) and graphitic carbon were generated heating up to 800 °C and higher than 800 °C, respectively (Piccirillo et al., 2017). This combination of OAp-carbon (bone char) revealed high efficiency for fluoxetine and diclofenac adsorption, whereas CAp is the best substrate for Pb(II) adsorption. In this context, when cod fish bones are initially washed with solutions of Ag or Fe(II) and then calcinated at T>650 °C, doped bioapatites were obtained (Piccirillo et al., 2015, 2014b). In the first case, Ag-HAp showed antibacterial and phytocatalytic activities against *Escherichia coli, Staphylococcus aureus*, and *Pseudomonas aeruginosa* under different conditions of luminosity. In the second case, Fe-HAp was an excellent photostable sunscreen protector with good absorption of whole UV range and without irritation and erythema problems.

On the other hand, hake bone is a source of calcium for efficient bone mineralization and bone deposition in comparison with Lithotame commercial supplement (Flammini et al., 2016). Porous HAp from fish scales was tested as scaffolds for cellular growth (Panda et al., 2014). HAp from codfish bones was applied as a material of construction for building facades indicating the need for lower levels of TiO_2 (titania) than commercial formulations (Saeli et al., 2018). Herring bone hydrolysate was an efficient P-source in feed for salmon (Albrektsen et al., 2018; Ytteborg et al., 2016). Finally, bone meal from fish showed effectivity as an ameliorant to reduce lead in contaminant soil (Gatimu et al., 2010) and water treatment for metals recovered (Pb(II), Zn(II), Cu(II), Cd(II), etc.) have been studied employing apatites from sardine, bogue, bluefish, seabream, and Atlantic cod (Kizilkaya et al., 2010; Oliva et al., 2010; Piccirillo et al., 2013b).

9 Silage, Biosilage, and Marine Peptones

The chemical treatment of fish wastes using mineral acids generates a pasty and cloudy liquid that is called silage (Bower and Hietala, 2008; Raa and Gildberg, 1982). The production of fish silage has been reported to be a feasible, simple, and lower-cost alternative (Olsen and Toppe, 2017; Vidotti et al., 2003) than fish meal, and it is a valuable option to stabilize wastes when the amount is not enough to be processed by fish meal plants or the distance of such plants is too far. This product is microbiologically established (final pH<3), is rich in free amino acids and broken peptides that after drying has potential application in farm animal feed (Bower and Hietala, 2008; Ganesan et al., 2009; Ramírez et al., 2016). Sulfuric and formic acids are the chemicals more commonly utilized in a process that entails the addition of acids to the wastes under agitation for homogenization of the bulk and developing acidity to preserve it (Özyurt et al., 2017; Pagarkar et al., 2006; Vizcarra-Magaña et al., 1999). Some reports have however indicated that this mixture is not very acceptable for animal fodders and suggests the use of organic acids such as lactic, acetic, or propionic acids as reagents of hydrolysis (Raghunath and Gopakumar, 2002), as well as autohydrolysis with endogenous enzymes present in viscera (Deraz, 2015). In this context, the application of LAB to ferment fish wastes has revealed to be the most efficient alternative (Faid et al., 1994; Lindgren and Pleje, 1983; Raghunath and Gopakumar, 2002; Rajendran et al., 2018; Vázquez et al., 2011). Biosilages have been produced by applying several lactic acid bacteria to different fish and seafood substrates from multiple species and experimental conditions (Góngora et al., 2012; Harrabi et al., 2017; Vázquez et al., 2011; Vázquez et al., 2010a). For instance, salmon head wastes were preserved by fermentation with LAB and discarded potatoes as a carbon source during 60 days at pH <4.8 and utilized for compost and animal and aquaculture feeds (Bower et al., 2011). Hake wastes from processing factory were successfully stabilized (pH 2) after 24 h of culture of *Lactobacillus arizonesis* (Góngora et al., 2012). Seafood processing by-products from shrimp head and octopus viscera were good substrates for terrestrial LAB such as *Enterococcus gallinarum* (Özyurt et al., 2017) and *L. plantarum* (Harrabi et al., 2017) but also for the growth marine bacteria such as *Exiguobacterium* sp. along with the concomitant production of antioxidant activities and proteases (Anil Kumar and Suresh, 2014). In all cases, an external source of sugar is needed for the glycolytic metabolism of LAB and must be added to the fish wastes (Shabani et al., 2018). Sucrose and glucose are the most conventional (Bower et al., 2011; Vázquez et al., 2011; Vázquez et al., 2010a), but sugarcane molasses have been extensively reported supplementing fish mince of silver bellies and fermented by *L. plantarum* (Rao et al., 2009), visceral wastes of freshwater carps (Ganesan et al., 2009) and offals, and fish by-catch inoculated with *L. plantarum* (Bridgewater and Lallo, 2005).

Peptones are protein hydrolysates, water soluble, and not thermally coagulable and are obtained by acid, alkaline, thermal, or enzymatic hydrolysis of raw materials, mainly by-products from animals, vegetal, or microbial origin. They are constituted of peptides of varied molecular weight, free amino acids, and a small concentration of nucleotides and carbohydrates. Commercially the name of peptones refers to the origin and enzyme used for producing them (soytone, casitone, tryptone, meat extract, etc.). Inexplicably those of marine origin are almost nonexistent even as viscera, heads, trimmings, and seafood effluents as an excellent substrate for marine peptone production (Dufosé et al., 1997; Horn et al., 2007; Ramkumar et al., 2016; Vázquez et al., 2015). Peptones are the most important and expensive source of organic nitrogen in the microbial culture media. In many cases, among 30–80% of the cost of the industrial production of vaccine, antibiotics, organic acids, etc., from microorganisms is due to the cost of peptones.

Marine peptones from fish and seafood by-products and effluents have been prepared mainly by hydrolysis with commercial protease (Djellouli et al., 2017; Safari et al., 2012; Safari et al., 2011) and also by boiling (Mohapatra et al., 2017; Rebah et al., 2008; Vázquez et al., 2008), autolytic process with endogenous enzymes (Deraz et al., 2011; Gildberg et al., 2010; Vázquez et al., 2004a), or acid/alkaline hydrolysis (Esakkiraj et al., 2009). LAB are the group of most employed microorganisms for testing the validity of marine peptones due to fastidious bacteria that require complex growth factors (Hammes and Hertel, 2009). *L. pentosus* have been grown on catfish manure waste as a source of carbon and nitrogen for the production of lactic acid (Shi et al., 2018). Biomasses of various *Lactobacillus* and *Lactococcus* were satisfactorily produced when peptones from tuna and salmon by-products were utilized (Aspmo et al., 2005a, 2005b, 2005c; Horn et al., 2007; Safari et al., 2012). Specifically, bacteriocins such as divergicin, pediocin, nisin, or sakacin produced from LAB have been obtained on media, including peptones from bolti fish viscera, tuna head and viscera, trout viscera, squid and shrimp FPH, octopus effluents, snow crab hepatopancreas, pelagic fish FPH and swordfish, shark and ray viscera (Beaulieu et al., 2009; Deraz et al., 2011; Shi et al., 2018; Tahiri et al., 2009; Vázquez et al., 2008; Vázquez et al., 2004a, 2006a; Vázquez and Murado, 2008a, 2008b). The growth of marine bacteria as *Vibrio anguillarum*, *Roseobacter* sp., *Carnobacterium divergens*, *Pseudomonas fluorescens*, and *Aeromonas salmonicida* was also studied on low-cost marine media (Gildberg et al., 2010; Safari et al., 2011; Vázquez et al., 2004b). In all cases mentioned, the corresponding growths and productions evaluated were similar or generally higher in media formulated with marine peptones than commercial ones.

Different microbial enzymes productions have been explored using peptones from marine origin with excellent outcomes: esterase from *Bacillus altitudinis* in fish processing waste meal (Esakkiraj et al., 2012), protease from *B. cereus* in sardinelle head and viscera powder (Sellami-Kamoun et al., 2011), fibrinolytic enzyme from *B. cereus* in cuttle fish waste (Biji et al., 2016), protease from *B. cereus* in tuna processing wastes (Esakkiraj et al., 2009), tranglutaminase from *Enterobacter* sp. and *Providencia* sp. in surimi wastewaters (H-Kittikun et al., 2012), lipases from *Staphylococcus xylosus* and *Rhizopus oryzae* in tuna by-products (Sellami et al., 2013), and proteolytic activities from *V. anguillaum* and *V. splendidus* in peptones from squid, swordfish, trout, and tuna viscera (Vázquez et al., 2006b). The production of other bacterial metabolites such as ramoplanin A2, polyhydroxybutyrate, and iturin A in fish protein were also reported (Erkan and Kayali, 2016; Mohapatra et al., 2017; Zohora et al., 2013). Finally, HA has been produced by *Streptococcus zooepidemicus* in culture media based on peptones extracted from various fish viscera wastes (Vázquez et al., 2009, 2010b; Vázquez et al., 2015).

10 Conclusion

In the present chapter, we have briefly reviewed a set of most important and added-value biocompounds and products that can be recovered, isolated, and produced from wastes generated by fishing activities, food processing of by-products, and discards. The studies reported in the references have demonstrated that using adequate processes and technologies, fish wastes are potential sources of proteins, biopolymers, bioactivities, oils, enzymes, minerals, and feed ingredients. The fields of application and potential market explored for those products deal with pharmacological, human-food, animal-feed, nutraceutical, cosmetic, biotechnological, clinical, and chemical sectors. From the perspective of the green and circular bioeconomy and the blue growth sustainability, the valorization of fish wastes is peremptory and the selective and optimal processes are fundamental to help moreover to the sustainability of fisheries resources.

Acknowledgments

The authors thank the projects IBEROS (0245_IBEROS_1_E, POCTEP 2015), CVMar+I (0302_CVMAR_I_1_P, POCTEP 2015), BLUEHUMAN (EAPA_151/2016, UE-INTERREG Atlantic Area Programme) and Xunta de Galicia (Grupos de Potencial Crecimiento, IN607B2018/19) for the economical financing of the present chapter.

References

Aam, B. B., Heggset, E. B., Norberg, A. L., Sørlie, M., Vârum, K. M., and Eijsink, V. G. H. 2010. Production of chitooligosaccharides and their potential applications in medicine. *Marine Drugs* 8:1482–517.

Abdelhedi, O., Jridi, M., Jemil, I., Mora, L., Toldrá, F., Aristoy, M., Boualga, A., Nasri, M., and Nasri, R. 2016. Combined biocatalytic conversion of smooth hound viscera: Protein hydrolysates elaboration and assessment of their antioxidant, anti-ACE and antibacterial activities. *Food Research International* 86:9–23.

Afonso, M. D., and Borquez, R. 2002. Review of the treatment of seafood processing wastewaters and recovery of proteins therein by membrane separation processes: Prospects of the ultrafiltration of wastewaters from the fish meal industry. *Desalination* 142:29–45.

Afonso, M. D., Ferrer, J., and Borquez, R. 2004. An economic assessment of proteins recovery from fish meal effluents by ultrafiltration. *Trends in Food Science & Technology* 15:506–12.

Agnello S., Gasperini L., Reis R. L., Mano J. F., Pitarresi G., Palumbo F. S., and Giammona G. 2016. Microfluidic production of hyaluronic acid derivative microfibers to control drug release. *Materials Letters* 182:309–13.

Ahn, C. B., Kim, J. G., and Je, J. Y. 2014. Purification and antioxidant properties of octapeptide from salmon byproduct protein hydrolysate by gastrointestinal digestion. *Food Chemistry* 147:78–83.

Albrektsen, S., Lock, E.-J., Bæverfjord, G., Pedersen, M., Krasnov, A., Takle, H., Veiseth-Kent, E., Ørnsrud, R., Waagbø, R., and Ytteborg, E. 2018. Utilization of H_2SO_4-hydrolysed phosphorus from herring bone by-products in feed for Atlantic salmon (*Salmo salar*) 0+ postsmolt. *Aquaculture Nutrition* 24:348–65.

Aldairi, A. F., Ogundipe, O. D., and Pye, D. A. 2018. Antiproliferative activity of glycosaminoglycan-like polysaccharides derived from marine molluscs. *Marine Drugs* 16:63.

Aleman, A., Giménez, B., Montero, P., and Gomez-Guillen, M. C. 2011. Antioxidant activity of several marine skin gelatins. *LWT-Food Science and Technology* 44:407–13.

Alves M. H. M. E., Nascimento, G. A., Cabrera, M. P., Da Cruz Silvério, S. I., Nobre, C., Teixeira, J. A., and de Carvalho L. B. Júnior 2017. Trypsin purification using magnetic particles of azocasein-iron composite. *Food Chemistry* 226:75–78.

Amado, I. R., González, M., Murado, M. A., and Vázquez, J. A. 2016. Shrimp wastewater as a source of astaxanthin and bioactive peptides. *Journal of Chemical Technology and Biotechnology* 91:793–805.

Amado, I. R., and Vázquez, J. A. 2015. Mussel processing wastewater: A low-cost substrate for the production of astaxanthin by *Xanthophyllomyces dendrorhous*. *Microbial Cell Factories* 14:177.

Amado, I. R., Vázquez, J. A., González, M. P., and Murado. M. A. 2013. Production of antihypertensive and antioxidant activities by enzymatic hydrolysis of protein concentrates recovered by ultrafiltration from cuttlefish processing wastewaters. *Biochemical Engineering Journal* 76:43–54.

Amado, I. R., Vázquez, J. A., Murado, M. A., and González, M. P. 2015. Recovery of astaxanthin from shrimp cooking wastewater: Optimization of astaxanthin extraction by response surface methodology and kinetic studies. *Food Bioprocess Technology* 8:371–81.

Ambati, R. R., Moi, P. H., Ravi, S., and Aswathanarayana, R. G. 2014. Astaxanthin: Sources, extraction, stability, biological activities and its commercial applications. A Review. *Marine Drugs* 12:128–52.

An, K., Liu, H., Guo, S., Kumar, D. N. T., and Wang, Q. 2010. Preparation of fish gelatine and fish gelatin/poly(l-lactide) nanofibers by electrospinning. *International Journal of Biological Macromolecules* 47:380–88.

Anil Kumar, P. K., and Suresh, P. V. 2014. Biodegradation of shrimp biowaste by marine *Exiguobacterium* sp. CFR26M and concomitant production of extracellular Protease and antioxidant materials: Production and process optimization by response surface methodology. *Marine Biotechnology* 16:202–18.

Arima, K., Fujita, H., Toita, R., Imazu-Okada, A., Tsutsumishita-Nakai, N., Takeda, N., Nakao, Y., Wang, H., Kawano, M., Matsushita, K., Tanaka, H., Morimoto, S., Nakamura, A., Kitagaki, M., Hieda, Y., Hatto, R., Watanabe, A., Yumura, T., Okuhara, T., Hayashi, H., Shimizu, K., Nakayama, K., Masuda, S., Ishihara, Y.,

Yoshioka, S., Yoshioka, S., Shirade, S., and Tamura, J.-I. 2013. Amounts and compositional analysis of glycosaminoglycans in the tissue of fish. *Carbohydrate Research* 366:25–32.

Arnesen, J. A., and Gildberg, A. 2007. Extraction and characterisation of gelatine from Atlantic salmon (*Salmo salar*) skin. *Bioresource Technology* 98:53–57.

Arvanitoyannis, I. S., and Kassaveti, A. 2008. Fish industry waste: Treatments, environmental impacts, current and potential uses. *International Journal of Food Science and Technology* 43:726–45.

Aspmo, S. I., Horn, S. J., and Eijsink, V. G. H. 2005a. Use of hydrolysates from Atlantic cod (*Gadus morhua* L.) viscera as a complex nitrogen source for lactic acid bacteria. *FEMS Microbiology Letters* 248:65–68.

Aspmo, S.I., Horn, S.J., and Eijsink, V. G. H. 2005b. Hydrolysates from Atlantic cod (*Gadus morhua* L.) viscera as components of microbial growth media. *Process Biochemistry* 40:3714–22.

Aspmo, S.I., Horn, S.J., and Eijsink, V. G. H. 2005c. Enzymatic hydrolysis of Atlantic cod (*Gadus morhua* L.) viscera. *Process Biochemistry* 40:1957–66.

Bajaj, M., Freiberg, A., Winter, J., Xu, Y., and Gallert, C. 2015. Pilot-scale chitin extraction from shrimp shell waste by deproteination and decalcification with bacterial enrichment cultures. *Applied Microbiology and Biotechnology* 99:9835–46.

Balti, R., Barkia, A., Bougatef, A., Ktari, N., and Nasri, M. 2009. A heat-stable trypsin from the hepatopancreas of the cuttlefish (Sepia officinalis): Purification and characterisation. *Food Chemistry* 113:146–54.

Barkia, A., Bougatef, A., Khaled, H. B., and Nasri, M. 2010. Antioxidant activities of sardinelle heads and/or viscera protein hydrolysates prepared by enzymatic treatment. *Journal of Food Biochemistry* 34:303–20.

Barros, M. P., Poppe, S. C., and Souza-Junior, T. P. 2011. Putative benefits of microalgal astaxanthin on exercise and human health. *Brazilian Journal of Pharmacognosy* 21:283–89.

Batista, I., Ramos, C., Coutinho, J., Bandarra, N. M., and Nunes, M. L. 2010. Characterization of protein hydrolysates and lipids obtained from black scabbardfish (*Aphanopus carbo*) by-products and antioxidative activity of the hydrolysates produced. *Process Biochemistry* 45:18–24.

Beaney, P., Lizardi-Mendoza, J., and Healy, M. 2005. Comparison of chitins produced by chemical and bioprocessing methods. *Journal of Chemical Technology and Biotechnology* 80:45–50.

Beaulieu, L., Desbiens, M., Thibodeau, J., and Thibault, S. 2009. Pelagic fish hydrolysates as peptones for bacterial culture media. *Canadian Journal of Microbiology* 55:1240–49.

Behçet, R. 2011. Performance and emission study of waste anchovy fish biodiesel in a diesel engine. *Fuel Processing Technology* 92:1187–94.

Bendiksen, E. Å., Johnsen, C. A., Olsen, H. J., and Jobling, M. 2011. Sustainable aquafeeds: Progress towards reduced reliance upon marine ingredients in diets for farmed Atlantic salmon (*Salmo salar* L.). *Aquaculture* 314:132–39.

Benjakul, S., Oungbho, K., Visessanguan, W., Thiansilakul, Y., and Roytrakul, S. 2009. Characteristics of gelatin from the skins of bigeye snapper, *Priacanthus tayenus* and *Priacanthus macracanthus*. *Food Chemistry* 116:445–51.

Bernocchi, B., Carpentier, R., and Betbeder, D. 2017. Nasal nanovaccines. *International Journal of Pharmaceutics* 530:128–38.

Betty, M., Awuor, O. L., Kirwa, M. E., and Jackim, M. F. 2014. Antioxidative and functional properties of *Rastrineobola argentea* (Dagaa) fish protein hydrolysate. *Discourse Journal of Agriculture and Food Science* 2:180–89.

Biji, G. D., Arun, A., Muthulakshmi, E., Vijayaraghavan, P., Arasu, M. V., and Al-Dhabi, N. A. 2016. Bio-prospecting of cuttle fish waste and cow dung for the production of fibrinolytic enzyme from *Bacillus cereus* IND5 in solid state fermentation. *3 Biotech* 6:231.

Bkhairia, I., Khaled, H. B., Ktari, N., Miled, N., Nasri, M., and Ghorbel, S. 2016. Biochemical and molecular characterisation of a new alkaline trypsin from *Liza aurata*: Structural features explaining thermal stability. *Food Chemistry* 196:1346–54.

Blanco, M., Fraguas, J., Sotelo, C. G., Pérez-Martín, R. I., and Vázquez. J. A. 2015a. Production of chondroitin sulphate from head, skeleton and fins of *Scyliorhinus canicula* by-products by combination of enzymatic, chemical precipitation and ultrafiltration methodologies. *Marine Drugs* 13:3287–308.

Blanco, M., Simpson, B. K., Pérez-Martín, R. I., Sotelo, C. G. 2014. Isolation and partial characterization of trypsin from pancreas of small-spotted catshark (*Scyliorhinus canicula*). *Journal of Food Biochemistry* 38:196–206.

Blanco, M., Sotelo, C. G., Chapela, M. J., and Pérez-Martín. R. I. 2007. Towards sustainable and efficient use of fishery resources: Present and future trends. *Trends in Food Science & Technology* 18:29–36.

Blanco, M., Sotelo, C. G., and Pérez-Martín, R. I. 2015b. Hydrolysis as a valorization strategy for unused marine food biomass: Boarfish and small-spotted catshark discards and by-products. *Journal of Food Biochemistry* 39:368–76.

Bougatef, A. 2013. Trypsins from fish processing waste: Characteristics and biotechnological applications-Comprehensive review. *Journal of Cleaner Production* 57:257–65.

Bougatef, A., Nedjar-Arroume, N., Manni, L., Ravallec, R., Barkia, A., Guillochon, D., and Nasri, M. 2010. Purification and identification of novel antioxidant peptides from enzymatic hydrolysates of sardinelle (*Sardinella aurita*) by-products proteins. *Food Chemistry* 118:559–65.

Bougatef, A., Nedjar-Arroume, N., Ravallec-Plé, R., Leroy, Y., Guillochon, D., Barkia, A., Nasri. M. 2008. Angiotensin I-converting enzyme (ACE) inhibitory activities of sardinelle (*Sardinella aurita*) by-products protein hydrolysates obtained by treatment with microbial and visceral fish serine proteases. *Food Chemistry* 111:350–56.

Bougatef, A., Souissi, N., Fakhfakh, N., Ellouz-Triki, Y., and Nasri, M. 2007. Purification and characterization of trypsin from the viscera of sardine (*Sardina pilchardus*). *Food Chemistry* 102:343–50.

Bougatef, H., Krichen, F., Capitani, F., Amor, I. B., Maccari, F., Mantovani, V., Galeotti, F., Volpi, N., Bougatef, A., and Sila, A. 2018. Chondroitin sulfate/dermatan sulfate from corb (*Sciaena umbra*) skin: Purification, structural analysis and anticoagulant effect. *Carbohydrate Polymers* 196:272–78.

Bourseau, P., Vandanjon, L., Jaouen, P., Chaplain-Derouiniot, M., Masse, A., Guerard, F., Chabeaud, A., Fouchereau-Peron, M., Gal, Y. L., Ravallec-Pie, R., Berge, J. P., Picot, L., Piot, J., Batista, I., Thorkelsson, G., Delannoy, C., Gakobsen, G., and Johansson, I. 2009. Fractionation of fish protein hydrolysates by ultrafiltration and nanofiltration: Impact on peptidic polulatoins. *Desalination* 244:303–20.

Boutinguiza, M., Pou, J., Comesaña, R., Lusquiños, F., De Carlos, A., and León, B. 2012. Biological hydroxyapatite obtained from fish bones. *Materials Science and Engineering C* 32:478–86.

Bower, C., Hietala, K., and Delaca, T. 2011. Bacterial preservation of pink salmon using potatoes as a carbohydrate source. *Journal of Food Processing and Preservation* 35:822–31.

Bower, C., Malemute, C., and Bechtel, P. 2011a. Endogenous protease activity in by-products of pink salmon (*Oncorhynchus gorbuscha*). *Journal of Food Biochemistry* 35:628–37.

Bower, C. K, and Hietala, K. A. 2008. Acidification methods for stabilization and storage of salmon by-products. *Journal of Aquatic Food Product Technology* 17:459–78.

Bowyer, J. N., Qin, J. G., and Stone, D. A. J. 2013. Protein, lipid and energy requirements of cultured marine fish in cold, temperate and warm water. *Reviews in Aquaculture* 5:10–32.

Bridgewater, S. T., and Lallo, C. H. O. 2005. Ensiling of fish waste with sugar cane molasses with and without inoculant as a protein supplement for livestock. *Tropical Agriculture* 82:300–06.

Bui, H. T. D., Khosravi, S., Fournier, V., Herault, M., and Lee, K. J. 2014. Growth performance, feed utilization, innate immunity, digestibility and disease resistance of juvenile red seabream (*Pagrus major*) fed diets supplemented with protein hydrolysates. *Aquaculture* 418–419:11–16.

Cai, L., Wu, X., Zhang, Y., Li, X., Ma, S., and Li, J. 2015. Purification and characterization of three antioxidant peptides from protein hydrolysate of grass carp (*Ctenopharyngodon idella*) skin. *Journal of Functional Foods* 16:234–42.

Cardoso, M. J., Costa, R. R., and Mano, J. F. 2016. Marine origin polysaccharides in drug delivery systems. *Marine Drugs* 14:34.

Carvalho, A. M., Teixeira, R., Novoa-Carballal, R., Pires, R. A., Reis, R. L., and Pashkuleva, I. 2018. Redox-responsive micellar nanoparticles from glycosaminoglycans for CD44 targeted drug delivery. *Biomacromolecules* 19:2991–99.

Castro-Ceseña, A. B., Del Pilar Sánchez-Saavedra, M., Márquez-Rocha, F. J. 2012. Characterisation and partial purification of proteolytic enzymes from sardine by-products to obtain concentrated hydrolysates. *Food Chemistry* 135:583–89.

Cerqueira M. T., Da Silva L. P., Santos T. C., Pirraco R. P., Correlo, V. M., Marques A. P., and Reis R. L. 2014. Human skin cell fractions fail to self-organize within a gellan gum/hyaluronic acid matrix but positively influence early wound healing. *Tissue Engineering Part A* 20:1369–78.

Chalamaiah, M., Kumar, B. D., Hemalatha, R., and Jyothirmayi, T. 2012. (FPH): Proximate composition, amino acid composition, antioxidant activities and applications: A review. *Food Chemistry* 135:3020–38.

Chatterjee, S., Adhya, M., Guha, A. K., and Chatterjee, B. P. 2005. Chitosan from *Mucor rouxii*: Production and physico-chemical characterization. *Process Biochemistry* 40:395–400.

Chatterjee, S., Chatterjee, B. P., and Guha, A. K. 2010. Kinetics of *Mucor rouxii* fermentation in relation to chitosan production. *Research Journal of Microbiology* 5:361–65.

Chen, X., Jiang, Q., Xu, Y., and Xia, W. 2017. Recovery of chitin from Antarctic krill (*Euphausia superba*) shell waste by microbial deproteinization and demineralization. *Journal of Aquatic Food Product Technology* 26:1210–20.

Chen, Y. C., Tou, J. C., and Jaczynski, J. 2007. Amino acid, fatty acid, and mineral profiles of materials recovered from rainbow trout (*Oncorhynchus mykiss*) processing by-products using isoelectric solubilization/precipitation. *Journal of Food Science* 72:527–35.

Cheng, C. Y., Duan, W. W., Duan, Z.-H. Hai, Y., Lei, X. G., Chang, H. 2013. Extraction of chondroitin sulfate from tilapia byproducts with ultrasonic-microwave synergistic. *Advanced Materials Research* 726–31:4381–85.

Cheng, C. Y., Duan, Z. H., Fan, J., Zhou, K. G., and Chang, H. 2011. Optimization of chondroitin sulfate extraction from tilapia byproduct with response surface methodology. *Food Science and Technology* 36:128–33.

Cheow, C. S., Norizah, M. S., Kyaw, Z. Y., and Howell, N. K. 2007. Preparation and characterisation of gelatins from the skins of sin croaker (*Johnius dussumieri*) and shortfin scad (*Decapterus macrosoma*). *Food Chemistry* 101:386–91.

Chi, C., Cao, Z., Wang, B., Hu, F., Li, Z., and Zhang, B. 2014. Antioxidant and functional properties of collagen hydrolysates from Spanish mackerel skin as influenced by average molecular weight. *Molecules* 19:11211–30.

Chi, C. F., Wang, B., Hu, F. Y., Wang, Y. M., Zhang, B., Deng, S. G., and Wu, C. W. 2015a. Purification and identification of three novel antioxidant peptides from protein hydrolysate of bluefin leatherjacket (*Navodon septentrionalis*) skin. *Food Research International* 73:124–29.

Chi, C. F., Wang, B., Wang, Y. M., Zhang, B., and Deng, S. G. 2015b. Isolation and characterization of three antioxidant peptides from protein hydrolysate of bluefin leatherjacket (*Navodon septentrionalis*) heads. *Journal of Functional Foods* 12:1–10.

Choonpicharn, S., Jaturasitha, S., Rakariyatham, N., Suree, N., and Niamsup, H. 2015. Antioxidant and antihypertensive activity of gelatin hydrolysate from Nile tilapia skin. *Journal of Food Science and Technology* 52:3134–39.

Chotikachinda, R., Tantikitti, C., Benjakul, S., Rustad, T., and Kumarnsit, E. 2013. Production of protein hydrolysates from skipjack tuna (*Katsuwonus pelamis*) viscera as feeding attractants for Asian seabass (*Lates calcarifer*). *Aquaculture Nutrition* 19:773–84.

Chuesiang, P., and Sanguandeekul, R. 2015. Protein hydrolysate from tilapia frame: Antioxidant and angiotensin I converting enzyme inhibitor properties. *International Journal of Food Science Amd Technology* 50:1436–44.

Coelho, R. C. G., Marques, A. L. P., Oliveira, S. M., Diogo, G. S., Pirraco, R. P., Moreira-Silva, J., Xavier, J. C., Reis, R. L., Silva, T. H., and Mano, J. F. 2017. Extraction and characterization of collagen from Antarctic and Sub-Antarctic squid and its potential application in hybrid scaffolds for tissue engineering. *Materials Science and Engineering C* 78:787–95.

Corinaldesi, C., Barone, G., Marcellini, F., Dell'Anno, A., and Danovaro, R. 2017. Marine microbial-derived molecules and their potential use in cosmeceutical and cosmetic products. *Marine Drugs* 15:118.

Costas, H. M., Freitas Júnior, A. C., Amaral, I. P., Hirata, I. Y., Paiva, P. M., Carvalho, L. B., and Bezerra, R. S. 2013. Metal-sensitive and termostable trypsin from the crevalle jack (Caranx hippos) pyloric caeca: Purification and characterization. *Chemistry Central Journal* 7:166.

Cuong, H. N., Minh, N. C., Van Hoa, N., and Trung, T. S. 2016. Preparation andcharacterization of high purity β-chitin from squid pens (*Loligo chenisis*). *International Journal of Biological Macromolecules* 93(Part A):442–47.

Da Silva, L. P., Pirraco, R. P., Santos, T. C., Novoa-Carballal, R., Cerqueira, M. T., Reis, R. L., Correlo, V. M., and Marques, A. P. 2016. Neovascularization induced by the hyaluronic acid-based spongy-like hydrogels degradation products. *ACS Applied Materials Interfaces* 8:33464–74.

da Trindade Alfaro, A., Simões Da Costa, C., Graciano Fonseca, G., Prentice, C. 2009. Effect of extraction parameters on the properties of gelatin from king weakfish (*Macrodon ancylodon*) Bones. *Food Science and Technology International* 15:553–62.

Dave, D., and Routray, W. 2018. Current scenario of Canadian fishery and corresponding underutilized species and fishery byproducts: A potential source of omega-3 fatty acids. *Journal of Cleaner Production* 180:617–41.

Del Campo, J. A., and García-González, M. 2007. Outdoor cultivation of microalgae for carotenoid production: Current state and perspectives. *Applied Microbiology and Biotechnology* 74:1163–74.

Deraz, S. F. 2015. Protein hydrolysate from visceral waste proteins of bolti fish (*Tilapia nilotica*): Chemical and nutritional variations as affected by processing pHs and time of hydrolysis. *Journal of Aquatic Food Product Technology* 24:614–31.

Deraz, S. F., El-Fawal, G. F., Abd-Ellatif, S. A., and Khalil, A. A. 2011. Autohydrolysed *Tilapia nilotica* fish viscera as a peptone source in bacteriocin production. *Indian Journal of Microbiology* 51:171–75.

Djellouli, M., Martínez-Álvarez, O., Arancibia, M. Y., Florez-Cuadrado, D., Ugarte-Ruíz, M., Domínguez, L., Zadi-Karam, H., Karam, N., Roudj, S., López-Caballero, M. E. 2017.

Effect of seafood peptones on biomass and metabolic activity by *Enterococcus faecalis* DM19. *LWT-Food Science and Technology* 81:94–100.
Dos Santos, S. A., Martins, V. G., Sallas-Mellado, M., and Prentice, C. 2011. Evaluation of functional properties in protein hydrolysates from bluewing searobin (*Prionotus punctatus*) obtained with different microbial enzymes. *Food Bioprocess Technology* 4:1399–406.
Duan, R., Zhang, J., Xing, F., Konno, K., and Xu, B. 2011. Study on the properties of gelatins from skin of carp (*Cyprinus carpio*) caught in winter and summer season. *Food Hydrocolloids* 25:368–73.
Duan, Z.-H., Cheng, C.-Y., Hai, Y., and Wang, J.-L. 2013. Determination and identification of chondroitin sulfate from tilapia byproducts. *Advanced Materials Research* 693:1318–21.
Dufossé, L., De La Broise, D., and Guerard, F. 1997. Fish protein hydrolysates as nitrogen sources for microbial growth and metabolite production. In: S. G. Pandalai (Ed.), *Recent Research Developments in Microbiology*, Vol. 1, pp. 365–81. Trivandrum, India: Research Signpost.
Elavarasan, K., Naveen Kumar, V., and Shamasundar, B. A. 2014. Antioxidant and functional properties of (FPH) from fresh water carp (*Catla catla*) as influenced by the nature of enzyme. *Journal of Food Processing and Preservation* 38:1207–14.
Enascuta, C. E., Stepan, E., Bolocan, I., Bombos, D., Calin, C., Oprescu, E.-E., and Lavric, V. 2018. Simultaneous production of oil enriched in ω-3 polyunsaturated fatty acids and biodiesel from fish wastes. *Waste Management* 75:205–14.
Ennaas, N., Hammami, R., Beaulieu, L., and Fliss, I. 2015. Purification and characterization of four antibacterial peptides fromprotamex hydrolysate of Atlantic mackerel (*Scomber scombrus*) by-products. *Biochemical and Biophysical Research Communications* 462:195–200.
Erkan, D., and Kayali, H. A. 2016. Replacement of soybean meal with animal origin protein meals improved ramoplanin A2 production by *Actinoplanes* sp. ATCC 33076. *Applied Biochemistry and Biotechnology* 180:306–21.
Esakkiraj, P., Immanuel, G., Sowmya, S.M., Iyapparaj, P., and Palavesam, A. 2009. Evaluation of protease-producing ability of fish gut isolate *Bacillus cereus* for aqua feed. *Food and Bioprocess Technology* 2:383–90.
Esakkiraj, P., Usha, R., Palavesam, A., and Immanuel, G. 2012. Solid-state production of esterase using fish processing wastes by *Bacillus altitudinis* AP-MSU. *Food and Bioproducts Processing* 90:370–76.
EU. 2018 https://ec.europa.eu/fisheries/cfp/fishing_rules/discards_en (consulted in May of 2018).
Eysturskarð, J., Haug, J., Elharfaoui, N., Djabourov, M., and Draget, I. 2009. Structural and mechanical properties of fish gelatin as a function of extraction conditions. *Food Hydrocolloids* 23:1702–11.
Fadhil, A. B., Al-Tikrity, E. T. B., and Albadree, M. A. 2017. Biodiesel production from mixed non-edible oils, castor seed oil and waste fish oil. *Fuel* 210:721–28.
Faid, M., Karani, H., Elmarrakchi, A., and Achkari-Begdouri, A. 1994. A biotechnological process for the valorization of fish waste. *Bioresource Technology* 49:237–41.
FAO. 2016. www.fao.org/3/a-i5555e.pdf.
Fernandes-Silva, S., Moreira-Silva, J., Silva, T. H., Perez-Martin, R. I., Sotelo, C. G., Mano, J. F. Duarte, A. R. C., and Reis, R. L. 2013. Porous hydrogels from shark skin collagen crosslinked under dense carbon dioxide atmosphere. *Macromolecular Bioscience* 13:1621–31.
Ferraro, V., Carvalho, A. P., Piccirillo, C., Santos, M. M, Castro, P. M. I., and Pintado, M. E. 2013. Extraction of high added value biological compounds from sardine, sardine-type fish and mackerel canning residues-A review. *Materials Science and Engineering C* 33:3111–20.

Ferraro, V., Cruz, I. B., Jorge, R. F., Malcata, F. X., Pintado, M. E., and Castro, P. M. L. 2010. Valorisation of natural extracts from marine source focused on marine by-products: A review. *Food Research International* 43:2221–33.

Ferreira, Isabel M. P. L. V. O., Eça, R., Pinho, O., Tavares, P., Pereira, A., and Roque, A. C. 2007. Development and valldation of an HPLC/UV method for quantification of bioactive peptides in fermented milk. *Journal of Liquid Chromatography & Related Technologies* 30:2139–47.

Fiori, L., Solana, M., Tosi, P., Manfrini, M., Strim, C., and Guella, G. 2012. Lipid profiles of oil from trout (*Oncorhynchus mykiss*) heads, spines and viscera: Trout by-products as a possible source of omega-3 lipids?. *Food Chemistry* 134:1088–95.

Flammini, L., Martuzzi, F., Vivo, V., Ghirri, A., Salomi, E., Bignetti, E., and Barocelli, E. 2016. Hake fish bone as a calcium source for efficient bone mineralization. *International Journal of Food Sciences and Nutrition* 67:265–73.

Foh, M. B. K., Kamara, M. T., Amadou, I., Foh, B. M., and Wenshui, X. 2011. Chemical and physicochemical properties of tilapia (*Oreochromis niloticus*) fish protein hydrolysate and concentrate. *International Journal of Biology and Chemistry* 5:21–36.

Freitas, A. C., Rodrigues, D., Rocha-Santos, T. A. P., Gomes, A. M. P., and Duarte, A. C. 2012. Marine biotechnology advances towards applications in new functional foods. *Biotechnology Advances* 30:1506–15.

Fu, S., Xia, J., and Wu, J. 2016. Functional chitosan nanoparticles in cancer treatment. *Journal of Biomedical Nanotechnology* 12:1585–603.

Galla, N. R., Pamidighantam, P. R., Akula, S., and Karakala, B. 2012. Functional properties and in vitro antioxidant activity of roe protein hydrolysates of *Channa striatus* and *Labeo horita*. *Food Chemistry* 135:1479–84.

Ganesan, P., Pradeep, M. G., Sakhare, P. Z., Suresh, P. V., and Bhaskar, N. 2009. Optimization of conditions for natural fermentation of freshwater fish processing waste using sugarcane molasses. *Journal of Food Science and Technology* 46:312–15.

Ganiari, S., Choulitoudi, E., and Oreopoulou, V. 2017. Edible and active films and coatings as carriers of natural antioxidants for lipid food. *Trends in Food Science & Technology* 68:70–82.

García-López, M., Pérez-Martín, R. I., and Sotelo, C. G. 2016. Carotenoid pigments composition of two commonly discarded decapod crustaceans in grand sole and the Galician-Northern Portugal coast fisheries. *Journal of Aquatic Food Product Technology* 25:114–21.

García-Moreno, P. J., Batista, I., Pires, C., Bandarra, N. M., Espejo-Carpio, F. J., Guadix, A., and Guadix, E. M. 2014. Antioxidant activity of protein hydrolysates obtained from discarded Mediterranean fish species. *Food Research International* 65:469–76.

García-Moreno, P. J., Pérez-Gálvez, R., Espejo-Carpio, F. J., Ruiz-Quesada, C., Pérez-Morilla, A. I., Martínez-Agustín, O., Guadix, A., and Guadix, E. M. 2017. Functional, bioactive and antigenicity properties of blue whiting protein hydrolysates: Effect of enzymatic treatment and degree of hydrolysis. *Journal of the Science of Food and Agriculture* 97:299–308.

Gatimu, E., Mwinyihja, M., and Killham, K. 2010. Determination of the effect of bone meal as an ameliorant on lead contaminated soils. *Research Journal of Environmental Toxicology* 4:237–47.

Geirsdottir, M. S., Sigurgisladottir, S., Hamaguchi, P. Y., Thorkelsson, G., Johannsson, R., Kristinsson, H. G., and Kristjansson, M. M. 2011. Enzymatic hydrolysis of Blue whiting (*Micromesistius poutassou*); functional and bioactive properties. *Journal of Food Science* 76: C14–C20.

Ghaly, A. E., Ramakrishnan, V. V., Brooks, M. S., Budge, S. M., and Dave, D. 2013. Fish processing wastes as a potential source of proteins, amino acids and oils: A critical review. *Journal of Microbial and Biochemical Technology* 5:107–29.

Gharehghani, A., Mirsalim, M., and Hosseini, R. 2017. Effects of waste fish oil biodiesel on diesel engine combustion characteristics and emission. *Renewable Energy* 101:930–36.

Ghorbel-Bellaaj, O., Hajji, S., Younes, I., Chaabouni, M., Nasri, M., and Jellouli, K. 2013. Optimization of chitin extraction from shrimp waste with *Bacillus pumilus* A1 using response surface methodology. *International Journal of Biological Macromolecules* 61:243–50.

Gildberg, A., Dahl, R., Mikkelsen, H., and Nilsen, K. 2010. Peptones from Atlantic cod stomach as nitrogen sources in growth media to marine bacteria. *Journal of Aquatic Food Product Technology* 19:75–83.

Giménez, B., Alemán, A., Montero, P., and Gómez-Guillén, M. C. 2009. Antioxidant and functional properties of gelatine hydrolysates obtained from skin of sole and squid. *Food Chemistry* 114:976–83.

Glauser, B. F., Mourão, P. A. S., and Pomin, V. H. 2013. Marine sulfated glycans with serpin-unrelated anticoagulant properties. *Advances in Clinical Chemistry* 62:269–303.

Gómez-Guillén, M. C., Giménez, B., López-Caballero, M. E., and Montero, M. P. 2011. Functional and bioactive properties of collagen and gelatin from alternative sources: A review. *Food Hydrocolloids* 25:1813–27.

Gómez-Guillén, M. C, Giménez, B., and Montero, P. 2005. Extraction of gelatin from fish skins by high pressure treatment. *Food Hydrocolloids* 19:923–28.

Gómez-Guillén, M. C., Ihl, M., Bifani, V., Silva, A., and Montero, P. 2007. Edible films made from tuna-fish gelatin with antioxidant extracts of two different murta ecotypes leaves (Ugni molinae Turcz). *Food Hydrocolloids* 21:1133–43.

Gómez-Guillén, M. C., and Montero, P. 2001. Extraction of gelatin from megrim (*Lepidorhombus boscii*) skins with several organic acids. *Journal of Food Science* 66:213–16.

Gómez-Guillen, M. C., Turnay, J., Fernández-Martín, M. D., Ulmo, N., Lizarbe, M., and Montero, P. 2002. Structural and physical properties of gelatin extracted from different marine species: A comparative study. *Food Hydrocolloids* 16:25–34.

Góngora, H. G., Ledesma, P., Valvo, V. R. L., Ruiz, A. E., and Breccia, J. D. 2012. Screening of lactic acid bacteria for fermentation of minced wastes of Argentinean hake (*Merluccius hubbsi*). *Food and Bioproducts Processing* 90:767–72.

Goosen, N. J., de Wet, L. F., Görgens, J. F., Jacobs, K., and de Bruyn, A. 2014. Fish silage oil from rainbow trout processing waste as alternative to conventional fish oil in formulated diets for Mozambique tilapia. *Oreochromis Mossambicus. Animal Feed Science and Technology* 188:74–84.

Gopakumar, K. 2012. Therapeutic applications of squalene—A review. *Fishery Technology* 49:1–9.

Grey, M., Forster, I., Dominy, W., Ako, H., and Giesen, A. F. 2009. Validation of a feeding stimulant bioassay using fish hydrolysates for the pacific white shrimp *Litopenaeus Vannamei*. *Journal of the World Aquaculture Society* 40:547–55.

Grossman, S., and Bergman, M. 1992. Process for the production of gelatin from fish skins. U.S. Patent No. 5093474.

Gu, R. Z., Li, C. Y., Liu, W. Y., Yi, W. X., and Cai, M. Y. 2011. Angiotensin I-converting enzyme inhibitory activity of lowmolecular-weight peptides from Atlantic salmon (*Salmo salar* L.) skin. *Food Research International* 44:1536–40.

Gudmundsson, M., and Hafsteinsson, H. 1997. Gelatin from cod skins as affected by chemical treatments. *Journal of Food Science* 62:37–39.

Hai-Lun, H., Xiu-La, C., Cai-Yun, S., Yu-Zhong, Z., and Bai-Cheng, Z. 2006. Analysis of novel angiotensin-I-converting enzyme inhibitory peptides from protease-hydrolyzed marine shrimp *Acetes Chinensis*. *Journal of Peptide Science* 12:726–33.

Halim, N. R. A., Yusof, H. M., and Sarbon, N. M. 2016. Functional and bioactive properties of fish protein hydolysates and peptides: A comprehensive review. *Trends in Food Science & Technology* 51:24–33.

Halldorsdottir, S. M., Sveinsdottir, H., Gudmundsdottir, A., Thorkelsson, G., and Kristinsson, H. G. 2014. High quality (FPH) prepared from by-product material with Fucus vesiculosus extract. *Journal of Functional Foods* 9:10–17.

Hammes, W. P., and Hertel, C. 2009. Genus I, Lactobacillus Beijerinck. In: De Vos, P., Garrity, G., Jones, D., Krieg, N. R., Ludwig, W., Rainey, F. A., Schleifer, K.-H. and Whitman, W. B. (Eds.), *Bergey's Manual of Systematic Bacteriology*, Vol. 3, pp. 465–511. New York: Springer.

Haq, M., Park, S.-K., Kim, M.-J., Cho, Y.-J., Chun, B.-S. 2018. Modifications of Atlantic salmon by-product oil for obtaining different -3 polyunsaturated fatty acids concentrates: An approach to comparative analysis. *Journal of Food and Drug Analysis* 26:545–56.

Harnedy, P. A., and FitzGerald, R. J. 2012. Bioactive peptides from marine processing waste and shellfish: A review. *Journal of Functional Foods* 24:6–24.

Harrabi, H., Leroi, F., Mihoubi, N. B., Chevalier, F., and Kechaou, N. 2017. Biological silages from Tunisian shrimp and octopus by-products. *Journal of Aquatic Food Product Technology* 26:279–95.

Hashiguchi, T., Kobayashi, T., Fongmoon, D., Shetty, A.K., Mizumoto, S., Miyamoto, N., Nakamura, T., Yamada, S., and Sugahara, K. 2011. Demonstration of the hepatocyte growth factor signaling pathway in the in vitro neuritogenic activity of chondroitin sulfate from ray fish cartilage. *Biochimica Et Biophysica Acta - General Subjects* 1810:406–13.

He, G., Yin, Y., Yan, X., and Yu, Q. 2014. Optimisation extraction of chondroitin sulfate from fish bone by high intensity pulsed electric fields. *Food Chemistry* 164:205–10.

He, S., Franco, C., and Zhang. W. 2013. Functions, applications and production of protein hydrolysates from fish processing co-products (FPCP). *Food Research International* 50:289–97.

Hermannsdottir, R., Johannsdottir, J., Smaradottir, H., Sigurgisladottir, S., Gudmundsdottir, B. K., and Bjornsdottir, R. 2009. Analysis of effects induced by a pollock protein hydrolysate on early development, innate immunity *and the bacterial community structure of first feeding of Atlantic halibut (Hippoglossus* hippoglossus L.) larvae. *Fish and Shellfish Immunology* 27:595–602.

Herpandi, H. N., and Adzitey, F. 2011. Fish bone and scale as a potential source of halal gelatin. *Journal of Fisheries and Aquatic Science* 6:379–89.

Hevrøy, E.M., Espe, M., Waagbø, R., Sandnes, K., Ruud, M., and Hemre, G. I. 2005. Nutrient utilization in Atlantic salmon (*Salmo salar* L.) fed increased levels of fish protein hydrolysate during a period of fast growth. *Aquaculture Nutrition* 11:301–13.

Himaya, S. W. A., Ngo, D. H., Ryu, B., and Kim, S. K. 2012. An active peptide purified from gastrointestinal enzyme hydrolysate of Pacific cod skin gelatin attenuates angiotensin-1 converting enzyme (ACE) activity and cellular oxidative stress. *Food Chemistry* 132:1872–82.

H-Kittikun, A., Bourneow, C., and Benjakul, S. 2012. Hydrolysis of surimi wastewater for production of transglutaminase by *Enterobacter* sp. C2361 and *Providencia* sp. C1112. *Food Chemistry* 135:1183–91.

Ho, T. C. W., Li-Chan, E. C. Y., Skura, B. J., Higgs, D. A., and Dosanjh, B. 2014. Pacific hake (*Merluccius productus* Ayres, 1855) hydrolysates as feed attractants for juvenile Chinook salmon (*Oncorhynchus tshawytscha* Walbaum, 1792). *Aquaculture Research* 45:1140–52.

Horn, S. J., Aspmo, S. I., and Eijsink, V. G. H. 2007. Evaluation of different cod viscera fractions and their seasonal variation used in a growth medium for lactic acid bacteria. *Enzyme and Microbial Technology* 40:1328–34.

Hsu, K., Li-Chan, E., and Jao, C. 2011. Antiploriferative activity of peptides prepared from enzymatic hydrolysates of tuna dark muscle on human breast cancer cell lines MCF-7. *Food Chemistry* 126:617–22.

Huang, D., Wang, W., Kang, Y., and Wang, A. 2012. Chitin and chitosan as multipurpose natural polymers for groundwater arsenic removal and As_2O_3 delivery in tumor therapy. *Journal of Macromolecular Science Pure and Applied Chemistry* 49:971–79.

Hung, C. C., Yang, Y. H., Kuo, P. F., and Hsu, K. C. 2014. Protein hydrolysates from tuna cooking juice inhibit cell growth and induce apoptosis of human breast cancer cell line MCF-7. *Journal of Functional Foods* 11:563–70.

IGE. 2018. www.ige.eu/web/mostrar_ficha_actividad.jsp?idioma=es&paxina=004005001&origen=org&operacion=65 (consulted in May of 2018).

Intarasirisawat, R., Benjakul, S., Vissessanguan, W., and Wu, J. 2014. Effects of skipjack roe protein hydrolysate on properties and oxidative stability of fish emulsion sausage. *LWT-Food Science and Technology* 58:280–86.

Ishak, N. H., and Sarbon, N. M. 2018. A review of protein hydrolysates and bioactive peptides deriving from wastes generated by fish processing. *Food Bioprocess Technology* 11:2–16.

Jamilah, B., and Harvinder, K. G. 2002. Properties of gelatins from skins of fish-black tilapia (*Oreochromis mossambicus*) and red tilapia (*Oreochromis nilotica*). *Food Chemistry* 77:81–84.

Jayasinghe, P., and Hawboldt, K. 2012. A review of bio-oils from waste biomass: Focus on fish processing waste. *Renewable and Sustainable Energy Reviews* 16:798–821.

Je, J. Y., Park, P. J., and Kim, S. K. 2005. Antioxidant activity of a peptide isolated from Alaska pollack (*Theragra chalcogramma*) frame protein hydrolysates. *Food Research International* 38:45–50.

Je, J. Y., Qian, Z. J., Byun, H. G., and Kim, S. K. 2007. Purification and characterization of an antioxidant peptide obtained from tuna backbone protein by enzymatic hydrolysis. *Process Biochemistry* 42:840–46.

Jemil, I., Jridi, M., Nasri, R., Ktari, N., Salem, R. B. S., Mehiri, M., Hjji, M., and Nasri, M. 2014. Functional, antioxidant and antibacterial properties of protein hydrolysates prepared from fish meat fermented by *Bacillus subtilis* A26. *Process Biochemistry* 49:963–72.

Jeon, Y. J., Byun, H. G., and Kim, S. K. 2000. Improvement of functional properties of cod frame protein hydrolysates using ultrafiltration membranes. *Process Biochemistry* 35:471–78.

Jeong, H. S., Venkatesan, J., and Kim, S. K. 2013. Isolation and characterization of collagen from marine fish (*Thunnus obesus*). *Biotechnology Bioprocress Engineering* 18:1185–91.

Jeraj, N., Kunic, B., Lenasi, H., and Breskvar, K. 2006. Purification and molecular characterization of chitin deacetylase from *Rhizopus Nigricans*. *Enzyme Microbial Technology* 39:1294–99.

Jesús-de la Cruz, K., Álvarez-González, C. A., Emyr Peña, E., Morales-Contreras, J. A., and Ávila-Fernández, A. 2018. Fish trypsins: Potential applications in biomedicine and prospects for production. *3 Biotech* 8:186.

Jonglareonrak, A., Benjakul, S., Visessanguan, W., Nagai, T., and Tanaka, M. 2005. Isolation and characterisation of acid and pepsin-solubilised collagens from the skin of brownstripe red snapper (*Lutjanus vitta*). *Food Chemistry* 93:475–84.

Jonglareonrak, A., Benjakul, S., Visessanguan, W., and Tanaka, M. 2006. Skin gelatin from bigeye snapper and brownstripe red snapper: Chemical compositions and effect of microbial transglutaminase on gel properties. *Food Hydrocolloids* 20:1216–22.

Joo, H. S., Park, G. C., Kim, K. M., Paik, S. R., and Chang, C. S. 2001. Novel alkaline protease from the polychaeta, *Periserrula leucophyrna*: Purification and characterization. *Process Biochemistry* 36:893–900.

Jung, W. K., Mendis, E., Je, J. Y., Park, P. J., Son, B. W., Kim, H. C., Choi, Y. K., and Kim, S.-K. 2006. Angiotensin Iconverting enzyme inhibitory peptide from yellow fin sole (*Limanda aspera*) frame protein and its antihypertensive effect in spontaneously hypertensive rats. *Food Chemistry* 94:26–32.

Kaczmarek, B., Sionkowska, A., Łukowicz, K., and Osyczka, M. A. 2017. The cells viability study on the composites of chitosan and collagen with glycosaminoglycans isolated from fish skin. *Materials Letters* 206:166–68.

Karaki, N., Aljawish, A., Humeau, C., Muniglia, L., and Jasniewski, J. 2016. Enzymatic modification of polysaccharides: Mechanisms, properties, and potential applications: A review. *Enzyme and Microbial Technology* 90:1–18.

Karayannakidis, P. D., and Zotos, A. 2016. Fish processing by-products as a potential source of gelatin: A review. *Journal of Aquatic Food Product Technology* 25:65–92.

Karnjanapratum, S., O'Callaghan, Y. C., Benjakul, S., O'Brien, N. 2016. Antioxidant, immunomodulatory and antiproliferative effects of gelatine hydrolysate from unicorn leatherjacket skin. *Journal of the Science of Food and Agriculture* 96:3220–26.

Kawahara, H., and Tanihata, T. 2005. U.S. Patent No. 10,683,586. Washington, DC: U.S. Patent and Trademark Office.

Kechaou, E. S., Dumay, J., Donnay-Moreno, C., Jaouen, P., Gouygou J. P., Bergé, J. P., and Amar, R. B. 2009. Enzymatic hydrolysis of cuttlefish (*Sepia officinalis*) and sardine (*Sardina pilchardus*) viscera using commercial proteases: Effects on lipid distribution and amino acid composition. *Journal of Bioscience and Bioengineering* 107:158–64.

Kerton, F. M., Liu, Y., Omaria, K. W., and Hawboldt, K. 2013. Green chemistry and the ocean-based biorefinery. *Green Chemistry* 15:860.

Ketnawa, S., Benjakul, S., Ling, T.C., Martínez-Álvarez, O., and Rawdkuen, S. 2013. Enhanced recovery of alkaline protease from fish viscera by phase partitioning and its application. *Chemistry Central Journal* 7:79–88.

Ketnawa, S., Martínez-Álvarez, O., Benjakul, S., and Rawdkuen, S. 2016. Gelatin hydrolysates from farmed Giant catfish skin using alkaline proteases and its antioxidative function of simulated gastro-intestinal digestion. *Food Chemistry* 192:34–42.

Khaled, H. B., Bougatef, A., Rafik Balti, A., Triki-Ellouz, Y., Souissi, N., and Nasri, M. 2008. Isolation and characterisation of trypsin from sardinelle (*Sardinella aurita*) viscera. *Journal of the Science of Food and Agriculture* 88:2654–62.

Khan, S. B., Qian, Z. J., Ryu, B., Kim, S. K. 2009. Isolation and biochemical characterization of collagens from seaweed pipefish *Syngnathus Schlegeli*. *Biotechnology and Bioprocess Engineering* 14:436–42.

Khandagale, A. S., Mundodi, L., and Sarojini, B. K. 2017. Isolation and characterization of trypsin from fish viscera of oil sardine (*Sardinella longiceps*). *International Journal of Fisheries and Aquatic Studies* 5:33–37.

Khangembam, B. K., and Chakrabarti, R. 2015. Trypsin from the digestive system of carp *Cirrhinus mrigala*: Purification, characterization and its potential application. *Food Chemistry* 175:386–94.

Khantaphant, S., and Benjakul, S. 2008. Comparative study on the proteases from fish pyloric caeca and the use for production of gelatin hydrolysate with antioxidative activity. *Comparative Biochemistry and Physiology Part B* 151:410–19.

Khiari, Z., Rico, D., Martin-Diana, A. B., and Barry-Ryan, C. 2013. Comparison between gelatines extracted from mackerel and blue whiting bones after different pretreatments. *Food Chemistry* 139:347–54.

Kim, H. J., Park, K. H., Shin, J. H., Lee, J. S., Heu, M. S., Lee, D. H., and Kim, J. S. 2011. Antioxidant ACE inhibiting activities of the rockfish *Sebastes hubbsi* skin gelatin hydrolysates produced by sequential two-step enzymatic hydrolysis. *Fisheries and Aquatic Science* 14:1–10.

Kim, S., Je, J., and Kim, S. 2007. Purification and characterization of antioxidant peptide from hoki (*Johnius belengerii*) frame protein by gastrointestinal digestion. *Journal of Nutritional Biochemistry* 18:31–38.

Kim, S. K. 2014. Marine cosmeceuticals. *Journal of Cosmetic Dermatology* 13:56–67.
Kittiphattanabawon, P., Benjakul, S., Visessanguan, W., Kishimura, H., and Shahidi, F. 2010a. Isolation and characterisation of collagen from the skin of brownbanded bamboo shark (*Chiloscyllium punctatum*). *Food Chemistry* 119:1519–26.
Kittiphattanabawon, P., Benjakul, S., Visessanguan, W., Nagai, T., and Tanaka, M. 2005. Characterisation of acid-soluble collagen from skin and bone of bigeye snapper (*Priacanthus tayenus*). *Food Chemistry* 89:363–72.
Kittiphattanabawon, P., Benjakul, S., Visessanguan, W., and Shahidi, F. 2010b. Isolation and characterization of collagen from the cartilages of brownbanded bamboo shark (*Chiloscyllium punctatum*) and blacktip shark (*Carcharhinus limbatus*). *LWT-Food Science and Technology* 43:792–800.
Kizilkaya, B., Tekinay, A. A., and Dilgin, Y. 2010. Adsorption and removal of Cu (II) ions from aqueous solution using pretreated fish bones. *Desalination* 264:37–47.
Klomklao, S., Benjakul, S., Visessanguan, W., Kishimura, H., and Simpson B. K. 2007. Purification and characterisation of trypsins from the spleen of skipjack tuna (*Katsuwonus pelamis*). *Food Chemistry* 100:1580–89.
Klomklao, S., Kishimura, H., and Benjakul, S. 2008. Endogenous proteinases in true sardine (*Sardinops melanostictus*). *Food Chemistry* 107:213–20.
Klompong, V., Benjakul, S., Kantachote, D., and Shahidi, F. 2007. Antioxidative activity and functional properties of protein hydrolysate of yellow stripe travally (*Selaroides leptolepis*) as influenced by the degree of hydrolysis and enzyme type. *Food Chemistry* 102:1317–27.
Kokou, F., and Fountoulaki, E. 2018. Aquaculture waste production associated with antinutrient presence in common fish feed plant ingredients. *Aquaculture* 495:295–310.
Kolodziejska, I., Skierka, E., Sadowska, M., Kolodziejski, W., and Niecikowska, C. 2008. Effect of extracting time and temperature on yield of gelatine from different fish offal. *Food Chemistry* 107:700–06.
Kousoulaki, K., Saether B. S., Albrektsen, S., Noble, C. 2015. Review on European sea bass (*Dicentrarchus labrax*, Linnaeus, 1758) nutrition and feed management: A practical guide for optimizing feed formulation and farming protocols. *Aquaculture Nutrition* 21:129–51.
Kralovec, J. A., Zhang, S., Zhang, W., and Barrow, C. J. 2012. A review of the progress in enzymatic concentration and microencapsulation of omega-3 rich oil from fish and microbial sources. *Food Chemistry* 131:639–44.
Krichen, F., Volpi, N., Sila, A., Maccari, F., Mantovani, V., Galeotti, F., Ellouz-Chaabouni, S., and Bougatef, A. 2017. Purification, structural characterization and antiproliferative properties of chondroitin sulfate/dermatan sulfate from tunisian fish skins. *International Journal of Biological Macromolecules* 95:32–39.
Ktari, N., Ben Khaled, H., Nasri, R., Jellouli, K., Ghorbel, S., and Nasri, M. 2012. Trypsin from zebra blenny (*Salaria basilisca*) viscera: Purification, characterisation and potential application as a detergent additive. *Food Chemistry* 130:467–74.
Kurita, K. 2006. Chitin and chitosan: Functional biopolymers from marine crustaceans. *Marine Biotechnology* 8:203–26.
Lamas, D. L. 2015. Characterization of proteolytic enzymes from *Sparus aurata* L. viscera, obtained by supercritical carbon dioxide, mechanical and n-hexane extraction methods. *Journal of Agriculture and Environmental Sciences* 4:8–17.
Lamas, D. L., Yeannes, M. I., and Massa, A. E. 2017. Alkaline trypsin from the viscera and heads of *Engraulis anchoita*: Partial purification and characterization. *Biotechnologia* 98:103–12.
Lanes, O., Leiros, I., Smalås, A. O., and Willassen, N. P. 2000. Purification and characterization of a cold-adapted uracil-DNA glycosylase from Atlantic cod. *Comparative Biochemistry and Physiology – Part B: Biochemistry & Molecular Biology* 127:399–410.

Lanes, O., Leiros, I., Smalås, A. O., and Willassen, N. P. 2002. Identification, cloning, and expression of uracil-DNA glycosylase from Atlantic cod (*Gadus morhua*): Characterization and homology modeling of the cold-active catalytic domain. *Extremophiles* 6:73–86.

Lassoued, I., Mora, L., Nasri, R., Jridi, M., Toldra, F., Aristoy, M., Barkia, A., and Nasri, M. 2015. Characterization and comparative assessment of antioxidant and ACE inhibitory activities of thornback ray gelatin hydrolysates. *Journal of Functional Foods* 13:225–38.

Lee, J. K., Jeon, J., and Byun, H. 2011. Effect of angiotensin I converting enzyme inhibitory peptide purified from skate skin hydrolysate. *Food Chemistry* 125:495–99.

Lee, S. H., Qian, Z. J., and Kim, S. K. 2010. A novel angiotensin I converting enzyme inhibitory peptide from tuna frame protein hydrolysate and its antihypertensive effect in spontaneously hypertensive rats. *Food Chemistry* 118:96–102.

Lewandowski, V., Decarli, J. A., de Araújo Pedron, F., Feiden, A., Signor, A., and Boscolo, W. R. 2013. Hidrolisados cárneos na alimentação do surubimdo Iguaçu (*Steindachneridion melanodermatum*). *Revista Brasileira De Ciência Veterinária* 20:222–26.

Liang, M., Wang, J., Chang, Q., and Mai, K. 2006. Effects of different levels of fish protein hydrolysate in the diet on the nonspecific immunity of Japanese sea bass, *Lateolabrax japonicus* (Cuvieret Valenciennes, 1828). *Aquaculture Research* 37:102–6.

Li-Chan, E. C. Y., Hunag, S. L., Jao, C. L., Ho, K. P., and Hsu, K. C. 2012. Peptides derived from atlantic salmon skin gelatin as dipeptidyl-peptidase IV inhibitors. *Journal of Agricultural and Food Chemistry* 60:973–78.

Limpisophon, K., Tanaka, M., Weng, W., Abe, S., and Osako, K. 2009. Characterization of gelatin films prepared from under-utilized blue shark (*Prionace glauca*) skin. *Food Hydrocolloids* 23:1993–2000.

Lin, C.-Y., and Li, R.-J. 2009a. Fuel properties of biodiesel produced from the crude fish oil from the soapstock of marine fish. *Fuel Processing Technology* 90:130–36.

Lin, C.-Y., and Li, R.-J. 2009b. Engine performance and emission characteristics of marine fish-oil biodiesel produced from the discarded parts of marine fish. *Fuel Processing Technology* 90:883–88.

Lin, L., Regenstein, J. M., Lv, S., Lu, J., Jiang, S. 2017b. An overview of gelatin derived from aquatic animals: Properties and modification. *Trends in Food Science & Technology* 68:102–12.

Lin, N. E, Mo, X., Yang, Y., and Zhang, H. 2017a. Purification and sequence characterization of chondroitin sulfate and dermatan sulfate from fishes. *Glycoconjugate Journal* 34:241–53.

Linder, M., Fanni, J., and Parmentier, M. 2005. Proteolytic extraction of salmon oil and PUFA concentration by lipases. *Marine Biotechnology* 7:70–76.

Lindgren, S., and Pleje, M. 1983. Silage fermentation of fish or fish waste products with lactic acid bacteria. *Journal of the Science of Food and Agriculture* 34:1057–67.

Liu, D., Nikoo, M., Boran, G., Zhou, P., and Regenstein, J. M. 2015. Collagen and gelatin. *Annual Review Food Science and Technology* 6:527–57.

Liu, H.-M., Ai, C.-Q., Liu, B., Ma, N., Li, Y.-L., and Song, S. 2017. Structure analysis of uronic acid-containing polysaccharides in viscera of four types of fish. *Modern Food Science and Technology* 33:33–38.

Liu, P., Liu, S., Guo, N., Mao, X., Lin, H., Xue, C., and Wei, D. 2014. Cofermentation of *Bacillus licheniformis* and *Gluconobacter oxydans* for chitin extraction from shrimp waste. *Biochemical Engineering Journal* 91:10–15.

Liu, Y., and Chen, C. 2016. Role of nanotechnology in HIV/AIDS vaccine development. *Advanced Drug Delivery Reviews* 103:76–89.

Lodish, H., Berk, A., Zipursky, S. L., Matsudaira, P., Baltimore, D., and Darnell, J. 2000. *Molecular Cell Biology*, 4th edition. Section 22.3. New York: W. H. Freeman; ISBN-10: 0-7167-3136-3.

Logithkumar, R., Keshavnarayan, A., Dhivya, S., Chawla, A., Saravanan, S., and Selvamurugan, N. 2016. A review of chitosan and its derivatives in bone tissue engineering. *Carbohydrate Polymers* 151:172–88.

Lopes, C., Antelo, L. T., Franco-Uría, A., Alonso, A. A., and Pérez-Martín, R. 2018. Chitin production from crustacean biomass: Sustainability assessment of chemical and enzymatic processes. *Journal of Cleaner Production* 172:4140–51.

López-Moreno, M. L., Cedeño-Mattei, Y., Bailón-Ruiz, S. J., Vazquez-Nuñez, E., Hernandez-Viezcas, J. A., Perales-Pérez, O. J., la Rosa, G. D, Peralta-Videa, J. R., and Gardea-Torresdey, J. L. 2018. Environmental behavior of coated NMs: Physicochemical aspects and plant interactions. *Journal of Hazardous Materials* 347:196–217.

Maccari, F., Galeotti, F., and Volpi, N. 2015. Isolation and structural characterization of chondroitin sulfate from bony fishes. *Carbohydrate Polymers* 129:143–47.

Mahmoodani, F., Ghassem, M., Babji, A. S., Yusop, S. M., and Khosrokhavar, R. 2014. ACE inhibitory activity of pangasius catfish (*Pangasius sutchi*) skin and bone gelatin hydrolysate. *Journal of Food Science and Technology* 51:1847–56.

Malde, M. K., Graff, I. E., Siljander-Rasi, H., Venäläinen, E., Julshamn, K., Pedersen, J. I., and Valaja, J. 2010. Fish bones - a highly available calcium source for growing pigs. *Journal of Animal Physiology and Animal Nutrition* 94:e66–e76.

Malerba, M., and Cerana, R. 2018. Recent advances of chitosan applications in plants. *Polymers* 10:118.

Mao, X., Guo, N., Sun, J., and Xue, C. 2017. Comprehensive utilization of shrimp waste based on biotechnological methods: A review. *Journal of Cleaner Production* 143:814–23.

Mapama. 2018. www.mapama.gob.es/es/estadistica/temas/estadisticas-pesqueras/pesca-maritima/estadistica-flota-pesquera/ (consulted in May of 2018).

Marcuschi, M., Espósito, T. S., Machado, M. F. M., Hirata, I. Y., and Machado, M. F. M. 2010. Purification, characterization and substrate specificity of a trypsin from the Amazonian fish tambaqui (*Colossoma macropomum*). *Biochemical and Biophysical Research Communications* 396:667–73.

Martínez-Álvarez, O., Chamorro, S., and Brenes, A. 2015. Protein hydrolysates from animal processing by-products as a source of bioactive molecules with interest in animal feeding: A review. *Food Research International* 73:204–12.

Martinou, A., Bouriotis, V., Stokke, B. T., and Vårum, K. M. 1998. Mode of action of chitin deacetylase from *Mucor rouxii* on partially N-acetylated chitosans. *Carbohydrate Research* 311:71–78.

Mata-Gómez, L. C., Montáñez, J. C., Méndez-Zavala, A. A., and Aguilar, C. N. 2014. Biotechnological production of carotenoids by yeasts: An overview. *Microbial Cell Factories* 13:1–11.

Mendis, E., Rajapakse, N., and Kim, S. K. 2005. Antioxidant properties of a radical-scavenging peptide purified from enzymatically prepared fish skin gelatin hydrolysate. *Journal of Agricultural and Food Chemistry* 53:581–87.

Minke, R., and Blackwell, J. 1978. The structure of β-chitin. *Journal of Molecular Biology* 120:167–81.

Mohapatra, S., Sarkar, B., Samantaray, D. P., Daware, A., Maity, S., Pattnaik, S., and Bhattacharjee, S. 2017. Bioconversion of fish solid waste into PHB using *Bacillus subtilis* based submerged fermentation process. *Environmental Technology* 38:3201–08.

Mohtar, N. F., Perera, C., and Quek, S.-Y. 2010. Optimisation of gelatine extraction from hoki (*Macruronus novaezelandiae*) skins and measurement of gel strength and SDS-PAGE. *Food Chemistry* 132:307–13.

Mondal, S., Mondal, B., Dey, A., and Mukhopadyay, S. S. 2012. Studies on processing and characterization of hydroxyapatite biomaterials from different bio wastes. *Journal of Minerals and Materials Characterization and Engineering* 11:55–67.

Mourão, P. A. S. 2015. Perspective on the use of sulfated polysaccharides from marine organisms as a source of new antithrombotic drugs. *Marine Drugs* 13:2770–84.

Mourya, V. K., and Inamdar, N. N. 2008. Chitosan-modifications and applications: Opportunities galore. *Reactive and Functional Polymers* 68:1013–51.

Murado, M. A., Fraguas, J., Montemayor, M. I., Vázquez, J. A., and González, P. 2010. Preparation of highly purified chondroitin sulphate form skate (*Raja clavata*) cartilage by-products. Process optimization including a new procedure of alkaline hydroalcoholic hydrolysis. *Biochemical Engineering Journal* 49:126–32.

Murado, M. A., González, M. P., and Vázquez, J. A. 2009. Recovery of proteolytic and collagenolytic activities from viscera by-products of rayfish (*Raja clavata*). *Marine Drugs* 7:803–15.

Murado, M. A., Montemayor, M. I., Cabo, M. L., Vázquez, J. A., and González, M. P. 2012. Optimization of extraction and purification process of hyaluronic acid from fish eyeball. *Food and Bioproducts Processing* 90:491–98.

Muralidharan, N., Shakila, R. J., Sukumar, D., and Jeyasekaran, G. 2013. Skin, bone and muscle collagen extraction from the trash fish, leather jacket (*Odonus niger*) and their characterization. *Journal of Food Science and Technology* 50:1106–13.

Muyonga, J. H., Cole, C. G. B., and Duodu, K. G. 2004. Extraction and physico-chemical characterisation of Nile perch (*Lates niloticus*) skin and bone gelatin. *Food Hydrocolloids* 18:581–92.

Nagai, T., Araki, Y., and Suzuki, N. 2002. Collagen of the skin of ocellate puffer fish (*Takifugu rubripes*). *Food Chemistry* 78:173–77.

Nagai, T., and Suzuki, N. 2000. Isolation of collagen from fish waste material - skin, bones and fins. *Food Chemistry* 68:277–81.

Najafian, L., and Babji, A. S. 2012. A review of fish-derived antioxidant and antimicrobial peptides: Their production, assessment, and applications. *Peptides* 33:178–85.

Nalinanon, S., Benjakul, S., Kishimura, H., and Shahidi, F. 2011. Functionalities and antioxidant properties of protein hydrolysates from the muscle of ornate threadfin bream treated with pepsin from skipjack tuna. *Food Chemistry* 124:1354–62.

Nalinanon, S., Benjakul, S., Visessanguan, W., and Kishimura, H. 2008. Improvement of gelatin extraction from bigeye snapper skin using pepsin-aided process in combination with protease inhibitor. *Food Hydrocolloids* 22:615–22.

Napavichayanun, S., and Aramwit, P. 2017. Effect of animal products and extracts on wound healing promotion in topical applications: A review. *Journal of Biomaterials Science*, Polymer Edition 28:703–29.

Narurkar, V. A., Fabi, S. G., Bucay, V. W., Tedaldi, R., Downie, J. B., Zeichner, J. A., Butterwick, K., Taub, A., Kadoya, K., Makino, E. T., Mehta, R. C., and Vega, V. L. 2016. Rejuvenating hydrator: Restoring epidermal hyaluronic acid homeostasis with instant benefits. *Journal of Drugs in Dermatology* 15:s24–s37.

Nasri, R., Sila, A., Ktari, N., Lassoued, I., Bougatef, A., Karra-Chaabouni, M., and Nasri, M. 2012. Calcium dependent, alkaline detergent-stable trypsin from the viscera of Goby (*Zosterisessor ophiocephalus*): Purification and characterization. *Process Biochemistry* 47:1957–64.

Naylor, R. L., Hardy, R. W., Bureau, D. P., Chiu, A., Elliott, M., Farrell, A. P., Forster, I., Gatlin, D. M., Goldburg, R. J., Hua, K., and Nichols, P. D. 2009. Feeding aquaculture in an era of finite resources. *Proceedings of the National Academy of Sciences of the United States of America* 106:15103–10.

Naziri, E., Mantzouridou, F., and Tsimidou M. Z. 2011. Squalene resources and uses point to the potential of biotechnology. *Lipid Technology* 23:270–73.

Nettleton, J. A. 1995. *Omega-3 Fatty Acids and Health*, pp. 64–76. Springer: New York (NY, U.S.). ISBN 978-1-4613-5860-2.

Ngo, D., Qian, Z., Ryu, B., Park, J. W., and Kim, S. 2010. In vitro antioxidant activity of a peptide isolated from Nile Tilapia (*Oreochromis niloticus*) scale gelatin in free radical-mediated oxidative systems. *Journal of Functional Foods* 2:107–17.

Ngo, D. H., Ryu, B., and Kim, S. K. 2014. Active peptides from skate (Okamejei kenojei) skin gelatin diminish angiotensin-I converting enzyme activity and intracellular free radical-mediated oxidation. *Food Chemistry* 143:246–55.

Nguyen, H. T. M., Perez-Galvez, R., and Berge, J. P. 2012. Effect of diets containing tuna head hydrolysates on the survival and growth of shrimp *Penaeus Vannamei*. *Aquaculture* 324:127–34.

Nikoo, M., Xu, X., Benjakul, S., Xu, G., Ramírez-Suárez, J. C., Ehsani, A., Kasankala, L. M., Duan, X., and Abbas, S. 2011. Characterization of gelatin from the skin of farmed Amur sturgeon *Acipenser schrenckii*. *International Aquatic Research* 3:135–45.

Novoa-Carballal, R., Pérez-Martín, R., Blanco, M., Sotelo, C. G., Fassini, D., Nunes, C., Coimbra, M. A., Silva, T. H., Reis, R. L., and Vázquez, J. A. 2017. By-products of *Scyliorhinus canicula, Prionace glauca* and *Raja clavata*: A valuable source of predominantly 6S sulphated chondroitin sulphate. *Carbohydrate Polymers* 157:31–37.

Okamoto, Y., Higashi, K., Linhardt, R. J., and Toida, T. 2018. Comprehensive analysis of glycosaminoglycans from the edible shellfish. *Carbohydrate Polymers* 184:269–76.

Oliva, J., De Pablo, J., Cortina, J.-L., Cama, J., and Ayora, C. 2010. The use of Apatite II™ to remove divalent metal ions zinc(II), lead(II), manganese(II) and iron(II) from water in passive treatment systems: Column experiments. *Journal of Hazardous Materials* 184:364–74.

Olsen, R. L., Johansen, A., and Myrnes, B. 1990. Recovery of enzymes from shrimp waste. *Process Biochemistry* 25:67–68.

Olsen, R. L., and Toppe, J. 2017. Fish silage hydrolysates: Not only a feed nutrient, but also a useful feed additive. *Trends in Food Science & Technology* 66:93–97.

Olsen, R. L., Toppe, J., and Karunasagar, I. 2014. Challenges and realistic opportunities in the use of by-products from processing of fish and shellfish. *Trends in Food Science & Technology* 36:144–51.

Ordoñez-Del Pazo, T., Antelo, L. T., Franco-Uría, A., Perez-Martin, R. I., Sotelo, C. G., and Alonso, A. A. 2014. Fish discards management in selected Spanish and Portuguese metiers: Identification and potential valorisation. *Trends in Food Science & Technology* 36:29–43.

Ovissipour, M., Abedian, A., Motamedzadegan, A., Rasco, B., Safari, R., and Shahiri, H. 2009. The effect of enzymatic hydrolysis time and temperature on the properties of protein hydrolysates from Persian sturgeon (*Acipenser persicus*) viscera. *Food Chemistry* 115:238–42.

Ovissipour, M., Abedian Kenari, A., Nazari, R., Motamedzadegan, A., and Rasco, B. 2014. Tuna viscera protein hydrolysate: Nutritive and disease resistance properties for Persian sturgeon (*Acipenser persicus* L.) larvae. *Aquaculture Research* 45:591–601.

Özyurt, G., Serhat Özkütük, A., Boğa, M., Durmuş, M., and Boğa, E. K. 2017. Biotransformation of seafood processing wastes fermented with natural lactic acid bacteria; the quality of fermented products and their use in animal feeding. *Turkish Journal of Fisheries and Aquatic Sciences* 17:543–55.

Pagarkar, A. U., Basu, S., Mitra, A., and Sahu, N. P. 2006. Preparation of bio-fermented and acid silage from fish waste and its biochemical characteristic. *Asian Journal of Microbiology, Biotechnology and Environmental Sciences* 8:381–87.

Pan, X., Zhao, Y. Q., Hu, F. Y., and Wang, B. 2016. Preparation and identification of antioxidant peptides from protein hydrolysate of skate (*Raja porosa*) cartilage. *Journal of Functional Foods* 25:220–30.

Pan, Y., Wang, P., Zhang, F., Yu, Y., Zhang, X., Lin, L., and Linhardt, R. J. 2018. Glycosaminoglycans from fish swim bladder: Isolation, structural characterization and bioactive potential. *Glycoconjugate Journal* 35:87–94.

Panagos, C. G., Thomson, D., Moss, C., Bavington, C. D., Ólafsson, H. G., and Uhrín, D. 2014. Characterisation of hyaluronic acid and chondroitin/dermatan sulfate from the lumpsucker fish *C. Lumpus*. *Carbohydrate Polymers* 106:25–33.

Panda, N. N., Pramanik, K., and Sukla, L. B. 2014. Extraction and characterization of biocompatible hydroxyapatite from fresh water fish scales for tissue engineering scaffold. *Bioprocess and Biosystems Engineering* 37:433–40.

Papakonstantinou, E., Roth, M., and Karakiulakis, G. 2012. Hyaluronic acid: A key molecule in skin aging. *Dermato-Endocrinology* 4:253–58.

Parrish, C. C. 1999. Determination of total lipid, lipid classes, and fatty acids in aquatic samples. In: Arts, M. T. and Wainman, B. C. (Eds.), *Lipids in Freshwater Ecosystems*, pp. 4–20. Springer: New York (NY, U.S.). ISBN 978-1-4612-0547-0.

Pavão, M. S. G. 2014. Glycosaminoglycans analogs from marine invertebrates: Structure, biological effects, and potential as new therapeutics. *Frontiers in Cellular and Infection Microbiology* 4:123.

Percot, A., Viton, C., and Domard, A. 2003. Optimization of chitin extraction from shrimp shells. *Biomacromolecules* 4:12–18.

Pereira, N. D. L. A., and Fernández-Gimenez, A. V. 2017. Exogenous enzymes in dairy technology: Acidic proteases from processing discards of shrimp *Pleoticus muelleri* and their use as milk-clotting enzymes for cheese manufacture. *International Journal of Food Science and Technology* 52:341–47.

Phanturat, P., Benjakul, S., Visessanguan, W., and Roytrakul, S. 2010. Use of pyloric caeca extract from bigeye snapper (*Priacanthus macracanthus*) for the production of gelatine hydrolysate with antioxidative activity. *LWT-Food Science and Technology* 43:86–97.

Piai J. F., Alves Da Silva, M. L., Martins, A., Torres, A., Faria, S., Reis, R. L., Muniz, E. C., and Neves, N. M. 2017. Chondroitin sulfate immobilization at the surface of electrospun nanofiber meshes for cartilage tissue regeneration approaches. *Applied Surface Sciences* 403:112–25.

Piccirillo, C., Moreira, I. S., Novais, R. M., Fernandes, A. J. S., Pullar, R. C., and Castro, P. M. L. 2017. Biphasic apatite-carbon materials derived from pyrolysed fish bones for effective adsorption of persistent pollutants and heavy metals. *Journal of Environmental Chemical Engineering* 5:4884–94.

Piccirillo, C., Pereira, S. I. A., Marques, A. P. G. C., Pullar, R. C., Tobaldi, D. M., Pintado, M. E., and Castro, P. M. L. 2013b. Bacteria immobilisation on hydroxyapatite surface for heavy metals removal. *Journal of Environmental Management* 121:87–95.

Piccirillo, C., Pinto, R. A., Tobaldi, D. M., Pullar, R. C., Labrincha, J. A., Pintado, M. M. E., Castro, P. M. L. 2015. Light induced antibacterial activity and photocatalytic properties of Ag_3PO_4-based material of marine origin. *Journal of Photochemistry and Photobiology A: Chemistry* 296:40–47.

Piccirillo, C., Pullar, R. C., Tobaldi, D. M., Castro, P. M. L., and Pintado, M. M. E. 2014a. Hydroxyapatite and chloroapatite derived from sardine by-products. *Ceramics International* 40:13231–40.

Piccirillo, C., Rocha, C., Tobaldi, D. M., Pullar, R. C., Labrincha, J. A., Ferreira, M. O., Castro, P. M. L., and Pintado, M. M. E. 2014b. A hydroxyapatite-Fe_2O_3 based material of natural origin as an active sunscreen filter. *Journal of Materials Chemistry B* 2:5999–6009.

Piccirillo, C., Silva, M. F., Pullar, R. C., Braga Da Cruz, I., Jorge, R., Pintado, M. M. E., and Castro, P. M. L. 2013a. Extraction and characterisation of apatite- and tricalcium phosphate-based materials from cod fish bones. *Materials Science and Engineering C* 33:103–10.

Pietsch, A., and Jaeger, P. 2007. Concentration of squalene from shark liver oil by short-path distillation. *European Journal of Lipid Science and Technology* 109:1077–82.

Pike, I. H., and Jackson, A. 2010. Fish oil: Production and use now and in the future. *Lipid Technology* 22:59–61.

Pomin, V. H. 2014. Anticoagulant motifs of marine sulfated glycans. *Glycoconjugate Journal* 31:341–44.

Pomin, V. H., and Mourão, P. A. S. 2014. Specific sulfation and glycosylation-a structural combination for the anticoagulation of marine carbohydrates. *Frontiers in Cellular and Infection Microbiology* 5:33.

Poonsin, T., Sripokar, P., Benjakul, S., Simpson, B. K., Visessanguan, W., and Klomklao, S. 2017. Major trypsin like-serine proteinases from albacore tuna (*Thunnus alalunga*) spleen: Biochemical characterization and the effect of extraction media. *Journal of Food Biochemistry* 41:1–9.

Popa, O., Băbeanu, N. E., Popa, I., Niţă, S., and Dinu-Pârvu, C. E. 2015. Methods for obtaining and determination of squalene from natural sources. *Biomedical Research International*, 2015:367202.

Port of Vigo. 2018. www.apvigo.es/es/paginas/estadisticas_puerto_pesquero (consulted in May of 2018).

Prabha, J., Narikimelli, A., Sajini, M. I., and Vincent, S. 2013. Optimization for autolysis assisted production of fish protein hydrolysate from underutilized fish *Pellona Ditchela*. *International Journal of Scientific and Engineering Research* 4:1863–69.

Raa, J., and Gildberg, A. 1982. Fish silage: A review. *CRC Critical Reviews in Food Science and Nutrition* 4:383–419.

Raghunath, M. R., and Gopakumar, K. 2002. Trends in production and utilization of fish silage. *Journal of Food Science and Technology* 39:103–10.

Rajendran, C. K. S. R., Mohan, A., Khiari, Z., Udenigwe, C. C., and Mason, B. 2018. Yield, physicochemical, and antioxidant properties of Atlantic salmon visceral hydrolysate: Comparison of lactic acid bacterial fermentation with Flavourzyme proteolysis and formic acid treatment. *Journal of Food Processing and Preservation* 42:e13620.

Ramírez, J. C., Espain, J. I. I., Leyva, R. G., Ulloa, J. A., and Ulloa, P. R. 2016. Use of biological fish silage in broilers feed: Effect on growth performance and meat quality. *Journal of Animal & Plant Sciences* 27:4293–304.

Ramkumar, A., Sivakumar, N., and Victor, R. 2016. Fish waste-potential low cost substrate for bacterial protease production: A brief review. *Open Biotechnology Journal* 10:335–41.

Ramrakhiani, L., Ghosh, S., and Majumdar, S. 2016. Surface modification of naturally available biomass for enhancement of heavy metal removal efficiency, upscaling prospects, and management aspects of spent biosorbents: A review. *Applied Biochemistry and Biotechnology* 180:41–78.

Rao, B. M., Babu, U. S. J., Khasim, D. I. I. 2009. Microbiological quality of fermented fish silage. *Journal of Food Science and Technology* 46:488–90.

Rao, M. S., and Stevens, W. F. 2005. Chitin production by *Lactobacillus* fermentation of shrimp biowaste in a drum reactor and its chemical conversion to chitosan. *Journal of Chemical Technology and Biotechnology* 80:1080–87.

Ratnasari, I., Yuwono, S. S., Nusyam, H., and Widjanarko, S. B. 2013. Extraction and characterization of gelatin from different fresh water fishes as alternative sources of gelatin. *International Food Research Journal* 20:3085–91.

Razali, A. N., Sarbon, N. M., and Amin, A. M. 2015. Antioxidant activity and functional properties of fractionated cobia skin gelatin hydrolysate at different molecular weight. *International Food Research Journal* 22:651–60.

Rebah, F. B., Frikha, F., Kamoun, W., Belbahri, L., Gargouri, Y., and Miled, N. 2008. Culture of *Staphylococcus xylosus* in fish processing by-product-based media for lipase production. *Letters in Applied Microbiology* 47:549–54.

Remme, J. F., Larsen W. E., and Stoknes I. S. 2005. Bioactive lipids in deep-sea sharks. *Report A0510 Project: Search for Bioactive Lipids in Internal Organs From Deep-Sea Sharks*. More Research, Ålesund, Norway.

Ren, J., Zhao, M., Shi, J., Wang, J., Jiang, Y., Cui, C., Kakuda, Y., and Xue, S. J. 2008. Purification and identification of antioxidant peptides from grass carp muscle hydrolysates by consecutive chromatography and electrospray ionization-mass spectrometry. *Food Chemistry* 108:727–36.

Rinaudo, M. 2006. Chitin and chitosan: Properties and applications. *Progress Polymer Science* 31:603–32.

Robert, M., Zatylny-Gaudin, C., Fournier, V., Corre, E., Le Corguillé, G., Bernay, B., and Henry, J. 2015. Molecular characterization of peptide fractions of a Tilapia (*Oreochromis niloticus*) by-product hydrolysate and in vitro evaluation of antibacterial activity. *Process Biochemistry* 50:487–92.

Roblet, C., Akhtar, M. J., Mikhaylin, S., Pilon, G., Gill, T., and Marette, A., Bazinet, L. 2016. Enhancement of glucose uptake in muscular cell by peptide fractions separated by electrodialysis with filtration membrane from salmon frame protein hydrolysate. *Journal of Functional Foods* 22:337–46.

Rubio-Rodríguez, N., Beltrán, S., Jaime, I., de Diego, S. M., Sanz, M. T., and Carballido, J. R. 2010. Production of omega-3 polyunsaturated fatty acid concentrates: A review. *Innovative Food Science and Emerging Technologies* 11:1–12.

Rubio-Rodríguez, N., de Diego, S. M., Beltrán, S., Jaime, I., Sanz, M. T., and Rovira, J. 2012. Supercritical fluid extraction of fish oil from fish by-products: A comparison with other extraction methods. *Journal of Food Engineering* 109:238–48.

Sadhasivam, G., Muthuvel, A., Pachaiyappan, A., and Thangavel, B. 2013. Isolation and characterization of hyaluronic acid from the liver of marine stingray *Aetobatus narinari*. *International Journal of Biological Macromolecules* 54:84–89.

Sae-Leaw, T., O'Callaghan, Y. C., Benjakul, S., and O'Brien, N. M. 2016. Antioxidant, immunomodulatory and antiproliferative effects of gelatine hydrolysates from seabass (*Lates calcarifer*) skins. *International Journal of Food Science and Technology* 51:1545–51.

Saeli, M., Piccirillo, C., Tobaldi, D. M., Binions, R., Castro, P. M. L., and Pullar, R. C. A. 2018. Sustainable replacement for TiO_2 in photocatalyst construction materials: Hydroxyapatite-based photocatalytic additives, made from the valorisation of food wastes of marine origin. *Journal of Cleaner Production* 193:115–27.

Safari, R., Motamedzadegan, A., Ovissipour, M., Regenstein, J. M., Gildberg, A., and Rasco, B. 2012. Use of hydrolysates from yellowfin tuna (*Thunnus albacares*) heads as a complex nitrogen source for lactic acid bacteria. *Food and Bioprocess Technology* 5:73–79.

Safari, R., Nasrollahzadeh Saravi, H., Pourgholam, R., Motalebi, A. A., and Ghoroghi, A. 2011. Use of hydrolysates from silver carp (*Hypophthalmichthys molitrix*) head as peptone for vibrio anguillarum and optimization using response surface method (RSM). *Journal of Aquatic Food Product Technology* 20:247–57.

Saha, A., Yadav, R., and Rajendran, N. 2014. Biomaterials from sponges, ascidians and other marine organisms. *International Journal of Pharmaceutical Sciences Review and Research* 27:100–9.

Sahena, F., Zaidul, I. S. M., Jinap, S., Jahurul, M. H. A., Khatib, A., and Norulaini, N. A. N. 2010. Extraction of fish oil from the skin of Indian mackerel using supercritical fluids. *Journal of Food Engineering* 99:63–69.

Sahena, F., Zaidul, I. S. M., Jinap, S., Saari, N., Jahurul, H. A., Abbas, K. A., and Norulaini, N. A. 2009. PUFAs in fish: Extraction, fractionation, importance in health. *Comprehensive Reviews in Food Science and Food Safety* 8:59–74.

Saidi, S., Deratani, A., Belleville, M.-P., Ben Amar, R. 2014. Antioxidant properties of peptide fractions from tuna dark muscle protein byproduct hydrolysate produced by membrane fractionation process. *Food Research International* 65:329–36.

Sai-Ut, S., Jongjareonrak, A., and Rawdkuen, S. 2012. Re-extraction, recovery and characteristics of skin gelatin from farmed giant catfish. *Food and Bioprocess Technology* 5:1197–205.

Salehi, E., Daraei, P., and Arabi Shamsabadi, A. 2016. A review on chitosan-based adsorptive membranes. *Carbohydrate Polymers* 152:419–32.

Salwowska, N. M., Bebenek, K. A., Żądło, D. A., Wcisło-Dziadecka, D. L. 2016. Physiochemical properties and application of hyaluronic acid: A systematic review. *Journal of Cosmetic Dermatology* 15:520–26.

Samaranayaka, A. G. P., and Li-Chan, E. C. Y. 2008. Autolysis-assisted production of fish protein hydrolysates with antioxidant properties from Pacific hake (*Merluccius productus*). *Food Chemistry* 107:768–76.

Sampath Kumar, N. S., Nazeer, R. A., and Jaiganesh, R. 2011. Purification and identification of antioxidant peptides from the skin protein hydrolysate of two marine fishes, horse mackerel (*Magalaspis cordyla*) and croaker (*Otolithes ruber*). *Amino Acids* 42:1641–49.

Sánchez, A., Blanco, M., Correa, B., Perez-Martin, R. I., Sotelo, C. G. 2018. Effect of fish collagen hydrolysates on type i collagen mRNA levels of human dermal fibroblast culture. *Marine Drugs* 16:144.

Sánchez, R., Stringari, G. B., Franco, J. M., Valencia, C., and Gallegos, C. 2011. Use of chitin, chitosan and acylated derivatives as thickener agents of vegetable oils for bio-lubricant applications. *Carbohydrate Polymers* 85:705–14.

Sayari, N., Sila, A., Haddar, A., Balti, R., Ellouz-Chaabouni, S., and Bougatef, A. 2016. Valorisation of smooth hound (*Mustelus mustelus*) waste biomass through recovery of functional, antioxidative and antihypertensive bioactive peptides. *Environmental Science and Pollution Research* 23:366–76.

Sears, A. V., Ho, C. K. M., and Walker, S. W. 2010. Omacor (omega-3-acid ethyl esters). *Practical Diabetes International* 27:38–39.

Sellami, M., Kedachi, S., Frikha, F., Miled, N., and Ben Rebah, F. 2013. Optimization of marine waste based-growth media for microbial lipase production using mixture design methodology. *Environmental Technology* 34:2259–66.

Sellami-Kamoun, A., Ghorbel-Frikha, B., Haddar, A., and Nasri, M. 2011. Enhanced Bacillus cereus BG1 protease production by the use of sardinelle (*Sardinella aurita*) powder. *Annals of Microbiology* 61:273–80.

Senaratne, L. S., Park, P. J., and Kim, S. K. 2006. Isolation and characterization of collagen from brown backed toadfish (*Lagocephalus gloveri*) skin. *Bioresource Technology* 97:191–97.

Senphan, T., and Benjakul, S. 2014. Antioxidative activities of hydrolysates from seabass skin prepared using protease from hepatopancreas of Pacific white shrimp. *Journal of Functional Foods* 6:147–56.

Shabani, A., Boldaji, F., Dastar, B., Ghoorchi, T., and Zerehdaran, S. 2018. Preparation of fish waste silage and its effect on the growth performance and meat quality of broiler chickens. *Journal of the Science of Food and Agriculture* 98:4097–103.

Shahidi, F. 2008. Omega-3 oils: Sources, applications, and health effects. In: Barrow, C. and Shahidi, F. (Eds.), *Marine Nutraceuticals and Functional Foods*, pp. 23–61. CRC Press: Boca Raton (Florida, U.S.). ISBN 1-57444487-5.

Shahidi, F., Kamil, Y. V. A. J. 2002. Enzymes from fish and aquatic invertebrates and their application in the food industry. *Food Science and Technology* 12:434–71.

Shahidi, F., and Miraliakbari, H. 2006. Marine oils: Compositional characteristics and health effects. In: Shahidi, F. (Ed.), *Nutraceutical and Specialty Lipids and their Co-Products*, pp. 227–47.CRC Taylor & Francis: Boca Raton (Florida, U.S.). ISBN 1-57444499-9.

Shakila, R. J., Jeevithan, E., Varatharajakumar, A., Jeyasekaran, G., and Sukumar, D. 2012. Functional characterization of gelatin extracted from bones of red snapper and grouper in comparison with mammalian gelatin. *Lebensmittel Wissenschaft Und Technologie* 48:30–36.

Shanmugam, V., Ramasamy, P., Subhapradha, N., Sudharsan, S., Seedevi, P., Moovendhan, M., Krishnamoorthy, J., Shanmugam, A., and Srinivasan, A. 2012. Extraction, structural and physical characterization of type I collagen from the outer skin of *Sepiella inermis* (orbigny, 1848). *African Journal of Biotechnology* 11:14326–37.

Shi, S., Li, J., and Blersch, D. M. 2018. Utilization of solid catfish manure waste as carbon and nutrient source for lactic acid production. *Applied Microbiology and Biotechnology* 102:4765–72.

Shuai, W., Das, R. K., Naghdi, M., Brar, S. K., and Verma, M. 2017. A review on the important aspects of lipase immobilization on nanomaterials. *Biotechnology and Applied Biochemistry* 64:496–508.

Shyni, K., Hema, G. S., Ninan, G., and Mathew, S. 2014. Isolation and characterization of gelatin from the skins of skipjack tuna (*Katsuwonus pelamis)*, dog shark (*Scoliodon sorrakowah)*, and rohu (*Labeo rohita*). *Food Hydrocolloids* 39:68–76.

Silva, J. F., Espósito, T. S., Marcuschi, M., Ribeiro, K., Cavalli, R. O., Oliveira, V., and Bezerra, R. S. 2011. Purification and partial characterisation of a trypsin from the processing waste of the silver mojarra (*Diapterus rhombeus*). *Food Chemistry* 129:777–82.

Silva, R. S. G., Bandeira, S. F., and Pinto, L. A. A. 2014b. Characteristics and chemical composition of skins gelatin from cobia (*Rachycentron canadum*). *LWT–Food Science and Technology* 57:580–85.

Silva, T. H., Alves, A., Ferreira, B. M., Oliveira, J. M., Reys, L. L., Ferreira, R. J. F., Sousa, R. A., Silva, S. S., Mano, J. F., and Reis, R. L. 2012. Materials of marine origin: A review on polymers and ceramics of biomedical interest. *International Materials Reviews* 57:276–307.

Silva, T. H., Moreira-Silva, J., Marques, A. L. P., Domingues, A., Bayon, Y., and Reis, R. L. 2014a. Marine origin collagens and its potential applications. *Marine Drugs* 12:5881–901.

Simpson, B. K. 2000. Digestive proteinases from marine animals. In: N. F. Haard and B. K. Simpson (Eds.), *Seafood enzymes: Utilization and influence on postharvest seafood quality*, pp. 531–40. New York: Marcel Dekker.

Souchet, N., and Laplante, S. 2011. Recovery and characterization of a serine collagenolytic extract from snow crab (*Chionoecetes opilio*) by-products. *Applied Biochemistry and Biotechnology* 163:765–79.

Souissi, N., Bougatef, A., Triki-Ellouz, Y., and Nasri, M. 2007. Biochemical and functional properties of sardinella (*Sardinella aurita*) by-product hydrolysates. *Food Technology and Biotechnology* 45:187–94.

Sousa, S. C., Vázquez, J. A., Pérez-Martín, R. I., Carvalho, A. P., and Gomes, A. M. 2017. Valorization of by-products from commercial fish species: Extraction and chemical properties of skin gelatins. *Molecules* 22:1545.

Sriket, C. 2014. Proteases in fish and shellfish: Role on muscle softening and prevention. *International Food Research Journal* 21:433–45.

Stefansson, B., Sandholt, G. B., and Gudmundsdottir, Á. 2017. Elucidation of different cold-adapted Atlantic cod (*Gadus morhua*) trypsin X isoenzymes. *Biochimica Et Biophysica Acta Proteins Proteomics* 1865:11–19.

Stephen, N. M., Bhaskar, N., Hosokawa, M., and Miyashita, K. 2010. Marine nutraceuticals: Sources, recovery and effective utilization. In: *Advances in Food Science and Technology*, pp. 1–17. Nova Science Publishers: Hauppauge (New York, U.S.). Ed. Haghi, A. K. ISBN: 978-1-61324-161-5.

Suwal, S., Roblet, C., Amiot, J., Doyen, A., Beaulieu, L., Legault, J., Bazinet, L. 2014. Recovery of valuable peptides from marine protein hydrolysate by electrodialysis with ultrafiltration membrane: Impact of ionic strength. *Food Research International* 65:407–15.

Synowiecki, J., and Al-Khateeb, N. A. 2003. Production, properties, and some new applications of chitin and its derivatives. *Critical Reviews in Food Science and Nutrition* 43:145–71.

Taheri, A., Anvar, S. A. A., Ahari, H., and Fogliano, V. 2013. Comparison the functional properties of protein hydrolysates from poultry by-products and rainbow trout (*Onchorhynchus mykiss*) viscera. *Iranian Journal of Fisheries Science* 12:154–69.

Tahiri, I., Desbiens, M., Lacroix, C., Kheadr, E., and Fliss, I. 2009. Growth of *Carnobacterium divergens* M35 and production of Divergicin M35 in snow crab by-product, a natural-grade medium. *LWT-Food Science and Technology* 42:624–32.

Takeda, N., Horai, S., and Tamura, J.-I. 2016. Facile analysis of contents and compositions of the chondroitin sulfate/dermatan sulfate hybrid chain in shark and ray tissues. *Carbohydrate Research* 424:54–58.

Tang, L., Chen, S., Su, W., Weng, W., Osako, K., and Tanaka, M. 2015. Physicochemical properties and film-forming ability of fish skin collagen extracted from different freshwater species. *Process Biochemistry* 50:148–55.

Tanuja, S., Viji, P., Zynudheen, A. A., and Joshy, C. G. 2012. Composition, functional properties and antioxidative activity of hydrolysates prepared from the frame meat of striped catfish (*Pangasianodon hypophthalmus*). *Egyptian Journal of Biology* 14:27–35.

Tavano, O. L. 2013. Protein hydrolysis using proteases: An important tool for food biotechnology. *Journal of Molecular Catalysis B: Enzymatic* 90:1–11.

Teves, J. F. C. and Ragaza, J. A. 2016. The quest for indigenous aquafeed ingredients: A review. *Reviews in Aquaculture* 8:154–71.

Theodore, A. E., Raghavan, S., and Kristinsson, H. G. 2008. Antioxidative activities of protein hydrolysates prepared from alkaline-aided channel catfish protein isolates. *Journal of Agricultural and Food Chemistry* 56:7459–66.

Thiansilakul, Y., Benjakul, S., and Shahidi, F. 2007. Compositions, functional properties and anti-oxidative activity of protein hydrolysates prepared from round scad (*Decapterus maruadsi*). *Food Chemistry* 103:1385–94.

Thuy, L. T. M., Okazaki, E., and Osako, K. 2014. Isolation and characterization of acid-soluble collagen from the scales of marine fishes from Japan and Vietnam. *Food Chemistry* 149:264–70.

Tolaimate, A., Desbrières, J., Rhazi, M., and Alagui, A. 2003. Contribution to the preparation of chitins and chitosans with controlled physico-chemical properties. *Polymers* 44:7939–52.

Trutnau, M., Suckale, N., Groeger, G., Bley, T., and Ondruschka, J. 2009. Enhanced chitosan production and modeling hyphal growth of *Mucor rouxii* interpreting the dependence of chitosan yields on processing and cultivation time. *Engineering Life Science* 9:437–43.

Valcarcel, J., Novoa-Carballal, R., Pérez-Martín, R. I., Reis, R. L., and Vázquez, J. A. 2017. Glycosaminoglycans from marine sources as therapeutic agents. *Biotechnology Advances* 35:711–25.

Valdez-Peña, A. U., Espinoza-Pérez, J. D., Sandoval-Fabian, G. C., Balagurusamy, N., Hernández-Rivera, A., De-la-Garza-Rodríguez, I. M., and Contreras-Esquivel, J. C. 2010. Screening of industrial enzymes for deproteinization of shrimp head for chitin recovery. *Food Science and Biotechnology* 19:553–57.

Vázquez, J., Caprioni, R., Nogueira, M., Menduiña, A., Ramos, P., and Pérez-Martín, R. 2016b. Valorisation of effluents obtained from chemical and enzymatic chitin production of *Illex argentinus* pen by-products as nutrient supplements for various bacterial fermentations. *Biochemical Engineering Journal* 116:34–44.

Vázquez, J. A., Blanco, M., Fraguas, J., Pastrana, L., and Pérez-Martín, R. I. 2016a. Optimisation of the extraction and purification of chondroitin sulphate from head by-products of *Prionace glauca* by environmental friendly processes. *Food Chemistry* 198:28–35.

Vázquez, J. A., Blanco, M., Massa, A. E., Amado, I. R., Pérez-Martín, R. I. 2017a. Production of fish protein hydrolysates from *Scyliorhinus canicula* discards with antihypertensive and antioxidant activities by enzymatic hydrolysis and mathematical optimization using response surface methodology. *Marine Drugs* 15:306.

Vázquez, J. A., Docasal, S. F., Mirón, J., González, M. P., and Murado, M. A. 2006b. Proteases production by two Vibrio species on residuals marine media. *Journal of Industrial Microbiology and Biotechnology* 33:661–68.

Vázquez, J. A., Docasal, S. F., Prieto, M. A., González, M. P., and Murado, M. A. 2008. Growth and metabolic features of lactic acid bacteria in media with hydrolysed fish viscera. An approach to bio-silage of fishing by-products. *Bioresource Technology* 99:6246–57.

Vázquez, J. A., González, M. P., and Murado, M.A. 2004a. Peptones from autohydrolysed fish viscera for nisin and pediocin production. *Journal of Biotechnology* 112:299–311.

Vázquez, J. A., González, M. P., and Murado, M. A. 2004b. A new marine medium. Use of different fish peptones and comparative study of the growth of selected species of marine bacteria. *Enzyme and Microbial Technology* 35:385–92.

Vázquez, J. A., González, M. P., and Murado, M. A. 2006a. Preliminary tests on nisin and pediocin production using waste protein sources. Factorial and kinetic studies. *Bioresource Technology* 97:605–13.

Vázquez, J. A., Montemayor, M. I., Fraguas, J., and Murado, M. A. 2009. High production of hyaluronic and lactic acids by *Streptococcus zooepidemicus* in fed-batch culture using commercial and marine peptones form fishing by-products. *Biochemical Engineering Journal* 44:125–30.

Vázquez, J. A., Montemayor, M. I., Fraguas, J., and Murado, M. A. 2010b. Hyaluronic acid production by *Streptococcus zooepidemicus* in marine by-products media from mussel processing wastewaters and tuna peptone viscera. *Microbial Cell Factories* 9:46.

Vázquez, J. A., and Murado, M. A. 2008a. Enzymatic hydrolysates from food wastewaters as a source of peptones for lactic acid bacteria productions. *Enzyme and Microbial Technology* 43:66–72.

Vázquez, J. A., and Murado, M. A. 2008b. Mathematical tools for objective comparison of microbial cultures. Application to evaluation of 15 peptones for lactic acid bacteria. *Biochemical Engineering Journal* 39:276–87.

Vázquez, J. A., Nogueira, M., Durán, A., Prieto, M. A., Rodríguez-Amado, I., Rial, D., González M. P., and Murado, M. A. 2011. Preparation of marine silage of swordfish, ray and shark visceral waste by lactic acid bacteria. *Journal of Food Engineering* 103:442–48.

Vázquez, J. A., Noriega, D., Ramos, P., Valcarcel, J., Novoa-Carballal, R., Pastrana, L., Reis, R. L., and Pérez-Martín, R. I. 2017c. Optimization of high purity chitin and chitosan production from Illex argentinus pens by a combination of enzymatic and chemical processes. *Carbohydrate Polymers* 174:262–72.

Vázquez, J. A., Pastrana, L., Piñeiro, C., Teixeira, J. A., Pérez-Martín, R. I., and Amado, I. R. 2015. Production of hyaluronic acid by Streptococcus zooepidemicus on protein substrates obtained from *Scyliorhinus canicula* discards. *Marine Drugs* 13:6537–49.

Vázquez, J. A., Ramos, P., Mirón, J., Valcarcel, J., Sotelo, C. G., and Pérez-Martín, R. I. 2017b. Production of chitin from *Penaeus vannamei* by-products to pilot plant scale using a combination of enzymatic and chemical processes and subsequent optimization of the chemical production of chitosan by response surface methodology. *Marine Drugs* 15:180.

Vázquez, J. A., Rodrigues, A. C., Fuciños, P., Pastrana, L., and Murado, M. A. 2010a. Biosilage of mussel work-processing wastes by lactobacilli on semi-solid culture. *Journal of Food Engineering* 97:355–59.

Vázquez, J. A., Rodríguez-Amado, I., Montemayor, M. I., Fraguas, J., González, M. P., and Murado, M. A. 2013. Chondroitin sulphate, hyaluronic acid and chitin/chitosan production using marine waste sources: Characteristics, applications and eco-friendly processes: A review. *Marine Drugs* 11:747–74.

Vidotti, R. M., Viegas, E. M. M., and Carneiro, D. J. 2003. Amino acid composition of processed fish silage using different raw materials. *Animal Feed Science and Technology* 105:199–204.

Villamil, O., Váquiro, H., and Solanilla, J. F. 2017. Fish viscera protein hydrolysates: Production, potential applications and functional and bioactive properties. *Food Chemistry* 224:160–71.

Vizcarra-Magaña, L., Avila, E., and Sotelo, A. 1999. Silage preparation from tuna fish wastes and its nutritional evaluation in broilers. *Journal of the Science of Food and Agriculture* 79:1915–22.

Volpi, N. 2006. *Chondroitin Sulfate, Volume 53 1st Edition. Structure, role and pharmacological activity* Ed. Nicola Volpi. eBook ISBN: 9780080471952.

Wald, M., Schwarz, K., Rehbein, H., Bußmann, B., and Beermann, C. 2016. Detection of antibacterial activity of an enzymatic hydrolysate generated by processing rainbow trout by-products with trout pepsin. *Food Chemistry* 205:221–28.

Wang, J., Wang, L., Yu, H., Zain-ul-Abdin, C.Y., Chen, Q., Zhou, W., Zhang, H., and Chen, X. 2016. Recent progress on synthesis, property and application of modified chitosan: An overview. *International Journal of Biological Macromolecules* 88:333–44.

Wang, T. Y., Hsieh, C. H., Hung, C. C., Jao, C. L., Chen, M. C., and Hsu, K. C. 2015. Fish skin gelatin hydrolysates as dipeptidyl peptidase IV inhibitors and glucagon-like peptide-1 stimulators improve glycaemic control in diabetic rats: A comparison between warm- and cold-water fish. *Journal of Functional Foods* 19:330–40.

Wang, Y., Zhu, F., Han, F., and Wang, H. 2008. Purification and characterization of antioxidative peptides from salmon protamine hydrolysate. *Journal of Food Biochemistry* 32:654–71.

Wasswa, J., Tang, J., Gu, X., and Yuan, X. 2007. Influence of the extent of enzymatic hydrolysis on the functional properties of protein hydrolysate from grass carp (*Ctenopharyngodon idella*) skin. *Food Chemistry* 104:1698–704.

Watanabe, T. 2002. Strategies for further development of aquatic feeds. *Fisheries Science* 68:242–52.

Weng, W., Tang, L.,Wang, B., Chen, J.,Su, W., Osako, K., and Tanaka, M. 2014. Antioxidant properties of fractions isolated from blue shark (*Prionace glauca*) skin gelatin hydrolysates. *Journal of Functional Foods* 11:342–51.

Win, N. N., and Stevens, W. F. 2001. Shrimp chitin as substrate for fungal chitin deacetylase. *Applied Microbiology and Biotechnology* 57:334–41.

Yang, J., Ho, H., Chu, Y., and Chow, C. 2008. Characteristic and antioxidant activity of retorted gelatin hydrolysates from cobia (*Rachycentron canadum*) skin. *Food Chemistry* 110:128–36.

You, L., Zhao, M., Regenstein, J. M., and Ren, J. 2010a. Changes in the antioxidant activity of loach (*Misgurnus anguillicaudatus*) protein hydrolysates during a simulated gastrointestinal digestion. *Food Chemistry* 120:810–16.

You, L., Zhao, M., Regenstein, J. M., and Ren, J. 2010b. Purification and identification of antioxidative peptides from loach (*Misgurnus anguillicaudatus*) protein hydrolysate by consecutive chromatography and electrospray ionization-mass spectrometry. *Food Research International* 43:1167–73.

Youn, D. K., No, H. K., and Prinyawiwatkul, W. 2013. Preparation and characterisation of selected physicochemical and functiional properties of β-chitosan from squid pen. *Food Science and Technology* 48:1661–69.

Younes, I., Nasri, R., Bkhairia, I., Jellouli, K., and Nasri, M. 2015. New proteases extracted from red scorpionfish (*Scorpaena scrofa*) viscera: Characterization and application as a detergent additive and for shrimp waste deproteinization. *Food and Bioproducts Processing* 94:453–62.

Ytteborg, E., Baeverfjord, G., Lock, E.-J., Pedersen, M., Takle, H., Ørnsrud, R., Waagbø, R., and Albrektsen, S. 2016. Utilization of acid hydrolysed phosphorous from herring bone by-products in feed for Atlantic salmon (*Salmo salar*) start-feeding fry. *Aquaculture* 459:173–84.

Yuan, G., Chen, X., and Li, D. 2016. Chitosan films and coatings containing essential oils: The antioxidant and antimicrobial activity, and application in food systems. *Food Research International* 89:117–28.

Yuan, P., Shen, X., Liu, J., Hou, Y., Zhu, M., Huang, J., and Xu, P. 2012. Effects of dentifrice containing hydroxyapatite on dentinal tubule occlusion and aqueous hexavalent chromium cations sorption: A preliminary study. *Plos One* 7:1–8.

Zamora-Sillero, J., Gharsallaoui, A., and Prentice, C. 2018. Peptides from fish by-product protein hydrolysates and its functional properties: An overview. *Marine Biotechnology* 20:118–30.

Zhang, F., Xu, S., Wang, Z. 2011. Pre-treatment optimization and properties of gelatin from freshwater fish scales. *Food and Bioproducts Processing* 89:185–93.

Zhang, M., Liu, W. T., and Li, G. Y. 2009. Isolation and characterisation of collagens from the skin of largefin longbarbel catfish (*Mystus macropterus*). *Food Chemistry* 115:826–31.

Zhang, Y., Duan, X., and Zhuang, Y. 2012. Purification and characterization of novel antioxidant peptides from enzymatic hydrolysates of tilapia (*Oreochromis niloticus*) skin gelatin. *Peptides* 38:13–21.

Zhao, T., Zhou, Y., Mao, G., Zou, Y., Zhao, J., Bai, S., Yang, L., and Wu, X. 2013. Extraction, purification and characterisation of chondroitin sulfate in Chinese sturgeon cartilage. *Journal of the Science of Food and Agriculture* 93:1633–40.

Zheng, K., Liang, M., Yao, H., Wang, J., and Chang, Q. 2012. Effect of dietary fish protein hydrolysate on growth, feed utilization and IGF-I levels of Japanese flounder (*Paralichthys olivaceus*). *Aquaculture Nutrition* 18:297–303.

Zheng, K., Liang, M., Yao, H., Wang, J., and Chang, Q. 2013. Effect of size-fractionated fish protein hydrolysate on growth and feed utilization of turbot (*Scophthalmus maximus* L.). *Aquaculture Research* 44:895–902.

Zhong, S., Ma, C., Lin, Y. C., and Luo, Y. 2011. Antioxidant properties of peptide fractions from silver carp (*Hypophthalmichthys molitrix*) processing by-product protein hydrolysates evaluated by electron spin resonance spectrometry. *Food Chemistry* 126:1636–42.

Zhou, P., and Regenstein, J. M. 2005. Effects of alkaline and acid pretreatments on Alaska pollock skin gelatin extraction. *Journal of Food Science* 70:c392–c396.

Zhou, Z., Ringø, E., Olsen, R. E., Song, S. K. 2018. Dietary effects of soybean products on gut microbiota and immunity of aquatic animals: A review. *Aquaculture Nutrition* 24:644–65.

Zhuang, Y., Li, B., and Zhao, X. 2009. The scavenging of free radical and oxygen species activities and hydration capacity of collagen hydrolysates from Walleye Pollock (*Theragra chalcogramma*) skin. *Journal of Ocean University of China* 8:171–76.

Zohora, U. S., Rahman, M. S., Khan, A. W., Okanami, M., and Ano, T. 2013. Improvement of production of lipopeptide antibiotic iturin A using fish protein. *Journal of Environmental Sciences* 25:S2–S7.

Zou, P., Yang, X., Wang, J., Li, Y., Yu, H., Zhang, Y., and Liu, G. 2016. Advances in characterisation and biological activities of chitosan and chitosan oligosaccharides. *Food Chemistry* 190:1174–81.

5

Recovery of Antioxidant Bioactive Compounds from Sweet Potato and By-Products

ZHENZHOU ZHU AND FANG WANG

College of Food Science and Engineering, Wuhan Polytechnic University, No.68 Xuefu South Road, Changqing Garden, 430023, Wuhan, China

ELENA ROSELLÓ-SOTO, FRANCISCO J. MARTÍ-QUIJAL, AND FRANCISCO J. BARBA

Nutrition and Food Science Area, Preventive Medicine and Public Health, Food Science, Toxicology and Forensic Medicine Department, Faculty of Pharmacy, Universitat de València, Avda. Vicent Andrés Estellés, s/n, 46100 Burjassot, València, Spain

KASHIF GHAFOOR

Department of Food Science and Nutrition, King Saud University, Riyadh, KSA

JOSÉ M. LORENZO

Centro Tecnológico de la Carne de Galicia, Adva. Galicia n° 4, Parque Tecnológico de Galicia, San Cibrao das Viñas, 32900 Ourense, Spain.

CYRIELLE GARCIA AND FABIENNE REMIZE

QualiSud, Université de La Réunion, CIRAD, Université Montpellier, Montpellier SupAgro, Université d'Avignon, 2 rue J. Wetzell, F-97490 Sainte Clotilde, France

Contents

1 Introduction

Sweet potato (*Ipomoea batatas* L.) is one of the most widely distributed food crops in the world, which is cultivated in more than 100 countries with a total production of around 111 million tons in 2013, with China's production alone representing around 71% of the production. It should be noted that the waste produced during sweet potato cultivation constitutes approximately 7% of total production (FAO 2018). Sweet potato exhibits an array of flesh and skin colors, including purple, orange, white, and yellow, depending on the variety (Xu et al. 2018). This variation is due to the contents of bioactive compounds. For instance, orange-fleshed varieties are rich in β-carotene, a precursor to vitamin A (Grace et al. 2014).

Carbohydrates make up 80–90% of sweet potato, of which 60–70% is starch. The crude protein, crude fat, crude fiber, carbohydrate, and ash contents range between 16.69–31.08, 2.08–5.28, 9.15–14.26, 42.03–61.36, and 7.39–14.66 g/100 g dry weight (DW), respectively (Sun et al. 2014). In addition, sweet potato contains important vitamins, minerals, especially K, P, Ca, Mg, Fe, Mn, and Cu, and antioxidant compounds such as carotenoids, phenolic compounds, flavonoids, and anthocyanins (Ji et al. 2015). The total antioxidant capacity of sweet potato has been reported to be 42.94% as compared to ascorbic acid (Pochapski et al. 2011). Bioactive compounds of sweet potato exhibit potential anti-inflammatory, antiproliferative, and anticancer activities, which make sweet potato a natural and valuable nutraceutical (Huang, Lin, Chen, and Lin 2004; Mohanraj and Sivasankar 2014; Sugata, Lin, and Shih 2015) (Figure 5.1).

Large amounts of waste and by-products are produced during cultivation and subsequent processing of sweet potatoes (e.g., root tuber in starch, alcoholic beverages, ethanol, flour, purees, and other products) (Zhu, Guan, Koubaa, Barba, and He 2017a). These by-products are both solid and liquid and do not have recognized added value (Akoetey, Britain, and Morawicki 2017). Solid residues originate from the skin and cuttings of the sweet potato root, as well as from leaves and vines. Despite the fact that vines and leaves are a good source of proteins, vitamins, minerals, and antioxidant bioactive compounds, such as carotenoids and polyphenols, these by-products are used only for feeding farm animals. Liquid waste, essentially in the form of a wastewater rich in nutrients, is also generated. Sweet potato by-products are therefore a good source of macro- (carbohydrates, proteins) and micronutrients (minerals, vitamins) as bioactive antioxidant compounds (e.g., polyphenols and carotenoids), which can be extracted and subsequently used for various subsequent processes. In this chapter, some of the potential applications of sweet potato by-products are reviewed.

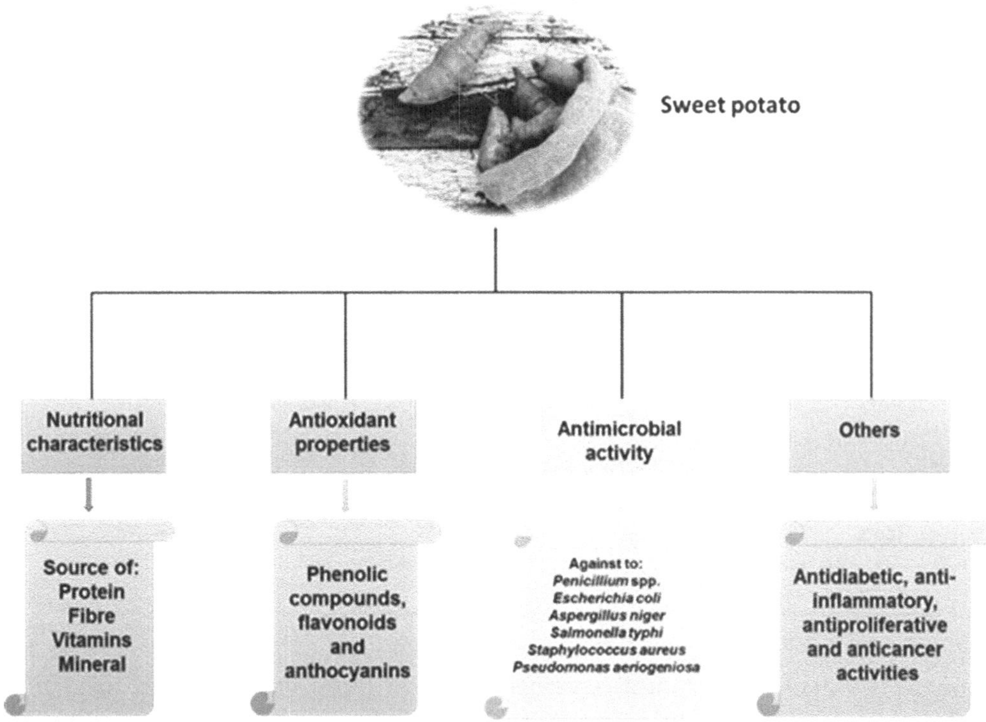

Figure 5.1 Main properties of sweet potato

2 Sweet Potato and By-Products as Source of Antioxidant Bioactive Compounds and Antimicrobials

2.1 Polyphenols

Phenolic compounds are the predominant antioxidant compounds in sweet potato leaves. In this sense, Sun et al. (2014) observed considerable amounts of different phenolic compounds (Table 5.1). Likewise, they also established a significant correlation between the values of total phenolic compounds (TPC) and the total antioxidant capacity measured with the ORAC method. The TPC content of sweet potato ranges from 1.42 to 17.1 g/100 g DW (Islam, Yoshimoto, Terahara, and Yamakawa 2002; Sun et al. 2014). These differences on the TPC content of sweet potato could be due to storage conditions, maturity, genotype, polyphenol oxidase activity, processing methods, variety of nutrient composition, differences in post-harvest, etc. (Sun et al. 2014).

In another subsequent study, the values of TPC of the leaves and cooked roots were compared, obtaining values up to 60 times higher for the leaves than for the cooked roots (Truong et al. 2007). Likewise, the values of antioxidant capacity were 10 times higher for the leaves extracts and themethanolic sweet potato

Table 5.1 Individual and total phenolic compounds obtained from leaves and roots of sweet potatoes and its total antioxidant capacity.

MATRIX	COMPOUNDS	REFERENCES
Leaves	5-O-caffeoylquinic acid 3-O-caffeoylquinic acid 4-O-caffeoylquinic acid Caffeic acid 4,5-di-O-caffeoylquinic acid 3,5-di-O-caffeoylquinic acid 3,4-di-O-caffeoylquinic acid 3,4,5-tri-O-caffeoylquinic acid Folin Ciocalteau method: 12.5 g chlorogenic acid/100 g DM ORAC: 1.28±0.07 μg TE/mg DM	(Sun et al. 2014)
	Folin Ciocalteau method: 1248.8 mg chlorogenic acid/100 g FW DPPH: 38.1 mM Trolox/g FW	(Truong et al. 2007)
Roots	Folin Ciocalteau method: 20.3 mg chlorogenic acid/100 g FW DPPH: 3.0 mM Trolox/g FW	(Truong et al. 2007)

extracts presented higher radical scavenging activity, reducing power and oxidation inhibition than α-tocopherol, and a higher iron-chelating capacity than ethylenediamine tetraacetic acid (EDTA).

Aqueous extracts of sweet potato leaves inhibit the production of nitric oxide in macrophages and protect mouse liposomes against oxidation (Huang, Chu, Juang, and Wang 2010).

Purple sweet potato (PSP) is a variety of sweet potatoes in which, according to literature, anthocyanin are considered to be the main group of phenolic compounds present and contribute to their characteristic color (Salawu et al. 2015). Anthocyanin concentration in PSP is among the highest anthocyanin production crops, such as blueberries, blackberries, pomegranate, or grapes (Barba et al. 2012, Barba, Esteve, and Frigola 2013, Brianceau, Turk, Boussetta, and Vorobiev 2015a, Barba, Galanakis, Esteve, Frigola, and Vorobiev 2015b; Putnik et al. 2019; Zhu et al. 2014). Moreover, PSP polyphenol and anthocyanins showed many functional abilities, such as antioxidant (Zhang et al. 2010), hepato-protection (Hwang et al. 2011), and memory enhancing (Lu et al. 2012). Because of their structures, identified as monoacylated or diacylated forms of cyanidin and peonidin, PSP anthocyanins have good heat stability (Montilla, Hillebrand, and Winterhalter 2011; Tiwari, O'Donnell, and Cullen 2009).

According to Grace et al. (2014) and Suda et al. (2003), anthocyanins detected in PSP are mainly cyanidin and peonidin glycosides acylated with p-hydroxybenzoic acids, ferulic acid, and caffeic acid. Grace et al. (2014) compared the effect of different processing methods (baking, boiling, and steaming) on the anthocyanins content of PSP. They observed that all processing methods increased the anthocyanin content from 1.24 mg CGE/g sample in raw samples to 2.05, 1.95, and 1.44 mg CGE/g sample, for steamed, baked, and boiled treatments, respectively. According to Ruttarattanamongkol et al. (2016), steaming is the most efficient way to inactivate enzymes such as polyphenoloxidase, peroxidase, and glycosidase that are able to degrade anthocyanin in fresh sweet potatoes. Therefore, steaming is the most effective heat treatment method to retain anthocyanin content in purple-fleshed sweet potato samples (Grace et al. 2014). A study carried out by Dong, Tai-hua and Hong-nan (2017) reported the importance of residual flour left during production as a source of dietary fiber (20.63–31.48 g per 100 g dry weight for sweet potato).

In another study, to intensify valuable compounds (polyphenols, especially anthocyanins, and proteins), extraction from PSP root, Zhu et al. (2017b) carried out an ultrasound-assisted extraction (UAE) pretreatment process by applying response surface methodology (RSM) with Box-Behnken design to optimize parameters in terms of total polyphenols (TP), total anthocyanins (TA), total proteins (TPr), and specific energy consumption (SEC). The maximal valuable compounds' (TP, TA, Tpr) extraction and minimal specific energy consumption were obtained as TP (3.877 mg/g), TA (0.293 mg/g), and TPr (0.753 mg/g), with minimal SEC (8406 J/mg) at conditions as follows: UAE time of 40 min, and a supplementary hot extraction at 80 °C up to 120 min; pH of the solution: 2.5 and ethanol concentration: 58%. Moreover, HPLC-DAD-ESI-MS^2 analysis was done to identify anthocyanin and non-anthocyanin polyphenols in PSP extract. The major polyphenol components identified were peonidin-3-caffeoyl-*p*-hydroxybenzoyl sophoroside-5-glucoside, peonidin-3-(6″-caffeoyl-6‴-feruloyl sophoroside)-5-glucoside, cyanidin-3-caffeoyl-*p*-hydroxybenzoyl sophoroside-5-glucoside for extract from UAE, whereas it was peonidin derivative 4, peonidin derivative 2, and peonidin-3-*p*-hydroxybenzoyl sophoroside-5-glucoside for extract from conventional solvent extraction (CSE) process without ultrasonic. The major non-anthocyanin polyphenols of UAE extract were chlorogenic acid and chlorogenic acid-3-glucose, whereas chlorogenic acid, caffeic acid, and quinic acid were found in the CSE extract. Phenolic acids were also identified in cooked PSP using HPLC-DAD-ESI-MS and proton nuclear magnetic resonance by Zhao et al. (2014) after semi-preparative HPLC purification from the acidified water extract. Five colorless caffeoyl compounds were the major types recovered, 5-caffeoylquinic acid (chlorogenic acid), 6-o′caffeoyl-β-D-fructofuranosyl-(2–1)-α-D-glucopyranoside, trans-4,5-dicaffeoylquinic acid, 3,5-dicaffeoylquinic acid, and 4,5-dicaffeoylquinic acid in this case. By-products of sweet potato from Shochu distillary were observed as an important source of phenolic compounds, whereas caffeic acid was identified as a dominant phenolic compound. Additionally, essential amino acids, minerals, citric acid, vitamins, and minerals have been detected, suggesting the utilization of sweet potato residue left after fermentation as food material (Yoshimoto et al. 2004).

Extraction of PSP anthocyanins using water as a solvent with ultrasonic circulating extraction equipment was studied by Zhu et al. (2016). After extraction at 840 W and 59 kHz, the extract was processed with ultrafiltration, using polyethersulfone membrane, to recover anthocyanins. Results showed that the recovery of polyphenol, protein, and anthocyanin was carried out. Pretreatment, via the centrifugation of PSP extracts, at 2,500 rpm over 6 min, led to an increased polyphenol recovery, with satisfactory protein removal (reused for future purposes), compared to PSP extract filtration without centrifugation. Results showed that anthocyanin could be recovered at an efficiency of 99% from permeate. Moreover, HPLC-DAD-ESI-MS^2 analysis evidenced that the main anthocyanin types and concentrations remained the same before and after ultrafiltration: peonidin-3-feruloylsophoroside-5-glucoside, peonidin-3-caffeoyl-*p*-hydroxybenzoylsophoroside—5-glucoside, and peonidin-3-caffeoyl-feruloyl sophoroside-5-glucoside.

Flavonoids can be purified from sweet potato leaves using ethanol after preliminary soxhlet extraction and homogenization. It is suggested that a pretreatment with dynamic high-pressure microfluidization homogenizer at 100 MPa can promote the solvent permeation and a 1.3-fold acceleration of the extraction of flavonoids from sweet potato leaves was found by comparison to a post-treatment or a conventional ethanol extraction, and led to a stronger radical scavenging activity in the purified samples (Huang et al. 2013).

2.2 Carotenoids

Carotenoids are found in variable concentration among the different colourfleshed sweet potatoes, ranging from 0.4 to 72.5 µg/g fresh weight. The amount found in orange-fleshed variety is the highest with a majority of trans β-carotene, ranging from 79.1 to 128.5mg/100 g DW, while xanthophylls, lutein, and zeaxanthin were identified as minor compounds (Wang et al. 2016). Carotenoids are generally altered by cooking. Boiling, in particular, was associated with an increase in *cis*-β-carotene and a decrease in the trans isomer in both in orange, yellow, and white-fleshed sweet potato roots (Islam et al. 2016), suggesting an isomerization that might influence β-carotene bioavailability. Sweet potatoes of the Suioh variety had a lutein content of 31.5–42.6 mg/100 g FW (Ishiguro and Yoshimoto 2006). Lutein is a yellow-to-orange non-provitamin A carotenoid playing an important role in the human retina protecting the macula; therefore, sweet potato leaves can be considered interesting for the prevention of eye diseases (Ishiguro and Yoshimoto 2006; Menelaou et al. 2006). In addition, lutein contained in sweet potato leaves could potentially be used as a substitute for the coloring tartrazine.

Hue (2014) isolated and purified the yellow fraction of sweet potato leaves using different solvents (acetone, methanol, tetrahydrofuran, and a mixture of petroleum ether:tetrahydrofuran 4:1). Once the extraction was carried out, the solvents were removed by means of a rotary evaporator, and the residue was saponified using different concentrations of potassium hydroxide with the objective of eliminating chlorophyll. Once the saponification was finished, the carotenoids were extracted with acetone. Of all the solvents used, acetone was the most effective for extracting β-carotene (10.8 mg/100 g fw) and lutein (10.0 mg/100 g fw), these amounts being up to 10 times higher than those found for β-carotene in pumpkin (1.2 mg/100 g fw) and in mango (*Magnifera indica*) (1.41 mg/100 g fw) (Bhaskarachary et al. 1995).

2.3 *Antimicrobial Activity of Sweet Potato Leaf Extracts*

Islam, Jalaluddin, Yoshimoto, and Yamakawa (2004) evaluated the use of lyophilized powder obtained from sweet potato to inhibit the growth of pathogenic bacteria. They observed inhibition of the growth of *Escherichia coli* and *Bacillus* spp. bacteria, which can be attributed to the polyphenols, but no growth inhibition of bifidobacteria, which can have a beneficial health effect.

Table 5.2 Antimicrobial activity of sweet potato leaf extracts (peptone, ethanol, and water extracts) using agar well diffusion technique zone of inhibition of organisms (mm) (Mbaeyi-Nwaoha and Emejulu 2013).

EXTRACTS	CONC. (MG/ML)	S. AUREUS	E. COLI	K. PNEU	S. TYPHI	P. AERO	A. NIGER	PENICIL
Water	100	15.7±0.6[a]	N.I.	12.0±0.0[a]	18.0±0.0[a]	13.7±0.6[a]	26.3±0.6[a]	N.I.
	50	12.7±0.6[b]	N.I.	9.3±00.6[b]	14.7±0.6[b]	11.0±0.0[b]	21.7±00.6[b]	N.I.
	25	9.0±0.0[c]	N.I.	N.I.	11.3±0.6[c]	8.7±0.6[c]	16.3±0.6[c]	N.I.
	12.5	N.I.	N.I.	N.I.	9.3±0.6[d]	N.I.	12.0±0.0[d]	N.I.
	6.25	N.I.	N.I.	N.I.	N.I.	N.I.	8.7±0.6[e]	N.I.
	3.125	N.I.	N.I.	N.I.	N.I.	N.I.	N.I.	N.I.
Peptone	100	N.I.	N.I.	N.I.	12.3±1.2[a]	N.I.	N.I.	N.I.
	50	N.I.	N.I.	N.I.	8.7±1.5[b]	N.I.	N.I.	N.I.
	25	N.I.	N.I.	N.I.	N.I.	N.I.	N.I.	N.I.
	12.5	N.I.	N.I.	N.I.	N.I.	N.I.	N.I.	N.I.
	6.25	N.I.	N.I.	N.I.	N.I.	N.I.	N.I.	N.I.
	3.125	N.I.	N.I.	N.I.	N.I.	N.I.	N.I.	N.I.
Ethanol	100	N.I.	N.I.	N.I.	12.0±1.0[a]	N.I.	N.I.	N.I.
	50	N.I.	N.I.	N.I.	8.7±0.6[b]	N.I.	N.I.	N.I.
	25	N.I.	N.I.	N.I.	N.I.	N.I.	N.I.	N.I.
	12.5	N.I.	N.I.	N.I.	N.I.	N.I.	N.I.	N.I.
	6.25	N.I.	N.I.	N.I.	N.I.	N.I.	N.I.	N.I.

Values are means ±standard deviation. [a–d] Means in the same row with different letters differ significantly ($P < 0.05$). S. *aureus* = *Staphylococcus aureus; E. coli* = *Escherichia coli; K. pneu* = *Klebsiella pneumoniae;* S. *typhi* = *Salmonella enterica* Typhi; *P. aero* = *Pseudomona aeroginosa; A. niger*= *Aspergillus niger, Penicil.* = *Penicillum* spp.; N.I. = No inhibition

Table 5.2 shows the antimicrobial activity of sweet potato leaves extracts (peptone, water, and ethanol extracts) (Mbaeyi-Nwaoha and Emejulu 2013). At the six different concentrations used, *Penicillium* spp. and *E. coli* were resistant to peptone, water, and ethanol extracts. On the contrary, *Aspergillus niger* was inhibited by water extract at concentrations of 6.25 mg/mL, *Salmonella enterica* Typhi was inhibited by water extract at concentrations of 12.5 mg/mL, *Staphylococcus aureus* and *Pseudomonas aeruginosa* were inhibited by water extract at concentrations of 25 mg/mL, whereas *Klebsiella pneumoniae* was inhibited by water extract at concentrations of 50 mg/mL. *S. enterica* Typhi was inhibited by peptone and ethanol extracts at 50 mg/mL. On the other hand, Pochapski et al. (2011) noticed that no concentration of sweet potato freeze dried extract (diluted up to 15% in water:DMSO 1:1) was able to inhibit growth of *S. aureus, Streptococcus mutans, S. mitis*, or *Candida albicans* in both agar disk and agar well diffusion tests. The antimicrobial properties of sweet potato leaves may depend on their antioxidative compound and/or pectin-like materials. In this regard, sweet potato leaves strongly suppressed the growth of both Gram-positive and Gram-negative bacteria: *E. coli 0157: H7, Bacillus cereus*, and *S. aureus* (Islam et al. 2004).

3 Conclusion

The richness and diversity of sweet potatoes in phenolic compounds and carotenoids are the cornerstones for recovering bioactive compounds. PSP is particularly relevant to obtain anthocyanins. Orange-fleshed variety can be used to recover β-carotene and other carotenoids. Depending on the method used for recovery, bioavailability could be increased by means of isomerization. Eventually, antimicrobial fractions are possibly recovered from leaves.

Acknowledgments

This work was supported by European Union project RE0017203.

References

Akoetey, W, M M Britain, and R O Morawicki 2017. "Potential Use of Byproducts from Cultivation and Processing of Sweet Potatoes [Potencial Para Utilização de Subprodutos Do Cultivo e Processamento de Batata-Doce]". *Ciencia Rural* 47: 5. doi:10.1590/0103-8478cr20160610

Barba, F J, S Brianceau, M Turk, N Boussetta, and E Vorobiev 2015a. "Effect of Alternative Physical Treatments (Ultrasounds, Pulsed Electric Fields, and High-Voltage Electrical Discharges) on Selective Recovery of Bio-Compounds from Fermented Grape Pomace". *Food and Bioprocess Technology* 8: 1139–48. doi: 10.1007/s11947-015-1482-3.

Barba, F J, M J Esteve, and A Frigola 2013. "Physicochemical and Nutritional Characteristics of Blueberry Juice after High Pressure Processing". *Food Research International* 50 (2): 545–49.

Barba, F J, C M Galanakis, M J Esteve, A Frigola, and E Vorobiev 2015b. "Potential Use of Pulsed Electric Technologies and Ultrasounds to Improve the Recovery of High-Added Value Compounds from Blackberries". *Journal of Food Engineering* 30 (4): 455–64.

Barba, F J, H Jäger, N Meneses, M J Esteve, A Frígola, and D Knorr 2012. "Evaluation of Quality Changes of Blueberry Juice during Refrigerated Storage after High-Pressure and Pulsed Electric Fields Processing". *Innovative Food Science and Emerging Technologies* 14: 18–24.

Bhaskarachary, K, D S S Rao, Y G Deosthale, and V Reddy 1995. "Carotene Content of Some Common and Less Familiar Foods of Plant Origin". *Food Chemistry* 54 (2): 189–93.

Dong JU, MU Tai-hua, SUN Hong-nan 2017. "Sweet Potato and Potato Residual Flours as Potential Nutritional and Healthy Food Material." *Journal of Integrative Agriculture* 16 (11): 2632–45.

FAO. 2018. "Statistical Website." *Statistical Website*. http://faostat3.fao.org/home.

Grace, M H, G G Yousef, S J Gustafson, V.-D. Truong, G C Yencho, and M A Lila 2014. "Phytochemical Changes in Phenolics, Anthocyanins, Ascorbic Acid, and Carotenoids Associated with Sweetpotato Storage and Impacts on Bioactive Properties". *Food Chemistry* 145: 717–24.

Huang, D -J, C -D Lin, H -J Chen, and Y -H Lin 2004. "Antioxidant and Antiproliferative Activities of Sweet Potato (Ipomoea Batatas [L.] Lam 'Tainong 57') Constituents". *Botanical Bulletin of Academia Sinica* 45 (3): 179–86.

Huang, M -H, H -L Chu, L -J Juang, and B -S Wang 2010. "Inhibitory Effects of Sweet Potato Leaves on Nitric Oxide Production and Protein Nitration". *Food Chemistry* 121 (2): 480–86.

Huang, X., Z. Tu, H. Xiao, Z. Li, Q. Zhang, H. Wang, Y. Hu, and L. Zhang. 2013. "Dynamic High Pressure Microfluidization-Assisted Extraction and Antioxidant Activities of Sweet Potato (*Ipomoea Batatas* L.) Leaves Flavonoid." *Food and Bioproducts Processing* 91 (1): 1–6.

Hue, S. 2014. "Extraction from Agricultural Waste, Ipomoea Batatas L. Lam Leaves as a Cheap Source of Natural Dye". *International Journal of Environmental Bioremediation & Biodegradation* 2 (1): 1–4.

Hwang, Y P, J H Choi, J M Choi, Y C Chung, and H G Jeong 2011. "Protective Mechanisms of Anthocyanins from Purple Sweet Potato against Tert-Butyl Hydroperoxide-Induced Hepatotoxicity". *Food and Chemical Toxicology: An International Journal Published for the British Industrial Biological Research Association* 49 (9): 2081–89.

Ishiguro, K, and M Yoshimoto 2006. "Content of an Eye-Protective Nutrient Lutein in Sweetpotato Leaves". *Acta Horticulturae* 703: 253–56.

Islam, M S, M Yoshimoto, N Terahara, and O Yamakawa 2002. "Anthocyanin Compositions in Sweetpotato (Ipomoea Batatas L.) Leaves". *Bioscience, Biotechnology and Biochemistry* 66 (11): 2483–86.

Islam, S, M Jalaluddin, M Yoshimoto, and O Yamakawa 2004. "Growth Inhibition of Pathogenic Bacteria by Artificially Grown Ipomoea Batatas (L.) Leaf". *Hort Science* 39 (4): 778B–778.

Islam, S. N., T. Nusrat, P. Begum, and M. Ahsan. 2016. "Carotenoids and β-Carotene in Orange Fleshed Sweet Potato: A Possible Solution to Vitamin A Deficiency." *Food Chemistry* 199: 628–31.

Ji, H, H Zhang, H Li, and Y Li 2015. "Analysis on the Nutrition Composition and Antioxidant Activity of Different Types of Sweet Potato Cultivars". *Food and Nutrition Sciences* 6 (1): 161.

Lu, J, D-M Wu, Y-L Zheng, B Hu, W Cheng, and Z-F Zhang 2012. "Purple Sweet Potato Color Attenuates Domoic Acid-Induced Cognitive Deficits by Promoting Estrogen Receptor-α-Mediated Mitochondrial Biogenesis Signaling in Mice". *Free Radical Biology & Medicine* 52 (3): 646–59.

Mbaeyi-Nwaoha, I E, and V N Emejulu 2013. "Evaluation of Phytochemical Composition and Antimicrobial Activity of Sweet Potato (Ipomoea Batatas) Leaf". *Pakistan Journal of Nutrition* 12 (6): 575–86.

Menelaou, E, A Kachatryan, J N Losso, M Cavalier, and D La Bonte 2006. "Lutein Content in Sweetpotato Leaves". *Hort Science* 41 (5): 1269–71.

Mohanraj, R, and S Sivasankar 2014. "Sweet Potato (Ipomoea Batatas [L.] Lam) – A Valuable Medicinal Food: A Review". *Journal of Medicinal Food* 17 (7): 733–41.

Montilla, E.C., Silke H., and Julián W. 2011. "Anthocyanins in Purple Sweet Potato (Ipomoea Batatas L.) Varieties". *Fruit, Vegetable and Cereal Science and Biotechnology* 5 (2): 19–24.

Pochapski, M T, E C Fosquiera, L A Esmerino, E B Dos Santos, P V Farago, F A Santos, and F C Groppo 2011. "Phytochemical Screening, Antioxidant, and Antimicrobial Activities of the Crude Leaves' Extract from Ipomoea Batatas (L.) Lam". *Pharmacognosy Magazine* 7 (26): 165–70.

Putnik, P, Ž Kresoja, T Bosiljkov, A Režek Jambrak, F J Barba, J M Lorenzo, S Roohinejad, D Granato, I Žuntar, and D Bursać Kovacević 2019. "Comparing the Effects of Thermal and Non-Thermal Technologies on Pomegranate Juice Quality: A Review". *Food Chemistry* 279: 150–61.

Ruttarattanamongkol, K, S Chittrakorn, M Weerawatanakorn, and N Dangpium 2016. "Effect of Drying Conditions on Properties, Pigments and Antioxidant Activity Retentions of Pretreated Orange and Purple-Fleshed Sweet Potato Flours". *Journal of Food Science and Technology* 53 (4): 1811–22.

Salawu, S O, E Udi, A A Akindahunsi, A A Boligon, and M L Athayde 2015. "Antioxidant Potential, Phenolic Profile and Nutrient Composition of Flesh and Peels from Nigerian White and Purple Skinned Sweet Potato (Ipomea Batatas L.)". *Asian Journal of Plant Science & Research* 5: 14–23.

Suda, I, T Oki, M Masuda, M Kobayashi, Y Nishiba, and S Furuta 2003. "Physiological Functionality of Purple-Fleshed Sweet Potatoes Containing Anthocyanins and Their Utilization in Foods". *Japan Agricultural Research Quarterly* 37 (3): 167–73.

Sugata, M, C -Y Lin, and Y -C Shih 2015. "Anti-Inflammatory and Anticancer Activities of Taiwanese Purple-Fleshed Sweet Potatoes (Ipomoea Batatas L. Lam) Extracts". *BioMed Research International* 2015: 768093.

Sun, H, T Mu, L Xi, M Zhang, and J Chen 2014. "Sweet Potato (Ipomoea Batatas L.) Leaves as Nutritional and Functional Foods". *Food Chemistry* 156: 380–89.

Tiwari, B K, C P O'Donnell, and P J Cullen 2009. "Effect of Sonication on Retention of Anthocyanins in Blackberry Juice". *Journal of Food Engineering* 93 (2): 166–71.

Truong, V-D, R F McFeeters, R T Thompson, L L Dean, and B Shofran 2007. "Phenolic Acid Content and Composition in Leaves and Roots of Common Commercial Sweetpotato (Ipomea Batatas L.) Cultivars in the United States". *Journal of Food Science* 72 (6): C343–49.

Wang, S., S. Nie, and F. Zhu. 2016. "Chemical Constituents and Health Effects of Sweet Potato." *Food Research International* 89: 90–116.

Xu, Y, A Cartier, A Porter, K Lalancette, Z Abraha-Eyob, E N Sismour, H Si, X Wang, N Rehmani, and L Githinji 2018. "Bioactive Compounds and Biological Activity of Extracts from Virginia-Grown Sweet Potatoes Affected by Different Cooking Methods". *Journal of Food Measurement and Characterization* 12 (4): 2591–97.

Yoshimoto, M., R. Kurata-Azuma, M. Fujii, D. -X Hou, K. Ikeda, T. Yoshidome, and M. Osako. 2004. "Phenolic Composition and Radical Scavenging Activity of Sweetpotato-Derived Shochu Distillery by-Products Treated with Koji." *Bioscience, Biotechnology and Biochemistry* 68 (12): 2477–83

Zhang, Z-F, J Lu, Y-L Zheng, B Hu, S-H Fan, D-M Wu, Z-H Zheng, Q Shan, and C-M Liu 2010. "Purple Sweet Potato Color Protects Mouse Liver against D-Galactose-Induced Apoptosis via Inhibiting Caspase-3 Activation and Enhancing PI3K/Akt Pathway". *Food and Chemical Toxicology: An International Journal Published for the British Industrial Biological Research Association* 48 (8–9): 2500–7.

Zhu, Z, Q Guan, M Koubaa, F J Barba, and J He 2017a. "Preparation of Highly Clarified Anthocyanin-Enriched Purple Sweet Potato Juices by Membrane Filtration and Optimization of Their Sensorial Properties". *Journal of Food Processing and Preservation* 41: 3. doi: 10.1111/jfpp.12929

Zhao, J. -G, Q. -Q Yan, R. -Y Xue, J. Zhang, and Y. -Q Zhang. 2014. "Isolation and Identification of Colourless Caffeoyl Compounds in Purple Sweet Potato by HPLC-DAD-ESI /MS and their Antioxidant Activities." *Food Chemistry* 161: 22–26.

Zhu, Z, Q Guan, M Koubaa, F J Barba, S Roohinejad, G Cravotto, X Yang, S Li, and J He 2017b. "HPLC-DAD-ESI-MS2 Analytical Profile of Extracts Obtained from Purple Sweet Potato after Green Ultrasound-Assisted Extraction". *Food Chemistry* 215: 391–400.

Zhu, Z, T Jiang, J He, F J Barba, G Cravotto, and M Koubaa 2016. "Ultrasound-Assisted Extraction, Centrifugation and Ultrafiltration: Multistage Process for Polyphenol Recovery from Purple Sweet Potatoes". *Molecules* 21: 11.

Zhu, Z, H Mhemdi, L Ding, O Bals, M Y Jaffrin, N Grimi, and E Vorobiev 2014. "Dead-End Dynamic Ultrafiltration of Juice Expressed from Electroporated Sugar Beets". *Food and Bioprocess Technology* 8 (3): 615–22. doi: 10.1007/s11947-014-1427-2.

6

Valuable Compounds Extraction from Cereal Waste and By-Products

MANUEL VIUDA-MARTOS, JUANA FERNÁNDEZ-LÓPEZ, AND JOSÉ ANGEL PÉREZ-ÁLVAREZ

Grupo de Industrialización de Productos de Origen Animal (IPOA), Universidad Miguel Hernández, Campus de Orihuela, Orihuela-03312, Alicante, Spain

Contents

1 Introduction

The cereals belong to the grass family (*Gramineae*) and they are cultivated mainly by the edible components that can be found in grains. From a strict point of view, cereals are formed by a caryopsis, which consists of fruit layer (pericarp) and a seed. The pericarp is firmly attached to the seed layer that surrounds the rest of the seed, which is formed by the germ and the endosperm. It is also possible to find another structure called aleurone layer, which is next to the pericarp. This layer is rich in proteins and minerals (Luithui et al., 2019). The endosperm constitutes the central part of the kernel and occupies most of it. This structure is basically composed of starch. On the other hand, the germ also called embryo is a small structure that is found in the lower part of the grain and it is mainly composed of lipids (Delcour and Hoseney, 2010).

For thousands of years, cereals and pulses have been produced and consumed as staple foods. The main cereals consumed by humans throughout the world are barley (*Hordeum vulgare*), corn (*Zea mays*), millet (*Pennisetum typhoides*), oat (*Avena sativa*), rice (*Oryza sativa*), rye (*Secale cereale*), and wheat (*Triticum aestivum* and *Triticum durum*). Cereals are cultivated in more than 73% of the total area harvested in the world. In this way, these types of products contribute to more than 60% of the world food production.

The cereal processing industries produce a great amount of waste and by-products, those might be a cheap source of compounds; in composition of these, it is possible to find a high concentration of bioactive compounds with a high added value, such as vitamins, minerals, polyphenolic compounds, and antioxidant dietary fiber (Ayala-Zavala et al., 2011). These by-products may be used as new ingredients for their application in the development of new foods, potentially with functional effects, since these bioactive compounds can exert positive effects on health and contribute in an effective way in the prevention of the development of various chronic diseases, such as cardiovascular diseases and diabetes, as well as different types of cancer, among others (Viuda-Martos et al., 2010a). The importance of the use of food additives of natural origin is increasing mainly due to the increasing use of natural compounds not only in the food industry but also in other industries, such as the cosmetics and pharmaceutical industries. This is done under the directives issued by the European Union in favor of the use of additives of natural origin instead of synthetic ones (Vági et al., 2007). Therefore, the recovery of compounds with high added value from natural sources is carried out today using the so called "5-stage universal recovery processing" as mentioned by Galanakis (2012). These stages are: (i) macroscopy pretreatment of samples; (ii) macro- and micromolecule separation; (iii) extraction of compounds with high added value; (iv) isolation and purification of products obtained; and (v) elaboration of the new product. In this way, the improvement of production processes could be realized following the short-, medium-, and long-term objectives as mentioned by Laufenberg et al. (2003):

(i) Short-term objectives consist of the reduction of waste generated through the reduction and recycling of valuable substances, obtaining, as a final result, by-products and wastes with lower emissions and environmental risks.
(ii) Medium-term objectives include the development of an efficient production process, adding value to by-products. The final result for the industries is their greater environmental responsibility accompanied by competitive advantages.
(iii) Long-term objectives include the step-by-step improvement in manufacturing process of a product that is environmentally friendly while also developing "innovative products." As a final result, the development of new foods with potentially functional properties may be the key that opens up new markets, in addition to meeting productivity objectives.

This chapter will examine current literature on both the recovery and use of valuable compounds obtained from the cereal waste and by-products, with a particular focus given to the main cereals consumed by humans around the world, such as rice, wheat, and corn.

2 Bioactive Compounds Present in Cereals and Its By-Products

At present, it is widely accepted by both the scientific community and the consumers that a diet rich in cereals has a beneficial effect on health since it has a very important influence on the prevention of the development of various chronic diseases, such as cardiovascular diseases and certain types of cancer. Likewise, it is widely demonstrated that this beneficial activity of the organism is due to its bioactive components present in its composition (Galaverna et al., 2008). The existence of great amounts, in cereals, of different bioactive compounds, such as vitamins, minerals, phenolic acids, flavonoids, sterols and stanols, tocopherols, and dietary fiber, ensures them a considerable nutritional value (Saura-Calixto and Goñi, 2005).

2.1 Dietary Fiber

Dietary fibers comprise a very diverse group of substances, which present the peculiarity that they can not be digested by the enzymes of the digestive apparatus of humans. Thus, World Health Organization and Food and Agriculture Organization (2008) defined fiber as carbohydrate polymers with 10 or more monomeric units, which are not hydrolyzed by the endogenous enzymes in the small intestine of humans.

2.1.1 Classification Dietary fiber or total dietary fiber (TDF), as it is known, consists of insoluble dietary fiber (IDF) as well as soluble dietary fiber (SDF) (Wang et al., 2002). The solubility refers to the ability of this fiber to disperse in water, although it must be taken into account that it is a somewhat inaccurate term (Figuerola et al., 2005).

When this form of classification was defined, it was thought that it would be a very simple way of predicting the physiological function that fiber would have, but really, in many cases, it has not always been so (Gallaher and Chneeman, 2003). Thus, SDF has been related to the decrease in the cholesterol levels in blood, as well as a decrease in the glucose absorption capacity of intestine. On the other hand, IDF has been associated with water absorption and intestinal regulating aspects (Angiolone and Collar, 2011). A classification of dietary fiber is shown in Figure 6.1.

Lignin, cellulose, and several hemicelluloses can be classified as IDFs and they have a structural function. All vegetables, as well as cereals, are a very important source of IDF. Among the cereals, corn and wheat have greater amount of this type of fiber (Dhingra et al., 2011). From a physiological point of view, insoluble dietary is responsible for a greater volume of feces and also responsible for helping the regulation of peristaltic movements of the large intestine. In cereals, the main components of the insoluble fiber can be found in cellulose. This compound is the main component of the cellular structure of plants providing strength and stability (Paster et al., 2003). The amount of cellulose present in a fiber influences their technofunctional, physiofunctional, and physiological properties, as well as the use of this fiber in several applications. Lignin is a highly complex molecular structure formed by a framework with an amorphous structure. Its main function is to act as a binder between the individual cells and between the fibrils that make up the cell wall (Mohanty et al., 2000). The

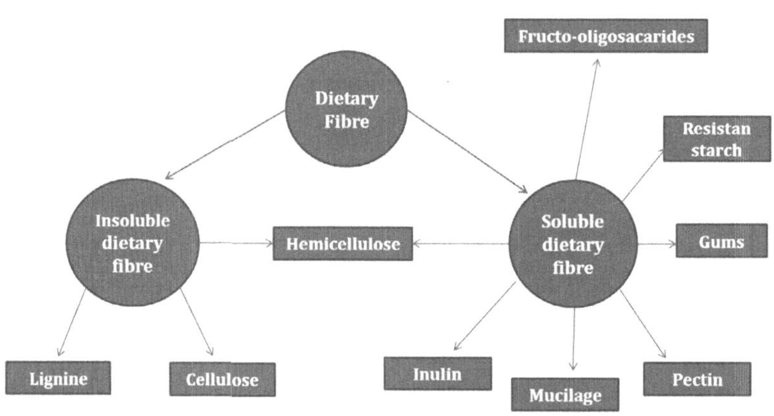

Figure 6.1 Classification of dietary fiber.

lignin is first formed between the neighboring cells in a "medium lamella," joining them firmly in a tissue and then extending toward the cell wall penetrating the hemicelluloses and joining the cellulose fibrils (Majumdar and Chanda, 2001).

The hemicellulose in the plants is formed by a very slightly branched structure composed basically of different polysaccharide polymers with a lower degree of polymerization and orientation than the cellulose or lignin may have (Jones et al., 2017). The main function of hemicellulose is to act as a filling agent for cellulose and lignin spaces. In its composition, you can find various sugars mainly glucose, xylose, galactose, arabinose, and mannose (Reddy and Yang, 2005). The main hemicellulose present in cereals and its by-products is arabinoxylan. Arabinoxylan is a hemicellulose comprising of backbone chains of β-(1-4)-linked D-xylopyranosyl residues to which α-l-arabinofuranose units are linked as side chains. Depending on the source from which the arabinoxylans are produced, there will be a significant variation in both the relative quantity as well as the distribution sequence of these structural elements. The main source of arabinoxylans is the bran of various cereals, such as rye, barley, oats, wheat, rice, sorghum, and millet (Izydorczyk and Dexter, 2008).

Pectins, gums, mucilages, algae polysaccharides, some polysaccharides, such as resistant starch, as well as some hemicelluloses are the main types of soluble fiber. These types of fibers have the ability to form gels and viscous structures (Perry and Ying, 2016). The main cereals rich in SDFs are oats, barley, as well as some fruits and vegetables (Lambo et al., 2005).

The main SDF found in cereal and cereal by-products is β-glucan. These compounds are the predominant constituents of cell walls of several cereal kernels, such as barley and oats (Brennan and Clearly, 2005). It is a linear unbranched polymer molecule comprising the β-(1-4) and (1-3) link units of glucopyranosyl. β-glucan is extremely soluble in water and loses its regularity as it dissolves in it (Havrlentova et al., 2011; Sidhu et al., 2007). The content of β-glucan in cereal bran ranges from 1 to 3% in wheat, 3 to 7% in oats, and 5 to 24% in barley (Skendi et al., 2003). As mentioned earlier, the main health benefits associated with an adequate intake of dietary fiber with an optimal ratio of IDF:SDF (3:1) include an increase in stool evacuation velocity, an increase in the feeling of satiety, and the reduction in the risk of developing various chronic diseases, such as type 2 diabetes, coronary heart disease, and, especially, certain types of cancer, such as colon cancer (Kendall et al., 2010; Viuda-Martos et al., 2010b).

2.2 Polyphenolic Compounds

Polyphenols are considered secondary metabolites of higher plants and due to their characteristics are considered as phytochemicals since they naturally give protection to plants against various pathogens, such as fungi and/or bacteria, against ultraviolet light.

They can also act as phytoalexins protecting the plant from predators or increasing the astringency of the fruits, which makes the food quite unpleasant for its consumption (Luithui et al., 2019). The polyphenolic compounds are mainly composed of phenolic acids and flavonoids. Phenolic acids are found in most of the plants and fruits, in which they contribute to their color and pleasant taste. In nature, it's possible to find two types of phenolic acids: those derived from benzoic acid and those derived from the cinnamic acid. Hydroxybenzoic acids, as shown in Figure 6.2a, are components of complex structures, such as hydrolyzable tannins (gallotannins or ellagitannins) (Juurlink et al., 2014), while hydroxycinnamic acids (Figure 6.2b) are more common in cereals than hydroxybenzoic acids. These compounds basically consist of *p*-coumaric, caffeic, ferulic, and sinapic acids (El Gharras, 2009). They possess nutraceutical potential as anti-inflammatory, anticarcinogenic, antimicrobial, and antioxidant activities. Furthermore, these substances have beneficial effects against diarrhea, ulcers, or cardiovascular problems. In cereals, these bioactive compounds are predominantly found in the outermost layers of the bran (Chandrasekara and Shahidi, 2010; Gorinstein et al., 2007; Sidhu et al., 2007).

Flavonoids are naturally occurring substances in cereals, vegetables, and fruits, and it is widely demonstrated that they can exert positive effects on various aspects of human health. Chemically, they are low molecular weight compounds consisting of 15 carbon atoms, arranged in a C_6–C_3–C_6 configuration. Flavonoids can be classified into seven subclasses depending on the type of heterocycle involved, such as flavonols, flavones, isoflavones, flavanones, anthocyanidins, and anthocyanins (Figure 6.3). These compounds have the ability to protect the body against the action produced by the reactive oxygen species in addition to showing a wide variety of biological effects, among which its antioxidant activity stands out (Sreeramulu and Raghunath, 2010). There are several studies, such as those conducted by Crespo et al. (2008), where it has been reported that flavonoids have the ability to inhibit a large number of enzymatic packages, mainly those related to oxidation systems, such as 5-lipoxygenase, cyclooxygenase, monooxygenase, or xanthine oxidase.

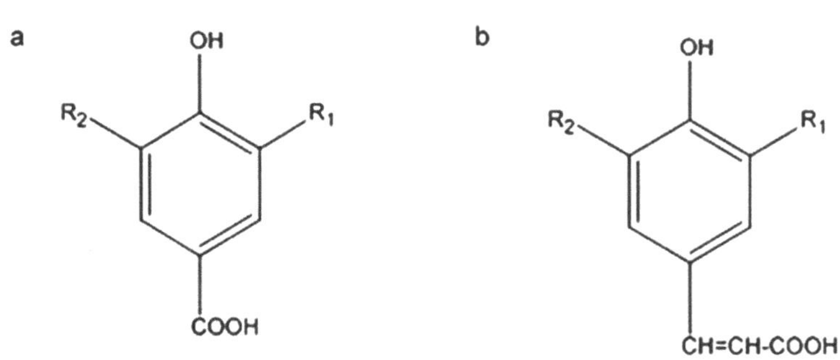

Figure 6.2 Phenolic acids structures as example of common fruit natural antioxidants.

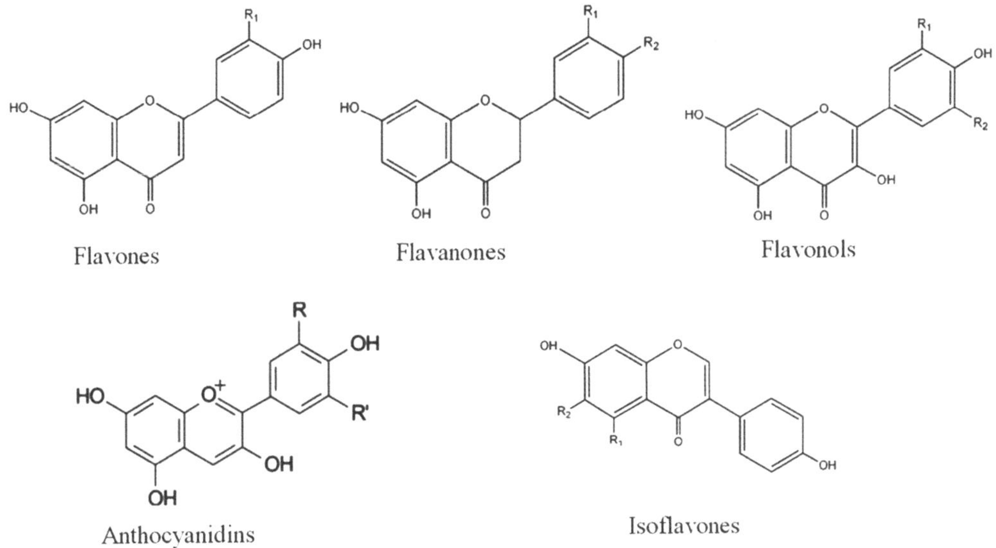

Figure 6.3 Structure of flavones, flavanones, flavonols, anthocyanidins, and isoflavones.

2.3 Tocopherols and Tocotrienols

Tocols, basically formed by tocopherols and tocotrienols, are a series of chemical compounds called vitamin E (Figure 6.4). Cereals are a very important source of vitamin E, since in its composition, it has both tocopherols and tocotrienols. However, it must be taken into account that its concentration and composition depend very much on the type of cereal as well as on the part of the cereal used, whether the endosperm, germ, or bran. This group includes eight naturally occurring isomers: four tocopherols (α-tocopherols, β-tocopherols, γ-tocopherols, and δ-tocopherols) and four tocotrienols (α-tocotrienols, β-tocotrienols, γ-tocotrienols, and δ-tocotrienols) (Moore et al., 2005; Nielsen and Hansen, 2008). The main source of tocopherols is oils obtained from various vegetables, although it is also true that it is possible to find considerable amounts of tocopherols in some cereals.

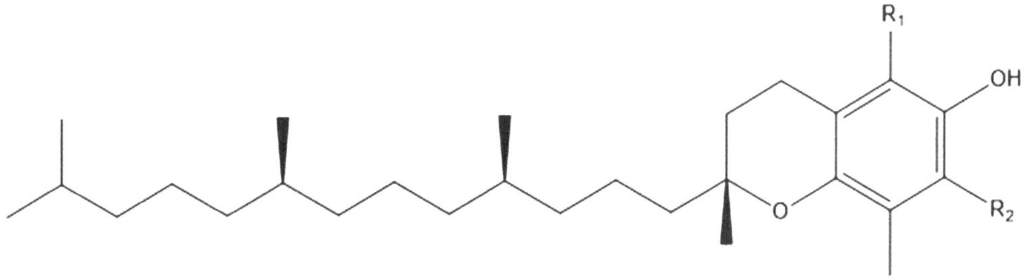

Figure 6.4 A tocopherol molecule, the side groups (R1 and R2) can be either hydrogen or a methyl group.

Thus, since cereals are a very important part of the diet of a large part of the world population; they, thus, become a very important source of tocopherols for these populations. Recent research shows that the considerable consumption of tocopherols exerts several beneficial effects on health, such as a reduction in the development of cardiovascular diseases, as well as various types of cancer (Das Gupta and Suh, 2016; Ju et al., 2009; Mathur et al., 2015).

2.4 Sterols and Stanols

Plant sterols are essential components of cell membranes and are present in all plants. They are structurally comparable to cholesterol, with differences in the lateral chain attached to the steroid ring (Carr and Jesch, 2006). All sterols have a chemical structure formed by four joined rings of 1,2-cyclopentanopehydrophenantren with hydroxyl group located in the position C3. In nature, it is possible to find more than 200 different compounds that can be classified as sterols. The difference between them is in the chemical structure of the side chain as well as the position in which the double bond is found in the fused ring (Bartłomiej et al., 2012). In cereal grains, the principal sterols include sitosterol, campesterol, stigmasterol, and brassicasterol (Figure 6.5a) (Nyström et al., 2008). Stanols are saturated sterols and are much less abundant in nature than the corresponding sterols. Plant stanols comprise about 5–10% of the total sterol/stanol mixture naturally present in the human diet (Valsta et al., 2004). The principal stanols of cereal grains (Figure 5b) are stigmastanol and campestanol, the concentration of both can reach up to 10–30% of the total of sterols present in the cereal (Piironen et al., 2002). In cereal grains, these compounds can be found mainly in the bran and the germ (Jiang and Wang, 2005; Nurmi et al., 2012). Many by-products generated in the industrialization of cereals present a considerable content of these bioactive compounds and could become potential sources of food ingredients for the development of functional foods that could have beneficial effects on health (Jiang and Wang, 2005; Wanyo et al., 2014).

The significance of phytosterols, as bioactive compounds, is in the ability of these compounds to reduce the levels of total cholesterol in blood, since these compounds compete with cholesterol for the formation of micelles in the intestinal lumen and inhibit the absorption of cholesterol (Liu, 2007; Pradeep et al., 2014). In addition, phytosterols have the ability to act as substances with anticarcinogenic, anti-inflammatory, free radical scavenging activities, and so on (Ribas et al., 2017).

2.5 Carotenoids

Carotenoids are a class of chemical compounds synthesized primarily by higher plants and some algae. In nature, it is possible to find more than 600 different

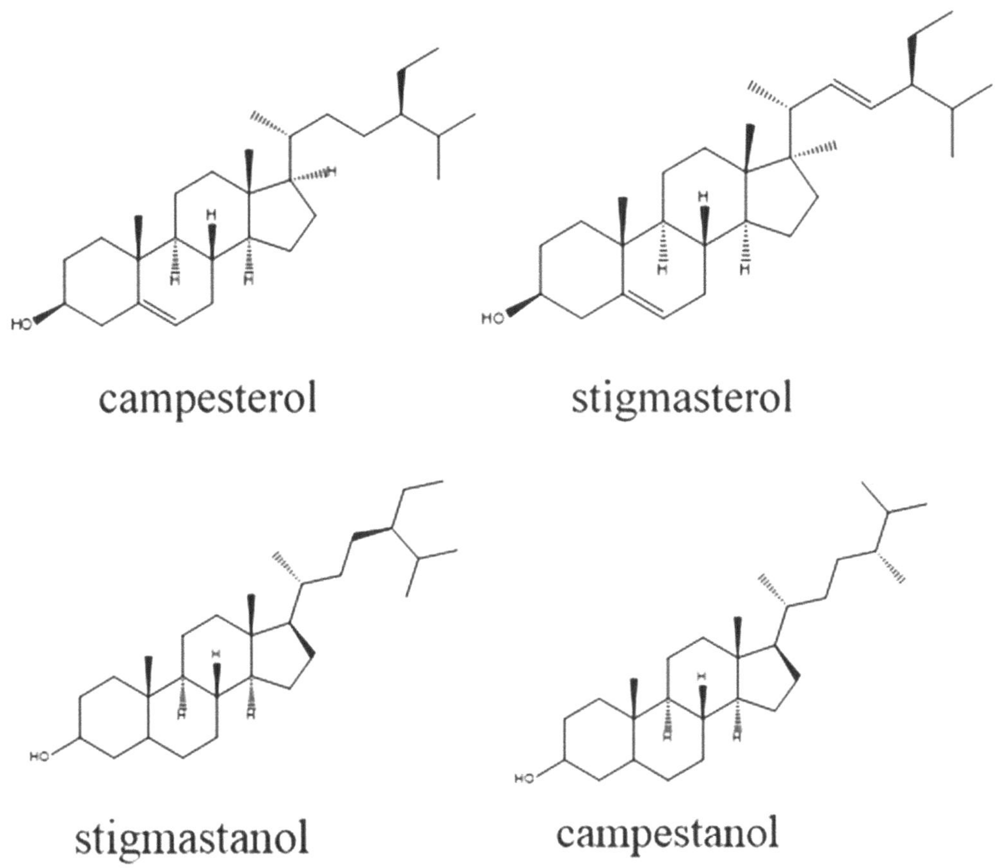

Figure 6.5 (a) Structure of the main stenols (campesterol and stigmasterol) and (b) structure of the main stanols (stigmastanol and campestanol) found in cereals.

types and its main function is to act as pigments in flowers, fruits, and leaves. The range of colors is very wide and ranges from red, orange red to yellow. In addition, these types of substances have provitamin A activity (Ribayo-Mercado et al., 2000). These compounds are characterized by a chemical structure of linear polyisoprene with conjugated double bonds, either as such (lycopene), or as derivatives of the cyclization of the two extremities, with oxidation (xanthophylls, such as lutein and zeaxanthin) or without oxidation (carotenes). Carotenoids are basically lipophilic substances, especially those that do not have hydroxyl groups in their structure (Noziere et al., 2006). Currently, there are more than 600 natural carotenoids identified in the nature. At least five of them, such as β-carotene, β-cryptoxanthin, α-carotene, lutein, and zeaxanthin, have been detected in the blood (Figure 6.6), which represents approximately more than 95% of the total carotenoids present in the blood (Maiani et al., 2009). These bioactive compounds due to a chemical structure with a large amount of conjugated double bonds show a series of very

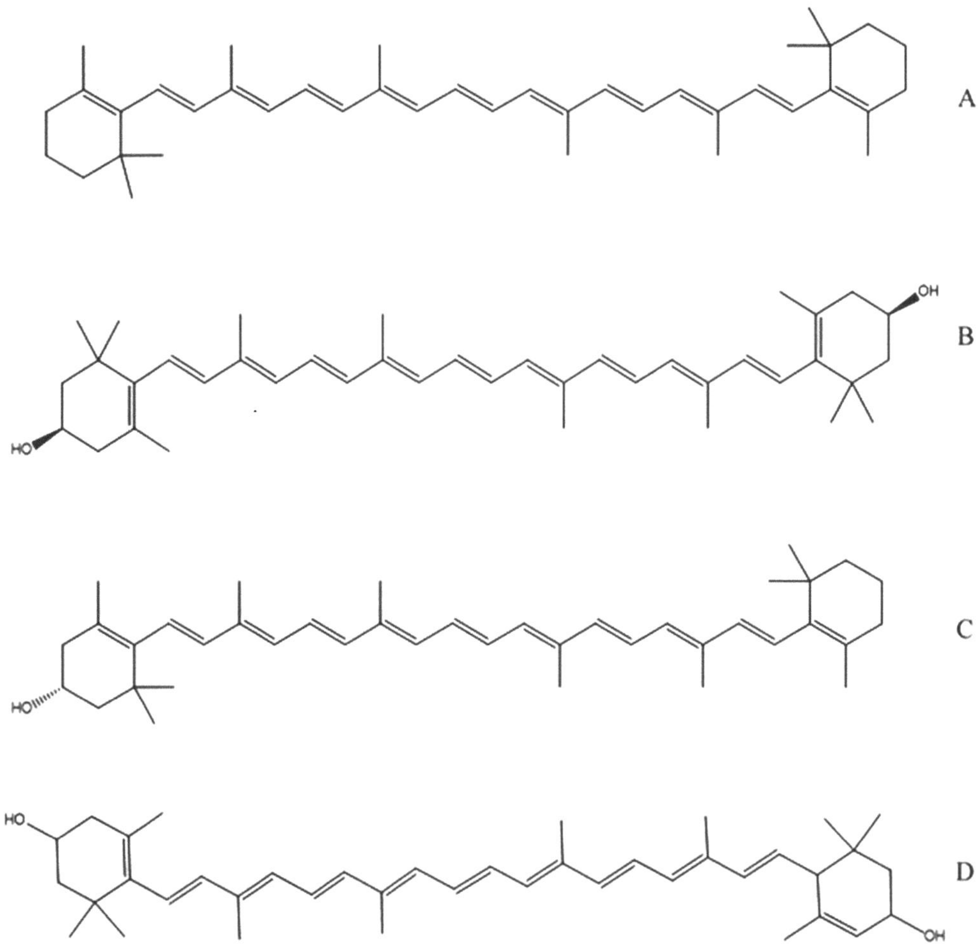

Figure 6.6 The main carotenes present in cereals: (a) β-carotene; (b) zeaxanthin; (c) criptoxanthin; and (D) lutein.

interesting properties, among which the antioxidant activity can be highlighted (Riccioni, 2009).

In cereals the type and concentration of carotenoids is very varied. In general, the main carotenoid found is lutein, followed by zeaxanthin and β-cryptoxanthin. All of them are considered as xanthophylls. It is also possible to find, although in much lower concentration, carotenoids, the main ones being α-carotene and β-carotene. In cereal grains, carotenoids are usually found in greater quantities, mainly in the embryo. The highest concentration of carotenoids can be found in the embryo, though this part of the seed only accounts for 3–5% of its total weight (Mellado-Ortega and Hornero-Méndez, 2015). On the other hand, in the endosperm, which holds between 80 and 85% of the weight of the cereal, the highest concentration of carotenoids can be found. The type and concentration of these compounds can vary depending on various factors such as different genotypes within the same type of cereal (Siebenhandl et al., 2007). In the same way, the distribution of the different types of carotenes within the cereal is very different, for example, carotenes, such as α-carotene, β-carotene, and zeaxanthin, are usually found in bran and germ, whereas lutein is distributed in a more homogeneous way (Borrelli et al., 2008; Konopka et al., 2004; Ndolo and Beta, 2013; Panfili et al., 2004).

3 Rice

Rice (*Oryza sativa* L.) is the main cereal crop grown across the world after wheat and corn (Altan et al., 2008). There are 18,000 varieties of rice and it is grown in more than 100 countries represented by around 25% of the world cereal production (Hernandez et al., 2000). In rice, it is possible to find bioactive compounds such as vitamins, minerals, phenolic acids, and flavonoids compounds such as anthocyanins, γ-oryzanol, tocopherols, tocotrienols, sterols, and dietary antioxidant-fibers, which have been shown to have various beneficial effects on human health (Mahanta and Sikia, 2016). The milling process of the rice grain (Figure 6.7) produces, as a main product, about 70–72% of rice formed, basically by the endosperm. While the by-products that are achieved are constituted by the rice husk that represents 20%, the rice bran with a content between 8 and 12%, and finally, the germ of rice that constitutes 2% of the total weight of the rice seed (Van Hoed et al., 2006).

3.1 Rice By-Products

3.1.1 Rice Husk Rice husk is the main by-product generated in the rice milling process. At present, most of the rice husk generated in the grinding of grains is not used for any other purpose or if it is then in a very low quantity. The reason why its use is almost null is due to the composition of this by-product where a hard surface is joined by a very low nutritive value; in addition, this by-product showed high

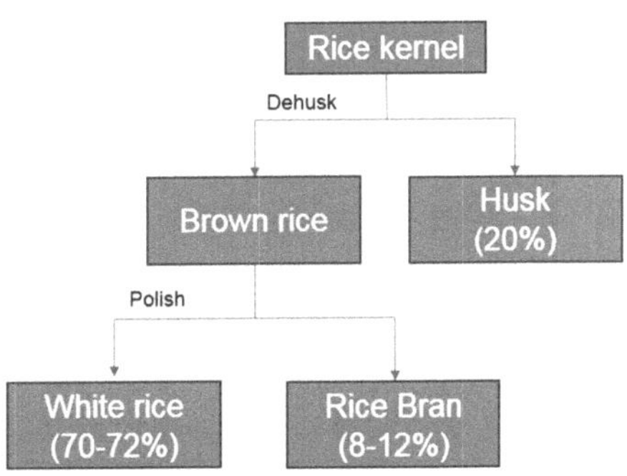

Figure 6.7 Milling process of rice kernel.

concentrations of silicon. On the other hand, rice husk had a low apparent and the decomposition produced by the bacteria discourages its use (Zou and Yang, 2019). The chemical composition of rice husk varies in different samples due to several factors, including season, geographical location, varieties, environmental conditions, soil, and fertilizer used. As mentioned earlier, the use of rice husk as potential ingredient for the development of new products is practically null.

3.1.2 Rice Bran Bran, combined with the husk, is the main by-product obtained in the rice milling process. Its by-product is formed by the inner pericarp, testa, aleurone, subaleurone, and occasionally, the embryo, small part of broken rice is also possible to find. In its composition, it is possible to find a large number of compounds with nutritional properties, among which are different types of lipids. In addition, it also presents a high concentration of bioactive compounds that could exert diverse health benefits (Khuwijitjaru et al., 2007). The chemical composition of rice bran varies, since it depends on many factors, such as the variety of the rice used, environmental conditions, weather, the process of collection and storage, the type of milling carried out, among others (Goufo and Trindade, 2017). In general, the rice bran is formed by proteins with 11–13 g/100 g; dietetic fiber with 11–12 g/100 g; fat with 15–20 g/100 g; and the rest between 40 and 45 g/100 g consists of carbohydrates (Moongngarm et al., 2012).

Rice bran also contains high concentrations of bioactive compounds. Among them, the most important is γ-oryzanol, which is formed by combination of ferulic acid esters of triterpene alcohols and sterols. Miller and Engel (2006) analyzed the concentration of γ-oryzanol in several rice samples obtained from different cultivars cultivated in Italy, France, and Spain during 2000, 2001, and 2002. These authors reported that the γ-oryzanol concentration was comprised between 262.0 and 627.0 mg/kg. They reported that this variation could be explained due to a different medioambiental factors rather than the stage of rice grain maturity. Similarly, γ-oryzanol content in different brown rice cultivars cultivated in several Brazilian locations was investigated by Carlos Iglesias Pascual et al. (2013). These authors reported that values of γ-oryzanol comprised between 183.2 and 192.2 mg/kg.

Other very important bioactive compounds present in rice bran are tocopherols and tocotrienols (Goufo and Trindade, 2014). Rice sterol and stanol content is reported to be in the range of 17–24 mg/kg (Heinemann et al., 2008), whereas rice bran oil contains higher tocol concentration (196–219 mg/kg) compared with brown rice (46.4–65.0 mg/kg) (Aguilar-Garcia et al., 2007). In rice bran, also, it is possible to find high concentration of polyphenols compounds, mainly flavonoids, anthocyanins, pro-anthocyanidins, and phenolic acids. Thus, the major phenolic acids in rice bran include ferulic and *p*-coumaric acids (Zhang et al., 2010). In this way, Tuncel and Yılmaz (2011) analyzed the rice variety "Osmancık-97", which is one of the principal rice variety cultivated in Turkey. These authors reported that the main phenolic acids found were ferulic and *p*-coumaric acids, and their concentrations were ranged between 119.98 and 18.10 μg/g, respectively. Similarly, Huang and Lai (2016) reported that the main bound phenolic acids in the rice brand obtained from black rice were ferulic acid (1,295–1,601 μg/g dry matter), protocatechuic acid (948–1,043 μg/g dry matter), vanillic acid (335–652 μg/g dry matter), and *p*-coumaric acid (140–232 μg/g dry matter). Anthocyanins and proanthocyanidins/condensed tannins are common pigments found in black or red rice, respectively. Thus, two major anthocyanins, such as cyaniding-3-glucoside and peonidin-3-glucoside, are present in pigmented rice (Hu et al., 2003). In a similar study, Hou et al. (2013) identified four different anthocyanins, namely cyanidin-3-glucoside, peonidin-3-glucoside, cyanidin-3,5-diglucoside, and cyanidin-3-rutinoside, in samples of black rice.

Regarding the carotenoid content, it is obvious that that the colored rice varieties are those that present some interest from the point of view of their composition in bioactive compounds (Belefant-Miller and Grace, 2010). In conventional rice grains, the concentration of compounds, such as carotenoids, is predominant in the bran. Thus, the grinding process in which the bran is eliminated results in the major loss of these compounds (Tan et al., 2005). As in the case with the rest of the cereals, the carotenoid that is in greater concentration is lutein (Belefant-Miller and Grace, 2010), although it is also possible to find zeaxanthin and/or β-carotene, especially in varieties of black or dark rice (Frei and Becker, 2005; Kim et al., 2010; Lamberts and Delcour, 2008; Nakornriab et al., 2008). In this way, Nakornriab et al. (2008) reported that the main carotenoid found in several cultivars of black rice cultivated in Thailand was β-carotene with concentrations up to 40 μg/g in its bran.

Other bioactive compounds found in rice bran are: γ-aminobutyric acid (GABA) (Kim et al., 2015), octacosanol (Chen et al., 2007), phytic acid (Canan et al., 2011), and sterols and stanols (Sugihara et al., 2010).

In this way, rice bran with a significant amount of bioactive compounds, such as polyphenols and/or tannins, makes it of considerable nutritional value. In addition, it is generally accepted that the beneficial effects for human health in the prevention of several illnesses are due to the amount of bioactive compounds of this by-product. Table 6.1 presented the main bioactive compounds present in rice bran and their health effects.

3.1.3 Rice Bran Oil The oil content present in the rice bran is approximately 12–18%. This oil is formed, in general terms, by monounsaturated fatty acids (47%), polyunsaturated fatty acids (PUFA) (33%), and saturated fatty acids (20%). It is also possible to find its composition in unsaponifiable fatty compounds (4.3%) (Yadav and Joyner, 2014). The rice oil has high concentration (1.21 mg/g) of vitamin E, mainly

Table 6.1 Main bioactive compounds present in rice bran and their health effects

FUNCTIONAL COMPONENT	HEALTH BENEFITS	REFERENCE
γ-oryzanol	Reduction of tumor incidence, inhibition of platelet aggregation, and anti-inflammatory activity	Lerma-Garcia et al. (2009)
	Antioxidant properties, total-and-LDL-cholesterol-lowering and HDL-cholesterol-increasing effects	Rong et al. (1997)
Tocotrienols and tocopherols	Inhibit cholesterol synthesis, lower serum cholesterol levels in various animal models, and suppress tumor cell proliferation	Qureshi et al. (2000)
	Antioxidant activity against cholesterol oxidation	Xu et al. (2001)
Phenolic acids	Exert cardioprotective effects on patients with coronary heart disease by improving plasma antioxidant levels and inhibiting inflammatory factors	Wang et al. (2007)
	Decrease atherosclerotic plaque formation and increase antioxidant levels in rabbits	Ling et al. (2001)
	Evaluation of gene expression in adipocytes involved in lipid accumulation was reduced, with a concomitant induction of the genes involved in lipid hydrolysis	Kim et al. (2013)
Anthocyanins	Suppresses both reactive oxygen species and nitric oxide in chemical and biological model systems	Hu et al. (2003)
	Inhibitory effect on cancer cell proliferation	Chen et al. (2006)
	High protection of endothelial cells from oxidative stress events	Zhang et al. (2006)
	Considerable protection against angiogenesis induced by vascular endothelial growth factor	Tanaka et al. (2011)

tocopherols and tocotrienols, as well as γ-oryzanol (3.83 g/100g), phytosterols (0.60 g/100g), and policosanol compounds (0.33 g/100g) (Wongwaiwech et al., 2019).

3.1.4 Defatted Rice Bran The rice bran, given the high content of lipids in its composition, is very easy to deteriorate by oxidative processes leading to a significant loss of bioactive compounds, as well as their nutritional value. Therefore, one way to avoid this degradative process is by defatting and drying the rice bran (Yadav and Joyner, 2014). After the extraction of oil, the defatted rice bran residual can be used for obtaining other products with a high added value, such as dietary fiber (Nandi and Ghosh, 2015). The chemical composition of dietary fiber obtained from rice bran after the oil extraction is mainly composed of cellulose, lignin, and hemicellulose (Chinma et al., 2015). Dietary fiber obtained from rice bran showed a very high water-holding capacity, as well as high oil-holding capacity (Daou and Zhang, 2014), showing its potential to be used as an ingredient in the development of new foods, such as bakery or meat products. It can also be used for its technofunctional properties, so it can be used for stabilizing structures or preventing syneresis or modifying the structural matrix of the food (Chinma et al., 2015; Kaur et al., 2012; Sairam et al., 2011).

4 Wheat

Wheat belongs to the genus *Triticum* and is classified into hard (*Triticum durum*) and soft (*Triticum aestivum*) types. Among the cereal grains, wheat accounts for around 29% of the total cereal production and is considered as the most popular cereal since 72% of the total production is destined for human consumption (Krishna and Chadrasekaran, 2013). Wheat grain have carbohydrate with values around 78.0%, protein with values around 14.0%, fat with values around 2.0%, and minerals with values around 2.0%, and considerable proportions of vitamins, mainly thiamine and vitamin B complex (Topping, 2007).

In the grain of integral wheat, it is basically possible to find three clearly differentiated structures, such as endosperm, bran, and germ. The endosperm, which is the most important part (around 80–85%), is formed mainly by starch and proteins and is surrounded by several external layers called "aleurone." A large number of bioactive compounds can be found in aleurone, as well as various dietary fibers, such as arabinoxylans and β-glucans, and some proteins (Hemery et al., 2010). The bran, which constitutes around 13–17%, is composed of pericarp and testa. The testa is the tissue rich in lignin, which is also characterized by a high content of lipid compounds, such as alkylresorcinols (Hemery et al., 2010). The pericarp lies in the outermost layers, which are composed of three layers, the epicarp, mesocarp, and endocarp (Yu et al., 2015), while the wheat germ that constitutes around 2–3% are formed of the embryo and the scutellum, which are separated from the endosperm by the aleurone. The wheat kernel milling (Figure 6.8) produced two main by-products, the brand and the germ.

4.1 Wheat By-Products

4.1.1 Wheat Bran After wheat grain milling, the bran obtained as a by-product is about 15%.

Wheat bran is a material composed of several layers, such as pericarp, which is rich in branched heteroxylans and cellulose; testa, which is rich in lipidic compounds;

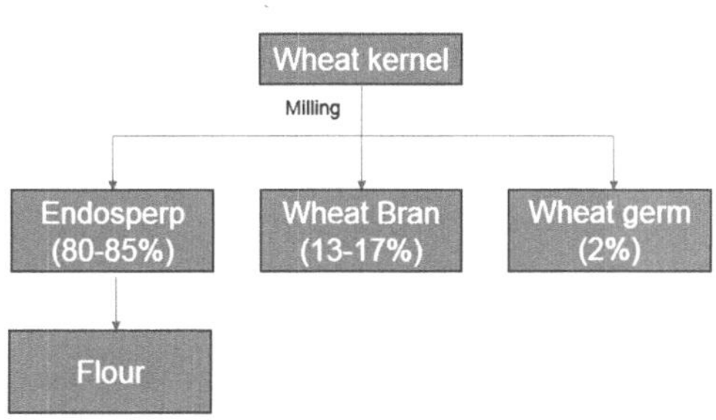

Figure 6.8 Milling process of wheat kernel.

and aleurone, which is rich in vitamins and tocols and germ (Šramková et al., 2009; Hemery et al., 2011a, 2011b), enclosed by thick cell walls formed of arabinoxylans and β-glucans (Luithui et al., 2019). In general terms, wheat bran is composed of water (12%), proteins (13–18%), fats (3.5%), and carbohydrates (56%) with an elevated content of bioactive compounds, such as dietary fibers, polyphenolic compounds, vitamins, and minerals (Apprich et al., 2014; Yan et al., 2015).

Dietary fiber is the main bioactive compound found in wheat bran. The chemical composition of wheat bran fiber is complex, but it contains, essentially, cellulose and pentosans, polymers based on xylose and arabinose, which are tightly bound to proteins. These substances are typical polymers present in the cell walls of wheat and layers of cells, such as aleurone layer (Šramková et al., 2009). Kahlon and Chow (2000) investigated the TDF, IDF, and SDF of wheat bran obtained from wheat cultivated in the United States. These authors reported a TDF content of 52.20 g/100 g with an insoluble IDF content of 48.40 g/100 g, while the SDF content was 3.80 g/100 g. In a similar study, Nandini and Salimath (2001) analyzed the TDF, IDF, and SDF of wheat bran obtained from wheat cultivated in India. These authors found that the wheat bran had a TDF content of 39.80 g/100 g with an IDF content of 35.80 g/100 g and SDF content of 4 g/100 g. In this way, Esposito et al. (2005) studied different fractions of wheat bran with different functional and nutritional characteristics obtained by an industrial milling process. These authors mentioned that the SDF content of the wheat bran ranged between 0.9 and 4.1 g/100 g sample, while that of IDF between 21 and 64 g/100 g sample.

The SDFs and IDFs found in wheat bran are composed, basically, of arabinoxylan (Du et al., 2009; Hollmann and Lindhauer, 2005) and β-glucan (Li et al., 2006). Arabinoxylans can be water-soluble or water-insoluble; flexibility of the arabinoxylan molecules is extremely dependent on xylose and arabinose subunits ratio. Only some portion of the arabinoxylans is soluble and majority of the arabinoxylans is insoluble. Du et al. (2009) analyzed the arabinoxylan content present in Consort, a soft wheat variety harvested in 2003 in the United Kingdom. These authors reported that the arabinoxylan content comprised between 16.6 and 26.80 g/100 g of sample.

Similarly, Hollmann and Lindhauer (2005) measured the arabinoxylan content present in wheat bran obtained from reins of the wheat cultivar Bussard, harvested in 2001 and cultivated in Switzerland. They reported a arabinoxylan content of 25 g/100 g. Wheat-β-D-glucans are mainly located in the aleurone layer of the wheat kernel. Hemery et al. (2007) reported concentrations around 2.9 and 4.3 g/100 g of β-glucans in wheat bran and aleurone, respectively, while Sovrani et al. (2012) reported values of 1.5–1.9 and 1.4–1.5 g/100 g of β-glucan in the pearling fractions of different wheat varieties.

Among phenolic acids, the bran layers contain the majority of phenolic acids. Most phenolic acids exist in three forms: (i) soluble free acids, (ii) soluble conjugated moieties esterified to sugars and other low molecular mass compounds, and (iii) insoluble bound moieties esterified to the cell wall structural components (Brouns et al., 2012). The predominant phenolics in wheat bran are benzoic acid derivatives, such as gallic, vanillic, and syringic acid, and cinnamic acid derivatives, such as ferulic, p-coumaric, and caffeic acid, of which, ferulic acid is the most potent in wheat bran and makes from 70 to 90% of the total derivatives (Moore et al., 2006). Li et al. (2008) informed that ferulic and *p*-coumaric acids are mostly found in the bound form (~92% and ~63%, respectively), whereas sinapic and vanillic acids are mostly found in the conjugated form (~69% and ~67%, respectively). Thus, Brewer et al. (2014) analyzed the total phenolic compounds present in the bran obtained after milling mixed Kansas (*Triticum aestivum* L.) hard red winter wheat. These authors reported a total phenolic content comprising between 4.73 mg ferulic acid equivalents/g of sample in coarse treatment and 6.72 mg ferulic acid equivalents/g sample in the fine treatment. Flavonoids are also present in wheat bran. Thus, Dinelli et al. (2009) informed that 32 flavones were identified in brans obtained from several wheat varieties, comprised of 21 C-glycosidic forms, 7 *O*-glycosidic derivatives, and 4 aglycone compounds.

Other bioactive compounds found in wheat bran are alkylresorcinol (Landberg et al., 2008; Rebolleda et al., 2013), vitamins (Brouns et al., 2012), tocopherols and tocotrienols (Engelsen and Hansen, 2009), or carotenoids (Brewer et al., 2014).

4.1.2 Wheat Germ Together with the bran, wheat germ is considered as the most important by-product of the dry milling industry. This wheat by-product is widely recognized as a nutritious raw material for incorporation into food product formulations or as a food in its own right. Typical applications are in germ-enriched bread, snack foods, and supplements to breakfast cereals (Mahmoud et al., 2015). The main bioactive compounds found in wheat germ are dietary fibers, amino acids, tocopherols, unsaturated fatty acids, and polyphenolic compounds.

In wheat germ, a significant amount of dietary fiber is also found. Therefore, Srivastava et al. (2006) reported a TDF content of 18.4 g/100 g, which is characterized by a greater amount of IDF (15.6 g/100 g) than SDF (2.8 g/100 g). In a similar study, Marti et al. (2014) informed that the TDF of wheat germ was 12.8 g/100 g with high content of IDF (11.3 g/100 g) than SDF (1.5 g/100 g). Bansal and Sudha (2011) reported TDF content in wheat germ as 16.47 g/100 g. Again, the IDF content (14.40 g/100 g) was higher than SDF (2.07 g/100 g).

Other very important bioactive compounds found in wheat germ are the polyphenolic compounds (phenolic acids and flavonoids). In a recent study, Liu et al. (2017) analyzed the total phenolic content of defatted wheat germ. These authors reported a total phenolic content of 10.55 mg gallic acid equivalents/g sample, of which 1.58 mg gallic acid equivalents/g sample were free phenolic compounds and 8.90 mg gallic acid equivalents/g sample were bound phenolic compounds. Zou et al. (2015) analyzed the total phenolic content of wheat germ from China. These authors reported a total phenolic content of 16.49 mg gallic acid equivalents/g sample. In the same study, the polyphenolic profile was determined. Six phenolic acids, including 3,4-dihydroxy benzoic, p-hydroxybenzoic, chlorogenic, caffeic, ferulic, and trans-2-hydroxy cinnamic acids, were found with ferulic acid having the highest concentration. In the same way, de Vasconcelos et al. (2013) analyzed the polyphenolic composition of wheat germ obtained from Porto (Portugal). These authors reported that the main low molecular weight phenolics found in the wheat germ soluble samples were gallic acid, chlorogenic acid, and caffeic acid with values of 8.29, 7.63, and 6.75 mg/kg sample, respectively.

Wheat germ, containing about 8–14% oil (average 10%), is mainly used in food, medical, and cosmetic industries. Physiological and biological characteristics of oils are influenced by the kinds and proportions of the fatty acid components and their position on the glycerol radical; thus, the stereospecific analysis of fatty acids in the triacylglycerol was considered important for the use of the lipid for both the dietary and industrial purposes (Yoon and Kim, 2003). A striking feature of wheat germ lipids was the relatively high level of PUFA, especially linoleic fatty acid, which was estimated at higher levels (55%). Thus, Dunford and Zhang (2003) reported that the wheat germ oil main fatty acids are C18:2, linoleic acid (55.2–56.9% of fatty acid content); C18:1, oleic acid (14.5–14.7% of fatty acid content); and C16:0, palmitic acid (16.4–16.6% of fatty acid content). Other reports stated that the crude fat in corn is predominated by two saturated fatty acids, palmitic (C16:0) and stearic acids (C18:0) and three unsaturated fatty acids, oleic (C18:1), linoleic (C18:2), and alpha-linolenic (C18:3) acids (White and Weber, 2003). In a similar study, Attia and Abou-Gharbia (2011) analyzed the fatty acid composition of wheat germ oil. These authors reported that linoleic acid was the major fatty acid, which accounted for 58.61% of the total fatty acids, whereas linolenic and oleic acids represent 9.58 and 11.21% of the total fatty acids, respectively. Small amount of palmitoleic acid (0.23%) was also present. In another study, the results reported by Srivastava et al. (2006) clearly indicated that germ samples are rich in palmitic acid (18.16–24.75%), apart from unsaturated fatty acids. Among the unsaturated fatty acids, content of linoleic acid and oleic acid were dominant at 46.09–57.63% and 15.94–22.80%, respectively. The variation between reported percentage compositions may be attributed to differences in wheat varieties, growth conditions, storage of the germ, and the extraction method of lipids.

Additionally, a significant quantities of bioactive compounds, such as tocopherols ranged between 300 and 740 mg/kg dry matter (Hidalgo and Brandolini, 2008), phytosterols comprised between 24 and 50 mg/kg (Nyström et al., 2007), policosanols 10 mg/kg (Irmak et al., 2006), and carotenoids 4–38 mg/kg (Hidalgo and Brandolini, 2008), were also found in wheat germ. Table 6.2 presents the principal bioactive compounds content in wheat bran and germ and their health effects.

Table 6.2 Main bioactive compounds present in wheat bran and germ and their health effects

BY-PRODUCT	FUNCTIONAL COMPONENT	HEALTH BENEFITS	REFERENCE
Wheat bran	Phenolic acids	Inhibition of oxidation of low-density lipoprotein cholesterol and deoxyribonucleic acid	Liyana-Pathirana and Shahidi (2007)
		Reduction in the thiobarbituric acid reactive substances (TBARS) production in LDL	Yu et al. (2005)
		Protect against free radical-induced oxidative damage in human erythrocytes	Wang (2009)
		Inhibits oxidation of LDL, does not initiate a chain reaction	Brouns et al. (2012)
		Reduced risk of colon cancer	Sang and Zhu (2014)
	Dietary fiber	Reduction in serum cholesterol from 5.57 to 4.385 mmol/L	Costabile et al. (2008)
		Decrease in serum total cholesterol, low density lipoprotein (LDL), triglycerides, and total serum cholesterol. LDL/HDL and total cholesterol/HDL ratios decreased significantly	Osfor et al. (2016)
		Control of glycemic index. Postprandial glucose and insulin responses were improved	Jensen et al. (2006)
	Alkylresorcinols	Inhibit platelet binding to fibrinogen, stimulate thromboxane production, and inhibit triglyceride formation, suggesting a potential role for phenolic compounds found in the bran fractions in CVD	Ross et al. (2004)
		In vitro and animal models have indicated that AR may have several bioactivities, including inhibition of some enzymes, delaying copper-induced LDL oxidation	Stasiuk and Kozubek (2010)
		Act as antioxidants since they are able to protect low-density lipoprotein from oxidation damage in vitro, but additionally exert biological effects, such as antimutagenic and antibacterial activity	Parikka and Wähälä (2009)
		High inhibitory effect against human prostrate adenocarcinoma cells	Luthria et al. (2015)
	Arabinoxylans	Fasting reduction in serum glucose, serum triglycerides, and apolipoprotein A-1	Garcia et al. (2006)
		Reduction in postprandial glycemic response due to cross-linked arabinoxylans to maintain their viscosity within the gastrointestinal tract, blunting the postprandial glucose response	Vogel et al. (2012)
		Reduction in the plasma total cholesterol and LDL-cholesterol concentrations by promoting the excretion of fecal lipids, regulating the activities of HMG-CoA reductase, and activity of cholesterol 7-α hydroxylase	Tong et al. (2014)
		Postprandial glucose and insulin responses were improved	Lu et al. (2000)
		Rats treated with the colon carcinogen 1,2-dimethylhydrazine reduced the occurrence of two types of preneoplastic lesions in the colon	
		Reduction in the postprandial glycemic response by maintaining viscosity in the gut, thereby reducing the risk of developing type 2 diabetes	Bernstein et al. (2013)
Wheat germ	Alkylresorcinol	Reduces human colon cancer cell growth	Zhu et al. (2012)
		Exerted high cytotoxic activity against the growth of human prostate adenocarcinoma (PC3) cells	Liu et al. (2012)
	β-sitosterols	The alleviation of physiological symptoms of diabetes, and reduction of fasting and postprandial glucose levels	Haripriya and Premakumari (2010)

5 Corn

Corn (*Zea mays*) is a crop that forms the basis of food for more than a billion people around the world. This crop is of utmost importance for the populations of Latin America and southern Africa where it offers almost one-third of its caloric needs (Díaz-Gómez et al., 2017).

For its use as food, the corn can be subjected to two different grinding processes, such as dry or wet milling. Dry milling is the traditional process of milling of clean grain, obtaining three types of products: endosperm, germ, and bran (Duensing et al., 2003). Dry milling's main objective is obtaining the endosperm of the grain as easily as possible to be used as corn meal and/or flour. On the other hand, separation of the germ can be used for obtaining oil. The last fraction of corn milling, that is, the bran, whose nutritional value is very low, is basically used for animal feed alone or in combination with the cake or corn germ meal (Rose et al., 2010). Wet milling of corn comprises, first, soaking the corn grain in a mix formed by water and sulfur dioxide. During this process, the moisture content increases to approximately 450 g/kg and the grains soften to facilitate the separation of the components: starch, gluten, fiber, and germ (Johnson and May, 2003). The main products obtained with the corn grain wet milling are: the endosperm from which flour or semolina can be obtained later, and the germ that, as in the case of dry milling, is used to obtain oil. The main by-products of this type of wet milling are formed by corn fiber, corn gluten, and soaking solids (Johnson and May, 2003).

Corn is a good source of many nutrients, including proteins, such as zein; vitamins, such as thiamine (vitamin B1), pantothenic acid (vitamin B5), vitamin C, and folate; and minerals, such as phosphorus and manganese. Apart from all these substances, from its composition, it is also possible to find bioactive compounds, such as carotenoids, polyphenolic compounds, antioxidant dietary fibers, and sterols and stanols, which contribute significantly to the multiple benefits that corn has for health (Yadav and Joyner, 2014).

5.1 Corn Dry-Milling By-Products

The main by-products obtained from corn dry milling process are the corn germ and the corn bran. Thus, the dry milling process of corn generated around 60–70 g/kg of corn bran.

5.1.1 Corn Bran The principal components of corn bran are dietary fiber, which contains heteroxylans and phenolic acids. Other compounds, such as starch with values 9–23%, proteins with values 10–13%, lipids with values 2–3%, and ash with values 2%, are too present in corn bran.

As mentioned earlier, the corn bran showed an elevated content of TDF. Thus, Kahlon and Chow (2000) investigated the TDF, IDF, and SDF content of corn bran obtained from corn cultivated in the United States. These authors reported a TDF content of 86.5 g/100 g with an IDF content of 86.3 g/100 g, while the SDF content was 0.2 g/100 g. In a similar study, Dikeman et al. (2006) analyzed the dietary fiber content of several corn bran samples. These authors reported values for TDF ranging between 52.6 and 73.50 g/100 g. Additionally, they reported that all samples are basically composed of IDF with values oscillating between 52.6 and 73.50 g/100 g. The IDF in corn bran is basically composed of cellulose approximately 28 g/100 g and hemicellulose 70 g/100 g, with only a low content of lignin around 1 g/100 g (Rose et al., 2010). On the other hand, the type of hemicelluloses that can be found is formed by a β-(1,4)-xylopyranosyl skeleton with α-L-arabinofuranosyl residues joined in the side chains (Gaspar et al., 2007). It is also possible to find D-glucuronic acid units in the side chains, as well as oligosaccharide side chains containing galactose, xylose, and arabinose.

Other very important bioactive compounds present in corn bran are polyphenolic compounds. Thus, corn bran is a very rich source of phenolic compounds, mostly in its bound form. Although we must not forget that it is also possible to find it, in smaller amount, in its free form. The main phenolic acid found in the bound form is ferulic acid, followed by *p*-coumaric and sinapic acids (Zhao et al., 2014). Depending on the type of corn used, the concentration and type of phenolic acids present varies. In general, the main phenolic acids found in this particular type of structure are ferulic acids in their conjugated form, as well as vanillic, caffeic, *p*-coumaric, and *p*-hydroxybenzoic acids (Das and Singh, 2016). It should be borne in mind that the type of corn has an effect in the concentrations obtained for total soluble phenolics present in the corn bran. In this way, Cabrera-Soto et al. (2009) reported that in dent corn grain a concentration comprising between 232.4 and 334.0 mg gallic acid equivalent/g of sample was obtained when seven different corn varieties were analyzed. In corn pop, Das and Singh (2016) reported a concentration of 0.013 mmol of ferulic acid equivalent/g of sample, while the concentration for quality protein in maize was 0.016 mmol of ferulic acid equivalent/g. Additionally, the insoluble phenolic content in this structure was higher (18–21 times) than soluble phenolics. Das and Singh (2016) informed that the values of soluble phenolics present in bran of two types of corn, such as dent and crystalline corn, were 0.012 and 0.011 mmol of ferulic acid equivalent/g, respectively. On the other hand, the concentrations of insoluble phenolics were 0.218 and 0.219 mmol of ferulic acid equivalent/g for the identical corn grains. In the soluble phenolic fraction, phenolic acids detected were vanillic, caffeic, ferulic, and *p*-coumaric acids, while in the insoluble fraction, the main phenolic acid found was ferulic acid, which was present in a concentration comprising between 40 and 50% (Salinas-Moreno et al., 2017).

One of the most important flavonoids present in the different structures of various types of corn is anthocyanins. These substances provide different color shades ranging from purple to pink depending on the pH and the concentration. Bran corn had the highest content of anthocyanins with values up to 50%, while aleurone comprises a low concentration of these compounds (Luna-Vital et al., 2017). In the bran obtained by milling purple corn, 23 anthocyanins were identified, including pelargonidin-3-glucoside, cyanidin-3-glucoside, delphinidin-3-glucoside, petunidin-3-glucoside, peonidin-3-glucoside, and malvidin-3-rutinoside (Ramos-Escudero et al., 2012). In the corn bran obtained from colored corn varieties, various compounds that can be classified as flavonoids were also found, among which include rutin, morin, quercetin, naringenin, or kaempferol (Salinas-Moreno et al., 2017).

5.2 Corn Wet-Milling By-Products

In the corn wet milling process, the main products obtained are corn fiber, corn gluten, and soaking solids. The most important of all of these and that may have a potency use is the corn fiber, which is composed of a mixture of corn husks and residual starch that is not extracted during the wet milling process. This part represents approximately 11% of the corn grain.

In general terms, in the composition of the corn fiber, it is possible to find a small amount of protein and some oil so it becomes a low-value food for animal feed. On the other hand, it is important to highlight the high content that is present in two types of fiber, such as cellulose and hemicellulose, which show antioxidant activity (Yadav and Joyner, 2014). Thus, Liang et al. (2014) reported that the main compounds present in corn fiber were carbohydrates, composed of 17.2% cellulose, 8.8% starch, 27.6% of xylan and galactan, and finally, 12.1% arabinan.

6 Conclusion

The valorization of food wastes and by-products has become a major subject of research for improving the sustainability of the food chain.

The valorization of the waste and by-products generated in the different agro-food industries has become one of the main topics of research by the scientific community with the sole purpose of implementing systems that improve the sustainability of the food chain. In this way, there are several by-products obtained from cereals after the milling process, which because of its composition is possible to find large quantity of bioactive compounds that can have a beneficial effect on health, among which are antioxidant dietary fibers, various minerals and vitamins, polyphenolic compounds, such as phenolic acids and flavonoids, as well as phytosterols. However, it must be borne in mind that all these residues generated in the cereal milling industry are underutilized and, as a general rule, are only used as livestock feed.

These valuable ingredients that are in high concentration in the by-products generated in the grain milling industry may be extracted from a low-cost material depending on the availability of adequate technology, converted into commercial products as raw materials for secondary processes called intermediate ingredients, or can also be used as ingredients for the advancement of new products, and thus taking advantage of the huge concentration of potentially beneficial compounds they contain.

References

Aguilar-Garcia, C., Gavino, G., Baragaño-Mosqueda, B., Hevia, P., and Gavino, V. C. 2007. Correlation of tocopherol, tocotrienol, goryzanol and total polyphenol content in rice bran with different antioxidant capacity assays. *Food Chemistry* 102:1228–1232.

Altan, A., McCarthy, K. L., and Maskan, M. 2008. Evaluation of snack foods from barley-tomato pomace blends by extrusion processing. *Journal of Food Engineering* 84 (2):231–242.

Angiolone, A., and Collar, C. 2011. Physicochemical and nutritional properties of reduced-caloric density high-fibre breads. *LWT-Food Science and Technology* 44:747–758.

Apprich, S., Tirpanalan, Ö., Hell, J., Reisinger, M., Böhmdorfer, S., Siebenhandl-Ehn, S., and Kneifel, W. 2014. Wheat bran-based biorefinery 2: valorization of products. *LWT-Food Science and Technology* 56(2):222–231.

Attia, R. S., and Abou-Gharbia H. A. 2011. Evaluation and stabilization of wheat germ and its oil characteristics. *Alexandria Journal of Food Science and Technology* 8(2):31–39.

Ayala-Zavala, J. F., Vega-Vega, V., Rosas-Domínguez, C., Palafox-Carlos, H., Villa-Rodriguez, J. A., Siddiqui, W., et al. 2011. Agro-industrial potential of exotic fruit byproducts as a source of food additives. *Food Research International* 44(7):1866–1874.

Bansal S., and Sudha M. L. 2011. Nutritional, microstructural, rheological and quality characteristics of biscuits using processed wheat germ. *International Journal of Food Science and Nutrition* 62(5):474–479.

Bartłomiej, S., Justyna, R. K., and Ewa N. 2012. Bioactive compounds in cereal grains – occurrence, structure, technological significance and nutritional benefits – a review. *Food Science and Technology International* 18(6):559–568.

Belefant-Miller, H., and Grace, S. C. 2010. Variations in bran carotenoid levels within and between rice subgroups. *Plant Foods for Human Nutrition* 65:358–363.

Bernstein, A. M., Titgemeier, B., Kirkpatrick K., Golubic, M., and Roizen, M. F. 2013. Major cereal grain fibers and psyllium in relation to cardiovascular health. *Nutrients* 5(5):1471–1487.

Borrelli, G. M., De Leonardis, A. M., Platani, C., and Troccoli, A. 2008. Distribution along durum wheat kernel of the components involved in semolina colour. *Journal of Cereal Science* 48:494–502.

Brennan, C. S., and Clearly L. J. 2005. The potential use of cereal (1-3, 1-4)-β-D-glucans as functional food ingredients. *Journal of Cereal Science* 42:1–13.

Brewer, L. R., Kubola, J., Siriamornpun, S., Herald, T. J., and Shi, Y. C. 2014. Wheat bran particle size influence on phytochemical extractability and antioxidant properties. *Food Chemistry* 152:483–490.

Brouns, F., Hemery, Y., Price, R., and Anson, N. M. 2012. Wheat aleurone: separation, composition, health aspects, and potential food use. *Critical Reviews in Food Science and Nutrition* 52:553–568.

Cabrera-Soto, M. L., Salinas-Moreno, Y., Velázquez-Cardelas, G. A., and Espinosa-Trujillo, E. 2009. Content of soluble and insoluble phenols in the structures of corn grain and their rela-tionship with physical properties. *Agrociencia* 43:827–839.

Canan, C., Lisboa Cruz, F. T., Delaroza, F., Delaroza, F., Casagrande, R., Sarmento, C. P. M., et al. 2011. Studies on the extraction and purification of phytic acid from rice bran. *Journal of Food Composition and Analysis* 24:1057–1063.

Carlos Iglesias Pascual, C. S., Louro Massaretto, I., Kawassaki, F., Cerdeira Barros, R. M., and Lanfer Marquez, U. M. 2013. Effects of parboiling, storage and cooking on the levels of tocopherols, tocotrienols and γ-oryzanol in brown rice (*Oryza sativa* L.). *Food Research International* 50:676–681.

Carr, T. P., and Jesch, E. D. 2006. Food components that reduce cholesterol absortion. *Advances in Food and Nutrition Research* 51:165–204.

Chandrasekara, A., and Shahidi, F. 2010. Content of insoluble bound phenolics in millets and their contribution to antioxidant capacity. *Journal of Agricultural and Food Chemistry* 57:6706–6714.

Chen, F., Wang, Z., Zhao, G., Liao, X., Ca, T., Guo, L., and Hu, X. 2007. Purification process of octacosanol extracts from rice bran wax by molecular distillation. *Journal of Food Engineering* 79:63–68.

Chen, P. N., Kuo, W. H,. Chiang, C. L., Chiou, H. L., Hsieh, Y. S., and Chu, S. C. 2006. black rice anthocyanins inhibit cancer cells invasion via repressions of MMPs and u-PA expression. *Chemico-Biological Interactions* 163:218–229.

Chinma C. E., Ramakrishnan, Y., Ilowefah, M., Hanis-Syazwani, M., and Muhammad K. 2015. Properties of cereal brans: a review. *Cereal Chemistry* 92:1–7.

Costabile, A., Klinder, A., Fava, F., Napolitano, A., Fogliano, V., and Leonard, C. 2008. Wholegrain wheat breakfast cereal has a prebiotic effect on the human gut microbiota. *British Journal of Nutrition* 99:110–120.

Crespo, I., García-Mediavilla, M. V., Almar, M., González, P., Tuñón, M. J., Sánchez-Campos, S., and González-Gallego, J. 2008. Differential effects of dietary flavonoids on reactive oxygen and nitrogen species generation and changes in antioxidant enzyme expression induced by pro-inflammatory cytokines in Chang Liver cells. *Food and Chemical Toxicology* 46(5):1555–1569.

Daou, C., and Zhang. H. 2014. Functional and physiological properties of total, soluble, and insoluble dietary fibres derived from defatted rice bran. *Journal of Food Science and Technology* 51(12):3878–3885.

Das, A. K., and Singh, V. 2016. Antioxidative free and bound phenolic constituents in botanical fractions of Indian specialty maize (*Zea mays* L.). *Food Chemistry* 201:298–306.

Das Gupta, S., and Suh, N. 2016. Tocopherols in cancer: an update. *Molecular nutrition & food research* 60(6):1354–1363.

de Vasconcelos, M. C. B. M., Bennett, R., Castro, C., Cardoso, P., Saavedra, M. J., and Rosa, E. A. 2013. Study of composition, stabilization and processing of wheat germ and maize industrial by-products. *Industrial Crops and Products* 42:292–298.

Delcour, J. A., and Hoseney, R. 2010. *Principles of cereal science and technology*. St. Paul, MN: AACC International Press.

Dhingra, D., Michael, M., Rajput, H., and Patil, R. T. 2011. Dietary fibre in foods: a review. *Journal of food science and technology* 49(3):255–266.

Díaz-Gómez J., Twyman, R. M., Zhu, C., Farré, G., Serrano, J. C., Portero-Otin, M., Muñoz, P., Sandmann, G., Capell, T., and Christou P. 2017. Biofortification of crops with nutrients: factors affecting utilization and storage. *Current Opinion in Biotechnololy* 44:115–123.

Dikeman, C. L., Murphy, M. R., and Fahey, G. C. 2006. Dietary fibers affect viscosity of solutions and simulated human gastric and small intestinal digesta. *Journal of Nutrition* 136:913–919.

Dinelli, G., Carretero, A. S., Di Silvestro, R., Marotti, I., Fu, S., Benedettelli, S., et al. 2009. Determination of phenolic compounds in modern and old varieties of durum wheat using liquid chromatography coupled with time-of-flight mass spectrometry. *Journal of Chromatography A* 1216(43):7229–7240.

Du, C., Campbell, G. M., Misailidis, N., Mateos-Salvador, F., Sadhukhan, J., Mustafa, M., et al. 2009. Evaluating the feasibility of commercial arabinoxylan production in the context of a wheat biorefinery principally producing ethanol. Part 1. Experimental studies of arabinoxylan extraction from wheat bran. *Chemical Engineering Research and Design* 87 (9):1232–1238.

Duensing, W. J., Roskens, A. B., and Alexander, R. J. 2003. Corn dry milling: processes, products, and applications. In: *Corn: Chemistry and Technology*. Edited by P. L. White and L. A.Johnson. St Paul, MN: American Association of Cereal Chemists. Pp. 407–447.

Dunford, N. T., and Zhang, M. 2003. Pressurized solvent extraction of wheat germ oil. *Food Research International* 36:905–909.

El Gharras, H. 2009. Polyphenols: food sources, properties and applications-a review. *International Journal of Food Science and Technology* 44:2512–2518.

Engelsen, M. M., and Hansen A. 2009. Tocopherol and tocotrienol content in commercial wheat mill streams. *Cereal Chemistry* 86(5):499–502.

Esposito, F., Arlotti, G., Bonifati, A. M., Napolitano, A., Vitale, D., and Fogliano, V. 2005. Antioxidant activity and dietary fibre in durum wheat bran by-products. *Food Research International* 38:1167–1173.

Figuerola, F., Hurtado, M. L., Estévez, A. M., Chiffelle, I., and Asenjo, F. 2005. Fibre concentrates from apple pomace and citrus peel as potential fibre sources for food enrichment. *Food Chemistry* 91:395–401.

Frei, M., and Becker, K. 2005. Fatty acids and all-trans-β-carotene are correlated in differently colored rice landraces. *Journal of the Science and Food Agricultural* 85:2380–2384.

Galanakis, C. M. 2012. Recovery of high added-value components from food wastes: conventional, emerging technologies and commercialized applications. *Trends in Food Science and Technology* 26(3):68–87.

Galaverna, G., Di Silvestro, G., Cassano, A., Sforza, S., Docena, A., Drioli, E., et al., 2008. A new integrated membrane process for the production of concentrated blood orange juice: effect on bioactive compounds and antioxidant activity. *Food Chemistry* 106:1021–1030.

Gallaher D., and Chneeman, B. O. 2003. Fibra alimentaria. In: *Conocimientos actuales sobre nutrición*. Edited by B. Bowman, R. Russel. Organizacion Panamericana de la Salud. Publicación Tecnica nº 59.

Garcia A. L., Steiniger, J., Reich, S. C., Weickert, M. O., Harsch I., Machowetz, A., et al. 2006. Arabinoxylan fiber consumption improved glucose metabolism, but did not affect serum adipokines in subjects with impaired glucose tolerance. *Hormone and Metabolic Research* 38:761–766.

Gaspar, M., Kalman, G., and Reczey, K. 2007. Corn fiber as a raw material for hemicellulose and ethanol production. *Process Biochemistry* 42:1135–1139.

Gorinstein, S., Vargas, O. J. M., Jaramillo, N. O., Salas, I. A., Ayala, A. L. M., Arancibia-Avila, P., et al. 2007. The total polyphenols and the antioxidant potentials of some selected cereals and pseudocereals. *European Food Research and Technolgoy* 225:321–328.

Goufo P., and Trindade H. 2014. Rice antioxidants: phenolic acids, flavonoids, anthocyanins, proanthocyanidins, tocopherols, tocotrienols, γ-oryzanol, and phytic acid. *Food Science & Nutrition* 2:75–104.

Goufo, P., and Trindade, H. 2017. Factors influencing antioxidant compounds in rice. *Critical Reviews in Food Science and Nutrition* 57:893–922.

Haripriya, S., Premakumari, S. 2010. Beta sitosterol of wheat germ reduces blood glucose in Humans. *Journal of Life Science* 2:87–92.

Havrlentova, M., Petrulakova, Z., Burgarova, A., Gago, F., Hlenkova, A., and Sturdik, E. 2011. Cereal β-glucans and their significance for the preparation of functional food—A review. *Czech Journal of Food Science* 29(1): 1–14.

Heinemann, R. J. B., Xu, Z. M., Godber, J. S., and Lanfer-Marquez, U. M. 2008. Tocopherols, tocotrienols, and g-oryzanol contents in Japonica and Indica subspecies of rice (*Oryza sativa* L.) cultivated in Brazil. *Cereal Chemistry* 85:243–247.

Hemery, Y., Chaurand, M., Holopainen, U., Lampi, A. M., Lehtinen, P., Piironen, V., et al. 2011a. Potential of dry fractionation of wheat bran for the development of food ingredients, part I: influence of ultra-fine grinding. *Journal of Cereal Science* 53:1–8.

Hemery, Y., Holopainen, U., Lampi, A. M., Lehtinen, P., Nurmi, T., Piironen, V., et al. 2011b. Potential of dry fractionation of wheat bran for the development of food ingredients, part II: electrostatic separation of particles. *Journal of Cereal Science* 53:9–18.

Hemery, Y., Rouau, X., Lullien-Pellerin, V., Barron,C., and Abecassis, J. 2007. Dry processes to develop wheat fractions and products with enhanced nutritional quality. *Journal of Cereal Science* 46(3):327–347.

Hemery, Y. M., Mateo Anson, N., Havenaar, R., Haenen, G. R. M. M., Noort, M. W., and Rouau, X. 2010. Dry-fractionation of wheat bran increases the bioaccessibility of phenolic acids in breads made from processed bran fractions. *Food Research International* 43 (5):1429–1438.

Hernandez, N., Rodriguez-Alegría, M. E., Gonzalez, F., and Lopez-Munguia, A. 2000. Enzymatic treatment of rice bran to improve processing. *Journal of the American Oil Chemists' Society* 77:177–180.

Hidalgo, A., and Brandolini, A. 2008. Protein, ash, lutein and tocols distribution in einkorn (*Triticum monococcum* L. subsp. *monococcum*) seed fractions. *Food Chemistry* 107:444–448.

Hollmann, J., and Lindhauer M. 2005. Pilot-scale isolation of glucuronoarabinoxylans from wheat bran. *Carbohydrate Polymers* 59(2):225–230.

Hou, Z., Qin, P., Zhang, Y., Cui, S., and Ren G. 2013. Identification of anthocyanins isolated from black rice *(Oryza sativa* L.) and their degradation kinetics. *Food Research International* 50:691–697.

Hu, C., Zawistowski, J., Ling, W., and Kitts, D. D. 2003. Black rice (*Oryza sativa* L. indica) pigmented fraction suppresses both reactive oxygen species and nitric oxide in chemical and biological model systems. *Journal of Agricultural and Food Chemistry* 51:5271–5277.

Huang, Y. P., and Lai, H. M. 2016. Bioactive compounds and antioxidative activity of colored rice bran. *Journal of Food and Drug Analysis* 24:564–574.

Irmak, S., Dunford, N. T., and Milligan, J. 2006. Policosanol contents of beeswax, sugar cane and wheat extracts. *Food Chemistry* 95:312–318.

Izydorczyk, M. S., and Dexter, J. E. 2008. Barley β-glucans and arabinoxylans: molecular structure, physicochemical properties, and uses in food products—A review. *Food Research International* 41:850–868.

Jensen, M. K., Koh-Banerjee, P., Franz, M., Sampson, L., Grønbaek, M., and Rimm E. B. 2006. Whole grains, bran, and germ in relation to homocysteine and markers of glycemic control, lipids, and inflammation 1. *The American Journal of Clinical Nutrition* 83(2):275–283.

Jiang, Y., and Wang, T. 2005. Phytosterols in cereal by-products. *Journal of American Oil Chemical Society* 86:439–445.

Johnson, L. A., and May, J. B. 2003. Wet milling: the basis for corn biorefineries. In: *Corn: Chemistry and Technology*. Edited by P. L.White and L. A. Johnson. St Paul, MN: American Association of Cereal Chemists. Pp. 449–494.

Jones, D., Ormondroyd, G. O., Curling, S. F., Popescu, C. M., and Popescu, M. C. 2017. Chemical compositions of natural fibres. In: *Advanced High Strength Natural Fibre Composites in Construction*. Edited by M. Fan and F. Fu. Sawston, Cambridge, UK: Woodhead Publishing. Pp. 23–58.

Ju, J., Picinich, S. C., Yang, Z., Zhao, Y., Suh, N., Kong, A. N., et al. 2009. Cancer-preventive activities of tocopherols and tocotrienols. *Carcinogenesis* 31(4):533–542.

Juurlink, B., Azouz, H., Aldalati, A., AlTinawi, B., and Ganguly, P. 2014. Hydroxybenzoic acid isomers and the cardiovascular system. *Nutrition journal* 13:63. Doi: 10.1186/1475-2891-13-63.

Kahlon, T. S., and Chow, F. 2000. In vitro binding of bile acids by rice bran, oat bran, wheat bran, and corn bran. *Cereal Chemistry* 77:518–521.

Kaur G., Sharma S., Nagi, H. P. S., and Dar, B. 2012. Functional properties of pasta enriched with variable cereal brans. J. *Journal of Food Science and Technology* 49(4):467–474.

Kendall, C. W. C., Esfahani, A., and Jenkins, D. J. A. 2010. The link between dietary fibre and human health. *Food Hydrocolloids* 24(1):42–48.

Khuwijitjaru, P., Nualchan, P., and Adachi, S. 2007. Foaming and emulsifying properties ofrice bran extracts obtained by subcritical water treatment. *Silpakorn University Science and Technology Journal* 1(1):7–12.

Kim, H. S., Lee, E. J., Lim, S. T., and Han, J. A. 2015. Self-enhancement of GABA in rice bran using various stress treatments. *Food Chemistry* 172:657–662.

Kim, H. W., Lee, A. Y., Yeo, S. K., Chung, H., Lee, J. H., Hoang, M. H., et al. 2013. Metabolic profiling and biological mechanisms of body fat reduction in mice fed the ethanolic extract of black-colored rice. *Food Research International* 53:373–390.

Kim, J. K., Lee, S. Y., Chu, S. M. Lim, S. H., Suh, S. C., Lee, Y. T., et al. 2010. Variation and correlation analysis of flavonoids and carotenoids in Korean pigmented rice (*Oryza sativa* L.) cultivars. *Journal of Agricultural and Food Chemistry* 58:12804–12809.

Konopka, I., Kozirok, W., and Rotkiewicz, D. 2004. Lipids and carotenoids of wheat grain and flour and attempt of correlating them with digital image analysis of kernel surface and cross-sections. *Food Research International* 37:429–438.

Krishna, J., and Chadrasekaran, M. 2013. Cereals. In: *Valorization of Food Processing by-Products*. Edited by M. Chadrasekaran. Boca Raton, FL: CRC Press. Pp. 303–330.

Lamberts, L., and Delcour, J. A. 2008. Carotenoids in raw and parboiled brown and milled rice. *Journal of Agricultural and Food Chemistry* 56:11914–11919.

Lambo, A. M., Öste, R., and Nyman, M. E. G. L. 2005. Dietary fibre in fermented oat and barley β-glucan rich concentrates. *Food Chemistry* 85(2):283–293.

Landberg, R., Eldin, K., Salmenkallio, A., Marttila, M., Rouau, X., and Aman, P. 2008. Localization of alkylresorcinols in barley, wheat and rye kernels. *Journal of Cereal Science* 48:401–406.

Laufenberg, G., Kunz, B., and Nystroem, M. 2003. Transformation of vegetable waste into value added products: (A) the upgrading concept; (B) practical implementations. *Bioresource Technology* 87:167–198.

Lerma-Garcia, M. J., Herrero-Martinez, J. M., Simó-Alfonso, E .F, Mendonça, C. R. B., and Ramis-Ramos, G. 2009. Composition, industrial processing and applications of rice bran γ-oryzanol. *Food Chemistry* 115:389–404.

Li, L., Shewry, P. R., and Ward, J. L. 2008. Phenolic acids in wheat varieties in the healthgrain diversity screen. *Journal of Agricultural and Food Chemistry* 56:9732–9739.

Li, W., Cui, S., and Kakuda Y. 2006. Extraction, fractionation, structural and physical characterization of wheat β-d-glucans. *Carbohydrate Polymers* 63(3):408–416.

Liang Y., Perez I., Goetzelmann K., and Trupia, S. 2014. Microbial lipid production from pretreated and hydrolyzed corn fiber. *Biotechnology Progress* 30(4):945–951.

Ling, W. H., Cheng, Q. X., Ma, J., and Wang, T. 2001. Red and black rice decrease atherosclerotic plaque formation and increase antioxidant status in rabbits. *Journal of Nutrition* 131:1421–1426.

Liu F., Chen Z., Shao, J., Wang, C., and Zhan, C. 2017. Effect of fermentation on the peptide content, phenolics and antioxidant activity of defatted wheat germ. *Food Bioscience* 20:141–148.

Liu, L., Winter, K. M., Stevenson, L., Morris, C., and Leach D. N. 2012. Wheat bran lipophilic compounds with in vitro anticancer effects. *Food Chemistry* 130(1):156–164.

Liu, R. H. 2007. Whole grain phytochemicals and health. *Journal of Cereal Science* 46:209–219.

Liyana-Pathirana, C. M., and Shahidi, F. 2007. The antioxidant potential of milling fractions from breadwheat and durum. *Journal of Cereal Science* 45:238–247.

Lu, Z. X., Walker, K. Z., Muir, J. G., Mascara, T., and O'Dea, K. 2000. Arabinoxylan fiber, a byproduct of wheat flour processing, reduces the postprandial glucose response in normoglycemic subjects. *The American Journal of Clinical Nutrition* 71(5):1123–1128.

Luithui, Y., Baghya Nisha, R., and Meera, M. S. 2019. Cereal by-products as an important functional ingredient: effect of processing. *Journal of Food Science and Technology* 56(1):1–11.

Luna-Vital, D., Li, Q., West, L., West, M., and de Mejia, E.G. 2017. Anthocyanin condensed forms do not affect color or chemical stability of purple corn pericarp extracts stored under different pHs. *Food Chemistry* 232:639–647.

Luthria, D. L., Lu, Y., and John K. M. 2015. Bioactive phytochemicals in wheat: extraction, analysis, processing, and functional properties. *Journal of Functional Foods* 18:910–925.

Mahanta, C. L., and Sikia, S. 2016. Bioactive compounds of rice as health promoters. In: *Natural Bioactive Compounds from Fruits and Vegetables as Health Promoters Part II*. Edited by L. Rodrigues Da Silva and B. M. Silva. Sharjah, United Arab Emirates: Bentham Science Publishers. Pp. 221–234.

Mahmoud A. A., Mohdaly A. A. A., and Elneairy, N. A. A. 2015. Wheat germ: an overview on nutritional value, antioxidant potential and antibacterial characteristics. *Food and Nutrition Sciences* 6:265–277.

Maiani, G., Caston, M. J., Catasta, G., Toti, E., Cambrodón, I. G., Bysted, A., et al. 2009. Carotenoids: actual knowledge on food sources, intakes, stability and bioavailability and their protective role in humans. *Molecular Nutrition and Food Research* 53(2):194–218.

Majumdar, P., and Chanda, S. 2001. Chemical profile of some lignocellulosic crop residues. *Indian Journal of Agriculture and Biochemistry* 14(1/2):29–33.

Marti, A., Torri, L., Cristina, M., Franzetti, L., Limbo, S., Morandin, F., Quaglia, L., and Pagani, M. 2014. Wheat germ stabilization by heat-treatment or sourdough fermentation: Effects on dough rheology and bread properties. *LWT-Food Science and Technology* 59: 1100-1106.

Mathur, P., Ding, Z., Saldeen T., and Mehta J. L. 2015. Tocopherols in the prevention and treatment of atherosclerosis and related cardiovascular disease. *Clinical Cardiology* 38 (9):570–576.

Mellado-Ortega, E., and Hornero-Méndez, D. 2015. Carotenoids in cereals: an ancient resource with present and future applications. *Phytochemical Reviews* 14(6):873–890.

Miller, A., and Engel, K. H. 2006. Content of γ-oryzanol and composition of steryl ferulates in brown rice (*Oryza sativa* L.) of European origin. *Journal of Agricultural and Food Chemistry* 54: 8127–8133.

Mohanty, A. K., Misra, M., and Hinrichsen, G. 2000. Biofibres, biodegradable polymers and biocomposites: an overview. *Macromolecular Materials and Engineering* 276/277:1–24.

Moongngarm, A., Daomukda, N., and Khumpika, S. 2012. Phytochemicals, and antioxidant capacity of rice bran, rice bran layer, and rice germ. *APCBEE Procedia* 2:73–79.

Moore, J., Hao, Z., Zhou, K., Luther, M., Costa, J., and Yu, L. 2005. Carotenoid, tocopherol, phenolic acid, and antioxidant properties of Maryland-grown soft wheat. *Journal of Agricultural and Food Chemistry* 53:6649–6657.

Moore, J., Liu, J. G., Zhou, K., and Yu, L. L. 2006. Effects of genotype and environment on the antioxidant properties of hard winter wheat bran. *Journal of Agricultural and Food Chemistry* 54(15):5313–5322.

Nakornriab, M., Sriseadka, T., and Wongpornchai, S. 2008. Quantification of carotenoid and flavonoid components in brans of some Thai black rice cultivars using supercritical fluid extraction and high-performance liquid chromatography-mass spectrometry. *Journal of Food Lipids* 15:488–503.

Nandi I., and Ghosh M. 2015. Studies on functional and antioxidant property of dietary fibre extracted from defatted sesame husk, rice bran and flaxseed. *Bioactive Carbohydrates and Dietary Fibre* 5(2):129–136.

Nandini, C. D., and Salimath, P. V. 2001. Carbohydrate composition of wheat, wheat bran, sorghum and bajra with good chapati/roti (Indian flat bread) making quality. *Food Chemistry* 73:197–203.

Ndolo, V. U., and Beta, T. 2013. Distribution of carotenoids in endosperm, germ, and aleurone fractions of cereal grain kernels. *Food Chemistry* 139:663–671.

Nielsen, M. M., and Hansen A. 2008. Rapid High-Performance Liquid Chromatography Determination of tocopherols and tocotrienols in cereals. *Cereal chemistry* 85(2):248–251.

Noziere, P., Graulet, B., and Lucas, A. 2006. Carotenoids for ruminants: from forages to dairy products. *Animal Feed Science and Technology* 131:418–450.

Nurmi, T., Lampi, A. M., Nyström, L., Hemery, Y., Rouau, X., and Piironen, V. 2012. Distribution and composition of phytosterols and steryl ferulates in wheat grain and bran fractions. *Journal of Cereal Science* 56:379–388.

Nyström, L., Lampi, A. M., Andersson, A. A., Kamal-Eldin, A., Gebruers, K., Courtin, C. M., et al. 2008. Phytochemicals and dietary fiber components in rye varieties in the HEALTHGRAIN Diversity Screen. *Journal of Agricultural and Food Chemistry* 56(21):9758–9766.

Nyström, L., Paasonen, A., Lampi, A. M., and Piironen, V. 2007. Total plant sterols, steryl ferulates and steryl glycosides in milling fractions of wheat and rye. *Journal of Cereal Science* 45:106–115.

Osfor M. M. H., Ashsh A. M., ElSawy N. A., Qusty N. F. H., and Alkushi A. G. 2016. Effect of wheat bran consumption on serum lipid profile of hypercholesterolemia patients residence in holly makah. *Asian Journal of Applied Sciences* 5(1):1–7.

Panfili, G., Fratianni, A., and Distaam, M. 2004. Improved normal-phase high-performance liquid chromatography procedure for the determination of carotenoids in cereals. *Journal of Agricultural and Food Chemistry* 52:6373–6377.

Parikka, K., and Wähälä, K. 2009. An expedient synthesis of 5-n-alkylresorcinols and novel 5-n-alkylresorcinol haptens. *Beilstein Journal of Organic Chemistry* 5:art.no. A22.

Paster, M., Pellegrino, J. L., and Carole, T. M. 2003. Industrial bioproducts: today and tomorrow. Report prepared for the US Department of Energy, Washington, DC.

Perry J. R., and Ying, W. 2016. A Review of Physiological Effects of Soluble and Insoluble Dietary Fibers. *Journal of Nutrition and Food Science* 6(2). Doi: 10.4172/2155-9600.1000476.

Piironen, V., Toivo, J., and Lampi A. M. 2002. Plant sterols in cereals and cereal products. *Cereal Chemistry* 79(1):148–154.

Pradeep, P. M., Jayadeep, A., Guha, M., and Singh V. 2014. Hydrothermal and biotechnological treatments on nutraceutical content and antioxidant activity of rice bran. *Journal of Cereal Science* 60:187–192.

Qureshi, A. A., Mo, H., Packer, L., and Peterson, D. M. 2000. Isolation and identification of novel tocotrienols from rice bran with hypocholesterolemic, antioxidant and antitumor properties. *Journal of Agricultural and Food Chemistry* 48:3130–3140.

Ramos-Escudero, F., Munoz, A. M., Alvarado-Ortiz, C., Alvarado, A., Yanez, J.A. 2012. Purple corn (*Zea mays* L.) phenolic compounds profile and its assessment as an agent against oxidative stress in isolated mouse organs. *Journal of Medical Food* 15:206–215.

Rebolleda, S., Beltrán, S., Sanz, M. T., González-Sanjosé, M. L., and Solaesa, Á. G. 2013. Extraction of alkylresorcinols from wheat bran with supercritical CO_2. *Journal of Food Engineering* 119:814–821.

Reddy, N., and Yang, Y. 2005. Biofibers from agricultural byproducts for industrial applications. *Trends in Biotechnology* 23(1):22–27.

Ribas, S. A., Sichieri, R., Moreira, A. S. B., Souza, D. O., Cabral, C. T. F., Gianinni, D. T., and Cunha, D. B. 2017. Phytosterol-enriched milk lowers LDL-cholesterol levels in Brazilian children and adolescents: double-blind, cross-over trial. *Nutrition, Metabolism and Cardiovascular Diseases* 27(11):971–977.

Ribayo-Mercado, J. D., Solon, S. F., Solon, M. A., Cabal-Barza, M. A., Perfecto, C. S., Tang, G., et al. 2000. Bioconversion of plant carotenoids to vit-A in Filipino school-aged children varies inversely with vit-A status. *American Journal of Clinical Nutrition* 72:455–465.

Riccioni G. 2009. Carotenoids and cardiovascular disease. *Current Atherosclerosis Report* 11:434–439.

Rong, N., Ausman, L. M., and Nicolosi, R. J. 1997. Oryzanol decreases cholesterol absorption and aortic fatty streaks in hamsters. *Lipids* 32:303–309.

Rose, D. J., Inglett G. E., and Liu, S. X. 2010. Utilisation of corn (Zea mays) bran and corn fiber in the production of food components. *Journal of the Science of Food and Agricultural* 90(6):915–924.

Ross, A. B., Kamal-Eldin, A., and Aman, P. 2004. Dietary alkylresorcinols: absorption, bioactivities and possible use as biomarkers of wholegrain wheat and rye rich foods. *Nutrition Reviews* 62:81–95.

Sairam S., Gopala Krishna, A. G., and Urooj, A. 2011. Physico-chemical characteristics of defatted rice bran and its utilization in a bakery product. *Journal of Food Science and Technology* 48(4):478–483.

Salinas-Moreno, Y. Garcia-Salinas, C., Ramírez-Díaz, J. L., and Alemán de la Torre, I. 2017. Phenolic Compounds in Maize Grains and Its Nixtamalized Products. In: *Phenolic Compounds – Natural Sources, Importance and Applications*. Edited by M. Soto-Hernandez. London, UK: IntechOpen Limited. Pp. 215–232.

Sang, S., and Zhu, Y. 2014. Bioactive phytochemicals in wheat bran for colon cancer prevention. In: *Wheat and Rice in Disease Prevention and Health: Benefits, Risks and Mechanisms of Whole Grains in Health Promotion*. Edited by R.R. Watson, R.P. Victor & Z. Sherma. Waltham, MA: Academic Press. Pp. 121–139.

Saura-Calixto, F., and Goñi, I. 2005. Alimentos funcionales: Fibra dietética y antioxidantes de la dieta española. *Alimentación, Nutricion y Salud* 12(4):132–149.

Sidhu, J. S., Kabir. Y., and Huffman, F. G. 2007. Functional foods from cereal grains. *International Journal of Food Properties* 10:231–244.

Siebenhandl, S., Grausgruber, H., Pellegrini, N., Del Rio, D., Fogliano, V., Pernice, R., et al. 2007. Phytochemical profile of main antioxidants in different factions of purple and blue wheat, and black barley. *Journal of Agricultural and Food Chemistry* 55:8541–8547.

Skendi, A., Biliaderis, C. G., Lazaridou, A., and Izydorczyk, M. S. 2003. Structure and rheological properties of water soluble b-glucans from oat cultivars of Avena sativa and Avena bysantina. *Journal of Cereal Science* 38:15–31.

Sovrani, V., Blandino, M., V. Scarpino, A. Reyneri, Coïsson, J. D., Travaglia, F., et al. 2012. Bioactive compound content, antioxidant activity, deoxynivalenol and heavy metal contamination of pearled wheat fractions. *Food Chemistry* 135(1):39–46.

Šramková, Z., Gregová E., and Šturdík E. 2009. Chemical composition and nutritional quality of wheat grain. *Acta Chimica Slovaca* 2:115–138.

Sreeramulu, D., and Raghunath, M. 2010. Antioxidant activity and phenolic content of roots, tubers and vegetables commonly consumed in India. *Food Research International* 43:1017–1020.

Srivastava A. K., Sudha, M. L. Baskaran, V., and Leelavathi, K. 2006. Studies on heat stabilized wheat germ and its influence on rheological characteristics of dough. *European Food Research and Technology* 224(3):365–372.

Stasiuk, M., and Kozubek A. 2010. Biological activity of phenolic lipids. *Cellular and Molecular Life Sciences* 67:841–860.

Sugihara N., Kanda, A., Nakano, T., Nakamura, T., Igusa, H., and Hara, S. 2010. Novel fractionation method for squalene and phytosterols contained in the deodorization distillate of rice bran oil. *Journal of Oleo Science* 59:65–70.

Tan, J., Baisakh, N., Oliva, N., Parkhi, V., Mayank Rai, M., Torrizo, L., et al. 2005. The screening of rice germplasm, including those transgenic rice lines which accumulate β-carotene in their polished seeds, for their carotenoid profile. *International Journal of Food Science and Technology* 40:563–569.

Tanaka J., Nakamura S., Tsuruma K., Shimazawa M., Shimoda, H., and Hara, H. 2011. Purple rice (*Oryza sativa* L.) extract and its constituents inhibit VEGF-induced angiogenesis. *Phytotherapy Research* 26(2):214–222.

Tong, L. T., Zhong, K., Liu, L., Qiu, J., Guo, L., Zhou, X., et al., 2014. Effects of dietary wheat bran arabinoxylans on cholesterol metabolism of hypercholesterolemic hamsters. *Carbohydrates Polymeres* 112:1–5.

Topping, D. 2007. Cereal complex carbohydrates and their contribution to human health. *Journal of Cereal Science* 46:220–229.

Tuncel, N. B, and Yılmaz, N. 2011. Gamma-oryzanol content, phenolic acid profiles and antioxidant activity of rice milling fraction. *European Food Research and Technology* 233:577–585.

Vági, E., Simándi, B., Vásárhelyiné, K. P., Daood, H., Kéry, A., Doleschall, F., et al. 2007. Supercritical carbon dioxide extraction of carotenoids, tocopherols and sitosterols from industrial tomato by-products. *Journal of Supercritical Fluids* 40:218–226.

Valsta, L. M., Lemström, A., Ovaskainen, M. L., Lampi, A.-M., Toivo, J., Korhonen, T., et al. 2004. Estimation of plant sterol and cholesterol intake in Finland: quality of new values and their effect on intake. *British Journal of Nutrition* 92:671–678.

Van Hoed, V., Depaemelaere, G., Villa Ayala, J., Santiwattana, P., Verhé, R., and De Greyt, W. 2006. Influence of chemical refining on the major & minor components of rice bran oil. *Journal of the American Oil Chemists' Society* 83:315–321.

Viuda-Martos, M., Fernández-López, J., and Pérez-Álvarez, J. A. 2010b. Pomegranate andits many functional components as related to human health: a review. *Comprehensive Review in Food Science and Food Safety* 9:635–654.

Viuda-Martos, M., López-Marcos, M. C., Fernández-López, J., Sendra, E., López-Vargas J. H., and Pérez-Álvarez J. A. 2010a. Role of fiber in cardiovascular diseases: a review. *Comprehensive Review in Food Science and Food Safety* 9(2):240–258.

Vogel, B., Gallaher, D. D., and Bunzel, M. 2012. Influence of cross-linked arabinoxylans on the postprandial blood glucose response in rats. *Journal of Agricultural and Food Chemistry* 60(15):3847–3852.

Wang J., Rosell, C. M., and Benedito de Barber C. 2002. Effect of the addition of different fibres on wheat dough performance and bread quality. *Food Chemistry* 79:221–226.

Wang, Q., Han, P. H., Zhang, M. W., Xia, M., Zhu, H. L., Ma, J., et al. 2007. Supplementation of black rice pigment fraction improves antioxidant and anti-inflammatory status in patients with coronary heart disease. *Asia Pacific Journal of Clinical Nutrition* 16:295–301.

Wang, Y. 2009. Prebiotics: present and future in food science and Technology. *Food Research International* 42:8–12.

Wanyo, P., Meeso, N., and Siriamornpun, S. 2014. Effects of different treatments on the antioxidant properties and phenolic compounds of rice bran and rice husk. *Food Chemistry* 157:457–463.

White, P. J., and Weber, E. J. 2003. Lipids of the kernel. In: *Corn Chemistry and Technology* 2nd ed. Edited by P. J. White, L. A. Johnson. Eagan, Minnesota. USA: American Association of Cereal Chemists, Inc. Pp. 355–405.

WHO/FAO. Codex Alimentarius Commission (CNFSDU). 2008. 30th Session. Available from: http://www.codexalimentarius.net/web/archives.jsp?yearZ09. ALINORM 09/32/26. Para 27-54 and page 46.

Wongwaiwech, D., Weerawatanakorn, M., Tharatha, S., and Ho, C. H. 2019. Comparative study on amount of nutraceuticals in by-products from solvent and cold pressing methods of rice bran oil processing. *Journal of Food and Drug Analysis* 27:71–82.

Xu, X., Hua, N., and Godber, S. 2001. Antioxidant activity of tocopherols, tocotrienols, and Û-oryzanol components from rice bran against cholesterol oxidation accelerated by 2,2'-azobis-(2-methylpropionamidine) dihydrochloride. *Journal of Agricultural and Food Chemistry* 49:2077–2081.

Yadav, B. K., and Joyner, J. J. 2014. Processing effects on functional components in cereals and grains. In: *Functional Foods and Dietary Supplements: Processing Effects and Health Benefits*, 1st edn. Edited by A. Noomhorm, I. Ahmad, and A. K. Anal. Hoboken, New Jersey, USA: John Wiley & Sons, Ltd. Published. Pp. 63–89.

Yan X., Ye, R., and Chen, Y. 2015. Blasting extrusion processing: the increase of soluble dietary fiber content and extraction of soluble-fiber polysaccharides from wheat bran. *Food Chemistry* 180:106–115.

Yoon, S. H., and Kim, E. 2003. Recent Progress in enzymatic production of structured lipids. *Food Science and Biotechnology* 12:721–726.

Yu, L., Zhou, K., and Parry, J. W. 2005. Inhibitory effect of wheat bran extracts on human LDL oxidation and free radicals. *LWT-Food Science and Technology* 38:463–470.

Yu, X., Li, B., Wang, L., Chen, X., Wang, W., Wang, Z., and Xiong, F. 2015. Systematic analysis of pericarp starch accumulation and degradation during wheat caryopsis development. *PLoS One* 10(9):e0138228.

Zhang, M. W., Zhang, R. F., Guo B. J., Chi J. W., Wei Z. C., and Xu, Z. H. 2006. The protective effects of anthocyanidin extracted from black rice fraction on endothelia cells injured by oxidative stress. *Acta Nutrimenta Sinica* 28:216–220.

Zhang, M. W. Zhang, R. F., Zhang, F. X., and Liu, R. H. 2010. Phenolic profiles and antioxidant activity of black rice bran of different commercially available varieties. *Journal of Agricultural and Food Chemistry* 58:7580–7587.

Zhao, S., Yao, S., Ou, S., Lin, J., Wang, Y., Peng, X., et al. 2014. Preparation of ferulic acid from corn bran: its improved extraction and purification by membrane separation. *Food and Bioproducts Processing* 92(3):309–313.

Zhu Y., Soroka, D. N. Sang, S. 2012. Synthesis and inhibitory activities against colon cancer cell growth and proteasome of alkylresorcinols. *Journal of Agricultural and Food Chemistry* 60(35):8624–8631.

Zou, Y., Yang, M., Zhang, G., He, H., and Yang, T. 2015. Antioxidant Activities and Phenolic Compositions of Wheat Germ as Affected by the Roasting Process. *Journal of American Oil Chemistry Society* 92:1303.

Zou, Y., and Yang, T. 2019. Rice Husk, Rice Husk Ash and Their Applications. In: *Rice Bran and Rice Bran Oil.* Edited by L. Z. Cheong and X. Xu. Elsevier Inc. Pp. 207–246.

7

Reuse of Tiger Nuts By-Products

Food Formulation, Clean Recovery of Oil and Bioactive Compounds, and Evaluation of the Cytotoxicity of the Oils

ELENA ROSELLÓ-SOTO, FRANCISCO J. BARBA AND FRANCISCO J. MARTÍ-QUIJAL

Nutrition and Food Science Area, Preventive Medicine and Public Health, Food Science, Toxicology and Forensic Medicine Department, Faculty of Pharmacy, Universitat de València, Avda. Vicent Andrés Estellés, s/n, 46100 Burjassot, València, Spain

MARIA G. DASKALAKI

School of Medicine, University of Crete, Heraklion, Greece

JOSÉ M. LORENZO

Centro Tecnológico de la Carne de Galicia, Adva. Galicia n° 4, Parque Tecnológico de Galicia, San Cibrao das Viñas, 32900 Ourense, Spain

CYRIELLE GARCIA AND FABIENNE REMIZE

QualiSud, Université de La Réunion, CIRAD, Université Montpellier, Montpellier SupAgro, Université d'Avignon, 2 rue J. Wetzell, F-97490 Sainte Clotilde, France

Contents

1 Introduction

Tiger nuts of Valencia and the "horchata" represent the cultural and traditional values in this territory, being a food of great nutritional value. Initially, the "horchata" was made for family consumption, but over time, it began to appear artisanal "horchaterías," later giving rise to the current horchata industries (Codina Torrella, 2014). The "horchata" is obtained from the tubers of the tiger nut (*Cyperus esculentus* L.), which is a refreshing drink with a milky appearance, white color, and pleasant flavor. The industrial production of horchata is of great economic importance in the Spanish food industry, and in particular, in the Valencian community, representing an annual consumption of ~50 million liters per year, and having a market value, at retail, estimated at 60 million euros (CRDO, 2018). Tiger nut is also consumed in other countries, such as West Africa, Egypt, and Korea, and more recently in the United States.

Currently, tiger nut and its derivatives have shown to be of a great culinary interest due to the organoleptic qualities that they confer to desserts, such as flan and creams, but also in main dishes or in starters, such as tiger nut soup (Roselló-Soto et al., 2018; Soriano del Castillo et al., 2014). It is becoming particularly popular in gluten-free diets and as an alternative to milk.

Before preparing "horchata," tiger nuts are first collected, and then are washed to remove impurities (stone, earth, and other remains contained in the tuber). The obtained tiger nuts contain between 45 and 50% humidity and are called "tender tiger nuts." To reduce the moisture content, a drying step is carried out, which consists of spreading the tiger nuts and stirring the tubers, so that drying is carried out uniformly. Drying lasts 3 months and humidity drops to 11%, resulting in dried tiger nuts (BOE, 1988; CRDO, 2018).

Once dried tiger nuts are available, they are washed to remove the dirt and impurities of the tuber using chlorinated drinking water (BOE, 1988). Next, the tiger nuts are selected by eliminating the tubers damaged by insects or microorganisms by saline flotation at a concentration between 15 and 17 °Baumé. The selected tubers are washed with drinking water to remove the remains of brine adhered on the surface. Then rehydration takes place by immersing them in drinking water for a certain period of time (8–24 h), where the tubers swell, decreasing its surface roughness and favoring the effectiveness of disinfection. The next step is disinfection using a solution of water and active chlorine (1%) with a time less than 30 minutes; then several washings are carried out to eliminate the remains of germicides (Martín-Esparza and González-Martínez, 2016; Soriano del Castillo et al., 2014).

Then, grinding is carried out in a mill, where water is added (3 liters of water/kg of dry chufa). After this, the mashed tiger nut mass is left to macerate with water for a short time (depending on the soaking time). The next step is pressing, which consists of separating the liquid from the solid waste. For this purpose, continuous pressing or pressing using a sieving device is used. After pressing them, the first extract is obtained and sifted. The residues of the sieve and the pressing are mixed, 2 liters of water/kg of tiger nuts are added to allow pressing and sieving and the formation of a second extract, which together with the one obtained above gives rise to the final extract. To the obtained liquid, 100–150 g sugar/liter is added. Finally, the "horchata" is cooled to a temperature of 0 °C and then the conservation is carried out at a temperature greater than or equal to 2 °C (Martín-Esparza and González-Martínez, 2016; Soriano del Castillo et al., 2014).

2 Direct Use in Food of the By-Products of "Horchata"

Because of large by-products generated in the food industry, measures have been established for the management of by-products, with the aim of reducing the environmental impact and promoting their use in other industrial processes, such as ingredients or raw materials (Granato et al., 2017; Roselló-Soto et al., 2015). That is why researchers have shown an increasing interest in the study of these by-products, due to their nutritional value, bioactive compounds, as well as the possible applications they may have.

During the production process of "Horchata de Chufa," a high amount of by-products is generated, which can represent up to 60% of the raw materials (Sánchez-Zapata et al., 2012b). Two kinds of by-products of "Horchata de Chufa," whose composition make them suitable for other uses, can be distinguished: solid and liquid by-products.

The reuse of the solid residue relies on its high content of dietary fiber, antioxidant compounds, and bioactive compounds (e.g., polyphenols and phytosterols), in addition to good technological properties (water retention capacity, emulsion capacity, and emulsion stability), according to Sánchez-Zapata et al. (2012a) (Table 7.1). Furthermore, the liquid by-product stands out for its high content of prebiotics, antioxidant compounds, and as a substitute for water in meat products (Sánchez-Zapata et al., 2012b). Sánchez-Zapata et al. (2012a) indicated that due to microbial load of the tiger nuts, it would be necessary to carry out disinfection treatment before the addition of by product in food products.

Several studies showed that the use of the by-products of "Horchata de chufa" might be useful in the meat industry from a nutritional point of view. A study carried out by Sánchez-Zapata et al. (2013) determined the effect of the addition of tiger nut

TABLE 7.1 Possible applications of tiger nut by-products.

APPLICATIONS	REFERENCES
Meat products with fiber source	Sánchez-Zapata et al. (2010, 2013c)
Production of gluten-free products	Aguilar et al. (2015)
Carbon source for the growth of probiotic bacteria	Sánchez-Zapata et al. (2013b)
Use of tiger nut oil as a source of biofuel	Barminas et al. (2001)
As an ingredient and substitute for water	Sánchez-Zapata et al. (2013a)

fiber (TNF) in percentages of 0.5 and 7.5% on physicochemical and sensory properties, as well as on the composition of dry and stored chorizo for 28 days. In addition, the oxidation reactions and the microbial quantity were also studied. These authors obtained a chorizo rich in fiber with the addition of TNF, with a lower percentage of fats and higher humidity. There were small changes in physicochemical properties, such as color changes; however, consumer acceptance was not modified.

Another study conducted by the same researchers evaluated the use of TNF in pork burgers. For this, increasing levels of TNF were used (5, 10, and 15%). The chemical composition, physicochemical properties, cooking characteristics, and sensory properties in pork burgers were determined. The results showed that hamburgers made with TNF had a higher fiber content. In addition, a higher cooking performance was obtained due to the water- and oil-fixing capacities. However, there were changes in color and texture due to the addition of TNF, being less fatty, less juicy, and with less meat flavor than the controls. There were no color changes noticed after the cooking process (Sánchez-Zapata et al., 2010).

Other authors added different combinations of TNF (1–2%) together with walnut oil (2.5–5%) in the "longaniza de pascua." "Longaniza de pascua" with walnut oil showed the highest moisture value; however, when walnut oil was added together with TNF, moisture decreased. The addition of dietary fiber in the meat products increased the water-retention capacity, while the walnut oil prevented the loss of water due to a waterproof film between the meat mass and the skin of the "longaniza de pascua." It was observed that "longaniza de pascua" with TNF (with or without walnut oil) showed lower pH values than the control. Regarding the color of the sausages, when both ingredients were added at the same time (TNF and walnut oil), the redness and the yellow tone increased. The oxidation of the lipids was probably reduced by the antioxidant compounds present in the TNF, such as polyphenols. Therefore, the use of by-products of "horchata" can be useful in the meat industry as a source of fiber (Sánchez-Zapata et al., 2013c).

Aguilar et al. (2015) studied the effect of the by-products of "horchata" in the preparation of gluten-free bread. These gluten-free products show a greater demand by the consumer due to the increased frequency of diagnosis of celiac diseases. Moreover, gluten-free diet is often associated with a low consumption of fibers (Jeske and Arendt, 2018; Wild et al., 2010). Breadcrumbs and darker rinds were obtained due to amino acids and sugars that contain by-products of "horchata," which favored the Maillard reaction.

Another application that can be found in the liquid by-products coming from the elaboration of the "horchata," is as an ingredient for replacing water. Sánchez-Zapata et al. (2013a) substituted water (50 or 100%) for the liquid by-product of "horchata de chufa" in pork liver pâté. The results show that these pâtés contain low moisture compared with control, higher heme iron content, and have a lower degree of metmyoglobin formation. However, the fat content increased. The addition of liquid by-products of horchata did not show differences in the physicochemical properties, in which pH, water activity, and color of the pâté were included.

On the other hand, Sánchez-Zapata et al. (2013b) studied the possible use of liquid by-products from the production of "horchata" as a carbon source for the growth of probiotic bacteria (*Lactobacillus acidophilus* and *Bifidobacterium animalis*). These liquid by-products contain approximately 94.44% of water and 5.56% of carbohydrates, proteins, fats, and ashes. The results indicated the potential of these liquid by-products of "horchata" as a carbon source and its possible utility in the food industry as a fermentable product, although more studies are required. Several studies emphasized the interest of lactic fermentation of tiger nut beverages and possible new product developments (Kizzie-Hayford et al., 2016; Roselló-Soto et al., 2018). Fermentation of liquid by-products also has many advantages with the gain of exopolysaccharides production (Roselló-Soto et al., 2018).

3 Antioxidant Compounds in Tiger Nut and By-Products

Polyphenols are compounds obtained from the secondary metabolism of plants. They constitute antioxidant compounds in our diet, with the consumption of fruits, vegetables, legumes, chocolates, and beverages, such as coffee, wine, and tea. Although the dietary intake is very variable, due to environmental factors, such as the degree of maturity, sun exposure, cooking methods used, bioavailability, among others (Quiñones et al., 2012), it is estimated that the intake is greater than 1 g/day approximately (Scalbert et al., 2005).

Polyphenols constitute a large group of substances. For example, flavonoids are divided into different subgroups depending on the degree of oxidation of the carbon ring, the main ones being flavonols, flavones, isoflavones, flavan-3-ols, flavanones, and anthocyanins (Santhakumar et al., 2018). In non-flavonoids group, we find phenolic acids, water-soluble tannins, and stilbenes, phenolic acids being the most common in the diet.

Polyphenolic compounds play an important role in the protection against reactive oxygen species, since the phenolic group acts by capturing unpaired electrons, limiting the risk of various neurodegenerative diseases associated with oxidative stress. Several studies have shown the beneficial properties of polyphenols in the prevention of cardiovascular diseases, cancer, osteoporosis, or diabetes (Costa et al., 2017; Santhakumar et al., 2018; Scalbert et al., 2005). A recent study with animal models has shown favorable results in the polyphenolic compounds of white wine in Alzheimer's disease (Mendes et al., 2018).

Currently, the number of research related to antioxidants has shown a growing interest in the benefits they produce in human health. However, more exhaustive studies are needed for determining both the specific compounds responsible for the prevention of diseases and their absorption and metabolism.

Tiger nuts are considered as a source of natural antioxidants. Luo et al. (2014) identified 20 flavonoids, including 6 new ones, called "eleocarinas" in the husks of the tiger nut (*Eleocharis tuberosa*). Koubaa et al. (2015) identified 78 polyphenols in tiger nut oil, in which 12 were common polyphenols (*p*-coumaric acid, ferulic acid, 3-phenylpropionic acid, cinnamic acid, vanillic acid, among others). Ezeh et al. (2014) indicated that tiger nut oil contains phenolic compounds and carotenoids. With regard to vitamins, it is worth mentioning vitamin E (10 mg/100 g), vitamin C (6 mg/100 g), and folic acid (141 μg/100 g) (BEDCA, 2018).

Liquid by-products showed a greater wealth in total phenolics, 169.8 ± 10.5 mg GAE/L in this case, than aqueous extracts of many fruit by-products; for instance, 46.00 mg GAE/g for apple or 12.15 mg GAE/g for tomato (Sánchez-Zapata et al., 2012). A value of total antioxidant capacity (TAC) of 1760 μM Trolox/g of dry matter was obtained by ethanolic extraction from liquid by-product (Roselló-Soto et al., 2018).

4 Extraction of Phenolic Compounds

The food and pharmaceutical industries show great interest in the recovery of bioactive compounds present in food by-products, due to the high added value and the great diversity of benefits they possess. These compounds can be extracted using different techniques, conventional or latest alternative extraction methods (Barba et al., 2016; Bursać-Kovačević et al., 2018; Koubaa et al., 2015). The extraction method, nature of the solvent, and its amount, in addition to the time and temperature applied, influence the extract composition (Roselló-Soto et al., 2016).

One of the most commonly used conventional techniques is solvent extraction (liquid–liquid or solid–liquid) using organic solvents. Many of these are toxic; methanol, hexane, or acetone, for instance, pose a risk to human health (Puértolas and Barba, 2016), that is why ethanol is preferred in the food industry. According to Mokrani and Madani (2016), acetone is the most efficient solvent for the extraction of proanthocyanidins and tannins, whereas ethanol is more efficient for the extraction of flavonoids and their glycosides. Phenolic acids and catechins are better extracted with methanol.

Roselló-Soto et al. (2018b) evaluated the impact of temperature (40–60 °C), extraction time (1–3 h), and solvent (mixtures of ethanol–water (v/v) at 0%, 25%, and 50%) on the recovery of total phenolic compounds (TPC), total flavonoids (TF), and TAC from "horchata" by-products. These authors observed a significant influence ($p < 0.05$) of all the investigated factors regarding the extraction of antioxidant bioactive compounds, obtaining the maximum TPC (16.02 mg GAE/100 g of dry matter [d.m.]) and TF (30.09 mg CE/100 g d.m.) when ethanol at 25% (*v/v*), during 3 h of extraction at 60 °C and 50 °C, respectively. However, the highest TAC values (1,759.81 μM Trolox equivalents/g d.m.) were found by these authors when they used 50% aqueous ethanol (v/v) at 60 °C for 3 h.

Currently, the need to obtain more efficient, ecological, and sustainable processes has led to innovative methodologies for obtaining compounds of interest due to their antioxidant or antimicrobial capacities, coming from waste or food by-products. Some of these technologies are pulsed electric fields, pressurized fluids (e.g., supercritical carbon dioxide [SC-CO_2] and gas-assisted mechanical expression [GAME]), ultrasound, microwave-assisted extraction (MAE), among others (Chemat et al., 2012; Roselló-Soto et al., 2015). These innovative or nonconventional methods have advantages over conventional methods, since extraction time is reduced, fewer toxic organic solvents and lower temperatures are used, contributing to greater efficiency in the extraction of antioxidant compounds and lower energy consumption. These innovative methods are called green extraction, since it is intended to preserve the environment, as well as consumer health (Chemat et al., 2012; Roselló-Soto et al., 2015).

5 Green Extraction of Oil and Polyphenol Composition

Commercial tiger nut oils are obtained from cold press processing. Tiger nut oil contains oleic acid as the most abundant fatty acid, but also phenolic compounds and vitamin E, mainly α-tocopherol at 86.7 mg/g, which confers high oxidative stability to the oil (Ezeh et al., 2014; Yeboah et al., 2012). Alternative extraction methods were used. A first study was carried out by Lasekan and Abdulkarim (2012) using SC-CO_2 for tiger nut oil extraction. The optimized parameters for oil extraction were calculated at a temperature of 60 °C, a pressure of 30.25 MPa, and an extraction time of 210 min. With the GAME method, a faster extraction, correlated with a higher polyphenol content was observed compared with that obtained by SC-CO_2 and MAE (Koubaa et al., 2015). Other researchers (Hu et al., 2018) used MAE for the extraction of tiger nut oil. The oil presented a higher phenolic content compared with the extraction with the conventional Soxhlet method, as well as a shorter extraction time and quantity of solvent used.

In a more recent study, Roselló-Soto et al. (2018a) evaluated the impact of SC-CO_2 (10–40 MPa) and conventional extraction (CE) (Folch and Soxhlet methods) to recover oil from "horchata" by-products. Moreover, they also studied the amount and profile of fatty acids, total polyphenols, α-tocopherol, and antioxidant capacity of the oils obtained, and established a relationship with lipid oxidation parameters. They found a linear relationship between extraction pressure and oil yield, obtaining the highest oil yield recovery when the highest pressure (40 MPa) was applied. However, when they compared the results obtained after SC-CO_2 and conventional Folch extraction, the highest oil yield was found after applying CE. It should also be noted that monounsaturated fatty acids were predominant and represented about 70% of total fatty acids in this "horchata" by-product, with slight nonsignificant differences regarding the fatty acid profile according to the treatment applied. Interestingly, lipid oxidation parameters were inversely correlated with TPC and TAC; thus, the lower oxidation indexes were obtained when SC-CO_2 pressure was augmented.

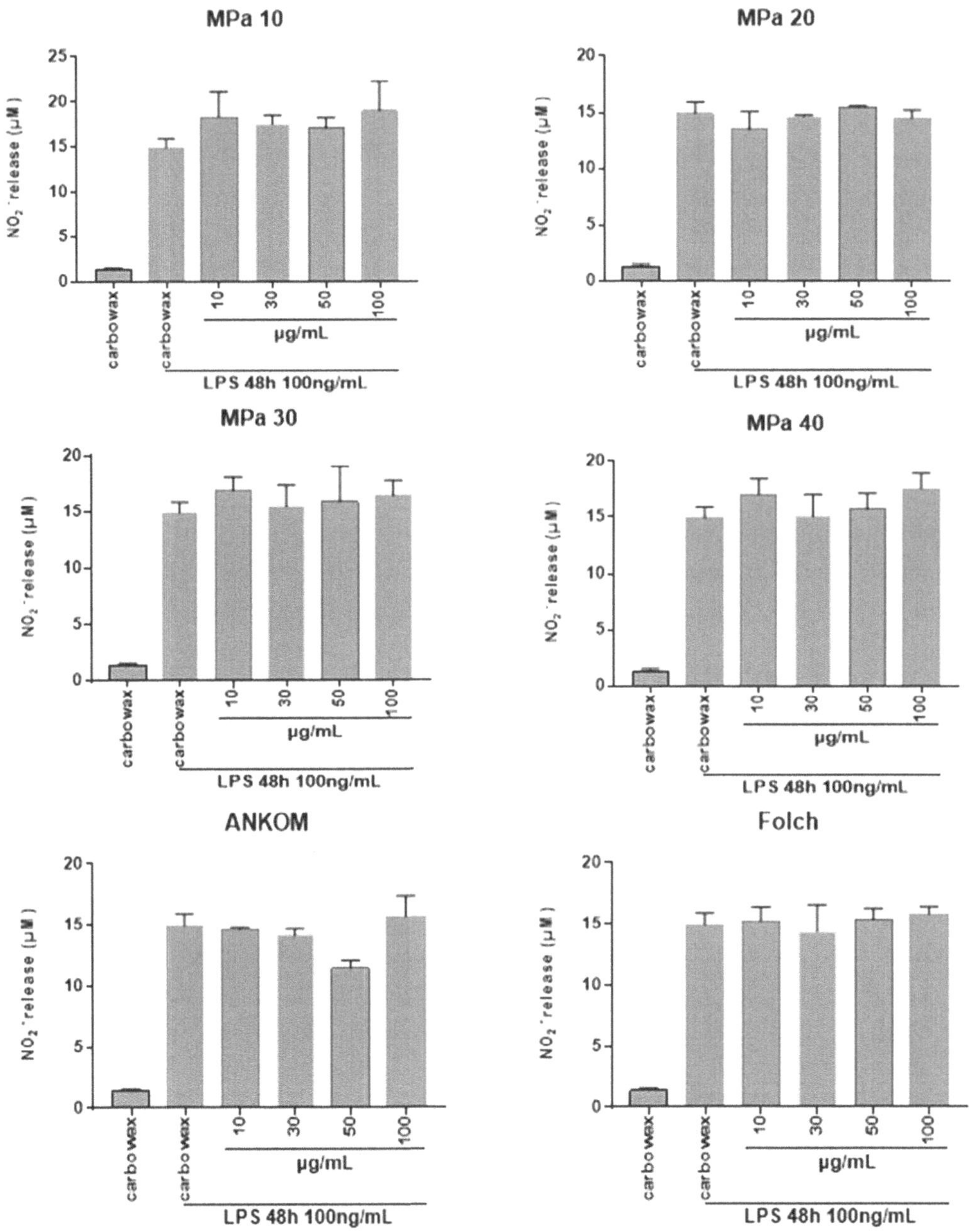

Figure 7.1 Determination of NO_2^- release of tiger nuts extracts obtained after supercritical carbon dioxide extraction at 10–40 MPa, and CE (ANKOM and Folch). Evaluation of the effect of tiger nuts extracts on RAW 264.7 LPS activated macrophages. Nitrite concentration was measured in cell culture supernatants using Griess reaction after 48 h of activation with 100 ng/mL LPS.

6 Evaluation of Inflammation Modulatory Effect of the Oil Extracts

Macrophages RAW 264.7 cell line stimulated with lipopolysaccharide (LPS) were used for evaluating the effect of oil extracts from "horchata" by-products on the production of inflammatory mediators. Different concentrations of each extract were tested, as well as the organic solvent used (carbowax). NO_2^- release was used as an indicator of inflammatory response, due to the activity of iNOS (inducible Nitric Oxide Synthase) as an important proinflammatory inductor, thus causing the activation of inflammatory cascade and other inflammatory associated effects, such as vasodilation. Nitrite concentration was measured in cell culture supernatants using Griess reaction after 48 h of activation with 100 ng/mL LPS.

As shown in Figure 7.1, the oil extracts did not trigger production of NO_2^- and did not affect NO_2^- release in response to LPS at the tested concentrations, compared with macrophages treated with LPS alone. It should also be noted that substantial differences do not appear in NO_2^- release between CE (Folch) and non-CE methods (SC-CO_2 and ANKOM), or by applying different pressure values to SC-CO_2 extraction (10, 20, 30, or 40 MPa). Finally, it is important to highlight that carbowax, the extracts solvent, did not have any influence in NO_2^- release by itself.

Another study on kidney slices by Olabiyi et al. (2018) found that incubation with hydroalcoholic extracts of tiger nuts could increase the NO levels in this tissue, and this could be attributed to the phenolic compounds. This apparent discrepancy could be explained regarding the nature of the extracts obtained, since Olabiyi et al. (2018) used an hydrophilic extract, while our extracts were lipophilic, with no wider presence of polar compounds. On the other hand, these authors also demonstrated using MTT assay that tiger nuts hydroalcoholic extracts did not affect cellular viability of kidney slices, foreseeing an absence of cytotoxicity.

7 Conclusion

During "horchata" production process, several by-products are generated, and they could be used as a source of fiber, antioxidant compounds, or even as preservatives when they are added to processed foods. They can also be used as prebiotics and for obtaining new compounds from lactic acid bacteria fermentation. However, CE of these antioxidant compounds represents serious safety problems due to the use of organic solvents. That is why it is interesting to use non-CE technologies, such as extraction with supercritical fluids (SC-CO_2), which reduce or eliminate the use of these harmful solvents. Our results confirmed that the application of SC-CO_2 could be a potential alternative to CE for obtaining oils from "horchata" by-products rich in high added value compounds without the use of organic solvents, which can be toxic.

Acknowledgments

This work was supported by the project GV/2018//040 "Implementación y optimización de procesos innovadores para la valorización de los subproductos obtenidos a partir del proceso de elaboración de la horchata" for emerging research groups from the Generalitat Valenciana. The authors would like to thank the Regulatory Council D.O. Tiger nut of Valencia (Valencia, Spain) for providing "horchata" by-products. Moreover, we also acknowledge Sotirios Kampranis and Christos Tsatsanis for their advice on measuring the effects of horchata on NO_2^- production.

References

Aguilar, N, E Albanell, B Miñarro, B Guamis, and M Capellas. 2015. "Effect of Tiger Nut-Derived Products in Gluten-Free Batter and Bread." *Revista de Agaroquimica y Tecnologia de Alimentos* 21 (5): 323–331.

Barba, F J, Z Zhu, M Koubaa, A S Sant'Ana, and V Orlien. 2016. "Green Alternative Methods for the Extraction of Antioxidant Bioactive Compounds from Winery Wastes and By-Products: A Review." *Trends in Food Science and Technology* 49: 96–109.

Barminas, J T, H M Maina, S Tahir, D Kubmarawa, and K Tsware. 2001. "A Preliminary Investigation into the Biofuel Characteristics of Tigernut (*Cyperus esculentus*) Oil." *Bioresource Technology* 79 (1): 87–89. https://doi.org/10.1016/S0960-8524(01)00026-8.

Base de Datos BEDCA. 2018. *Base de Datos Española de Composición de Alimentos*. Red-BEDCA & AESAN: Spain.

BOE. 1988. *REAL DECRETO 1338/1988 de 28 de Octubre Por El Que Se Aprueba La Reglamentación Técnico-Sanitaria Para La Elaboración y Venta de Horchata de Chufa*. Madrid, Spain: Ministerio de Relaciones Con Las Cortes y de La Secretaría Del Gobierno.

Bursać-Kovačević, D, F J Barba, D Granato, C M Galanakis, Z Herceg, V Dragović-Uzelac, and P Putnik. 2018. "Pressurized Hot Water Extraction (PHWE) for the Green Recovery of Bioactive Compounds and Steviol Glycosides from *Stevia rebaudiana* Bertoni Leaves." *Food Chemistry* 254: 150–157.

Chemat, F, M A Vian, and G Cravotto. 2012. "Green Extraction of Natural Products: Concept and Principles." *International Journal of Molecular Sciences* 13 (7): 8615–8627.

Codina Torrella, I. 2014. "Optimización Del Proceso de Elaboración y Aplicación de La Homogenización a Ultra Alta Presión Como Tecnología de Conservación de Licuado de Chufa." Universitat Autònoma de Barcelona.

Costa, C, A Tsatsakis, C Mamoulakis, M Teodoro, G Briguglio, E Caruso, D Tsoukalas, et al. 2017. "Current Evidence on the Effect of Dietary Polyphenols Intake on Chronic Diseases." *Food and Chemical Toxicology* 110 (April): 286–299.

CRDO. 2018. "Consejo Regulador de La D.O. Chufa de Valencia. Horchata de Chufa de Valencia." *Chufa de Valencia*.

Ezeh, O, M H Gordon, and K Niranjan. 2014. "Tiger Nut Oil (*Cyperus esculentus* L.): A Review of Its Composition and Physico-Chemical Properties." *European Journal of Lipid Science and Technology* 116 (7): 783–794.

Granato, D, D S Nunes, and F J Barba. 2017. "An Integrated Strategy between Food Chemistry, Biology, Nutrition, Pharmacology, and Statistics in the Development of Functional Foods: A Proposal." *Trends in Food Science & Technology* 62: 13–22. https://doi.org/10.1016/j.tifs.2016.12.010.

Hu, B, K Zhou, Y Liu, A Liu, Q Zhang, G Han, S Liu, Y Yang, Y Zhu, and D Zhu. 2018. "Optimization of Microwave-Assisted Extraction of Oil from Tiger Nut (*Cyperus esculentus* L.) and Its Quality Evaluation." *Industrial Crops and Products* 115: 290–297.

Jeske, S, and E K Arendt. 2018. "Past, Present and Future: The Strength of Plant-Based Dairy Substitutes Based on Gluten-Free Raw Materials." *Food Research International* 110 (August): 42–51. https://doi.org/10.1016/J.FOODRES.2017.03.045.

Kizzie-Hayford, N, D Jaros, S Zahn, and H Rohm. 2016. "Effects of Protein Enrichment on the Microbiological, Physicochemical and Sensory Properties of Fermented Tiger Nut Milk." *LWT – Food Science and Technology* 74 (December): 319–324. https://doi.org/10.1016/J.LWT.2016.07.067.

Koubaa, M, F J Barba, H Mhemdi, N Grimi, W Koubaa, and E Vorobiev. 2015. "Gas Assisted Mechanical Expression (GAME) as a Promising Technology for Oil and Phenolic Compound Recovery from Tiger Nuts." *Innovative Food Science & Emerging Technologies* 32: 172–180.

Koubaa, M, E Roselló-Soto, J Šic Žlabur, A Režek Jambrak, M Brncić, N Grimi, N Boussetta, and F J Barba. 2015. "Current and New Insights in the Sustainable and Green Recovery of Nutritionally Valuable Compounds from *Stevia rebaudiana* Bertoni." *Journal of Agricultural and Food Chemistry* 63 (31): 6835–6846. https://doi.org/10.1021/acs.jafc.5b01994.

Lasekan, O, and S M Abdulkarim. 2012. "Extraction of Oil from Tiger Nut (Cyperus Esculentus L.) with Supercritical Carbon Dioxide (SC-CO 2)." *LWT – Food Science and Technology* 47 (2): 287–292.

Luo, Y, X Li, J He, J Su, L Peng, X Wu, R Du, and Q Zhao. 2014. "Isolation, Characterisation, and Antioxidant Activities of Flavonoids from Chufa (Eleocharis Tuberosa) Peels." *Food Chemistry* 164 (April): 30–35. https://doi.org/10.1016/j.foodchem.2014.04.103.

Martín-Esparza, E, and C González-Martínez. 2016. "Horchata de Chufa: A Traditional Spanish Beverage with Exceptional Organoleptic, Nutritive, and Functional Attributes." In: *Functional Properties of Traditional Foods*, eds. K Kristbergsson and S Ötles, 371–375. Springer US: New York, USA.

Mendes, D, M M Oliveira, P I Moreira, J Coutinho, F M Nunes, D M Pereira, P Valentão, P B Andrade, and R A Videira. 2018. "Beneficial Effects of White Wine Polyphenols-Enriched Diet on Alzheimer's Disease-like Pathology." *The Journal of Nutritional Biochemistry* 55 (April): 165–177. https://doi.org/10.1016/j.jnutbio.2018.02.001.

Mokrani, A, and K Madani. 2016. "Effect of Solvent, Time and Temperature on the Extraction of Phenolic Compounds and Antioxidant Capacity of Peach (*Prunus persica* L.) Fruit." *Separation and Purification Technology* 162 (April): 68–76. https://doi.org/10.1016/j.seppur.2016.01.043.

Olabiyi, A A, F B Carvalho, N B Bottari, V M Morsch, A F Morel, G Oboh, and M R Schetinger. 2018. "Tiger Nut and Walnut Extracts Modulate Extracellular Metabolism of ATP and Adenosine through the NOS/CGMP/PKG Signalling Pathway in Kidney Slices." *Phytomedicine* 43: 140–149.

Puértolas, E, and F J Barba. 2016. "Electrotechnologies Applied to Valorization of By-Products from Food Industry: Main Findings, Energy and Economic Cost of Their Industrialization." *Food and Bioproducts Processing* 100: 172–184.

Quiñones, M, M Miguel, and A Aleixandre. 2012. "Los Polifenoles, Compuestos de Origen Natural Con Efectos Saludables Sobre El Sistema Cardiovascular." *Nutrición Hospitalaria* 27 (1): 76–89.

Roselló-Soto, E, F J Barba, J M Lorenzo, R Dominguez, M Pateiro, J Mañes, and J C Moltó. 2018a. "Evaluating the Impact of Supercritical-CO_2 Pressure on the Recovery and Quality of Oil from 'Horchata' by-Products: Fatty Acid Profile, α-Tocopherol,

Phenolic Compounds, and Lipid Oxidation Parameters." *Food Research International* 120: 888-894.

Roselló-Soto, E, M Koubaa, A Moubarik, R P Lopes, J A Saraiva, N Boussetta, N Grimi, and F J Barba. 2015. "Emerging Opportunities for the Effective Valorization of Wastes and By-Products Generated during Olive Oil Production Process: Non-Conventional Methods for the Recovery of High-Added Value Compounds." *Trends in Food Science and Technology* 45 (2): 296–310. https://doi.org/10.1016/j.tifs.2015.07.003.

Roselló-Soto, E, O Parniakov, Q Deng, A Patras, M Koubaa, N Grimi, N Boussetta, et al. 2016. "Application of Non-Conventional Extraction Methods: Toward a Sustainable and Green Production of Valuable Compounds from Mushrooms." *Food Engineering Reviews* 8 (2): 214–234.

Roselló-Soto, E, F Barba, P Putnik, D B Kovačević, J Lorenzo, Y Cantavella-Ferrero 2018b. "Enhancing Bioactive Antioxidants' Extraction from 'Horchata de Chufa' By-Products." *Foods* 7 (10): 161. https://doi.org/10.3390/foods7100161.

Roselló-Soto, E, C Garcia, A Fessard, F Barba, P Munekata, J Lorenzo, F Remize, et al. 2018c. "Nutritional and Microbiological Quality of Tiger Nut Tubers (*Cyperus esculentus*), Derived Plant-Based and Lactic Fermented Beverages." *Fermentation* 5 (1): 3. https://doi.org/10.3390/FERMENTATION5010003.

Sánchez-Zapata, E, J Díaz-Vela, M L Pérez-Chabela, J A Pérez-Alvarez, and J Fernández-López. 2013. "Evaluation of the Effect of Tiger Nut Fibre as a Carrier of Unsaturated Fatty Acids Rich Oil on the Quality of Dry-Cured Sausages." *Food and Bioprocess Technology* 6 (5): 1181–1190.

Sánchez-Zapata, E, E Fuentes-Zaragoza, M Viuda-Martos, J Fernández-López, E Sendra, E Sayas, and J A Pérez-Alvarez. 2012a. "Reclaim of the By-Products from 'Horchata' Elaboration Process." *Food and Bioprocess Technology* 5 (3): 954–963.

Sánchez-Zapata, E, Muñoz MC, E Fuentes, J Fernández-López, E Sendra, E Sayas, C Navarro, and J A Pérez-Alvarez. 2010. "Effect of Tiger Nut Fibre on Quality Characteristics of Pork Burger." *Meat Science* 85 (1): 70–76.

Sánchez-Zapata, E, E Sayas-Barberá, J A Pérez-Alvarez, and J Fernández-López. 2013a. "The Effect of Replacing Water with Tiger Nut Milk (Horchata) Liquid Coproduct on the Physicochemical Properties and Oxidation (Haemopigments and Lipids) of a Cooked Pork Liver Meat Product." *Food Technology and Biotechnology* 51 (2): 422–428.

Sánchez-Zapata, E, V Zunino, J A Pérez-Alvarez, and J Fernández-López. 2013c. "Effect of Tiger Nut Fibre Addition on the Quality and Safety of a Dry-Cured Pork Sausage ('Chorizo') during the Dry-Curing Process." *Meat Science* 95 (3): 562–568.

Sánchez-Zapata, E, J Fernández-López, J A Pérez-Alvarez, J Soares, S Sousa, A M P Gomes, and M M E Pintado. 2013b. "In Vitro Evaluation of 'Horchata' Co-Products as Carbon Source for Probiotic Bacteria Growth." *Food and Bioproducts Processing* 91 (3): 279–286. https://doi.org/10.1016/J.FBP.2012.11.003.

Sánchez-Zapata, E, J Fernández-López, and J A Pérez-Alvarez. 2012b. "Tiger Nut (Cyperus Esculentus) Commercialization: Health Aspects, Composition, Properties, and Food Applications." *Comprehensive Reviews in Food Science and Food Safety* 11 (4): 366–377. https://doi.org/10.1111/j.1541-4337.2012.00190.x.

Santhakumar, A B, M Battino, and J M Alvarez-Suarez. 2018. "Dietary Polyphenols: Structures, Bioavailability and Protective Effects against Atherosclerosis." *Food and Chemical Toxicology* 113 (April): 49–65. https://doi.org/10.1016/j.fct.2018.01.022.

Scalbert, A, C Manach, C Morand, C Rémésy, and L Jiménez. 2005. "Dietary Polyphenols and the Prevention of Diseases." *Critical Reviews in Food Science and Nutrition* 45 (4): 287–306. https://doi.org/10.1080/1040869059096.

Soriano del Castillo, J M, G Alcayde, J Blesa, L Bosch, A I Catalá, P García, C Martínez, et al. 2014. *El Gran Libro de La Horchata y La Chufa de Valencia*. 1st ed. Universitat de València: Col.lecions de les Clíniques de la Universitat de València: Valencia, Spain.

Wild, D, G G Robins, V J Burley, and P D Howdle. 2010. "Evidence of High Sugar Intake, and Low Fibre and Mineral Intake, in the Gluten-Free Diet." *Alimentary Pharmacology & Therapeutics* 32 (4): 573–581. https://doi.org/10.1111/j.1365-2036.2010.04386.x.

Yeboah, S O, Y C Mitei, J C Ngila, L Wessjohann, and J Schmidt. 2012. "Compositional and Structural Studies of the Oils from Two Edible Seeds: Tiger Nut, Cyperus Esculentum, and Asiato, Pachira Insignis, from Ghana." *Food Research International* 47 (2): 259–266.

8

Green Extraction of Nutritional and Antioxidant Valuable Compounds from Wine By-Products

FRANCISCO J. BARBA

Nutrition and Food Science Area, Preventive Medicine and Public Health, Food Science, Toxicology and Forensic Medicine Department, Faculty of Pharmacy, Universitat de València, Avda. Vicent Andrés Estellés, s/n, 46100 Burjassot, València, Spain

BELÉN GÓMEZ

Centro Tecnológico de la Carne de Galicia, rúa Galicia n° 4, Parque Tecnológico de Galicia, San Cibrao das Viñas, Ourense, Spain

GABRIELA I. DENOYA

Instituto Nacional de Tecnología Agropecuaria (INTA), Instituto Tecnología de Alimentos (Argentina), Hurlingham, Buenos Aires, Argentina
Consejo Nacional de Investigaciones Científicas y Técnicas (CONICET), Argentina

MLADEN BRNČIĆ AND SUZANA RIMAC-BRNČIĆ

University of Zagreb Faculty of Food Technology and Biotechnology, Pierottijeva ulica 6, 10000, Zagreb, Croatia

JOSE M. LORENZO

Centro Tecnológico de la Carne de Galicia, rúa Galicia n° 4, Parque Tecnológico de Galicia, San Cibrao das Viñas, Ourense, Spain

ANDRÉS MORENO

University of Castilla-La Mancha, Faculty of Chemical Sciences and Technologies (San Alberto Magno Building), Department of Organic Chemistry, Av. Camilo José Cela, 10, Ciudad Real, 13071, Spain

Contents

1 Introduction

The world wine-growing area in 2017 stands at 7.5 million hectares, with Spain having the largest number (967,000 ha), around 13% of the world surface, ranking ahead of countries, such as China (870,000 ha) and France (786,000 ha) (OIV [The International Organization of Vine and Wine], 2019). In terms of world's wine production, Italy is the leading producer (48.5 million hectoliters), followed by France (46.4 million hectoliters) and Spain (40.9 million hectoliters) (OIV, 2019).

During the wine production process, a large number of wastes and by-products are generated (Figure 8.1), having a large number of valuable compounds with a high nutritional value and biological properties (e.g., proteins and bioactive compounds).

Among the bioactive compounds present in the vinification products, polyphenols represent one of the most important groups of natural products associated with healthy benefits (Haminiuk et al., 2012). In the last two decades, there has been a growing interest in these compounds by the scientific community due to their numerous beneficial properties, many of them associated with their antioxidant capacity. Up to now, the nature of all these compounds and the mechanism of action are not completely known, although it has been shown that, depending on the chemical structure, they have different antioxidant properties. Then, their antioxidant activity not only depends on the quantity but also on the type of phenolic compound that is obtained. For this reason, the selective extraction of these compounds acquires a relevant interest. Particularly, some phenolic compounds were associated with antiallergenic, anti-inflammatory, antimicrobial, antioxidant, antithrombotic, and cardioprotective properties (Beres et al., 2017).

One of the main objectives in Europe is the elimination of waste in a sustainable way. Traditionally, many of the residues from the agri-food industry have been discarded without further consideration, without taking into account that they could represent a valuable source of compounds with a high nutritional value. However, in recent decades, there has been a growing interest from both the food industry and the scientific community to add value to these residues. One of the solutions is the extraction of

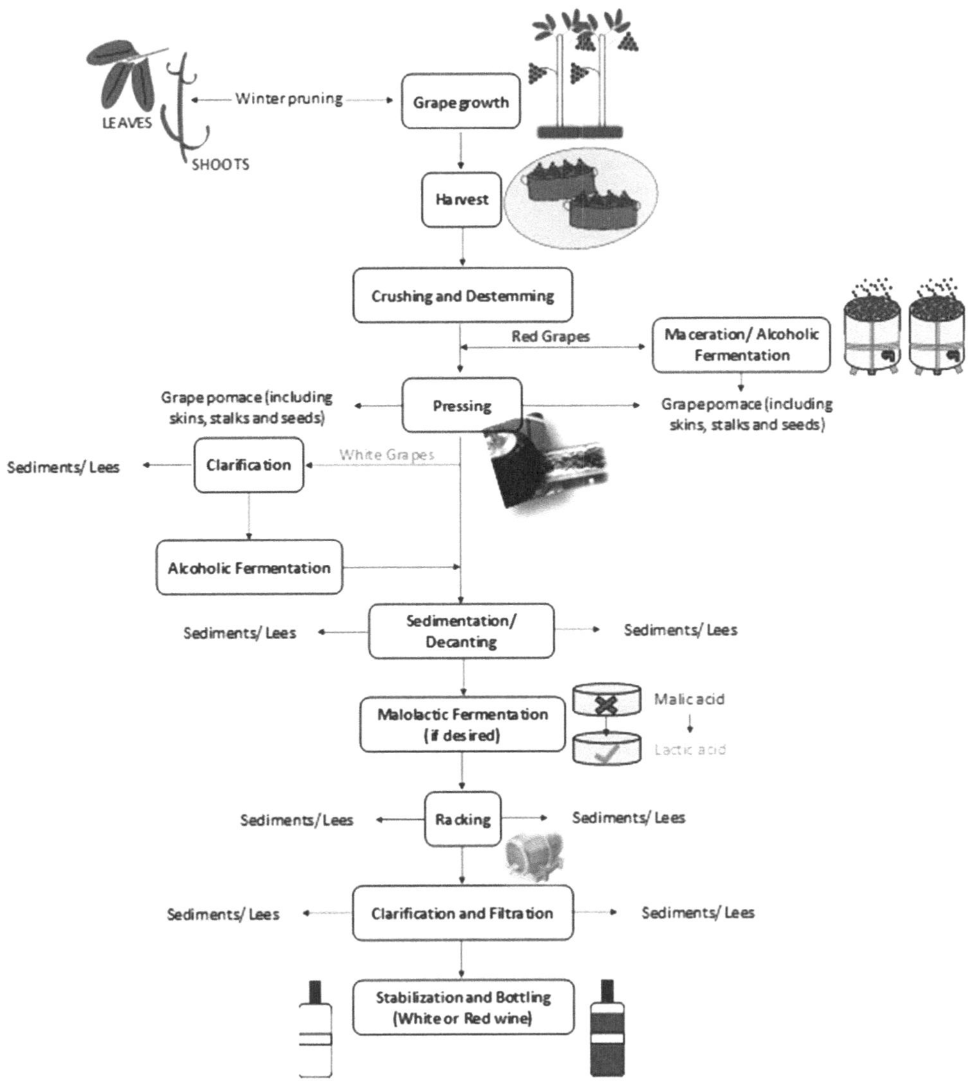

Figure 8.1 General scheme of the winemaking process with the by-products generated in each step.

compounds that can be used in different applications, such as pharmaceutical, food, and so on, as food additives, or nutraceuticals, among others. For instance, this year, Baca-Bocanegra et al. (2019) contributed to the global idea of waste reduction and by-product valorization, studying the potential use of wine by-products as a source of copigments for improving the quality of wines.

Conventional solid–liquid extraction (SLE) has been the main way to extract these compounds. However, in many cases, this involves the use of solvents that can be toxic and the use of temperatures above 60 °C, with the subsequent degradation of thermolabile compounds. Thus, the use of new green technologies that can partially or totally replace conventional methodologies is being investigated to reduce solvent consumption,

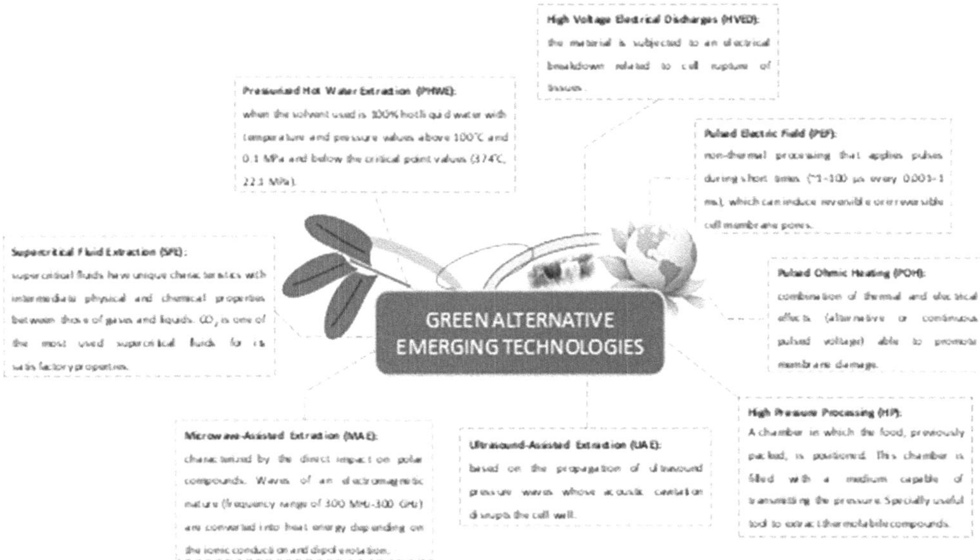

Figure 8.2 Short description of some nonconventional eco-friendly extraction technologies.

as well as to reduce the temperature and the treatment time. The purpose is to achieve more efficient and sustainable processes from the economic and environmental point of view. The application of conventional and innovative extraction processes (Figure 8.2) on the different wine industry wastes and by-products for obtaining valuable products is discussed in the different sections of this chapter.

2 Pomace

Grape pomace is an abundant by-product from the winemaking industry, which includes seeds, and the remaining skin and stalks (Figure 8.3) and it can represent around 20% of the total grape weight used in the winemaking process (Bordiga et al., 2019). In addition, it is interesting to take into account that in Italy, France, and Spain, countries with a relevant production of wine, the annual grape pomace generation can reach nearly 1,200 tons per year (Beres et al., 2017). This fact highlights the need to implement a waste reduction policy.

Traditional extraction using a mixture of water and organic solvent provides high yields (Schieber, 2017). However, the current challenge consists in reducing the use of organic solvents, as well as the temperature and the extraction time, looking for more sustainable processes. In this sense, several authors have evaluated the use of pulsed electric fields (PEF) treatment to extract polyphenols from the grape pomace. For instance, Brianceau et al. (2015) found that a PEF treatment (1.2 kV/cm, 18 kJ/kg) combined with a hydroalcoholic extraction at different temperatures with a pomace density of 1 g/cm^3 allowed extraction of more

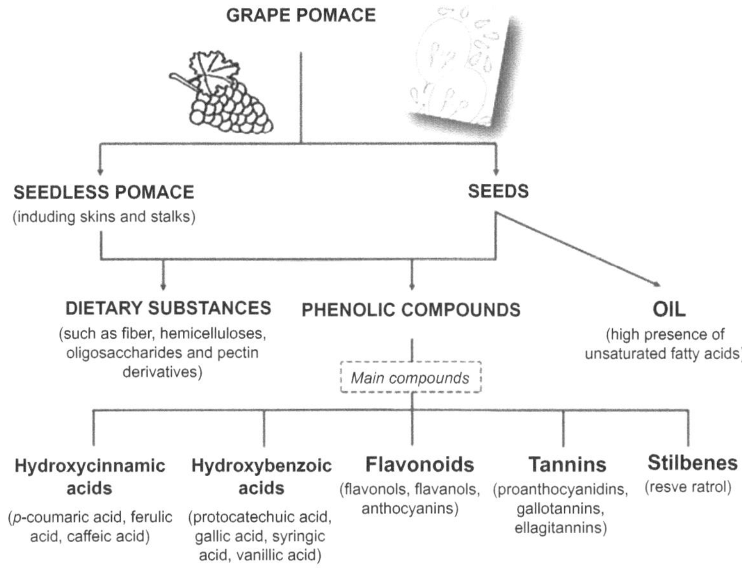

Figure 8.3 Fractions obtained from grape pomace and main components found in them.

polyphenols from grape pomace than the conventional extraction alone. They also obtained a selective extraction of anthocyanins with PEF treatment (ratio anthocyanins: total flavan-3-ols at 20 °C of 7.1 and 9.0 for the control and PEF-treated samples, respectively). In another study, the impact of PEF (13.3 kV/cm/0–564 kJ/kg) on the selective recovery of high added value compounds from fermented grape pomace (variety *Dunkelfelder*) was evaluated (Barba et al., 2015). The authors established different cell disruption levels to evaluate if the extraction assisted by PEF was correlated with the degree of cell disruption, and they observed an increased anthocyanin extraction when the cell disruption level was increased.

In some cases, the use of pulsed ohmic heating (POH), a procedure that is able to increase the temperature by ionic movements following the application of PEF treatment could be a potential strategy to obtain intracellular compounds at moderate temperatures (20–50 °C) and electric field (<1,000 V/cm). For instance, the consequences of using POH (20–800 V/cm) with different proportions of ethanol–water (0–50% w/w) on the extraction of polyphenols from red grape pomace were investigated (El Darra et al., 2013). POH treatment was able to promote cell membrane disruption, thus increasing polyphenols recovery from grape pomace, obtaining 36% more polyphenol content than in the untreated samples, when POH process was performed at 400 V/cm followed by a diffusion step for 60 min at 50 °C and with a hydroalcoholic mixture of 30% ethanol in water. On the other hand, the application of high voltage electrical discharges (HVED) for polyphenols extraction from grape pomace was also investigated. After the HVED treatment, the extraction of total polyphenols from grape pomace, seeds, skins, and stems increased up to seven times. Nevertheless, it should be considered that a pilot-scale (35 L) needs higher treatment energies than the laboratory scale to obtain similar results (Boussetta et al., 2012).

Another nonconventional technique that has been reported as appropriate for the recovery of phenolic compounds from grape pomace is ultrasound-assisted extraction (UAE). Corrales et al. (2008) evaluated the effects of UAE combined with thermal treatment (70 °C) to obtain antioxidant bioactive compounds from grape by-products. After applying ultrasound (USN) treatments, authors detected a substantial increase in total phenolic compounds of the extracts in comparison to the recoveries observed with conventional water extraction. In the same way, Vilkhu et al. (2008) found a significant improvement in the extraction yield of polyphenols (6–35%) after red grape marc was subjected to UAE.

In the same way, supercritical fluid extraction (SFE) is a fairly novel technology to pull out analytes of interest from solid matrices, including grapes-derived products (Mohamed et al., 2016). Supercritical CO_2 extraction (SC-CO_2) can offer many advantages in comparison to SLE, such as it is cheap and nontoxic. Oliveira et al. (2013) studied the composition and the antimicrobial activity of extracts recovered from Merlot and Syrah grape pomace after being treated with SC-CO_2 and SC-CO_2 + co-solvent at temperatures of 50 and 60 °C and pressures up to 300 bars. Although they observed low extraction yields of polyphenols (gallic acid, *p*-OH-benzoic acid, vanillic acid, and epicatechin), the extracts under study showed interesting antibacterial (*Bacillus cereus, Escherichia coli, Staphylococcus aureus*, and *Pseudomonas aeruginosa*) and antifungal (*Candida parapsilosis, Candida albicans*, and *Candida krusei*) activities.

Other authors evaluated the impact of pressurized hot water extraction (PHWE), a nonconventional system consisting of the use of hot water as a solvent to extract molecules. In particular, Vergara-Salinas et al. (2013) employed certain conditions (50–200 °C, 103 bars, and 5–30 min) to recover antioxidants (anthocyanins, tannins, and polyphenols) from grape pomace before and after fermentation. They observed that the recovery of antioxidant compounds was increased proportionally with the increase in water temperature, finding the highest anthocyanin extraction yields at 100 °C and the maximum yield for tannins and tannin-anthocyanin adducts at 150 °C. Nevertheless, assays with temperatures higher than 100 °C and longer extraction times showed significant decreases in the polyphenol extraction yields. Furthermore, the authors also detected that some thermolabile proanthocyanidins were extracted only at 50 and 100 °C. Additionally, they did not find a specific relationship between the main polyphenol content and the antioxidant recovery after applying PHWE. This issue could be explained by the extraction of other antioxidant compounds (e.g., vitamin E), which could be present in grape pomace but not analyzed.

3 Skins

Several studies have demonstrated the potential of the grape skins as a source of phenolics and antioxidants. For instance, PEF was used for pretreatment for enhancing recovery of tannins and anthocyanins from this by-product (Cholet et al., 2014; Delsart et al., 2013) obtaining interesting results. In fact, Delsart et al. (2013) found extraction of tannins (34%) after using PEF (0.7 kV/cm/200 ms/31 Wh/kg), while they obtained the higher anthocyanin extraction after 4 kV/cm/1 ms/4 Wh/kg. The authors attributed this enhanced extraction to the ability of PEF to promote modifications in the grape skin structure, as well as to the depolymerization of skin tannins after applying PEF treatment. To verify the biochemical changes occurring in grape skin after PEF, Cholet et al. (2014) evaluated the impact of different PEF extraction parameters (4 and 0.7 kV/cm during 1 ms and 200 ms) on the grape skin structure. They observed a little effect of PEF (4 kV/cm for 1 ms) on the polyphenol structure, while PEF (0.7 kV/cm for 200 ms) promoted marked changes in the skin cell wall organization.

Another technique studied for obtaining anthocyanins from grape skins was the microwave-assisted extraction (MAE) (Liazid et al., 2011). MAE application allowed a significant reduction in the processing time from 5 h (with SLE) to 5 min. Moreover, MAE demonstrated another advantage compared with the conventional method, which consisted of recovering three additional acyl derivatives at measurable levels. In the same line, Yu et al. (2014) reported that the optimal conditions for extracting polyphenols from grape skin were 1 g grape skin, 40 mL of extraction of aqueous solution with 50% ethanol, 540 W microwave power, and 3 min of extraction time.

In the same way, SFE was also evaluated and optimized for obtaining resveratrol from grape skin of *Vitis vinifera* (Pascual-Marti et al., 2001). They observed that the best SFE conditions to recover 100% of resveratrol were at 40 °C, 110 bars, 7.5% ethanol, and extraction time of 15 min. In another study, the impact of SC-CO_2 at 35–55 °C and 100–400 bars, and the addition of a modifier (5% [v/v] of ethanol) on the recovery of resveratrol from grape seeds, pomace, stems, and skin of the Palomino fine grape variety was investigated (Casas et al., 2010). It was reported that the highest recovery of resveratrol from skins (49.1 mg/100 g dry sample) was obtained when SC-CO_2 was applied at 35 °C and 400 bars, and 5% (v/v) of ethanol as co-solvent. Likewise, high pressure/high temperature (HPHT) was checked as another potential procedure to recover phenolics from grape by-products (Casazza et al., 2010a, 2012), and particularly resveratrol from grape skins (Casazza et al., 2010b), finding promising results. These authors evaluated the extraction yields and antioxidant activities of phenolic compounds from grape seeds and skins obtained by a variety of nonconventional (UAE, MAE, and HPHT) and traditional SLE methods. Specifically, ethanol seemed to be a more suitable solvent than methanol for both the seed and skin extractions. In addition, results revealed that HPHT working in a Parr reactor provided the maximum content in total polyphenols, *o*-diphenols, and flavonoids, from both seeds (108.3, 47 mg GAE/g and 47.2 mg CE/g dry weight) and skins (34.2, 10.1 mg GAE and 21.6 mg CE/g dry weight), respectively. Casazza et al. (2012) identified the phenolic compounds of grape skin extracts using HPLC, showing gallic acid, protocatechuic acid, catechin, vanillic acid, syringic acid, coumaric acid, *trans*-resveratrol, and quercetin as the major phenolics with 163.2, 69.9, 420, 20.6, 603, 20.1, 42.4, and 117.1 mg/100 g, respectively.

Likewise, the effects of the application of PHWE at 40–120 °C and 150 bars for the extraction of antioxidant compounds (polyphenols and anthocyanins) was reported from skins of two different wine grape varieties (Alibernet and St. Laurent). The results were compared with the ones obtained with the same grape skin samples extracted with both methanol and ethanol under pressurized fluid extraction conditions. The authors concluded that PHWE was the most appropriate and efficient process for obtaining promising phenolic antioxidants from grape skins (Stavikova et al., 2011).

On the other hand, an important step in the manufacture of new products is their purification. In this sense, Bleve et al. (2008) reported a new method using CO_2 to carry out the purification of anthocyanins from grape skin extracts. Previously, they obtained anthocyanins by treating grape skins with a mixture of ethanol/water with 0.2% of trifluoroacetic acid. After the purification stage, two differentiated phases were recovered: one enriched in pure anthocyanins, and the second containing ethanol and other grape skin components. In sum, under optimal conditions (30–40 °C, 100–130 bars, pH 2–4, 25–30% ethanol, 25–50 mL/min CO_2 flow, and 3–10% extract flow ratio), the purification process was suitable for removing ethanol without any further heat treatment and to produce a nondegradable anthocyanin extract.

4 Seeds

Seeds are one of the main components of grapes and were reported as an important source of oil, which are widely used in the cosmetic industry (Beres et al., 2017). The recovery of polyphenols from grape seeds was evaluated by Boussetta et al. (2013). For this purpose, the authors evaluated and compared the effects of HVED (40 kV), PEF (8–20 kV/cm, 0–20 μs), and grinding (180 W, 40 s), obtaining the optimal yield (9 g of GAE/100 g of dry matter) after using HVED or grinding plus supplementary ethanol extraction (30 g ethanol/100 g solid).

On the other hand, other authors used conventional solvent extraction (Soxhlet) and USN for extracting oil from grape seed rich in polyphenols (Da Porto et al., 2013). They obtained a similar extraction yield (14 g/100 g) with Soxhlet's extraction (6 h) compared with UAE (150 W/20 kHz/30 min). They also observed a decrease in the time (15 min) needed for extracting a similar amount of polyphenols when USN was used compared with maceration (12 h).

In another study, grape seed oil was subjected to SC-CO_2, and the conditions of the process were optimized by the response surface methodology. Both the effect of pressure and temperature on the recovered oil yield and the extract's antioxidant activity was evaluated. The highest antioxidant activity (DPPH inhibition of 37.1%) and yield (14.5%) were obtained under the following optimum experimental conditions: 41 °C and 400 bars (Jokić et al., 2016).

Palma and Taylor (1999) developed a new methodology using near critical carbon dioxide to extract polyphenols from white grape seeds. For this purpose, they optimized the processing conditions (CO_2 density, organic modifier, percentage of modifier, and extraction temperature), and obtained an extract with gallic acid, catechin, and epicatechin as the predominant phenolic compounds.

5 Wine Lees

Lees are yeast residues collected after different stages of the winemaking process, such as filtration, centrifugation, or even deposits in wine vessels. The red wine yeast lees involve around 12 kg of pigments per ton (Hogervorst et al., 2017). In fact, these by-products have been characterized for their rich composition consisting of yeasts, tartaric acid, polyphenols, and inorganic compounds. Moreover, lees incorporate enzymes that can change polyphenols into simple phenolic components (Vattem & Shetty, 2003), as well as are usually employed in the aging process for improving the wine quality by reducing bitterness (Tao et al., 2014). With the aim of incrementing its economic value, wine lees are being used as raw material for the recovery of polyphenols. For instance, Tao et al. (2014) compared the extraction of antioxidant compounds from wine lees using UAE and conventional maceration, finding an approximate recovery percentage of 20% lower in the second case, for both the total phenolic content (58.76 vs. 49.04 mg/g) and total anthocyanin concentration (6.69 vs. 5.55 mg/g). Particularly, the optimal UAE conditions for the highest total phenolic content, considering the sonication time, temperature, liquid–solid ratio, and ethanol proportion were 25 min, 60 °C, 60:1, and ~44%, respectively. While the maximum anthocyanin yield was observed after selecting these values for these parameters: 36.3 min, 59.9 °C, and 51.5% of ethanol concentration.

In another study (Pérez-Serradilla & De Castro, 2011), the control of the quantity (phenolic index) and the antioxidant ability (ORAC data) of the extracts from wine lees allowed reaching optimal MAE of polyphenols. In particular, the optimal parameters were 200 W, 75% of ethanol as solvent, and 17 min of processing duration. This treatment time was significantly lower than that of conventional extraction (24 h). Meanwhile, Farías-Campomanes et al. (2015) carried out the SFE of polyphenols from pisco-making lees working at two pressures 20 and 35 MPa and with 10% ethanol as solvent. The lower pressure (20 MPa) leads to higher polyphenol recovery compared with 35 MPa, including phenolic acids (vanillic, gallic, protocatechuic, ferulic, syringic, and *p*-coumaric acids derivatives) and flavonoids (quercetin). Despite Soxhlet method resulting in higher polyphenol extraction yields, the high purity of the recovered components and the low solvent consumption of SFE showed that this technology could be a better polyphenol extraction method than the conventional one.

6 Leaves

In the winemaking process, in addition to the grapes, we must consider olive tree that needs attention similar to pruning for its well-being and optimal production. In this context, it has been reported that the pruning of this plant involves around 7 kg of waste by year, in which the leaves represent a high percentage (Cárcel et al., 2010). For this reason, the authors used olive leaves for extracting antioxidant compounds through the application of high-intensity USNs. Data showed that extracts from leaves contained higher antioxidant ability (16,146 ± 1,116 μmol trolox/L) than grape stalks (9,327 ± 1,697 μmol trolox/L). In addition, recently, grape leaves have also been described as a source of antioxidant compounds with a complex chemical composition, including enzymes, lipids, sugars, vitamins, organic acids, flavanols, tannins, procyanidins, and anthocyanins (Hogervorst et al., 2017).

7 Vine Shoots

Vine shoots are an important waste of viticulture, obtained from the traditional pruning practice. Before being exploited as an acceptable source of phenolic compounds, vine shoots were habitually burned or abandoned on the ground. There are several studies that have identified a great variety of polyphenols in this by-product, including caffeic acid, gallic acid, catechin, epicatechin, luteolin, pyrogallol, hydroxybenzoic acid, syringic acid, vanillic acid, ferulic acid, procyanidol B1, procyanidol B2, procyanidol B3, procyanidol B4, procyanidol B7, kaempferol, resveratrol, viniferin, piceatannol, vitisin A, vitisin B, ampelopsin A, or hopeaphenol (Maroun et al., 2017).

Rajha et al. (2014a) investigated the potential of PEF technology for extracting polyphenols and proteins from vine shoots, observing an increased recovery yield of these compounds after PEF treatment (50 kJ/kg). The same research group also observed an important polyphenols' yield (34.5 mg GAE/g dry matter) and high purity (89%) after applying HVED (10 kJ/kg), which is the significantly lower energy required compared with USN (1,010 kJ/kg) treatment (Rajha et al., 2014b). Meanwhile, Rajha et al. (2015a) also found protein concentrations of 0.097 and 0.082 mg/mL, and phenolic concentrations of 0.4 and 0.3 mg/mL, after grinding (360 kJ/kg) and HVED (242 kJ/kg) treatments, respectively.

On the other hand, Rajha et al. (2015b) assayed the capacity of β-cyclodextrin (β-CD), a hydrotropic solvent, to extract polyphenols from vine shoots. Authors opted for the response surface methodology with the following factors for optimizing the recovery conditions: β-CD concentration, time, and temperature. The obtained results were compared with those observed with conventional hydroethanolic treatments. As conclusion to stand out that, under β-CD-assisted extraction conditions (37.7 mg/ml β-CD, 66.6 °C, and 48 h), it was possible to obtain 8 mg of GAE/g of dry matter, as well as an antioxidant capacity of 3,146 βM trolox equivalent/mL. They detected similar polyphenol amounts after 1.65 h of hydroethanolic (50%, v/v) treatment, while the antioxidant ability was substantially lower (2,000 βM trolox equivalent) than with β-CD.

8 Possible Applications of the Extracted Compounds

Nowadays, the so-called food wastes can be considered as by-products susceptible to be manufactured for the recovery of cheap and functional components (Galanakis, 2012), taking advantage of the numerous emerging green technologies cited in the previous sections of this chapter. The overall process consists of: (i) classification of the food by-product, (ii) investigation in which high added value components could be obtained, (iii) exploration of the most appropriate techniques for their extraction and purification, (iv) consideration of the future applications, and (v) discussion of the costs and environmental safety factors.

During the processing of these materials, some undesirable substances in liquid or solid form are also recuperated. However, discarding certain streams should not affect the potentiality of reutilizing others with promising market value inside the food chain. For instance, valuable compounds derived from food wastes, including the phenolic antioxidants from wine by-products, can be commercialized within the following approaches:

- Improving the organoleptic characteristics of food products (odor, color, and taste).
- Preserving and extending the shelf-life of final products by delaying rancidity and off-flavors.
- Food safety (prevent microbial deterioration and/or avoid toxic and carcinogenic effects).
- Stabilizing, gelling, emulsifying, and texturizing (or with other technological properties) agents.
- Prebiotics.
- Cosmetic purposes.

9 Conclusion

Similar to other agro-industrial processes, the wine industry generates a considerable number of wastes and by-products (especially grape pomace) whose valorization is important considering the eco-friendly policies and for maintaining the sustainability of this practice. In this sense, natural bioactive compounds are an industrial requirement, not only for the current strict laws but also for offering safer and higher quality products to the consumers. These types of ingredients can exert a variety of healthy properties, as well as they can be used for different technological purposes, replacing the common synthetic additives. In this line, the methodologies chosen for obtaining valuable substances will play a key role in relation to the yields, purity, costs, and environmental contaminations. Moreover, in addition to the origin of the raw material and the part of the plant (skin, seeds, and leaves), the pretreatment applied, and extraction technique and solvent used will notably influence the composition and quality of the extracts.

References

Baca-Bocanegra, B., J. Nogales-Bueno, I. García-Estévez, M. T. Escribano-Bailón, J. M. Hernández-Hierro, & F. J. Heredia. 2019. "Screening of wine extractable total phenolic and ellagitannin contents in revalorized cooperage by-products: Evaluation by micro-NIRS technology." *Food and Bioprocess Technology* 12(3): 477–485.

Barba, F. J., Brianceau, S., Turk, M., Boussetta, N., & Vorobiev, E. 2015. Effect of alternative physical treatments (ultrasounds, pulsed electric fields, and high-voltage electrical discharges) on selective recovery of bio-compounds from fermented grape pomace. *Food and Bioprocess Technology* 8(5): 1139–1148. doi:10.1007/s11947-015-1482-3

Beres, C., Costa, G. N., Cabezudo, I., Da Silva-James, N. K., Teles, A. S., Cruz, A. P., … Freitas, S. P. 2017. Towards integral utilization of grape pomace from winemaking process: a review. *Waste Management, 68*: 581–594.

Bleve, M., Ciurlia, L., Erroi, E., Lionetto, G., Longo, L., Rescio, L., … Vasapollo, G. 2008. An innovative method for the purification of anthocyanins from grape skin extracts by using liquid and sub-critical carbon dioxide. *Separation and Purification Technology, 64*(2): 192–197.

Bordiga, M., Travaglia, F., & Locatelli, M. 2019. Valorisation of grape pomace: an approach that is increasingly reaching its maturity–a review. *International Journal of Food Science & Technology* 54: 933-942. doi:10.1111/ijfs.14118

Boussetta, N., Turk, M., De Taeye, C., Larondelle, Y., Lanoisellé, J. L., & Vorobiev, E. 2013. Effect of high voltage electrical discharges, heating and ethanol concentration on the extraction of total polyphenols and lignans from flaxseed cake. *Industrial Crops and Products, 49*: 690–696.

Boussetta, N., Vorobiev, E., Reess, T., De Ferron, A., Pecastaing, L., Ruscassié, R., & Lanoisellé, J.-L. 2012. Scale-up of high voltage electrical discharges for polyphenols extraction from grape pomace: effect of the dynamic shock waves. *Innovative Food Science & Emerging Technologies, 16*: 129–136.

Brianceau, S., Turk, M., Vitrac, X., & Vorobiev, E. 2015. Combined densification and pulsed electric field treatment for selective polyphenols recovery from fermented grape pomace. *Innovative Food Science & Emerging Technologies, 29*: 2–8.

Cárcel, J. A., García-Pérez, J. V., Mulet, A., Rodríguez, L., & Riera, E. 2010. Ultrasonically assisted antioxidant extraction from grape stalks and olive leaves. *Physics Procedia, 3*(1): 147–152.

Casas, L., Mantell, C., Rodríguez, M., Ossa, E. J. M.de la, Roldán, A., Ory, I. De, … Blandino, A. 2010. Extraction of resveratrol from the pomace of Palomino fino grapes by supercritical carbon dioxide. *Journal of Food Engineering*, *96*(2): 304–308.

Casazza, A. A., Aliakbarian, B., Mantegna, S., Cravotto, G., & Perego, P. 2010a. Extraction of phenolics from Vitis vinifera wastes using non-conventional techniques. *Journal of Food Engineering*, *100*(1): 50–55.

Casazza, A. A., Aliakbarian, B., Montoya, E. Y. O., & Perego, P. 2010b. t-resveratrol recovery from grape skins using high pressure and temperature extraction. *Journal of Biotechnology*, *150*(Suppl 0): S333.

Casazza, A. A., Aliakbarian, B., Sannita, E., & Perego, P. 2012. High-pressure high-temperature extraction of phenolic compounds from grape skins. *International Journal of Food Science and Technology*, *47*(2): 399–405.

Cholet, C., Delsart, C., Petrel, M., Gontier, E., Grimi, N., L'Hyvernay, A., … Gény, L. 2014. Structural and biochemical changes induced by pulsed electric field treatments on cabernet sauvignon grape berry skins: impact on cell wall total tannins and polysaccharides. *Journal of Agricultural and Food Chemistry*, *62*(13): 2925–2934.

Corrales, M., Toepfl, S., Butz, P., Knorr, D., & Tauscher, B. 2008. Extraction of anthocyanins from grape by-products assisted by ultrasonics, high hydrostatic pressure or pulsed electric fields: a comparison. *Innovative Food Science and Emerging Technologies*, *9*(1): 85–91.

Da Porto, C., Porretto, E., & Decorti, D. 2013. Comparison of ultrasound-assisted extraction with conventional extraction methods of oil and polyphenols from grape (*Vitis vinifera* L.) seeds. *Ultrasonics Sonochemistry*, *20*(4): 1076–1080.

Delsart, C., Cholet, C., Ghidossi, R., Grimi, N., Gontier, E., Geny, L., … Mietton-Peuchot, M. 2013. Effects of pulsed electric fields on Cabernet Sauvignon grape berries and on the characteristics of wines. *Food and Bioprocess Technology*, 7: 424-436.

El Darra, N., Grimi, N., Vorobiev, E., Louka, N., & Maroun, R. 2013. Extraction of polyphenols from red grape pomace assisted by pulsed ohmic heating. *Food and Bioprocess Technology*, *6*(5): 1281–1289.

Farías-Campomanes, A. M., Rostagno, M. A., Coaquira-Quispe, J. J., & Meireles, M. A. A. 2015. Supercritical fluid extraction of polyphenols from lees: overall extraction curve, kinetic data and composition of the extracts. *Bioresources and Bioprocessing*, *2*(1): 45.

Galanakis, C. M. 2012. Recovery of high added-value components from food wastes: conventional, emerging technologies and commercialized applications. *Trends in Food Science & Technology*, *26*(2): 68–87.

Haminiuk, C. W., Maciel, G. M., Plata-Oviedo, M. S., & Peralta, R. M. 2012. Phenolic compounds in fruits–an overview. *International Journal of Food Science & Technology*, *47*(10), 2023–2044.

Hogervorst, J. C., Miljić, U., & Puškaš, V. 2017. Extraction of bioactive compounds from grape processing by-products. In Galanakis, C. M. (ed.), *Handbook of Grape Processing By-Products* (pp. 105–135). Academic Press: Cambridge, Massachusetts, USA.

Jokić, S., Bijuk, M., Aladić, K., Bilić, M., & Molnar, M. 2016. Optimisation of supercritical CO 2 extraction of grape seed oil using response surface methodology. *International Journal of Food Science & Technology*, *51*(2): 403–410.

Liazid, A., Guerrero, R. F., Cantos, E., Palma, M., & Barroso, C. G. (2011). Microwave assisted extraction of anthocyanins from grape skins. *Food Chemistry*, *124*(3): 1238–1243.

Maroun, R. G., Rajha, H. N., Vorobiev, E., & Louka, N. 2017. Emerging technologies for the recovery of valuable compounds from grape processing by-products. In Galanakis, C. M. (ed.), *Handbook of Grape Processing By-Products* (pp. 155–181). Academic Press: Cambridge, Massachusetts, USA.

Mohamed, H. B., Duba, K. S., Fiori, L., Abdelgawed, H., Tlili, I., Tounekti, T., & Zrig, A. 2016. Bioactive compounds and antioxidant activities of different grape (*Vitis vinifera* L.) seed oils extracted by supercritical CO_2 and organic solvent. *LWT*, *74*: 557–562.

OIV (The International Organization of Vine and Wine). (2019). www.oiv.int/en/the-international-organisation-of-vine-and-wine

Oliveira, D. A., Salvador, A. A., Smania, A., Smania, E. F. A., Maraschin, M., & Ferreira, S. R. S. 2013. Antimicrobial activity and composition profile of grape (Vitis vinifera) pomace extracts obtained by supercritical fluids. *Journal of Biotechnology*, *164*(3): 423–432.

Palma, M., & Taylor, L. T. 1999. Extraction of polyphenolic compounds from grape seeds with near critical carbon dioxide. *Journal of Chromatography A*, *849*(1): 117–124.

Pascual-Marti, M. C., Salvador, A., Chafer, A., & Berna, A. 2001. Supercritical fluid extraction of resveratrol from grape skin of vitis vinifera and determination by HPLC. *Talanta*, *54* (4): 735–740.

Pérez-Serradilla, J. A., & De Castro, M. L. 2011. Microwave-assisted extraction of phenolic compounds from wine lees and spray-drying of the extract. *Food Chemistry*, *124*(4): 1652–1659.

Rajha, H. N., Boussetta, N., Louka, N., Maroun, R. G., & Vorobiev, E. 2014b. A comparative study of physical pretreatments for the extraction of polyphenols and proteins from vine shoots. *Food Research International*, *65*: 462–468.

Rajha, H. N., Boussetta, N., Louka, N., Maroun, R. G., & Vorobiev, E. 2015a. Effect of alternative physical pretreatments (pulsed electric field, high voltage electrical discharges and ultrasound) on the dead-end ultrafiltration of vine-shoot extracts. *Separation and Purification Technology*, *146*: 243–251.

Rajha, H. N., Chacar, S., Afif, C., Vorobiev, E., Louka, N., & Maroun, R. G. 2015b. β-Cyclodextrin-assisted extraction of polyphenols from vine shoot cultivars. *Journal of Agricultural and Food Chemistry*, *63*(13): 3387–3393.

Rajha, H. N., Ziegler, W., Louka, N., Hobaika, Z., Vorobiev, E., Boechzelt, H. G., & Maroun, R. G. 2014a. Effect of the drying process on the intensification of phenolic compounds recovery from grape pomace using accelerated solvent extraction. *International Journal of Molecular Sciences*, *15*(10): 18640–18658.

Schieber, A. 2017. Side streams of plant food processing as a source of valuable compounds: selected examples. *Annual Review of Food Science and Technology*, *8*: 97–112.

Stavikova, L., Polovka, M., Hohnova, B., Karasek, P., & Roth, M. 2011. Antioxidant activity of grape skin aqueous extracts from pressurized hot water extraction combined with electron paramagnetic resonance spectroscopy. *Talanta*, *85*(4): 2233–2240.

Tao, Y., García, J. F., & Sun, D. W. 2014. Advances in wine aging technologies for enhancing wine quality and accelerating wine aging process. *Critical Reviews in Food Science and Nutrition*, *54*(6): 817–835.

Vattem, D. A., & Shetty, K. 2003. Ellagic acid production and phenolic antioxidant activity in cranberry pomace (Vaccinium macrocarpon) mediated by Lentinus edodes using a solid-state system. *Process Biochemistry*, *39*(3): 367–379.

Vergara-Salinas, J. R., Bulnes, P., Zuniga, M. C., Perez-Jimenez, J., Torres, J. L., Mateos-Martin, M. L., … Perez-Correa, J. R. 2013. Effect of pressurized hot water extraction on antioxidants from grape pomace before and after enological fermentation. *Journal of Agricultural and Food Chemistry*, *61*(28): 6929–6936.

Vilkhu, K., Mawson, R., Simons, L., & Bates, D. 2008. Applications and opportunities for ultrasound assisted extraction in the food industry – a review. *Innovative Food Science and Emerging Technologies*, *9*(2): 161–169.

Yu, H.-B., Ding, L.-F., Wang, Z., & Shi, L.-X. 2014. Study on extraction of polyphenol from grape Peel microwave-assisted activity. *Advanced Materials Research*, *864–867*: 520–525.

9

Valorization of Olive Oil and Oilseed By-Products through Green Extraction Techniques

SONIA BARBA-ORELLANA

Centro Sanitario Integrado de Xirivella, Consorci Hospital General Universitari València, Xirivella, Spain

ELENA ROSELLÓ-SOTO

Nutrition and Food Science Area, Preventive Medicine and Public Health, Food Science, Toxicology and Forensic Medicine Department, Faculty of Pharmacy, Universitat de València, Avda. Vicent Andrés Estellés, s/n, 46100 Burjassot, València, Spain

PAULO E.S. MUNEKATA AND JOSÉ M. LORENZO

Meat Technological Center of Galicia, Ourense, Spain

KASHIF GHAFOOR

Department of Food Science and Nutrition, King Saud University, Riyadh, Saudi Arabia

CÁTIA DOURADO AND JORGE A. SARAIVA

QOPNA & LAQV-REQUIMTE, Department of Chemistry, University of Aveiro, 3810-193 Aveiro, Portugal

FRANCISCO J. BARBA

Nutrition and Food Science Area, Preventive Medicine and Public Health, Food Science, Toxicology and Forensic Medicine Department, Faculty of Pharmacy, Universitat de València, Avda. Vicent Andrés Estellés, s/n, 46100 Burjassot, València, Spain

Contents

1 Introduction

Edible oil production is crucial for the agricultural sector, especially in the olive oil production in the countries of the Mediterranean region, as olive oil constitutes one of the main components of the Mediterranean diet. One of the main reasons that justifies the central role of olive oil is the high content of natural antioxidants (Tsarouhas et al., 2015; Žugčić et al., 2019). Similarly, edible vegetable oils are important sources of nutrients (particularly unsaturated fatty acids), vitamins, and bioactive compounds that contribute to reduce the risk of noncommunicable diseases (Johnson and Fritsche, 2012; Lin et al., 2013; Schwingshackl and Hoffmann, 2012).

The conventional method for olive oil extraction consists of grinding, mechanical pressing in hot or cold conditions, and separation of oil by a centrifuge (two- or three-phase extraction processes can be carried out, but the latter is becoming the norm) (Roselló-Soto et al. 2015a). Similarly, the extraction of oil from oilseeds involves the mechanical expression followed by solvent extraction and finally refining for improving oil stability during storage (Koubaa et al., 2016). However, huge amounts of by-products from edible oil extraction are produced depending on the extraction method used. This is one of the main reasons that support the research and application of innovative edible oil extraction methods that are superior to conventional ones. Optimization of processing variables, such as the use of lower extraction time and temperature than conventional methods, is being studied in the context of environmental friendly extraction technologies. Another relevant aspect of green technologies is the lower consumption of solvents, which can reduce the environmental impact of vegetable oil processing (Roselló-Soto et al. 2015b).

Particularly for olive oil processing, the main by-products generated are the olive wood and leaves, pomace, olive seeds, the water used during the milling process, among others. These by-products contain polyphenols, volatile compounds, aromatics, phytosterols, and pigments, such as chlorophyll and carotenoids (Roselló-Soto et al. 2015a). Similarly, oilseeds generate residues, such as hulls (Boussetta et al., 2014), skin (Ballard et al., 2010), seed cake/meal (Dubie et al., 2013; Sarkis et al., 2015a; Teh et al., 2015; Yan et al., 2015; Zago et al., 2015), and fronds (Ofori-Boateng and Lee, 2013) rich in polyphenols, lignans, carotenoids, phytosterols, and tocopherols.

The recovery of high-value compounds from vegetable oil processing by-products is an effective strategy to reduce the environmental impact associated with the vegetable oil industry and generates supplies for food, cosmetic, and biofuel production. For example, phenolic compounds are associated with beneficial effects on health as they can reduce the risk of neurodegenerative diseases, ulcerous, and anti-inflammatory and anticancer diseases (Barba et al., 2016). This chapter aims to review and discuss the extraction of high-value compounds from olive oil and oilseed processing by-products by green extraction technologies: pulsed electric fields (PEF); high voltage electrical discharges (HVED); ultrasound (USN); microwaves (MW); sub- (SubC) and supercritical (SC) fluid, and high pressure (HP) extraction.

2 Green Extraction Technologies

The database *Science direct* (sciencedirect.com) was searched with the aim of finding innovative methods for the extraction of waste and food by-products, in terms of their usefulness and application at an industrial level. An initial search involved the use of keywords related to innovative technologies and high-value compounds, such as PEF, polyphenols, and extraction. The results were further filtered through segregation of review articles from research articles and also through limiting the date from 2010 to the present. In addition, a few relevant articles published before 2010 have also been used.

The selection of articles was based on the abstracts, and several exclusion criteria were applied, namely, the absence of at least one innovative technology used as an extraction method and the application in matrices other than olives or oilseed. The information obtained is structured in two ways in the current chapter: first, the different innovative methods that can be used for the extraction of polyphenols and second, the reports that explain the use of these methods for the extraction of polyphenols from olive and other vegetable oil production wastes. Some of the innovative extraction methods (Barba et al., 2016) have been elaborated here.

2.1 *Pulsed Electric Fields*

Extraction of bioactive components from plant materials using PEF is an innovative method that can induce important changes in the structure of plant matrix and can facilitate the extraction of high added value compounds. The process occurs in microseconds when the matrix is placed between two electrodes and subjected to electrical pulse discharge. Once cell membranes are exposed to electrical fields, an electric potential is formed that leads to the separation of membrane components according to individual molecular charge. Consequently, pores are formed, and intracellular compounds can easily be extracted by solvents. Although cell membrane composition is influenced by several factors, all cell membranes have a critical electrical field value. Once this value is exceeded, irreversible pores are formed. Inducing specific membrane effects is necessary for successful recovery of high added value compounds (Azmir et al., 2013; Puértolas and Barba, 2016).

Controlling external variables (such as temperature, solvent composition, and concentration) is also relevant for achieving optimal extraction yields (Barba et al., 2016). In this sense, PEF has great utility in the extraction of compounds, such as polyphenols. Interestingly, PEF has also been very useful for improving food preservation (Barba et al., 2015a; Chemat et al., 2015; Roselló-Soto et al., 2015a).

2.2 High Voltage Electrical Discharges

In HVED treatment, the matrix is placed within a treatment chamber (composed of two electrodes) with a high local electric field that causes electrical discharges. Because of the passage of electrical discharges, secondary effects are produced, such as the formation of cavitation bubbles and alterations in the cell membranes, which can be ruptured. Mass transfer of intracellular compounds is enhanced and diffused into the solvent augmented (Li, Fan, and Xi, 2019). It is also important to highlight that HVED is considered as a more efficient technique to remove intercellular compounds than PEF (Barba et al., 2016; Roselló-Soto et al., 2015c).

Because of the advantages of this technology, important advances about the equipment design and operational conditions, mainly at the laboratory level, have been made in the last decade to extract polyphenols, sugars, oils, and proteins (Puértolas, Koubaa, and Barba, 2016; Li, Fan, and Xi, 2019). Although this technology has been showing promising results, more information about the effects on matrix and target compounds is still required (Barba et al., 2016; Roselló-Soto et al., 2015c).

2.3 Sub- (SubC) and Supercritical (SC) Fluid Extraction

The extraction with solvents can be greatly influenced by changing the solvent properties, especially pressure and temperature. Among the several possible combinations of pressure and temperature, two important approaches can be highlighted: SubC and SC fluid extraction. The SubC fluid extraction is characterized by combining temperature and pressure to keep the solvent in the liquid state between its evaporation and critical points. The solvent properties of water (the main fluid used in this technique) are changed with this technique. Compounds that once were poorly extracted with water at room temperature and atmospheric pressure can be removed from a complex matrix. This effect is associated with the reduction of the dielectric constant of water around 80 at room temperature and atmospheric pressure (polar solvent) to lower values as temperature rises (Herrero, Cifuentes, and Ibañez, 2006).

In a similar way, SC fluid extraction explores combinations of pressure and temperature above the solvent's critical point. In this condition, fluids display interesting combinations of physical properties, such as density values, expected for fluids in the liquid state, while the viscosity value observed is for the liquid in the gas state and the diffusivity value between liquid and gas states. Among the several fluids used for carrying out SC extraction, the most common choice is CO_2 because of its low viscosity, low surface tension, and high diffusivity compared with other solvents. These characteristics are useful for the recovery of high-value compounds, such as polyphenols (Barba et al., 2016; Roselló-Soto et al., 2015c).

2.4 Ultrasound

USN extraction technology is a technique that involves the propagation of USN waves through a liquid and/or solid media. During the passage of these waves, a phenomenon known as cavitation occurs leading to the production of intense shear forces in the media and formation of bubbles. The eventual implosion of bubbles on the surface of solids generates microinjection of fluid. Consequently, the surface of solids can be peeled, and particles are disintegrated (Chemat et al., 2017).

The application of USNs is a method that decreases extraction time, uses lower solvents amounts, and requires less energy during the process, resulting in greater performance in the recovery of products compared with conventional methods. This method of extraction can be employed using both the raw materials and the final products. In addition, the pharmaceutical industry has been showing great interest in UNS technology (Amofa-Diatuo et al., 2017; Barba et al., 2016; Roselló-Soto et al., 2015a, 2015b, 2015c).

2.5 Microwaves

The principle of MW includes exposing a matrix to nonionizing electromagnetic radiation in the range of 0.3–300 GHz. When polar compounds are exposed to MW, two phenomena can occur, ionic conduction and dipole rotation. The electric field caused by MW induces the migration of ions and realignment of the dipoles. Eventually, migrating ions cause friction, and the rapid change of the electrical field heats the matrix. Moreover, both effects can occur simultaneously. Because of these phenomena, the efficiency of recovering soluble compounds in a polar solvent is increased (Mandal, Mohan, and Hemalatha, 2006). The increase of the internal pressure and temperature facilitates the breakage of the cell walls, favoring the release of trapped solutes. Other relevant characteristics of MW are the low energy consumption, the short extraction time, and the high extraction efficiency (Barba et al., 2016; Chemat et al., 2015).

2.6 High Pressure

HP processing is a new technique in which a matrix is subjected to elevated pressure for a short time. Research interest has increased very substantially in the study of this technology since it has shown important results by destructing allergenic compounds, improving the extraction of several classes of compounds, as well as the bioavailability of bioactive compounds, and increasing the shelf-life of several foods (Barba et al. 2015b). One of the effects associated with HP is the disruption of cell membranes and walls, which facilitates the extraction of trapped compounds (Chanioti and Tzia, 2018). However, more research is required to increase the current knowledge about the mechanisms of this extraction method.

3 Recovery of High-Value Compounds by Green Technologies

The need to improve extraction methods has led researchers to carry out experimental studies using new technologies. Figure 9.1 displays the current green extraction technologies used for extracting high-value compounds from olive oil and oilseed processing residues.

3.1 PEF and HVED Extraction

Among the several promising green technology approaches, PEF and HVED were explored to obtain polyphenols, phytosterols, tocopherols, and carotenoids (Table 9.1). In case of extraction of total polyphenols from olives, extraction conditions of 2 kV/cm and 11.25 kJ/kg were used for increasing polyphenols extraction from Arroniz olive variety by 11.5% (Puértolas and Martínez de Marañón, 2015). The extraction of phytosterols (9.9%) and tocopherols (15%) were also improved by applying the same PEF treatment. However, a different outcome was reported for the extraction of polyphenols and α-tocopherol (2 kV/cm at 5.22 kJ/kg) from Arbequina olive variety (Abenoza et al., 2013). While the extraction of α-tocopherol was improved to 252.0 mg/kg oil, the contents of phenolic compounds and carotenoids in the oil were slightly reduced to 112.2 mg gallic acid equivalent (GAE)/kg and 3.94 mg lutein/kg compared with control extraction (247.84 mg α-tocopherol/kg oil, 148.94 mg GAE/kg, and 4.36 mg lutein/kg).

The impact of energy input and electrical field intensity in the range used in these studies revealed contrasting results, indicating that the successful extraction of high added value compounds from olive requires more intense treatments. In this regard, PEF treatments in the range 19–109 kJ/kg were used to obtain polyphenols and carotenoids from olive kernels of the Cornicabra and Koroneiki varieties (Roselló-Soto et al. 2015a). As a result, a greater polyphenols quantity was obtained, particularly for Cornicabra variety (618.9 mg GAE/L). However, no effect on carotenoids extraction was associated with PEF treatments. Similarly, the experiment carried out using Tsounati, Amfissis, and Manaki olive varieties

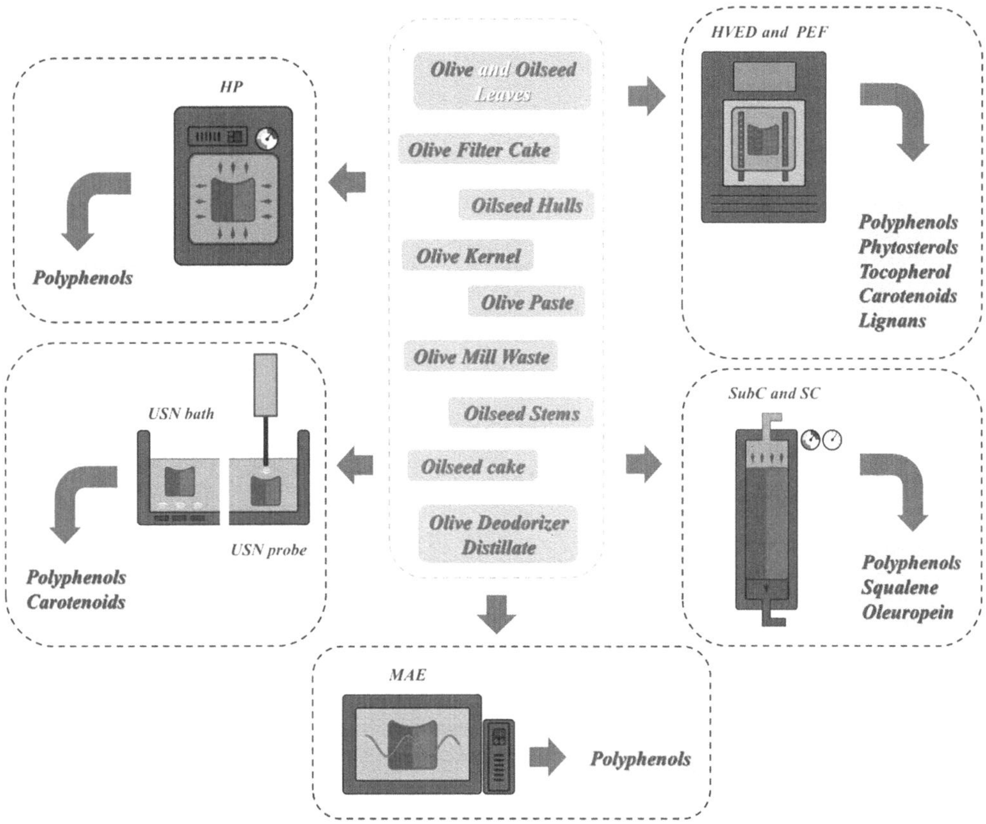

Figure 9.1 Extraction of high-value compounds from olive and oilseed processing by-products by green technologies.

revealed that optimum PEF treatment could be obtained at specific energy input values (Andreou et al., 2017). While the highest polyphenol extraction yield for Amfissis variety was obtained at 70 kJ/kg (176 mg GAE/kg), Manaki and Tsounati varieties required 20 (341 mg GAE/kg) and 13.3 (988.5 mg GAE/kg) kJ/kg, respectively.

The extraction of polyphenols from rapeseed stems and leaves was improved by increasing PEF treatment intensity. The highest polyphenol extraction yield for both the rapeseed stems and leaves was obtained at 20 kV/cm and 160 kJ/kg than at 0.8 kV/cm and 6.4 kJ/kg or 5 kV/cm and 40 kJ/kg (Yu et al., 2015a). The use of PEF technology for recovering polyphenols from rapeseed press-cake was also influenced by input energy (Barba, Boussetta, and Vorobiev, 2015). Contrary to other studies, the highest polyphenol recovery rate was achieved by PEF treatment at 80 kJ/kg, which indicated an optimum input energy value. In addition to treatment intensity, the liquid–solid ratio effect was also evaluated by the authors, and the highest polyphenol extraction yield was obtained by the liquid–solid ratio of 20.

Table 9.1 Experimental values of studies using pulsed electric field (PEF) and high voltage electrical discharge (HVED) as extraction methods

MATRIX	EXTRACTED COMPOUNDS	TREATMENT CONDITIONS	EFFECT	REFERENCE
Olive paste (Arroniz variety)	Polyphenols, phytosterols, and tocopherols	PEF: 2 kV/cm and 11.25 kJ/kg	Increased extraction yield by 11.5 (polyphenols), 9.9 (phytosterols), and 15% (tocopherol)	Puértolas and Martínez de Marañón (2015)
Olive paste (Arbequina variety)	Polyphenols and tocopherols	PEF: 1–2 kV/cm and input energy (1.47–5.22 kJ/kg)	Increased α-tocopherol extraction (252.0 mg/kg oil) and slight reduction in polyphenol extraction (112.2 mg GAE/kg) compared with control; *optimized extraction condition: 2 kV/cm and 5.22 kJ/kg*	Abenoza et al. (2013)
Olive kernel (Cornicabra and Koroneiki variety)	Polyphenols and carotenoids	PEF and HVED: 13.3 kV/cm, 0–300 pulses, 10 μs, 0.5 Hz, and input energy (18, 55, or 109 kJ/kg); further extraction with pH (2.5, 7.5, or 12.5) and ethanol (0, 25, or 50%)	The highest polyphenol extraction yield was obtained with HVED treatment (618.9 mg GAE/L); no effect on carotenoid extraction for both technologies; *optimized extraction condition: HVED at 81 kJ/kg, 47% ethanol and pH 2.5 for Cornicabra var.*	Roselló-Soto et al. (2015a)
Olive kernel (*Tsounati, Amfissis,* and *Manaki var.*)	Polyphenols	PEF: 1.8 kV/cm, 15 μs, 300 Hz, and input energy (1.6, 13.3, 20, 50, or 70 kJ/kg)	Increased from 106 to 176 (at 70 kJ/kg, *Amfissis var.*), 271 to 341 (at 20 kJ/kg, *Manaki var.*), and 768.5 to 988.5 (at 13.3 kJ/kg, *Tsounati var.*) mg GAE/kg	Andreou et al. (2017)
Rapeseed stems and leaves	Polyphenols	PEF: electrical field intensity (0.8, 5, or 20 kV/cm) and input energy (6.4, 40, or 160 kJ/kg)	The highest extraction yield was obtained at 20 kV/cm and 160 kJ/kg for both steam and leaf; *optimized extraction condition: 20 kV/cm, 160 kJ/kg, and 20 (leaves) and 30 (stems) min*	Yu et al. (2015a)
Rapeseed press-cake	Polyphenols	PEF: input energy (up to 400 kJ/kg) and liquid–solid ratio (5–20)	Increased from 143.1 (liquid–solid ratio of 5) to 559.2 (liquid–solid ratio of 20) mg GAE/100 g both; *optimized extraction condition: 80 kJ/kg and liquid–solid ratio of 20*	Barba et al. (2015a)
Hulled and non-hulled rapeseed	Polyphenols and tocopherols	PEF: 30 μs, electrical field intensity (5–7 kV/cm), number of pulses (60–120), and input energy (21.4–84.0 kJ/kg)	Increased α-tocopherol extraction by 20%, regardless of PEF treatment (hulled rapeseed); no effect on γ-tocopherol extraction; increased polyphenol extraction by both PEF treatments (non-hulled rapeseed); *optimized*	Guderjan et al. (2007)

(*Continued*)

Table 9.1 (Cont.)

MATRIX	EXTRACTED COMPOUNDS	TREATMENT CONDITIONS	EFFECT	REFERENCE
			extraction condition: 7 kV/cm, 120 pulses, and 84 kJ/kg for hulled rapeseed and 5 kV/cm, 60 pulses, and 42 kJ/kg for non-hulled rapeseed	
Sesame seed cake	Polyphenols and lignans	PEF and HVED: 40 kV, input energy (0–291 kJ/kg), time (up to 60 min), ethanol (10, 30, or 50%), and temperature (20, 40, or 60 °C)	PEF and HVED were more efficient than control extraction, regardless of time, ethanol concentration, and temperature for polyphenols and lignans; *optimized extraction condition: 83 kJ/kg, 10 (polyphenols) and 50% (lignans) ethanol, and 60 °C*	Sarkis et al. (2015a)
Flaxseeds hulls	Polyphenols	PEF: 300 kJ/kg, 10 ms, electrical field intensity (10, 15, or 20 kV/cm), further extraction with ethanol (0, 20, or 50%), and citric acid (0.05, 0.1, or 0.3 M), or NaOH (0.05, 0.1, or 0.3 M)	Between three- and fourfold increase as electrical field intensity rises; electrical field intensity improved extraction regardless of ethanol, citric acid, and NaOH concentration; *optimized extraction condition: 20 kV, 20% ethanol, and 0.3 M NaOH*	Boussetta et al. (2014)
Sesame seed (*Schistosoma indicum*)	Polyphenols	PEF and HVED: energy input (40, 80, 160, or 240 kJ/Kg)	HVED was more efficient for extracting polyphenols than PEF, regardless of the energy input	Sarkis et al. (2015b)
Maize germ	Phytosterols	PEF: 91.4 kJ/kg and electrical field intensity (0.6 or 7.3 kV/cm)	Increased extraction yield (1039.4 mg/100 g oil); *optimized extraction condition: 0.6 kV/cm*	Guderjan et al. (2005)

PEF application to recover compounds from rapeseed matrix can also be customized for tocopherols. An experiment with hulled and non-hulled rapeseed indicated that α-tocopherol extraction was increased by 20% (particularly for hulled rapeseed), but no effect on γ-tocopherol extraction (for both the hulled and non-hulled rapeseed) was reported by Guderjan et al. (2007). In the case of polyphenols, extraction yield was increased by PEF treatment, regardless of electrical field intensity, the number of pulses, and input energy. In addition, combining PEF with pressing technology caused up to a fourfold increase in the recovery of polyphenols from non-hulled rapeseed.

An interesting comparison between PEF and HVED effect on the extraction of polyphenols from sesame seed was reported in the literature. Two recent studies indicated that HVED was more efficient for improving the recovery of phenolic compounds than PEF. In the first study, sesame cake seed was subjected to PEF and HVED (Sarkis et al., 2015a). Both technologies improved the release of polyphenols and lignans compared with conventional extraction, but the highest extraction yield was obtained from samples subjected to HVED treatment, regardless of input energy (0–291 kJ/kg), time (0–60 min), ethanol concentration (10, 30, or 50%), and temperature (20, 40, or 60 °C). The second study on the extraction of polyphenols from sesame matrix also supports the use of HVED compared with PEF, under the same treatment conditions (Sarkis et al., 2015b). Although both technologies displayed an input energy-dependent effect on the recovery of polyphenols, the extraction with HVED was more efficient than PEF.

A similar study about the impact of PEF electrical field intensity (10, 15, or 20 kV/cm) on the extraction of polyphenols from flaxseeds hulls revealed that the extraction yield increased with the electrical field intensity (Boussetta et al., 2014). A fourfold increase in polyphenol extraction was obtained from flaxseeds hulls subjected to PEF treatment at 20 kV/cm compared with untreated samples. In addition, the authors also optimized the solvent composition after PEF treatment with 20% ethanol and 0.3 M NaOH. Similarly, an interesting experiment with corn revealed that PEF (0.6 kV/cm) improved the extraction of phytosterols (Guderjan et al., 2005). This experiment also revealed that the treatment of more intense electrical field (7.3 kV/cm) reduced the extraction yield of phytosterols.

3.2 SubC, SC, and GAME Extraction

The extraction of compounds with technological applications can be carried out by exploring the physical properties of the extracting solvent. Some interesting techniques explored in the last decades are the pressurized liquid extraction (PLE), which includes SubC fluid extraction, SC fluid extraction, and combined techniques, such as gas-assisted mechanical expression (GAME). Table 9.2 displays the studies carried out to characterize and optimize the use of these technologies for recovering high added value compounds from by-products of olive and oilseed processing.

The extraction of polyphenols from by-products of olive and oilseed processing can be performed by SC-CO_2 extraction. For instance, increasing the extraction pressure from 10 to 40 MPa (40 °C) improved the recovery of polyphenolic compounds from olive kernels (Koubaa et al., 2017). However, the extraction of polyphenols may not achieve similar extraction yields than other extraction techniques, such as conventional solvent extraction. This outcome was reported in a study that compared the SC-CO_2 with conventional solvent extraction technique to recover polyphenols from olive oil mill waste (Lafka et al., 2011). In this experiment, the solvent composition was associated with different extraction yields, wherein alcoholic solvents (mainly methanol and ethanol) were indicated as the most efficient. Moreover, SC-CO_2 extraction (solvent flux of 2 g/min at 40 °C and 350 bar for 60 min) displayed intermediate extraction yield compared with extractions carried out with 50% ethanol, isopropanol, and ethyl acetate. Another relevant outcome

Table 9.2 Impact of SubC, PLE, SC, and GAME technologies on the extraction of high added value compounds from olive and oilseed processing by-products

MATRIX	EXTRACTED COMPOUNDS	EXTRACTION CONDITIONS	EFFECT	REFERENCE
Olive kernel (*Olea europaea*)	Polyphenols	SC-CO_2: 40 °C and pressure (10–40 MPa)	Increased extraction yield in a pressure-dependent manner; *optimized extraction condition: 40 MPa*	Koubaa et al. (2017)
Olive oil mill waste	Polyphenols	SC-CO_2: 2 g/min, 40 °C, 350 bar, and 60 min	Lower polyphenol extraction yield and antioxidant activity than ethanolic extraction	Lafka et al. (2011)
Olive leaves	Polyphenols	SC-CO_2: 2 mL/min, methanol as co-solvent (2, 5, and 10%), temperature (80, 100, and 120 °C), pressure (155, 207, 256, and 334 bar), and extraction time (0–160 min)	Extraction of polyphenols increased in a cosolvent-, temperature-, pressure-, and time-dependent manner; *optimized extraction condition: 10% methanol, 100 °C, 334 bar, and 140 min*	Floch et al. (1998)
Olive leaves	Polyphenols	SC-CO_2: 300 bar, 40 °C, total flow rate 2.5 mL/min, and ethanol as cosolvent (0.3 mL/min)	Reduced extraction yield compared with hydroalcoholic extraction	Jimenez et al. (2011)
Olive kernel (*Olea europaea*)	Polyphenols	GAME: 40 °C, 1 h, 30 MPa (mechanical expression) + 10 MPa (SC-CO_2); mechanical expression (30 MPa) followed by GAME treatment (40 °C, 1 h, 30 MPa [mechanical expression] + 10 MPa [SC-CO_2])	Reduced extraction yield by carrying out GAME extraction compared with mechanical expression; similar extraction yield to that obtained by mechanical expression	Koubaa et al. (2017)
Olive leaves	Polyphenols	PLE: 20 min, solvent (water and ethanol), and temperature (50, 100, 150, and 200 °C)	Increasing temperature and using water improved the extraction of polyphenols and total extract yield; *optimized extraction condition: water at 200 °C*	Herrero et al. (2011)
Olive oil filter cake	Selected polyphenols	PLE: ethanol % (0, 15, 50, and 85%) and temperature (40, 60, 120, and 175 °C)	Decarboxymethyl-oleuropein aglycone and luteolin were the major polyphenols; *optimized extraction condition: 50% ethanol at 120 °C*	Lozano-Sánchez et al. (2014)
Olive fruit dreg	Polyphenols	SubC: water, temperature (100–180 °C), solvent/solid ratio (20–60), and time (5–60 min)	Increased polyphenol extraction compared with conventional solvent extraction with methanol, ethanol, and acetone; *optimized extraction condition: 160 °C, solvent/solid ratio of 50, and 30 min*	Yu et al. (2015b)

(*Continued*)

Table 9.2 (Cont.)

MATRIX	EXTRACTED COMPOUNDS	EXTRACTION CONDITIONS	EFFECT	REFERENCE
Olive leaves	Oleuropein	PLE: temperature (40, 50, 115, and 150 °C) and ethanol % (0, 60, and 100%)	Extraction yield was improved by increasing temperature and ethanol percentage in the solvent; *optimized extraction condition: 100% ethanol at 115 °C*	Xynos et al. (2012)
Olive leaves	Oleuropein	PLE: temperature (40–190 ° C), ethanol (0–100%), and number of cycles (1 and 3)	Extraction was improved by increasing temperature and reducing the number of cycles; ethanol % displayed an optimum value; *optimized extraction condition: 56.04% ethanol, 1 cycle, and 190 °C*	Xynos et al. (2014)
Olive leaves	Oleuropein	SC-CO_2: solvent/feed ratio (120 and 290) and ethanol as cosolvent (5 and 20%)	Extraction was possible only with a high proportion of cosolvent; *optimized extraction condition: 20% cosolvent and solvent/feed ratio of 290*	Xynos et al. (2012)
Olive leaves	Oleuropein	PLE: after SC-CO_2, ethanol % (0 and 60%) and temperature (50 and 150 °C)	Increasing temperature and reducing ethanol percentage in the solvent; *optimized extraction condition: 50% ethanol at 150 °C*	Xynos et al. (2012)
Olive oil deodorizer distillate	Squalene	SC-CO_2: pressure (110, 130, 150, and 170 bar) and temperature (40 and 60 °C)	The highest extraction yield was 83.7%; *optimized extraction condition: 104.8 bar, 52.05 °C, and 180 min*	Bondioli et al. (1993)
Olive oil deodorizer distillate	Squalene	SC-CO_2: solvent/feed ratio (13–60), temperature (313, 323, and 343 K), and pressure (150, 180, 200, and 230 bar)	Squalene recovery was improved by increasing temperature and solvent/feed ratio while decreasing pressure; *optimized extraction condition: solvent/feed ratio of 51, 351.1 K, and 176.8 bar*	Vázquez et al. (2007)
Olive oil deodorizer distillate	Squalene	SC-CO_2: pressure (120, 150, and 180 bar) and temperature (40, 50, and 60 °C)	The highest extraction yield was 73.57%; maximum yield from theoretical estimation: 74.08%; *optimized extraction condition: 105 bar, 50 °C, and 180 min*	Akgu (2011)
Olive leaves	Tocopherols	SC-CO_2: pressure (25, 35, and 45 bar), temperature (313, 323, and 333 K), particle size (0.25, 0.5, and 1.5 mm), and solvent flow (0.5, 1.0, and 1.5 L/min)	Tocopherol extraction was improved by increasing pressure and particle size; no effect was associated with temperature and solvent flow; *optimized extraction condition: 313 K, 25 MPa, 1.5 mm, 1 L/min, and 2 h*	de Lucas et al. (2002)

reported by the authors is an intermediate antioxidant activity compared with other conventional solvent extractions.

In this sense, SC-CO_2 extraction must be optimized for solvent–solute interaction and the properties of the by-product matrix. The use of cosolvents for improving the recovery of polyphenols by SC-CO_2 extraction is a common and effective approach. Optimizing the cosolvent percentage, extraction temperature, and time (in addition to pressure) was associated with increased polyphenol recovery. An experiment with olive leaves revealed that polyphenol extraction yield was increased in a cosolvent percentage-, temperature-, pressure-, and time-dependent manner (Floch et al., 1998). The optimized SC-CO_2 conditions were 10% methanol at 100 °C and 334 bar for 140 min. However, the polyphenol extracts obtained by SC-CO_2 technique may display a lower yield than conventional solvent extractions. Nevertheless, a study with olive leaves indicated that hydroalcoholic extraction was more efficient to remove polyphenols and to preserve antioxidant activity than SC-CO_2 technique (at 300 bar and 40 °C with total flow rate 2.5 mL/min) even by using ethanol as cosolvent (0.3 mL/min) (Jimenez et al., 2011).

An interesting approach by combining the SC-CO_2 technique with mechanical expression (GAME) was recently explored (Koubaa et al., 2017). The combined effect of carrying out SC-CO_2 and mechanical expression extractions together (at 300 bar and 40 °C with total flow rate 2.5 mL/min and 0.3 mL/min of ethanol as cosolvent) resulted in lower polyphenol extraction than mechanical expression (30 MPa) alone. However, the authors also explored the combination of mechanical expression for 1 h, followed by GAME extraction for 1 h. In this case, GAME approach increased the polyphenol percentage in the final extract compared with mechanical expression for 2 h.

The use of PLE for obtaining polyphenols from olive processing is seen as a promising technique. The solvent composition and the extraction temperature are important variables for improving the recovery of polyphenols. This outcome was explored with polyphenols extracted from olive leaves that were obtained with pressurized ethanol and water at 50, 100, 150, and 200 °C (Herrero et al., 2011). According to the authors, water was the most efficient solvent, and higher extraction yields were obtained by increasing temperature, particularly at 200 °C (13.42 vs 7.83 mg/g extract for aqueous and ethanolic extracts, respectively). An interesting outcome was highlighted in a later study. Although the extraction yield of the main identified phenolic compound (hydroxytyrosol) was improved by temperature with both solvents, the recoveries of other identified phenolic compounds (such as oleuropein and apigenin-7-glucoside) were gradually reduced as temperature increased, particularly in the case of aqueous extraction.

Similarly, another study indicated that polyphenol extraction yield from the olive oil filter cake by PLE showed a solvent composition and temperature dependency (Lozano-Sánchez et al., 2014). The most efficient solvent composition for the recovery of selected polyphenols (hydroxytyrosol, decarboxymethyl-oleuropein aglycone, decarboxymethyl-ligstroside aglycone, oleuropein aglycone, luteolin, and apigenin) was 50% ethanol at 120 °C. Increasing temperature from 40 to 120 °C led to an increased release of all selected phenolic compounds by carrying out the extraction with 50% ethanol. In addition, the effect of solvent composition on extractions performed at 120 °C revealed that water reduced the recovery of selected polyphenols, for example, for hydroxytyrosol.

The thermodynamic condition under which PLE is carried out (using water heated above 100 °C) reveals that this process can be called SubC extraction (Yu et al., 2015b). This approach was optimized (temperature, solvent/solid ratio, and time) to obtain polyphenols from olive fruit dreg (a by-product of olive soft drink processing). The authors indicated that SubC water extraction recovered more polyphenols than conventional solvent extraction with methanol, ethanol, and acetone. Moreover, the optimum extraction conditions were solvent/solid ratio of 50 at 160 °C for 30 min.

To obtain oleuropein, some research groups evaluated the extracting conditions of PLE, SC-CO_2, and the possible sequential use of both technologies. Regarding the extraction yield obtained by PLE approach, this technology was optimized at temperature (40–150 °C) and solvent concentration (0, 60, and 100% ethanol) (Xynos et al., 2012). This experiment revealed that high extraction yields were obtained with high ethanol percentages in the solvent. A similar effect was associated with increasing temperature, with the maximum oleuropein extraction yield being obtained with 100% ethanol at 115 °C (16.4%). A similar outcome was reported in a posterior study that evaluated the number of extraction cycles (1 or 3) in addition to temperature (40–190 °C) and ethanol content (0–100%) in the solvent (Xynos et al., 2014). In the latter study, the predicted optimization revealed that the optimum extraction yield of 26.1% was obtained by carrying out PLE with 56.04% ethanol for 1 cycle at 190 °C.

In addition to PLE, oleuropein can also be isolated from olive leaves by SC-CO_2 technology. However, this technology requires further studies related to the extraction of oleuropein, since contrasting results were reported from the evaluation of solvent/solid ratio (120 vs 290) and the percentage of ethanol as cosolvent (5 vs 20%) (Xynos et al., 2012). While oleuropein was not extracted by SC-CO_2 with a solvent/solid ratio of 120 and 5% ethanol as co-solvent, carrying out the extraction with a ratio of 290 and 20% generated an extract with 30% of oleuropein. Interestingly, these authors also explored the sequential extraction of oleuropein by subjecting olive leaves to, first, SC-CO_2 approach, and then PLE technology. In this condition, the effects of ethanol percentage (0 and 60%) and temperature (50 and 150 °C) in the PLE stage were explored, and an optimum extraction yield of 10.4% was obtained with 50% ethanol at 150 °C.

Olive oil processing generates residues rich in squalene, a compound with potential use in the cosmetic industry (Huang, Lin, and Fang, 2009). This compound can be extracted from olive processing by-products using SC-CO_2 technology, which is influenced by some extraction variables, such as pressure and temperature. For instance, the maximum extraction yield reported for SC-CO_2 in the range of 40–60 °C temperature and 110–170 bar pressure was 83.7% at 104.8 bar, 52.05 °C, and 180 min (Bondioli et al., 1993). A similar experiment indicated that maximum extraction yield was 73.57% at 105 bar and 50 °C for 180 min from olive oil deodorizer distillate (Akgu, 2011). Interestingly, another study about the recovery of squalene from olive oil deodorizer distillate reported that a countercurrent SC-CO_2 extraction process was optimized for recovering 92% of squalene (Vázquez et al., 2007). The experiment evaluated the impact of solvent/feed ratio (13–60), temperature (313, 323, and 343 K), and pressure (150, 180, 200, and 230 bar) wherein the optimized extraction conditions were solvent/feed ratio of 51 at 351.1 K and 176.8 bar.

SC-CO_2 technology was also used for extracting tocopherols from olive leaves. The impact of pressure (25, 35, and 45 MPa), temperature (313, 323, and 333 K), particle size (0.25, 0.5, and 1.5 mm), and solvent flow (0.5, 1.0, and 1.5 L/min) were evaluated (Lucas et al., 2002). Although the results indicated that tocopherol extraction yield could be improved by increasing pressure and particle size, the temperature and the solvent flow had a nonsignificant effect. The authors indicated that optimum extracting conditions (10.10 mg/100 g leaves) were 313 K with 25 MPa, particle size of 1.5 mm, and solvent flow of 1 L/min for 2 h of extraction.

3.3 MW Extraction

The impact of MW on the extraction of high added value compounds extracted from by-products of olive and oilseed processing is presented in Table 9.3. The use of MW, compared with other technologies, can improve the recovery of polyphenols from olive leaves. This outcome was obtained by comparing MW (1.200 W with 80% methanol at 80 °C for 6 min) with SC-CO_2 (150 bar at 40 °C for 2 h with 6.6% of ethanol as cosolvent) and PLE (100 bar at 150 °C for 20 min) (Taamalli et al., 2012a). The authors also indicated that the main phenolic compounds obtained by MW were 10-hydroxy-oleuropein and 2″-methoxyoleuropein. Similarly, an experiment reported higher phenolic extraction yield by MW extraction than conventional solvent extraction (Taamalli et al., 2012b). In this case, the MW was particularly efficient to recover 2″-methoxyoleuropein and isomers (30.3 vs 26.1% of peak area, respectively), oleuropein and isomers (12.5 vs 6.3% of peak area, respectively), and apigenin-7-o-glucoside (7.4 vs 1.4% of peak area, respectively).

MW technology was also successfully applied to recover valuable compounds from oilseed processing by-products. A recent study explored the impact of MW extraction variables (MW power, liquid/solid ratio, and time) on polyphenols recovery from canola seed cake (Teh et al., 2015). The optimized conditions were

Table 9.3 Impact of MW technology on the extraction of high added value compounds from olive and oilseed processing by-products

MATRIX	EXTRACTED COMPOUNDS	EXTRACTION CONDITIONS	EFFECT	REFERENCE
Olive leaves	Polyphenols	1.200 W, 80% methanol, 80 °C, and 6 min	Most efficient technology to extract total phenolic compounds in comparison to conventional solvent, supercritical, and pressurized liquid extraction protocols	Taamalli et al. (2012a)
Olive leaves (Croniky, Roghani, and Mission cultivar)	Polyphenols	Time (2–15 min) and solvent (water, 80% methanol, 50% ethanol, and acetone)	Higher extraction yields were obtained with ethanol and methanol; *optimized extraction condition: 80% methanol or 50% ethanol and 15 min*	Rafiee et al. (2011)
Olive leaves	Apigenin-7-o-glucoside, methoxyoleuropein and isomers, and luteolin-glucoside and isomers	1.200 W, 80% methanol, 80 °C, and 6 min	Higher extraction yield than conventional solvent extraction method for all compounds	Taamalli et al. (2012b)
Canola seed cake	Polyphenols	Microwave power (440, 770, and 1,100 W), liquid/solid ratio (4, 5, and 6 mL/g), and time (1, 3, and 5 min)	Extraction yield improved by increasing extraction time and liquid/solid ratio; optimum microwave power was achieved; *optimized extraction condition: 633.33 W, liquid/solid ratio of 6, and 5 min*	Teh et al. (2015)
Peanut skin	Polyphenols	Microwave power (10, 50, and 90%), time (30, 90, 150 s), and sample mass (1.5, 2.5, and 3.5 g peanut skins)	Extraction yield was improved by increasing microwave power while decreasing the sample mass and time; *optimized extraction condition: 90% power, 30 s, and 1.5 g of skins*	Ballard et al. (2010)
Rapeseed seed meal	Polyphenols	Microwave roasting (2.45 GHz, 10 kW, 160 °C, and 50 kg/h) or followed by superheated steam (2 kg/h at 180 °C)	Both approaches partially prevented the thermal degradation of phenolic compounds (4% decrease); microwave roasting facilitated the release of canolol but combining superheated steam reduced canolol extraction	Zago et al. (2015)

achieved by increasing the liquid/solid ratio to 6 mL/g and extraction time to 5 min while an optimum MW power was observed as 633.33 W. Moreover, this experiment also explored the recovery of flavonoids, which were also increased by the optimized extraction conditions. Likewise, the extractions of phenolic compounds from peanut skins, a residue from peanut processing, were optimized by MW technology (Ballard et al., 2010). They studied the effects of MW power (10, 50, and 90%), extraction time (30, 90, and 150 s), and sample mass (1.5, 2.5, and

3.5 g peanut skins) on phenolics recovery. It was observed that polyphenol extraction yield was improved (90%) by increasing MW power. Interestingly, the highest extraction yield was obtained by reducing the sample mass to 1.5 g of peanut skin and extraction time to 30 s.

Finally, an experiment about the effect of MW roasting alone (2.45 GHz, 10 kW, 160 °C, and 50 kg/h) or followed by superheated steam (2 kg/h at 180 °C) was explored to improve the recovery of polyphenols from rapeseed meal. In this case, the MW roasting alone or combined with superheated steam partially prevented the thermodegradation of phenolic compounds wherein 4% decrease was indicated by the authors. Additionally, both the MW roasting approaches displayed equal or superior capacity to reduce the thermal degradation of polyphenols in comparison to simple roasting (Zago et al., 2015).

3.4 USN Extraction

USN is an interesting approach to recover high added value compounds from residues of olive and oilseed processing. The USN-assisted extraction can be carried out using either sonication probes or bath devices (Table 9.4). It is worth mentioning that USN probes and baths can result in different extraction yields. Particularly for olive kernel, both the equipment can improve the extraction of the main polyphenols, but the highest extraction yield was obtained using the ultrasonic probe. For instance, the extraction yield of oleuropein (the main quantified phenolic compound) was higher in extracts obtained using USN probe in comparison to USN bath and conventional solvent extraction (16.5 vs 12.3 and 9.2 mg/g dw, respectively) (Jerman, Trebše, and Vodopivec, 2010).

Regarding the use of USN probes to improve the extraction of phenolic compounds from olive processing residues, Ahmad-Qasem et al. (2013) explored the effect of electric power (40, 60, 80, and 100%), emitter surface (1.5, 3.8, and 12.6 cm^2), and temperature (25–50 °C). The authors observe that the increase in electrical power (100%) was associated with high polyphenol extraction yield, while an optimum quantity (66 mg GAE/g dw) of phenolics was obtained for 3.8 cm^2 emitter surface (25 °C temperature). Moreover, the application of USN-assisted extraction at optimized conditions resulted in obtaining an optimum yield in 15 min, which was much lower than that of the conventional solvent extraction method (24 h).

Similarly, Japón-Luján et al. (2006) also optimized the extraction process for recovering specific polyphenols (apigenin-7-glucoside, luteolin-7-glucoside, oleuropein, and verbacoside) from olive leaves. The authors indicated that the highest extraction yields of polyphenols (488, 970, 1,072, and 22,610 mg/kg for verbacoside, luteolin-7-glucoside, apigenin-7-glucoside, and oleuropein, respectively) were achieved at 30% radiation amplitude, 70% duty cycle, 40 °C temperature, probe positioned at 4 cm, 5 mL/min ethanol (59%), and 25 min extraction time. In addition, the authors

Table 9.4 Impact of USN technology on the extraction of high added value compounds from olive and oilseed processing by-products

MATRIX	EXTRACTED COMPOUNDS	EXTRACTION CONDITIONS	EFFECT	REFERENCE
Olive kernel	Caffeoyl-6′-secologanoside, comselogoside, demethyloleuropein, hydroxytyrosol glucoside, luteolin-7-*o*-glucoside, oleuropein, quercitrin, rutin, and verbascoside	Methanol, USN probe (100 W and 30 kHz), and USN bath (400 W and 30 kHz)	Both equipment increased the extraction of main polyphenols; the highest extraction yield was obtained with the ultrasound probe	Jerman et al. (2010)
Olive leaves	Polyphenols	USN probe, electric power (40, 60, 80, and 100%), emitter surface (1.5, 3.8, and 12.6 cm^2), and temperature (25–50 °C)	Extraction yield was improved by increasing electric power; *an optimum value was obtained for emitter* surface and temperature; higher extraction yield than static and agitated conventional extraction; *optimized extraction condition: 100% electric power, emitter surface 3.8* cm^2*, and 25 °C*	Ahmad-Qasem et al. (2013)
Olive leaves	Apigenin-7-glucoside, luteolin-7-glucoside, oleuropein, and verbacoside	USN probe, radiation amplitude (10–50%), duty cycle (30–70%), irradiation time (6–30 min), extractant flow-rate (4–6 mL/min), ethanol % (50–90%), probe position (0–4 cm), and temperature (25–40 °C)	Extraction yield increased as radiation amplitude, duty cycle, probe position, and temperature increases; optimum values for irradiation time, extractant flow-rate, and ethanol %; *optimized extraction condition: 30% of radiation amplitude, 70% of duty cycle, 25 min, 5 mL/min, 59% ethanol, and probe position at 4 cm, 40 °C*	Japón-Luján et al. (2006)
Olive mill wastewater	Polyphenols	USN probe, 100 W, 30 kHz, pretreatment (fresh and freeze-dried), time (5, 10, 15, and 20 min), and solvent (ethyl acetate and diethyl ether)	Higher extraction yield than conventional solvent extraction for both the total and selected polyphenols; the highest extraction yield from freeze-dried samples; *optimized extraction condition: 10 min, 5 cycles, and ethyl acetate*	Klen and Mozeti (2011)
Olive oil alperujo	Polyphenols	USN probe, 20 kHz, 450 W, room temperature, radiation amplitude (10–50%), duty cycle (10–50%), time (1–13 min), extractant flow-rate (1–2 mL/min), methanol %	Main extracted compounds were tyrosol, catechin, and tannic acid; *optimized extraction condition: radiation amplitude of 10%, duty cycle of 50%, 13 min, 2 mL/min, 25% methanol, probe position at 3 mm, and 6 mL of extractant*	Priego-Capote et al. (2004)

		(25–100%), probe position (1–3 mm), and extractant volume (2–6 mL)		
Mustard seed meal	Polyphenols	USN probe, ethanol % (0–95%), solvent/solid ratio (10–60), temperature (20–75 °C), and time (10–60 min)	Extraction yield was improved by increasing temperature; optimized optimum extraction conditions were achieved for ethanol %, solvent/solid ratio, and time; *optimized extraction condition: 70% ethanol, solvent/solid ratio of 40, 40 °C, and 30 min*	Dubie et al. (2013)
Olive leaves	Polyphenols	USN bath, 50 Hz, solid/solvent ratio (500/20, 500/15, and 500/10 mg/mL), time (20, 40, and 60 min), and ethanol % (0, 50, and 100%)	Extraction yield was improved by increasing solid/solvent ratio and extraction time while an optimum ethanol % was achieved; *optimized extraction condition: solid/solvent ratio of 500/10, 50% ethanol, and 60 min*	Şahin and Şamli (2013)
Olive leaves	Polyphenols	USN bath, pH (3, 7, and 11), solid/solvent ratio (500/20, 500/15, and 500/10 mg/mL), time (20, 40, and 60 min), temperature (30, 45, and 60 °C); response surface methodology vs artificial neural network approach	Extraction yield was associated with low pH and high extraction temperature, time and solid/solvent ratio; artificial neural network showed the highest predication accuracy; *optimized extraction condition: pH 3.52, 59.87 °C, 59.57 min, and solid/solvent ratio of 500/19.78 mg/mL*	İlbay et al. (2014)
Olive leaves	Polyphenols	USN bath, 50 Hz, and 25 °C	Lower extraction yield than homogenized-assisted extraction	Bilgin and Selin (2013)
Olive leaves and kernel	Phytosterols	USN probe, 20 kHz, 450 W, radiation amplitude (10–50%), duty cycle (10–50%), time (5–20 min), dichloromethane–hexane ratio (33.3–100), volume (5–10 mL), number of cycles (1–5), temperature (15–25 °C), and probe position (1–3 cm)	USN treatment improved the extraction of phytosterols from both olive kernel and leaves; olive kernels had higher sterol content than leaves; *optimized extraction condition: radiation amplitude of 10%, duty cycle of 10%, 10 min, dichloromethane–hexane ratio of 66, 10 mL, 5 cycles, 20 °C, and probe position at 1 cm*	Orozco-Solano et al. (2010)

(Continued)

Table 9.4 (Cont.)

MATRIX	EXTRACTED COMPOUNDS	EXTRACTION CONDITIONS	EFFECT	REFERENCE
Rapeseed meal	Carotenoids	USN bath, USN power (120 and 180 W), temperature (20 and 50 °C), liquid/solid ratio (15 and 30 mL/g), time (10 and 45 min), and extraction cycles (1 and 3)	Extraction yield was improved by increasing USN power, temperature, liquid/solid ratio, and time; *optimized extraction condition: 252.9 W, 49.6 °C, liquid/solid ratio of 41.4 mL/g, 48.5 min, and 1 cycle*	Yan et al. (2015)
Oil palm fronds	β-carotene, lutein, and zeaxanthin	USN bath, solvent (ethanol, acetone, and hexane), solvent/solid ratio (10–50), time (10–50 min), and temperature (30–70 °C)	Ethanol was the most efficient solvent to recover carotenoids; optimum extraction conditions were achieved for all carotenoids; *optimized extraction condition: ethanol, 30 °C, 37 min, and solvent/solid ratio of 22 for β-carotene; ethanol, 30°C, 37 min, and solvent/solid ratio of 23 for lutein; and ethanol, 30°C, 39 min, and solvent/solid ratio of 20 for zeaxanthin*	Ofori-Boateng and Lee (2013)

also stated that the same extraction yield obtained from USN extraction was achieved by conventional solvent extraction using 59% ethanol for 24 h.

Another similar experiment was carried out by Priego-Capote et al. (2004), who improved the extraction of phenolic compounds from olive alperujo, a semisolid residue obtained from olive oil processing. The USN method was optimized for radiation amplitude (10%), duty cycle (50%), time (13 min), extractant flow rate (2 mL/min), methanol % (25%), probe position (3 mm), and extractant volume (6 mL). Additionally, the authors indicated that the main extracted compounds were tyrosol, catechin, and tannic acid.

Klen and Mozeti (2011) obtained interesting results by exploring the use of pretreatment and USN extraction variables for recovering polyphenols from olive mill wastewater. In this study, the USN extraction protocol was optimized for pretreatment (fresh or freeze-dried), extraction time (5–20 min), and solvent composition (ethyl acetate or diethyl ether). The authors argued that freeze-drying of olive mill wastewater was the most efficient method, as pretreatment, to preserve phenolic compounds, compared with fresh samples. Regarding the effect of USN treatment on the extraction of polyphenols, the optimum conditions indicated by the authors consisted in five consecutive USN treatments (10 min each) using ethyl acetate, which was mainly composed of *p*-HPEA-DEDA. The authors also compared the optimized USN extraction method with conventional solvent extraction (with agitation), solid-phase extraction, and direct filtration of olive mill wastewater. In this case, the optimized USN treatment and filtration displayed the highest extraction yields, but a significant increase in recovery of hydroxytyrosol glucoside, hydroxytyrosol, and caffeoyl-60-secologanoside by USN treatment was highlighted by the authors. Additionally, USN extraction with the probe was studied to recover polyphenols from mustard seed meal (Dubie et al., 2013). In this experiment, the ethanol content in the solvent (0–95%), solvent/solid ratio (10–60), temperature (20–75 °C), and extraction time (10–60 min) were selected for optimization. The maximum extraction yield was 8.85 mg sinapic acid equivalents/g mustard seed meal by carrying out the extraction with 70% ethanol and solvent/solid ratio of 40 at 40 °C for 30 min.

Conversely, the extraction of polyphenols using USN baths was also studied in the last decade. For instance, the protocol proposed by Şahin and Şamli (2013) explored the impact of solid/solvent ratio (500/20, 500/15, and 500/10 mg/mL), extraction time (20, 40, and 60 min), and ethanol content in solvent (0, 50, and 100%) revealing that the extraction yield was improved by increasing solid/solvent ratio and extraction time. The authors also reported a maximum polyphenol recovery of 25.06 mg GAE/g while USN-assisted extraction was carried out at the solid/solvent ratio of 500/10 and 50% ethanol for 60 min. Likewise, an experiment compared the extraction of polyphenols from olive leaves using two approaches: response surface methodology and artificial neural network (İlbay, Şahin, and Büyükkabasakal, 2014). Both optimization strategies displayed good agreement with experimental results regarding the pH (3, 7, and 11), solid/solvent ratio (500/20, 500/15, and 500/10 mg/mL), time (20, 40, and 60 min), and temperature (30, 45, and 60 °C). The optimization of USN extraction by response surface methodology indicated that at pH 3.52 and 59.87 °C for 59.57 min with a solid/solvent ratio of 500/19.78 mg/mL could achieve a maximum polyphenols extraction yield of 56.2 mg GAE/g dw. Additionally, the authors also stated that artificial neural network showed higher predication accuracy than response surface methodology.

However, USN bath may not cause significant changes in the residue matrix that lead to higher extraction yields compared with homogenized conventional solvent extraction. This outcome was reported by Bilgin and Selin (2013), who explored the extraction of polyphenols from olive leaves collected from Bursa, Mardin Ayvalik, Kas Tekirdag, and Canakkale regions of Turkey. All extracts displayed lower polyphenol content than their counterparts obtained by conventional solvent extractions.

The extraction of other antioxidant and bioactive compounds from oilseed by-products was also carried out using USN technology. An interesting example is the recovery of phytosterols from olive leaves and kernels using USN treatment (Orozco-Solano, Ruiz-Jiménez, and Luque de Castro, 2010). The extraction was optimized for radiation amplitude (10%), duty cycle (10%), extraction time (10 min), dichloromethane–hexane ratio (66), extraction volume (10 mL), extraction repetition (5 cycles), temperature (20 °C), and probe position (1 cm) to achieve 369.1 and 3063.3 μg/g in olive kernel and leaves, respectively. According to the authors, β-sitosterol (2,594.6 and 204.6 μg/g in olive kernel and leaves, respectively) was the main compound extracted using the optimized protocol.

Another relevant group of bioactive compounds extracted by USN are carotenoids. The extraction of carotenoids was carried out by Yan et al. (2015) from rapeseed meal by evaluating the effect of USN power (120 and 180 W), temperature (20 and 50 °C), liquid/solid ratio (15 and 30 mL/g), time (10 and 45 min), and extraction cycles (1 and 3). The highest carotenoid yield obtained at optimized conditions (252.9 W and 49.6 °C with a liquid/solid ratio of 41.4 mL/g during 48.5 min and 1 cycle) was 0.158 mg/g. Likewise, Ofori-Boateng and Lee (2013) optimized the solvent composition (ethanol, acetone, and hexane), solvent/solid ratio (10–50), extraction time (10–50 min), and temperature (30–70 °C) for the recovery of β-carotene, lutein, and zeaxanthin from oil palm fronds. Interestingly, optimum USN extraction conditions were compound-dependent, except for temperature (30 °C) and solvent (ethanol): 37 min and solvent/solid ratio of 22 for β-carotene (17.95 μg/g dw); 37 min and solvent/solid ratio of 23 for lutein (261.99 μg/g dw); and 30 °C, 39 min, and solvent/solid ratio of 20 for zeaxanthin (29.99 μg/g dw).

3.5 HP Extraction

The use of HP to obtain high-value compounds from olive oil processing by-products was carried out by Chanioti and Tzia (2018), who explored the effects of pressure (300 and 600 MPa), time (5 and 10 min), and solvent composition (water, ethanol, and natural deep eutectic solvents) on the polyphenols recovery, with temperature motorization. The virgin olive pomace was packaged with selected solvent and subjected to HP treatment. The authors observed that the increase of temperature and extraction time was associated with a higher polyphenol recovery (600 MPa for 10 min) with natural deep eutectic solvents, particularly for lactic acid in the deep eutectic solvent. Moreover, under optimized processing conditions, the comparison of natural deep eutectic solvents with water and 70% ethanol was carried out. The results revealed that using lactic acid in the natural deep eutectic solvent was more efficient to recover polyphenols than 70% ethanol and water, particularly hydroxytyrosol (2.57 vs 0.41 and 0.25 mg/g dw, respectively). Interestingly, 70% ethanol was highly efficient to recover oleuropein compared with lactic acid in the deep eutectic solvent (4.42 vs 1.94 mg/g dw).

4 Conclusion

During the production of food products, such as olive oil, a series of residues and by-products are produced with high amounts of polyphenols, carotenoids, phytosterols, and tocopherols. The pharmaceutical and food industries have been attracted by these, showing great interest in the recovery of these compounds due to their different economic benefits. Innovative methods, such as HVED, USN, HP, and MW, are promising alternative extraction techniques for recovering of these compounds. The optimization of extraction variables is of great importance since significant differences are observed in several studies, particularly for extracting solvent composition, temperature, time, and energy inputs. Consequently, distinct advantages related to these new methodologies are the reduced extraction time, amount of extracting solvent, energy consumption, and environmental impact compared with conventional extraction methods.

However, it is not possible to state which method is the most favorable among all innovative methods, since it depends on the matrix used, and the type of solvent and its concentration. Moreover, advances and further studies in the application of each technology are still required. For instance, both the HVED and PEF are promising technologies with important results reported in recent years, which must be tested at pilot and industrial scales; SC-CO_2 extraction requires major advances in improving its efficiency for recovering polyphenols (which normally is carried out with solvents of intermediate polarity) and then enhance the extraction yield; the GAME approach demands more studies to optimize the extraction of polyphenols and to explore the extraction of other high-value compounds; and an improvement in the current level of information about the impact of HP on extraction of polyphenols is also needed. The other technologies discussed in this chapter must be tested on a pilot and then on an industrial scale.

Finally, the stability of bioactive compounds after extraction and their further utilization in a complex matrix (e.g., food systems, cosmetics, and packaging) demands more attention by researchers and industrial professionals to evaluate the influence on product characteristics and viability of new products. It is important to explore the extraction of other less studied high-value compounds, particularly carotenoids, phytosterols, and tocopherols, that have already been associated with important health benefits, and cosmetic and technological applications. Therefore, the research work is essential for the process optimization to ensure the recovery of optimum yields for high-quality bioactive compounds.

Acknowledgments

Paulo E. S. Munekata acknowledges postdoctoral fellowship support from the Ministry of Economy and Competitiveness (MINECO, Spain) "Juan de la Cierva" program (FJCI-2016-29486).

References

Abenoza, M, M Benito, G Saldana, I Alvarez, J Raso, and A C Sanchez-Gimeno 2013. "Effects of Pulsed Electric Field on Yield Extraction and Quality of Olive Oil". *Food and Bioprocess Technology* 6 (6): 1367–1373.

Ahmad-Qasem, M H, J Cánovas, E Barrajón-Catalán, V Micol, J Andrés, and J V García-Pérez 2013. "Kinetic and Compositional Study of Phenolic Extraction from Olive Leaves (Var. Serrana) by Using Power Ultrasound". *Innovative Food Science and Emerging Technologies* 17: 120–29. doi: 10.1016/j.ifset.2012.11.008.

Akgu, N A 2011. "Separation of Squalene from Olive Oil Deodorizer Distillate Using Supercritical Fluids". *European Journal of Lipid Science and Technology* 113: 1558–1565. doi: 10.1002/ejlt.201000466..

Amofa-Diatuo, T, D M Anang, F J Barba, and B K Tiwari 2017. "Development of New Apple Beverages Rich in Isothiocyanates by Using Extracts Obtained from Ultrasound-Treated Cauliflower by-Products: Evaluation of Physical Properties and Consumer Acceptance". *Journal of Food Composition and Analysis* 61: 73–81. doi: 10.1016/j.jfca.2016.10.001.

Andreou, V, G Dimopoulos, Z Alexandrakis, G Katsaros, D Oikonomou, S Toepfl, V Heinz, and P Taoukis 2017. "Shelf-Life Evaluation of Virgin Olive Oil Extracted from Olives Subjected to Nonthermal Pretreatments for Yield Increase". *Innovative Food Science and Emerging Technologies*, Special Issue from the '29th EFFoST International Conference 40 (Supplement C): 52–57. doi: 10.1016/j.ifset.2016.09.009.

Azmir, J, I S M Zaidul, M M Rahman, K M Sharif, A Mohamed, F Sahena, M H A Jahurul, K Ghafoor, N A N Norulaini, and A K M Omar 2013. "Techniques for Extraction of Bioactive Compounds from Plant Materials: A Review". *Journal of Food Engineering* 117 (4): 426–436. doi: 10.1016/j.jfoodeng.2013.01.014.

Ballard, T S, P Mallikarjunan, K Zhou, and S O Keefe 2010. "Microwave-Assisted Extraction of Phenolic Antioxidant Compounds from Peanut Skins". *Food Chemistry* 120 (4): 1185–192. doi: 10.1016/j.foodchem.2009.11.063.

Barba, F J, N Boussetta, and E Vorobiev 2015. "Emerging Technologies for the Recovery of Isothiocyanates, Protein and Phenolic Compounds from Rapeseed and Rapeseed

Press-Cake: Effect of High Voltage Electrical Discharges". *Innovative Food Science and Emerging Technologies* 31: 67–72. doi: 10.1016/j.ifset.2015.06.008.

Barba, F J, O Parniakov, S A Pereira, A Wiktor, N Grimi, N Boussetta, J A Saraiva et al. 2015a. "Current Applications and New Opportunities for the Use of Pulsed Electric Fields in Food Science and Industry". *Food Research International* 77. doi: 10.1016/j.foodres.2015.09.015.

Barba, F J, N S Terefe, R Buckow, D Knorr, and V Orlien 2015b. "New Opportunities and Perspectives of High Pressure Treatment to Improve Health and Safety Attributes of Foods. A Review". *Food Research International* 77 (November): 725–742.

Barba, F J, Z Zhu, M Koubaa, A S Sant'Ana, and V Orlien 2016. "Green Alternative Methods for the Extraction of Antioxidant Bioactive Compounds from Winery Wastes and By-Products: A Review". *Trends in Food Science and Technology* 49: 96–109.

Bilgin, M, and Ş Selin 2013. "Effects of Geographical Origin and Extraction Methods on Total Phenolic Yield of Olive Tree (Olea Europaea) Leaves". *Journal of the Taiwan Institute of Chemical Engineers* 44: 8–12. doi: 10.1016/j.jtice.2012.08.008.

Bondioli, P, C Mariani, A Lanzani, E Fedeli, and A Muller 1993. "Squalene Recovery from Olive Oil Deodorizer Distillates". *Journal of the American Oil Chemists' Society* 70 (8): 763–766.

Boussetta, N, E Soichi, J -L Lanoiselle, and E Vorobiev 2014. "Valorization of Oilseed Residues: Extraction of Polyphenols from Flaxseed Hulls by Pulsed Electric Fields". *Industrial Crops and Products* 52 (0): 347–353.

Chanioti, S, and C Tzia 2018. "Extraction of Phenolic Compounds from Olive Pomace by Using Natural Deep Eutectic Solvents and Innovative Extraction Techniques". *Innovative Food Science and Emerging Technologies* 48: 228–239. doi: 10.1016/j.ifset.2018.07.001.

Chemat, F, A S A Fabiano-Tixier, M A A Vian, B Allaf, and E C Vorobiev 2015. "Solvent-Free Extraction of Food and Natural Products". *TrAC - Trends in Analytical Chemistry* 71: 157–168. doi: 10.1016/j.trac.2015.02.021.

Chemat, F, N Rombaut, A-G Sicaire, A Meullemiestre, A-S Fabiano-Tixier, and M Abert-Vian 2017. "Ultrasound Assisted Extraction of Food and Natural Products. Mechanisms, Techniques, Combinations, Protocols and Applications. A Review". *Ultrasonics Sonochemistry* 34: 540–560.

Dubie, J, A Stancik, M Morra, and C Nindo 2013. "Antioxidant Extraction from Mustard (Brassica Juncea) Seed Meal Using High-Intensity Ultrasound". *Journal of Food Science* 78 (4): 542–548. doi: 10.1111/1750–3841.12085.

Floch, F L, M T Tena, A Ríos, and M Valcárcel. 1998. "Supercritical Fluid Extraction of Phenol Compounds from Olive Leaves". *Talanta* 46: 1123–30.

Guderjan, M, P Elez-Martinez, and D Knorr 2007. "Application of Pulsed Electric Fields at Oil Yield and Content of Functional Food Ingredients at the Production of Rapeseed Oil". *Innovative Food Science and Emerging Technologies* 8 (1): 55–62.

Guderjan, M, S Töpfl, A Angersbach, and D Knorr 2005. "Impact of Pulsed Electric Field Treatment on the Recovery and Quality of Plant Oils". *Journal of Food Engineering* 67 (3): 281–287. doi: 10.1016/j.jfoodeng.2004.04.029.

Herrero, M, A Cifuentes, and E Ibañez. 2006. "Sub- and Supercritical Fluid Extraction of Functional Ingredients from Different Natural Sources: Plants, Food-by-Products, Algae and Microalgae: A Review". *Food Chemistry* 98 (1): 136–148.

Herrero, M, T N Temirzoda, A Segura-Carretero, R Quirantes, M Plaza, and E Ibañez. 2011. "New Possibilities for the Valorization of Olive Oil By-Products". *Journal of Chromatography A* 1218: 7511–7520. doi: 10.1016/j.chroma.2011.04.053.

Huang, Z-R, Y-K Lin, and J-Y Fang 2009. "Biological and Pharmacological Activities of Squalene and Related Compounds: Potential Uses in Cosmetic Dermatology". *Molecules* 14: 540–554. doi: 10.3390/molecules14010540.

İlbay, Z, S Şahin, and K Büyükkabasakal. 2014. "A Novel Approach for Olive Leaf Extraction through Ultrasound Technology: Response Surface Methodology versus Artificial Neural Networks". *Korean Journal of Chemical Engineering* 31 (9): 1661–1667. doi: 10.1007/s11814-014-0106-3.

Japón-Luján, R, J M Luque-Rodríguez, and M D Luque de Castro 2006. "Dynamic Ultrasound-Assisted Extraction of Oleuropein and Related Biophenols from Olive Leaves". *Journal of Chromatography A* 1108: 76–82. doi: 10.1016/j.chroma.2005.12.106.

Jerman, T, P Trebše, and B M Vodopivec 2010. "Ultrasound-Assisted Solid Liquid Extraction (USLE) of Olive Fruit (Olea Europaea) Phenolic Compounds". *Food Chemistry* 123: 175–182. doi: 10.1016/j.foodchem.2010.04.006.

Jimenez, P, L Masson, A Barriga, J Chávez, and P Robert 2011. "Oxidative Stability of Oils Containing Olive Leaf Extracts Obtained by Pressure, Supercritical and Solvent-Extraction". *European Journal of Lipid Science and Technology* 113: 497–505. doi: 10.1002/ejlt.201000445.

Johnson, G H, and K Fritsche 2012. "Effect of Dietary Linoleic Acid on Markers of Inflammation in Healthy Persons: A Systematic Review of Randomized Controlled Trials". *Journal of the Academy of Nutrition and Dietetics* 112 (7): 1029–41. doi: 10.1016/j.jand.2012.03.029.

Klen, T J, and B Mozeti 2011. "Ultrasonic Extraction of Phenols from Olive Mill Wastewater: Comparison with Conventional Methods". *Journal of Agricultural and Food Chemistry* 59: 12725–12731. doi: 10.1021/jf202800n.

Koubaa, M, L Lepreux, F J Barba, H Mhemdi, and E Vorobiev 2017. "Gas Assisted Mechanical Expression (GAME) for the Selective Recovery of Lipophilic and Hydrophilic Compounds from Olive Kernel". *Journal of Cleaner Production* 166: 387–394.

Koubaa, M, H Mhemdi, F J Barba, S Roohinejad, R Greiner, and E Vorobiev 2016. "Oilseed Treatment by Ultrasounds and Microwaves to Improve Oil Yield and Quality: An Overview". *Food Research International* 85: 59–66.

Lafka, T-I, A E Lazou, V J Sinanoglou, and E S Lazos 2011. "Phenolic and Antioxidant Potential of Olive Oil Mill Wastes". *Food Chemistry* 125 (1): 92–98. doi: 10.1016/j.foodchem.2010.08.041.

Li, Z, Y Fan, and J Xi 2019. "Recent Advances in High Voltage Electric Discharge Extraction of Bioactive Ingredients from Plant Materials". *Food Chemistry* 277: 246–260. doi: 10.1016/j.foodchem.2018.10.119.

Lin, L, H Allemekinders, A Dansby, L Campbell, S Durance-Tod, A Berger, and P J H Jones 2013. "Evidence of Health Benefits of Canola Oil". *Nutrition Reviews* 71 (6): 370–385. doi: 10.1111/nure.12033.

Lozano-Sánchez, J, M Castro-Puyana, J A Mendiola, A Segura-Carretero, A Cifuentes, and E Ibañez. 2014. "Recovering Bioactive Compounds from Olive Oil Filter Cake by Advanced Extraction Techniques". *International Journal of Molecular Sciences* 15: 16270–16283. doi: 10.3390/ijms150916270.

Lucas, A de, E M de la Ossa, J Rincón, M A Blanco, and I Gracia 2002. "Supercritical Fluid Extraction of Tocopherol Concentrates from Olive Tree Leaves". *Journal of Supercritical Fluids* 22: 221–228.

Mandal, V, Y Mohan, and S Hemalatha 2006. "Microwave Assisted Extraction - An Innovative and Promising Extraction Tool for Medicinal Plant Research". *Pharmacognosy Reviews* 1 (1): 7–18.

Ofori-Boateng, C, and K T Lee 2013. "Response Surface Optimization of Ultrasonic-Assisted Extraction of Carotenoids from Oil Palm (Elaeis Guineensis Jacq.) Fronds". *Food Science & Nutrition* 1 (3): 209–221. doi: 10.1002/fsn3.22.

Orozco-Solano, M I, J Ruiz-Jiménez, and M D Luque de Castro 2010. "Ultrasound-Assisted Extraction and Derivatization of Sterols and Fatty Alcohols from Olive Leaves and

Drupes Prior to Determination by Gas Chromatography-Tandem Mass Spectrometry". *Journal of Chromatography A* 1217: 1227–35. doi: 10.1016/j.chroma.2009.12.040.

Priego-Capote, F, J Ruiz-Jiménez, and M D Luque de Castro 2004. "Fast Separation and Determination of Phenolic Compounds by Capillary Electrophoresis-Diode Array Detection Application to the Characterisation of Alperujo after Ultrasound-Assisted Extraction". *Journal of Chromatography A* 1045: 239–246. doi: 10.1016/j.chroma.2004.06.019.

Puértolas, E, and F J Barba 2016. "Electrotechnologies Applied to Valorization of By-Products from Food Industry: Main Findings, Energy and Economic Cost of Their Industrialization". *Food and Bioproducts Processing* 100: 172–184.

Puértolas, E, M Koubaa, and F J Barba 2016. "An Overview of the Impact of Electrotechnologies for the Recovery of Oil and High-Value Compounds from Vegetable Oil Industry: Energy and Economic Cost Implications". *Food Research International* 80 (February): 19–26.

Puértolas, E, and I Martínez de Marañón 2015. "Olive Oil Pilot-Production Assisted by Pulsed Electric Field: Impact on Extraction Yield, Chemical Parameters and Sensory Properties". *Food Chemistry* 167 (Supplement C): 497–502. doi: 10.1016/j.foodchem.2014.07.029.

Rafiee, Z, M Aalami, and M Khomeiri 2011. "Microwave-Assisted Extraction of Phenolic Compounds from Olive Leaves; a Comparison with Maceration". *The Journal of Animal & Plant Sciences* 24 (4): 738–45.

Roselló-Soto, E A, F J A Barba, O B C Parniakov, C M D Galanakis, N C Lebovka, N B Grimi, and E B Vorobiev 2015a. "High Voltage Electrical Discharges, Pulsed Electric Field, and Ultrasound Assisted Extraction of Protein and Phenolic Compounds from Olive Kernel". *Food and Bioprocess Technology* 8 (4): 885–94. doi: 10.1007/s11947-014-1456-x.

Roselló-Soto, E A, C M Galanakis, M Brnčić, V Orlien, F J Trujillo, R Mawson, K Knoerzer, B K Tiwari, and F J Barba 2015b. "Clean Recovery of Antioxidant Compounds from Plant Foods, by-Products and Algae Assisted by Ultrasounds Processing. Modeling Approaches to Optimize Processing Conditions". *Trends in Food Science & Technology* 42 (2): 134–49.

Roselló-Soto, E A, M Koubaa, A Moubarik, R P Lopes, J A Saraiva, N Boussetta, N Grimi, and F J Barba 2015c. "Emerging Opportunities for the Effective Valorization of Wastes and By-Products Generated during Olive Oil Production Process: Non-Conventional Methods for the Recovery of High-Added Value Compounds". *Trends in Food Science and Technology* 45 (2): 296–310. doi: 10.1016/j.tifs.2015.07.003.

Şahin, S, and R Şamli. 2013. "Optimization of Olive Leaf Extract Obtained by Ultrasound-Assisted Extraction with Response Surface Methodology". *Ultrasonics Sonochemistry* 20: 595–602. doi: 10.1016/j.ultsonch.2012.07.029.

Sarkis, J R, N Boussetta, C Blouet, I C Tessaro, L D F Marczak, and E Vorobiev 2015a. "Effect of Pulsed Electric Fields and High Voltage Electrical Discharges on Polyphenol and Protein Extraction from Sesame Cake". *Innovative Food Science and Emerging Technologies* 29: 170–77. doi: 10.1016/j.ifset.2015.02.011.

Sarkis, J R, N Boussetta, I C Tessaro, L D F Marczak, and E Vorobiev 2015b. "Application of Pulsed Electric Fields and High Voltage Electrical Discharges for Oil Extraction from Sesame Seeds". *Journal of Food Engineering* 153: 20–7. doi: 10.1016/j.jfoodeng.2014.12.003.

Schwingshackl, L, and G Hoffmann 2012. "Monounsaturated Fatty Acids and Risk of Cardiovascular Disease: Synopsis of the Evidence Available from Systematic Reviews and Meta-Analyses". *Nutrients* 4: 1989–2007. doi: 10.3390/nu4121989.

Taamalli, A, D Arráez-Román, E Barrajón-Catalán, V Ruiz-Torres, A Segura-Carretero, and A Fernández-Gutiérrez. 2012a. "Use of Advanced Techniques for the Extraction of Phenolic Compounds from Tunisian Olive Leaves: Phenolic Composition and Cytotoxicity against Human Breast Cancer Cells". *Food and Chemical Toxicology* 50 (6): 1817–25. doi: 10.1016/j.fct.2012.02.090.

Taamalli, A, D Arráez-Román, E Ibañez, M Zarrouk, A Segura-Carretero, and A Fernández-Gutiérrez. 2012b. "Optimization of Microwave-Assisted Extraction for the Characterization of Olive Leaf Phenolic Compounds by Using HPLC-ESI-TOF-MS/IT-MS2". *Journal of Agricultural and Food Chemistry* 60: 791–8. doi: 10.1021/jf204233u.

Teh, S-S, B E Niven, A E A Bekhit, A Carne, and E John Birch 2015. "Microwave and Pulsed Electric Field Assisted Extractions of Polyphenols from Defatted Canola Seed Cake". *International Journal of Food Science and Technology* 50: 1109–15. doi: 10.1111/ijfs.12749.

Tsarouhas, P, C. Achillas, D Aidonis, D Folinas, and V Maslis 2015. "Life Cycle Assessment of Olive Oil Production in Greece". *Journal of Cleaner Production* 93 (Supplement C): 75–83. doi: 10.1016/j.jclepro.2015.01.042.

Vázquez, L, C F Torres, T Fornari, F J Señoráns, and G Reglero 2007. "Recovery of Squalene from Vegetable Oil Sources Using Countercurrent Supercritical Carbon Dioxide Extraction". *Journal of Supercritical Fluids* 40: 59–66. doi: 10.1016/j.supflu.2006.04.012.

Xynos, N, G Papaefstathiou, E Gikas, A Argyropoulou, N Aligiannis, and A-L Skaltsounis 2014. "Design Optimization Study of the Extraction of Olive Leaves Performed with Pressurized Liquid Extraction Using Response Surface Methodology". *Separation and Purification Technology* 122: 323–30. doi: 10.1016/j.seppur.2013.10.040.

Xynos, N, G Papaefstathiou, M Psychis, A Argyropoulou, N Aligiannis, and A-L Skaltsounis 2012. "Development of a Green Extraction Procedure with Super/Subcritical Fluids to Produce Extracts Enriched in Oleuropein from Olive Leaves". *The Journal of Supercritical Fluids* 67: 89–93. doi: 10.1016/j.supflu.2012.03.014.

Yan, F, K Fan, J He, and M Gao 2015. "Ultrasonic-Assisted Solvent Extraction of Carotenoids from Rapeseed Meal: Optimization Using Response Surface Methodology". *Journal of Food Quality* 38: 377–86.

Yu, X, O Bals, N Grimi, and E Vorobiev 2015a. "A New Way for the Oil Plant Biomass Valorization: Polyphenols and Proteins Extraction from Rapeseed Stems and Leaves Assisted by Pulsed Electric Fields". *Industrial Crops & Products* 74: 309–18. doi: 10.1016/j.indcrop.2015.03.045.

Yu, X-M, P Zhu, Q-P Zhong, M-Y Li, and H-R Ma 2015b. "Subcritical Water Extraction of Antioxidant Phenolic Compounds from XiLan Olive Fruit Dreg". *Journal of Food Science and Technology* 52 (8): 5012–20. doi: 10.1007/s13197-014-1551–z.

Zago, E, J Lecomte, N Barouh, C Aouf, P Carré, F Fine, and P Villeneuve 2015. "Influence of Rapeseed Meal Treatments on Its Total Phenolic Content and Composition in Sinapine, Sinapic Acid and Canolol". *Industrial Crops & Products* 76: 1061–70. doi: 10.1016/j.indcrop.2015.08.022.

Žugčić, T, R Abdelkebir, C Alcantara, M C Collado, J V García-Pérez, A J Meléndez-Martínez, A Jambrak Režek, J M Lorenzo, and F J Barba 2019. "From Extraction of Valuable Compounds to Health Promoting Benefits of Olive Leaves through Bioaccessibility, Bioavailability and Impact on Gut Microbiota". *Trends in Food Science & Technology* 83: 63–77. doi: 10.1016/j.tifs.2018.11.005.

10

Research, Development, and Innovation in Dairy and Meat-Based Foods Using Valued Added Compound Obtained from Mediterranean Fruit By-Products

JOSÉ ANGEL PÉREZ-ALVAREZ, MANUEL VIUDA-MARTOS, AND JUANA FERNÁNDEZ-LÓPEZ

IPOA Research Group, Agrifood Technology Department, Orihuela Polytechnical High School, Miguel Hernández University, Carretera a Beniel km 3.2, Orihuela 03312, Alicante, Spain

Contents

1 Introduction

In the context of food in the 21st century, the perceptions of consumers concerning the importance of functional foods have changed. Their trust and willingness to pay for the added value, which is involved in purchasing these foods, are of great interest for the food industry (Mirosa and Mangan-Walker, 2018), especially when it comes to identifying the "target" population (Maeda-Yamamoto, 2017; Predieri et al., 2017) they consider important. For example, knowing the nutrient or bioactive compound responsible for producing a beneficial effect on health (as announced by the corresponding health claims), such as dietary fibers (Viuda-Martos et al., 2010a, 2010b), bioactive compounds (Pogorzelska-Nowicka et al., 2018), among others.

A new strategy for developing and innovating functional foods, in general, is to apply the "5Stars Foods" concept. Each star corresponds to the following characteristics of the food, which must be healthy, safety, tasty, sustainable, and socially accepted (Pérez-Alvarez, 2016). This new food concept has a strong ally in the use of functional and potentially functional ingredients (dietary fibers, antioxidants, etc.), obtained through the valorization of by-/coproducts. Following the 5S foods concept, functional foods should comply with the following criteria: (i) healthy: reduced calorie intake, either by reducing the fat content of the product (Kumar et al., 2017; Belc et al., 2018), modification of its lipid profile by including "healthy" fats, such as omega-3 fatty acids (Ospina-E et al., 2015; Bolger et al., 2018), reduced salt content (Zapata-López, 2016; Belc et al., 2018), or reduced sugar content (Belc et al., 2018; Gabriel et al., 2018); (ii) safety: the producer must have followed the criteria of traceability of raw materials, as required by legislation, developed a plan of Hazards and Critical Control Points to the process of elaboration of functional foods and should have followed hygiene and food safety protocols for raw materials, processes, and products, as well as should have trained various members of the team in food handling (Roselino et al., 2017; Halagarda et al., 2018); (iii) tasty: food products, in general, and functional food products, in particular, must be attractive visually and flavor wise. They should at least taste like conventional products. For this reason, it is very important that texture, color, aroma, and taste be "pleasant" for the consumer. According to Gabriel et al. (2018), flavor becomes a very relevant quality criterion for virtually all functional foods. According to Delwiche (2004), flavor is an emergent phenomenon that arises from the combination of taste, odor, and chemesthesis, or a combination of multiple sensations, while Sellimi et al. (2017) also mentioned that color plays an important role in functional foods. The appearance and colors of a foodstuff are among the first sensory stimuli to be noticed when eating food (Delwiche, 2012); (iv) sustainable: throughout the elaboration process, criteria of eco-efficiency should be taken into account (Martín-Sánchez et al., 2009; Latiffi et al., 2016; Sonesson et al., 2017; Faucher et al., 2018; Gabriel et al., 2018). According to the United Nations, the growing use in foods of legume proteins (Schutyser et al., 2015; Vinauskiene et al., 2015; Baugreet et al., 2016; Tulbek et al., 2016; Rocchetti et al., 2019) and pseudocereals, such as quinoa (Ruiz et al., 2017; Opazo-Navarrete et al., 2018; Pellegrini et al., 2018) and amaranth (Barba de la Rosa et al., 2009; Cabanillas et al., 2017), or oleaginous plants (Brückner et al., 2010; Mireles-Arriaga et al., 2017), such as sunflower or chia (Zettel and Hitzmann, 2018), favor sustainable food production since they allow crop rotation, require less fertilizer than others, and have low carbon emissions or, in the case of insects as alternative sources of animal protein (FAO, 2013; Smetana et al., 2015). Buss (2019) mentioned that food companies are applying circular economy principles as they strive to operate more sustainably, responsibly, and creatively. Finally, (v), foods must be socially accepted: this is a complex matter because of the huge number of factors that may affect the acceptance of a product. For example, the following criteria may influence consumers: (i) type and origin of raw materials (Gutiérrez, 2018; Leidi et al., 2018; Silva et al., 2018; Żyżelewicz et al., 2018)—in this respect, fruits and vegetables are well accepted (Nunes, 2015; Schifferstein et al., 2019; van Stokkom et al., 2019); (ii) "new supply species" (game animals, insects, etc.) (Kim et al., 2016, 2017; Schnettler et al., 2018); (iii) nonconventional sources of proteins with excellent nutritional value, such as chia (*Salvia hispanica*), quinoa (*Chenopodium quinoa* Willd.), or amaranth (*Amaranthus caudatus*) (Boukid et al., 2018; López et al., 2018); (iv) religious rites (Evivie et al., 2017; Sharif et al., 2018); (v) food taboos (Köhler et al., 2019); (vi) food "philosophies" (vegetarian, vegan, reducterians, etc.) (Goetzke et al., 2014; Sobiecki et al., 2016); (vii) comfort (Spence, 2017); (viii) neophobia and neophilia (Giordano et al., 2018; Laureati et al., 2018; Rioux et al., 2018); (ix) in vitro meat (Acevedo et al., 2018; Stephens et al., 2018); (x) the use of blood and heme derivatives (Stephens et al., 2018), among others.

In general, for the development of functional foods, a conventional food is chosen and subjected to "functionalization," much practice leading to the creation of a "prototype" (Olsen, 2015). Once a food is selected to "functionalize," there are several recommendations for designing healthy foods. One of them, described by Decker and Park (2010), applies fruit value-added compounds rich in extracts, dietary fibers obtained with eco-efficient technologies from fruits, some of which have been studied by Lario et al. (2004), Aleson-Carbonell et al. (2005), Fernández-Ginés et al. (2003), Fernández-López et al. (2009), Viuda-Martos et al. (2009), Sánchez-Zapata et al. (2011), López-Vargas et al. (2013), Martín-Sánchez et al. (2014b), and López-Marcos et al. (2015). Development is based on the identification of the following factors or criteria: (i) Is the bioactive compound consumed by the general population? (ii) Is the bioactive compound consumed by a target population (children, teenagers, millennials, men, women, elderly people, etc.)? (iii) What is the scientific evidence supporting its efficacy in humans? What are the health claims, if any? (iv) Is the bioactive compound bioavailable and bioaccessible? (v) How stable is the bioactive compound in the food of interest, both in its processing and its conservation and useful life? This criterion must also take into account oxidative stability to light, the enzymes endogenous in the food (their own or from the microbiota present), pH, and thermal stability; (vi) How do bioactive compounds affect the sensory quality of food, considering the color, aroma, texture, and taste? (Adding it must at least maintain the "original flavor" of the food or improve it); (vii) Does the incorporation of the bioactive compound change the food's ultrastructure and, with it, the process of elaboration? (viii) Does the bioactive compound lose its effectiveness (health properties) during the digestion process? (ix) Is the added compound or ingredient economically viable? (x) Is there a reliable and continuous commercial supply of the compound or ingredient to be added? (xi) Is the ingredient or compound to be added recognized as safe? (xii) Are there legal restrictions to incorporating bioactive compounds in food products in general or particularly in some? Here, it must be remembered that a given bioactive compound may be permitted in one food group (bakery products, dairy products, etc.) and be prohibited in others (e.g., the use of chitin and chitosan as dietary fiber in foods); (xiii) Does its incorporation have any real interest? What does it actually do? (xiv) Are there criteria or quality standards for the ingredient or bioactive compound? (xv) Can we make any health claims about the ingredient once it has been added to the food product? (xvi) What health benefits of the product can be communicated to the consumer? (xvii) Is there a truthful communicator and a community manager to defend the product on social networks? In short, industries have to offer accurate information, which has been proven and is easily understandable by the consumer.

2 Definitions

2.1 Valorization

"Valorization" is the process of creating value from knowledge by making it suitable and/or available for economic and/or societal use and translating it into competitive products, services, processes, and entrepreneurial activity (de Jong, 2015). Compounds obtained as a result of valorization are well regarded by consumers as they have a natural origin, which consumers appreciate and are beneficial for the environment. The compounds obtained from the valorization of fruits and vegetables have a high biological, technofunctional, and physiofunctional value, and may help in the economic activity of the companies concerned. Such valorization practices are considered pivotal as an effective bio-economy strategy for rural development. The distance between the citrus processors and the citrus by-product plant is one of the main criteria for choosing alternative valorization pathways (Raimondo et al., 2018). Without valorization, the great loss of valuable materials that may result may give rise to serious management problems, both from the economic and environmental point of view (Damian, 2018).

2.2 Waste

Waste is defined as any substance or object that the holder discards or intends to discard or is required to discard (Directive, 2008). According to Barbosa-Pereira and coworkers (2014), food "wastes" are products with special characteristics, as they have a high degree of biological instability, a potentially pathogenic nature, high water content, a potential for rapid auto-oxidation, and high amounts of enzymes and high enzymatic activity.

The generation of "wastes" mainly consisting of organic residues of processed raw materials (animal and vegetal origin) by the food industry is unavoidable. According to the EU norms (Directive, 2008), the quality of by-/coproducts must meet certain microbiological, chemical, physiochemical, and other criteria for their valorization. For example, their composition must be standardized as much as possible. However, such standardization is very difficult and complex, and thus potentially expensive. The richness of bioactive compounds, dietetic fibers, and compounds of high added value much appreciated in the food, pharmaceutical, and chemical industries, and as biocomposites, has led to new approaches to valorization. In general, vegetal residues are easily fermentable so that they should be dehydrated before attempts at valorization. In this respect, the costs of drying, storing, and transport are considerable and may well limit such attempts (Oreopoulou and Tzia, 2007).

2.3 By-Product

Article 5 on by-products EU Directive (Directive, 2008) mentions, "substances or objects resulting from a production process not primarily aimed at producing such substances or objects are by-products and not waste." A by-product is not a planned product and is produced after carrying out a given process. From an economic point of view, in some food industries, by-products have no value, a negative value (elimination costs money!), or little economic value. Ningrum et al. (2018) mentioned that most food by-products or food waste are produced by households, while approximately 39% of losses occur in the food manufacturing industry. For these reasons, by-products in some sectors are considered as a "waste" with negative connotations. The decision that a substance is not a waste can only be taken based on a coordinated approach, has to be regularly updated, and should be consistent with the protection of the environment and human health. The use of a by-product to obtain value added products must be legally permitted and any processing must have no adverse environmental or human health impacts.

Martins and coworkers (2017) reported that food industrial by-products differ in terms of their origin and include a wide range of components: peel, stem, leaf, seed, shell, bran, kernel, pomace, and oil cake. Figure 10.1, for example, shows the fruit by-product (peel) that remains after extracting juice from citrus fruits, which normally constitutes about 20–25% of the fresh fruit weight, with a high moisture content and carbohydrates, mainly sugars and dietary fibers (Paraman et al., 2015). Fruit by-products are treated as an agro-industrial waste with very little or no economic value, and are considered as a big environmental problem since they contain large amount of water [66.4–78.2%, wet basis (wb)] and fermentable sugars (3.6%, wb), and, as a result, their direct disposal in soil creates environmental concerns due to uncontrolled fermentation and high chemical oxygen demand (COD) (300 g COD/kg pomace) during degradation. Alternative techniques to avoid such damage are expensive and may discourage companies from applying them except if they can obtain high value-added compounds in return for use as bioactive substances in the food, chemical, or pharmaceutical industries.

Figure 10.1 Citrus peel as by-product from the citrus juice industry.

2.4 Food Coproduct

Several definitions of this term can be summarized as follows: a product obtained from the main product (juice, fruit pulp, etc.) and having an equally important economic value. It is valorization of by-products to obtain valued-added products, such bioactive compounds and dietary fibers that led to their being designated coproducts. For example, in pomegranate juice processing, juice is the product and bagasses, pulp, and husk are the coproducts, which are used as raw material for the extraction of high valued-added compounds (antioxidants, food coloring, antimicrobials, dietetic fiber, pro/prebiotics, etc.). Indeed, such is the importance of these products that we may even speak of joint production.

Pérez-Chabela and Hernández-Alcántara (2018) defined coproducts as "wastes" or residues with a high organic load, which are obtained or derived from the processing of foods (fruit, cereals, meat, and dairy) or during postharvest handling or storage; Hall and Kononoff (2011) included, in this definition, dietary fiber and bioenergy products for human consumption. These authors mention that coproducts generally have high amounts of fiber, are rich in protein, and high in energy. According to Pérez-Chabela and Hernández-Alcántara (2018), agri-industrial coproducts are cheap and an excellent source of functional ingredients, which, after suitable processing, can be used for human consumption. The use of these coproducts as food ingredients decreases the environmental impact if they are not correctly disposed of, which consumers find praiseworthy organic matter. In the same manner, fruit peels are cheaper underutilized resources. Coproducts can be an important source of income for companies which otherwise would be expensive to eliminate.

2.5 Bioactive Compounds

According to Amiot and coworkers (2012), "bioactive" or "bioactive compounds" can be defined as compounds that can have a positive effect on chronic pathologies, although different levels of proof have been offered.

Several authors mentioned that bioactive compounds might have antioxidant properties. Dangles (2012) mentioned that it is generally recognized that the health effects of bioactive compounds are mediated by specific interactions of their circulating metabolites with cellular targets but not antioxidant activity per se. However, the antioxidant properties may be relevant in food or cosmetic stabilization, because they can be used for replacing synthetic antioxidants, such as butylated hydroxytoluene, or for the prevention of oxidation of polyunsaturated fatty acids in the gut (Brewer, 2011).

The recovery of such commodities from by-products appears advantageous and generates economic and environmental benefits (Rabetafika et al., 2014). Several researchers (Baiano and Del Nobile, 2016) describe methods that are suitable for obtaining bioactive or functional components for food uses—mostly molecules with functional (color, antioxidant, aroma, or taste), physiofunctional, and technofunctional properties. However, Renard (2018) mentioned that the use of bioactive compounds as isolated molecules has proven disappointing.

2.6 Intermediate Food Products

Intermediate food products (IFP) are those ingredients or complementary products that fulfil specific functions, for which they are incorporated into food during processing. From a technological point of view, most of them can be considered as food (e.g., date paste) (Martín-Sánchez et al., 2014b). Their application in food represents high added value and clearly differentiates food products from functional food. IFP are edible products, not necessarily nutritious, which are not basic raw materials of industrialized foods (meat, milk, fruit, and eggs), nor are they consumed directly, but they come from transformations of these basic raw materials (from by-/coproducts) and are used for facilitating the industrial elaboration of potentially functional foods (Bonet et al., 2000). Figure 10.2 depicts a food intermediate product obtained from the valorization of citrus coproducts (IPOA-UMH orange fiber) Fernández-López et al. (2004).

2.6.1 Processes for the Valorization of Co/By-Products The importance of food fiber has led to the development of a large and potential market for fiber-enriched foods (King et al., 2012). There is ongoing interest to find new sources of dietary fiber and fiber-

Figure 10.2 IPOA-UMH orange intermediate food product (IFP) rich fiber.

associated bioactive compounds, thus contributing to their valorization (Banerjee et al., 2017) and improving the profitability of industries—such as the agro-industrial industry, where coproducts have traditionally been undervalued (López-Marcos et al., 2015a). Many of the fruit bioactive compounds useful for the development of functional foods are found as secondary metabolites in these coproducts, where they tend to be concentrated in the outer, least palatable parts of fruits and vegetables and are, therefore, present in high concentrations when the coproducts are processed (Renard, 2018). Both researchers and the food industry need to take into account the high variability of such resources when attempting to recover these valued-added compounds. Garcia and coworkers (2009) reported that concentrations of target compounds are likely to vary, sometimes more than 10-fold, depending on the fruit exact variety and its maturity stage. To obtain bioactive compounds from fruit coproducts, the choice of extraction method is very important, and this will depend on many factors. Ramandi and coworkers (2012) reported that these factors might include effectiveness, capital cost, operating cost, simplicity of operation, and waste production. The same authors mention that in the past, many extraction methods, such as water (cold and hot) extraction (Contreras-López and Salvá-Ruiz, 2018), hot expelling, cold pressing, Soxhlet, or extraction by organic solvents, were widely used for these purposes. However, these techniques are time-consuming and require large volumes of expensive organic solvents, so that they cannot be considered environmentally friendly.

3 Citrus Fruits

Citrus is the most abundant crop in the world, and one-third of the crop is processed. Oranges, lemons, grapefruits, and mandarins represent approximately 98% of the entire industrialized crop, oranges being the most abundant (approximately 82% of the total) (Marín et al., 2007). A possible way to valorize citrus peels and seeds, which are by-products of the juice extraction industry, is to use them as natural antioxidants. Safari et al. (2018) suggested other possibilities for valorizing citrus by-products in the form of nanoparticles, biosorbents, or biofertilizer.

The processing of by-products into fiber also affects those parameters known to determine the nutritional, functional value of a product. In the studied samples, losses in soluble dietary fibers, flavonoids, ascorbic acid, and total antioxidant activity were observed, while insoluble dietary fiber levels increased.

Although more research is needed into the processing of citrus waste into fibers to optimize their properties, the broad composition of the different citrus industrial wastes, together with the changes produced during processing and an appropriate blending, might allow the design of specific fibers for different purposes. For example, pectin (from citrus valorization) content of food industry by-products was 2- to 10-fold higher than that obtained from the chemical industry. In CIB-1, 2, 3, and 4, pectin contributed 4 to 16% to the total amount of chemical fibers, while in food industry by-products, the percentage ranged from 25 to 30% (Marin et al., 2005).

The process of obtaining citrus intermediate products can be summarized in these steps: reception of the raw materials, mainly from citrus juice industries, grinding, washing, extraction, purification/concentration, and drying at low temperatures (less that 60 °C). The use of a suitable method for citrus extraction is important to maximize the yield and the quality. In this regard, factors, such as extraction conditions (temperature, pressure, etc.), solid/liquid ratio, type of solvent and polarity, and particle size, are known to affect the extraction kinetics (Grassino et al., 2018). Lario and coworkers (2004b) and Sendra et al. (2008a) described methods to obtain citrus fibers as IFP from citrus by-products. Fernandez-López and coworkers (2009) developed a stable high dietary fiber powder from orange by-products. This product has large amounts of dietary fiber (71.62 ± 0.24 g 100 g^{-1} dry sample) and high polyphenol content (40.67 ± 0.45 mg g^{-1} dry sample). Hesperidin was the major phenolic compound (higher than 50%). Another by-product for citrus is the orange juice wastewater generated in the process of obtaining citrus fiber (Viuda-Martos et al., 2011), which is rich in bioactive compounds. The most representative of these bioactive compounds contained in washing waters are narirutin and hesperidin (phenolic compounds), glucose (sugars), and succinic acid (organic acids). Bitter citrus (orange and pomelo) have also been studied as a source of dietary fiber and bioactive compounds (Coksever and Saricoban, 2010; Soriao et al., 2015).

3.1 Technofunctional Properties

O'Shea and coworkers (2015) described orange pomace as a fibrous ingredient with applications suited to products requiring improved water/oil-holding and binding properties, that is, a high water hydration capacity (4.40 ml/g). It had a favorable nutritional composition: high dietary fiber (40.47%), low fat (2.14%), and high mineral content.

Water- and oil-holding capacities are two easy-to-measure concepts for use in dietary fiber application and in the development of new foods. Marin and coworkers (2007) studied the relation of both properties and found that the water-holding capacity is related to the soluble dietary fiber content of citrus fibers, and that the oil-holding capacity is correlated with lignin content. Similar results were obtained by Saikia and Mahanta (2016), but their studies also included swelling (which were associated with insoluble dietary fiber) and oil-holding capacities. Canela-Xandri and coworkers (2018) reported that when citrus fruit by-products are enzymatically treated, the water-holding and swelling capacities increase, while they also show greater viscosity.

Lario and coworkers (2004b) mentioned that water-holding capacity was enhanced by washing (7 g water/g non-washed fiber powder; 12.6 g water/g washed fiber powder) and slightly decreased by a reduction in fiber particle size.

Oil-holding capacity (6.7 g oil/g fiber powder) was not affected by these factors. Washing prevented fiber browning during drying as reflected in color parameters.

Yi et al. (2014b) described how the fermentation of citrus by-products improved the dietary fiber recovery process (77.91 g/100 g). These authors mentioned that fermentation favors the enhancement of water retention (13.31 mL water/g powder), swelling capacity (8.55 mL/g powder), and oil-holding (8.37 g oil/g powder) capacity, while improving the heavy metal-binding ability (Cu^{2+}, 27.54 μmol/g; Pb^{2+}, 42.85 μmol/g; Cd^{2+}, 30.92 μmol/g).

Knowing these properties and their correlation with the citrus compounds present in dietary fibers (obtained from by-products), these results may help in the design of better foods in the future by adjusting the contents of these constituents. The fiber samples exhibited antioxidant activity.

3.2 Physiofunctional Properties

Citrus fruit peels showed a significant decrease in blood cholesterol, triacylglyceride, and glucose. Mandarin peels decreased cholesterol more than other citrus fruits (Fayek et al., 2017). Saikia and Mahanta (2016) reported that citrus fiber showed good results for glucose adsorption and amylase activity inhibition, the glucose diffusion rate index, in the digestive system.

3.3 Bioactive Compounds

Citrus fruit and juices represent one of the main sources of compounds with high health-promoting properties. Fresh lime by-products contained significant amounts of vitamin C, phenolic compounds, and flavonoids (Jongaroontaprangsee et al., 2014). Hesperidin was a major flavonoid, and only one polymethoxyflavone (tangeretin) was detected in small amounts in oranges. Omar et al. (2016) mentioned that hesperidin, a naturally occurring flavonoid, exerts many clinically appreciable effects, including antioxidant, anti-allergic, and anti-inflammatory actions. As regards hesperidin, Li and Schluesener (2017) mentioned that this flavanone (also its aglycone hesperetin) has pharmacological activities and widespread application in diseases, such as neurological disorders, psychiatric disorders, and cardiovascular diseases and others, due to its anti-inflammatory, antioxidant, lipid-lowering, and insulin-sensitizing properties. Barreca et al. (2017) reported that citrus bioactive compounds (hesperetin, naringenin, eriodictyol, isosakuranetin, and their respective glycosides) also enhance the body's defenses against oxidative stress.

Louati et al. (2019) mentioned that to increase the availability of these active molecules the cell walls should be broken during valorization of by-products. Also, Jongaroontaprangsee and coworkers (2014) mentioned that vacuum drying (80 °C) is the most suitable method for preparing dietary fiber powder from lime residues due to the short drying time required and its ability to retain bioactive compounds.

Vacuum drying lime fiber has the potential to reduce blood glucose and cholesterol levels by exhibiting high glucose retardation index and bile acid retardation index.

3.4 Healthy Citrus-Based Ingredients: Application to Dairy- and Muscle-Based Foods

The demand for healthier foods by consumers has driven the food industry to develop foods that contain less fat and sugars while adding dietary fibers and increasing the content of healthy bioactive compounds without altering the sensory characteristics and nutritional properties. The need to increase fiber intake and the image of a fiber-added food as a healthy food has good acceptance. To satisfy these demands, citrus is a good option.

3.4.1 Dairy Products The ice cream industry uses high quantities of fat in their formulations, on average about 10% (Boff et al., 2013). These authors mentioned that ice cream with added citrus peel (0.74 and 1.10%) obtained acceptance rates of 78% and obtained a fat reduction of more than 25%, by claiming these products were "light." Citrus-based fibers, mainly from orange and lemon, have been applied to several products; for example, orange fiber was applied to yogurt (Lario et al., 2004a) studied its influence on physical properties, such as rheology was studied (Garcia-Pérez et al., 2006). In the mentioned study, the fiber was added and its concentration rheology was modified. Thus, the perception of creaminess was reduced and acidity perception increased. The addition of 1% fiber-enriched yogurt reduced syneresis and improved the textural properties of yogurt (gel firmness increased, stickiness), color (Garcia-Pérez et al., 2004), and microbiology (Lario et al., 2004a). The effect of fiber pasteurization was clear, since pasteurization provided counts of total viable microorganisms, streptococci, molds, and yeasts similar to those without fiber, the absence of enterobacteria, and populations of lactobacilli were lower (0.2 log units) in fiber yogurt. Non-pasteurized fiber caused unacceptable counts of molds and yeasts after 21 days of cold storage (Sendra et al., 2008b). In this work, authors mentioned that citrus fiber in fermented milks enhanced bacterial growth and the survival of probiotic bacteria (*Lactobacillus acidophilus* CECT 903 and *Lactobacillus casei* CECT 475). The study indicated that citrus fiber-enriched fermented milk has good acceptability and is a good vehicle for a variety of commercial probiotics. Yi and coworkers (2014) reported that micronized orange fiber DF showed better physicochemical and functional properties than other orange fiber, and that 2% of this micronized fiber can act as a fat replacer in yogurt, retaining most of the textural and sensory properties of full-fat yogurt, without sacrificing good taste and other qualities of full-fat yogurt.

Other findings (unpublished data) mention that the addition of lemon fiber to rennet coagulated fresh cheese and cream cheese increased cheese yield and reduced syneresis but did not affect cheese shelf-life. The overall acceptability of these cheeses was better for cream cheese than for rennet coagulated cheese.

3.4.2 Citrus-Derived Compounds in Meat Products There is more information about the use of citrus-derived compounds in meat products, where they are sometimes used for fat replacement, fresh, cooked, and dry-cured meat products. (i) Fresh meat products: Baioumy and coworkers (2014) used orange albedo as fat replacer and thus decreased the caloric value of meat products. The same study mentioned that the technofunctional properties, water-holding capacity, and cooking yield increased as the level of added albedo increased. In a sensory evaluation, acceptability was improved by using 5 and 7.5% of orange albedo as a disadvantage; burger with added albedo was less hard than control treatment. Alesón-Carbonell and coworkers (2005) reported that lemon albedo (cooked or/and dried and minced) could be added to beef burger. The type of albedo slightly influenced the lipid oxidation and pH, while the cooking properties were significantly improved by some of the added albedo samples (increased yield between 5.5 and 11%) compared with control samples. (ii) cooked meat products: Fernández-Ginés et al. (2003) used orange citrus fiber as an ingredient in a bologna sausage elaboration process. The same authors (2004) also incorporated lemon albedo (fresh, cooked, dehydrated fresh, and dehydrated cooked albedo) as coproduct in bologna sausages. The use of lemon albedo might be considered as a potential dietary fiber source for this type of meat product since it enhances the functional and technological properties (Sariçoban et al., 2008). A very interesting finding for the application of citrus by-products or coproducts in the meat industry was reported by Viuda-Martos and coworkers (2010), who mentioned that citrus industry coproducts (albedo, high dietary fiber powder, and citrus washing water, from citrus fiber processing by-product) are all good sources of biocompounds. All of them may help to reduce the residual nitrite level in meat products. Of these coproducts, citrus fiber shows the highest potential for reducing the residual nitrite level, followed by the albedo and finally, the washing water. The combination of essential oils (oregano) with orange dietary fiber improves the shelf-life of cooked meat products (e.g., Bologna type sausages) (Viuda-Martos et al., 2010a). Similar results were obtained in Mortadella type sausage; a sensorial analysis showing that the sausage containing orange dietary fiber and rosemary essential oil was more appreciated by consumers (Viuda-Martos et al., 2010b). For low-fat cooked meat products, citrus fiber improved the water-binding ability by increasing the hydrophobic interactions (myofibrillar proteins) and changing the sausage microstructure (Song et al., 2016). Gupta and coworkers (2015) reported a disadvantage associated with citrus fiber addition to emulsified meat products, citing that the addition of orange pulp decreased the emulsion stability and cooking yield. But this problem could be avoided by incorporating 1% carrageenan in the formula. (iii) Dry-cured meat products: There are several studies related to the incorporation of citrus fibers in this type of meat product. Fernández-López et al. (2007) manufactured dry-cured sausages with orange fiber obtained from juice industry by-products at different concentrations (2.5, 5.0, 7.5, and 10%). The use of this fiber significantly reduced the residual nitrite level. In another work, Fernández-López and coworkers (2007) reported that

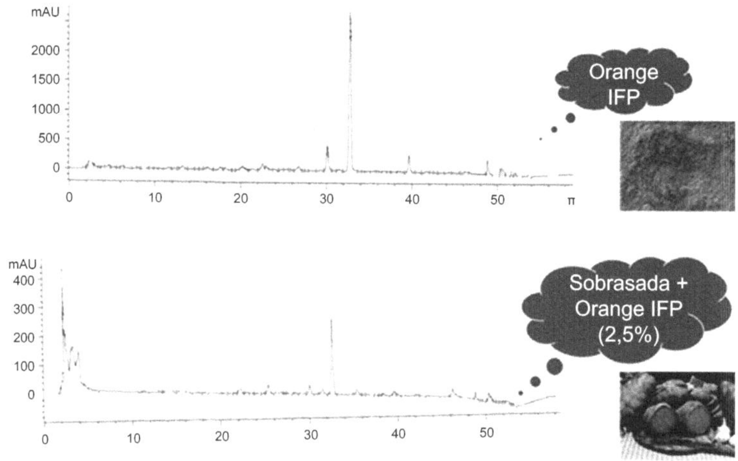

Figure 10.3 Chromatograms of citrus intermediate food products (CIFP) and Sobrasada (paprika-added dry-cured sausage) added with CIFP.

hesperidin is the most suitable compound for monitoring changes in polyphenols in sausages with orange fiber added during processing (Figure 10.3). In a traditional Spanish dry-cured sausage (*Salchichón*), the addition of albedo improved the nutritional properties as a result of fiber addition, also delayed oxidation (based on TBARS values), and the sensory properties of sausages contained up to 50 g/kg dehydrated raw albedo and 75 g/kg dehydrated cooked albedo, were well accepted by untrained panel. Aleson-Carbonell et al. (2003) found similar results in *Longaniza de Pascua*, another Spanish dry-cured sausage specialty (Aleson-Carbonell et al., 2004). Sayas and coworkers (2012) added orange dietary fiber powder (1%) with probiotic lactic acid bacteria (*L. casei* CECT 475) to a dry-cured sausage (Longaniza de Pascua). The combination of orange fiber and probiotic starter culture showed several advantages including a faster fall in pH (reducing dry-maturation stage) and the release of organic acids and glycerol. The resulting sausage obtained the best sensory scores and provided the best guarantees of food safety, low pH and water activity, decreased counts of enterobacteria, and reduced residual nitrite compared with other types of sausage.

In another type of dry-cured sausage, Sucuk, a Turkish dry-fermented sausage, 2% of added orange fiber improved the texture, color, odor, taste, and general acceptability (Yalinkiliç et al., 2012). For this type of meat product, Coksever and Saricoban (2010) reported that low concentrations (<5%) of bitter orange fiber enhance the nutritional quality characteristics as it acts as a source of fiber.

4 Pomegranate (*Punica Granatum* L)

Pomegranate is a primeval, mystical, and distinctive fruit of heuristic medicinal value (Chaudhari et al., 2013). In several cultures, pomegranate is considered as a "superfruit." The importance of superfruits as a necessary part of the human diet and as cash crops is well recognized. This term was originally used by the food industry to indicate fruit with exceptional nutrient richness and nutritional qualities, including a high concentration of phytochemicals (Sau et al., 2016). Abid et al. (2017) and Yang et al. (2018) reported a method to obtain pectin from pomegranate peels. It is well known in pharmaceutical sciences that the phenolic compounds from pomegranate have been documented as possessing numerous prophylactic and therapeutic qualities against various pathological infections as well as noninfectious disorders (Saeed et al., 2018). Pomegranate possesses antioxidant, anti-inflammatory, antiproliferative, antiangiogenic, anti-invasive, and antimetastatic properties, and it induces apoptosis (Khwairakpam et al., 2018). Punicalagin present in several pomegranate compounds (among others) is used as a drug.

Pomegranate husk, a by-product of the pomegranate juice industry, is an inexpensive and abundant source of ellagic acid; 3.5% of ellagic acid can be obtained from this by-product. This acid, obtained by the extraction of tannins, is widely used as a functional food for its physiological functions, and is suitable for wide applications in the food industry (Jingjing and Qipeng, 2008). Singh and coworkers (2018) mentioned that pomegranate peel, a juice by-product, represents around 30–40% of the fruit. The same authors mentioned that pomegranate peel contains the main phenolic compounds reported in the literature: flavonoids (anthocyanins, such as pelargonidin, delphinidin, and cyanidin along with their derivatives, and anthoxanthins, such as catechin, epicatechin, and quercetin), tannins (ellagitannins and ellagic acid derivatives, such as punicalagin, punicalin, and pedunculagin), and phenolic acids (such as chlorogenic, caffeic, syringic, sinapic, *p*-coumaric, ferulic, ellagic, gallic, and cinnamic acid).

The commercial promotion of antioxidants of plant origin, which can be used for improving the properties of food, both for nutritional and preservation purposes, is of growing interest (Rice-Evans et al., 1997).

4.1 Pomegranate-Based Health-Related Ingredients in Muscle-Based Foods

The addition of nonmeat ingredients to food products can modify the technological, nutritional, and sensory characteristics of these products. As a new ingredient is added, it can disrupt the normal evolution or processing because it can change emulsion stability, protein precipitation, color, and so on. This is especially important in polyphenol-rich fiber since they can act as pro-oxidants (Perez-Alvarez, 2008).

Lopez-Marcos and coworkers (2015b) reported that pomegranate fiber-rich powder (IFP) obtained from valorization processes showed good cholesterol retention (32% cholesterol retained) due to its healthy properties and ability to adsorb cholesterol. Recent studies related to pomegranate seed (20% w/w of the whole fruit) pointed to their potential as a good source of nutrients and antioxidants. It has been suggested that dietary supplementation with pomegranate seeds may prevent DNA damage (Guo et al., 2007), reduce the risk of cancer, and alleviate menopausal symptoms (Keşkekoğlu and Üren, 2014).

4.2 Pomegranate-Derived Compounds in Meat Products

Devatkal et al. (2010) reported that pomegranate rind and seed powders have the potential to be used as natural antioxidants in meat products and can be used in goat meat patties. These authors reported that both the pomegranate powders are bioactive-rich compounds (phenolics) with free radical scavenging activity. The TBARS values during refrigerated storage (4 ± 1 °C) were significantly lower in goat meat patties with added pomegranate rind powder and higher in those with pomegranate seed powder added. Goat meat patties with both pomegranate powders added were darker than the control. Similar results were obtained in salted chicken patties (Devatkal et al., 2011). Another study (Naveena et al., 2008a, 2008b) using pomegranate rind powder extract in cooked chicken patties found that 10 mg equivalent phenolics/100 g meat would be sufficient to protect against oxidative rancidity for longer periods than the most commonly used synthetic antioxidant, such as BHT. Bouarab-Chibane and coworkers (2017) used pomegranate extract for beef burger preservation. These authors found that pomegranate peel extracts retarded lipid oxidation (reduced TBARS values level) during 12 days' storage, diminished color variations, and maintained the redness index of meat. Andrés et al. (2017) mentioned that red grape and olive pomace extracts have better antioxidant activity than pomegranate. Shah et al. (2014) reported that pomegranate extracts exhibited better antioxidant properties than some synthetic compounds (butylated hydroxyanisole, butylated hydroxytoluene, tert-butylhydroquinone, and propyl gallate) commonly used in meat products. This extract improved the overall sensory and nutritional quality of meat and meat products and their shelf-life through the inhibition of lipid oxidation and degradation of meat pigments, while also delaying the onset of rancid flavors and stabilizing the color of meat. Other pomegranate products (juice) and coproducts (rind and seed extracts) were also added to beef patties, without any changes in sensorial scores (Ibrahim et al., 2012).

Pomegranate seed extract (0.5%) added to beef and chicken meatballs during cooking (oven roasting, pan cooking, charcoal-barbecue, and deep-fat frying) inhibit (68%) the formation of heterocyclic aromatic amines. The total heterocyclic aromatic amine formation was reduced by 39 and 46% in beef meatballs cooked by charcoal-barbecue and deep-fat frying, respectively. In deep-fat-fried chicken meatballs with pomegranate added, the total heterocyclic aromatic amine formation was inhibited by 49%, but increased by 70% after oven roasting (Keşkekoğlu and Üren, 2014). Narsaiah and coworkers (2011) used a blade tenderizer and pomegranate seed powder for improving the textural properties of goat meat, where toughness is a major problem with regard to consumer acceptance. However, combining blade tenderization and pomegranate seed extract (4%) improved tenderization but had the disadvantage of a slight adverse color change and taste. The authors mentioned that pomegranate seed powder might be considered for mixing with other spices to marinate goat meat, mainly for its beneficial effects.

Qin et al. (2014) studied a new process for the meat industry, involving extracts, bioactive compounds, and antioxidant extracts in biofilms. For example, a chitosan film containing pomegranate rind powder was used as an active packaging material for pork meat patties prior to storage (4 ± 1 °C, 20 days). These authors achieved a microbiological shelf-life extension of 8 days. Wrapping with chitosan-pomegranate rind powder retarded increase in TBARS value and metmyoglobin content during storage. The same samples also gradually released phenolic compounds during storage, enhancing the antioxidative and antimicrobial activities of the film and maintaining the quality and shelf-life of pork meat patties. The samples wrapped in this way kept an acceptable sensory quality throughout the storage.

The effect of vacuum packaging and pomegranate peel extract on ground goat meat and cooked nuggets during refrigerated storage (4 ± 1 °C) was studied by Devatkal and coworkers (2012). These authors reported that vacuum packaging along with 1% pomegranate peel extract resulted in a more stable color than meat stored in atmospheric packaging. Also, TBARS values were significantly lower in samples with vacuum and pomegranate extract. This study, therefore, showed that vacuum packaging with pomegranate peel extract has a synergistic antioxidant effect and vacuum packaging extended the refrigerated shelf-life of goat meat and nuggets.

5 Persimmon (*Diospyros Kaki* Thunb.)

This fruit has recently aroused great interest for its health-related properties. As a minority fruit crop in the Mediterranean area, there has been less research into the composition of persimmon (fibers, vitamins, minerals, antioxidants, and other phytonutrients that can improve health) (Sau et al., 2016). As a colored fruit, persimmons have bitter and astringent compounds that may make them undesirable, especially when processed as totally immature fruit (Beaulieu and Stein-Chisholm, 2016).

Consumer demand for natural and naturally processed food products is increasing worldwide. From this point of view, fruits and vegetables and their products are one of the most important food groups. Persimmons and derived products with this fruit are attractive to many people due to their unique flavor and aroma (Kursun and Karaca, 2018). Persimmon is rich in carbohydrates, carotenoids, phenolic compounds, such as condensed tannins, vitamins A and C, minerals, and dietary fiber components with high antioxidant (reducing oxidative stress) characteristics. To obtain better quality dietary fibers from persimmon, Akter and coworkers (2010) recommended the use of hot water washing.

Lucas-González et al. (2017) obtained a coproduct from the persimmon juice industry that was processed as flour. This persimmon IFP is rich in carbohydrates (38.07–46.98 g/100 g) and total dietary fibers (32.07–43.57 g/100 g), insoluble dietary fiber representing more than 68% of the total dietary fiber content.

5.1 Technofunctional Properties

The persimmon IFP showed high values for water-holding capacity (9.47–12.19 g water/g sample) and emulsifying activity (69.33–74.00 mL/mL), data that are interesting for dairy- and muscle-based food products.

5.2 Physiofunctional Properties

Persimmon has many beneficial effects mainly for the digestive and cardiovascular systems (Kursun and Karaca, 2018), since persimmon peel might possess a potential anti-inflammatory (Rashed and Pan, 2014) and antiatherogenic effect (Son et al., 2013). Matsumoto et al. (2014) reported that persimmon dietary fiber is rich in tannin and has bile acid-binding properties. The bile-holding capacity and fat/oil-binding values were analyzed by Lucas and coworkers (2017), who obtained a persimmon IFP with a higher bile-holding capacity and fat/oil-binding capacity (9.78–19.61 g water/g sample) than THF (16.12–12.40 g bile/g sample and 9.78–7.96), respectively.

Heat treatment of persimmon dietary fibers reduced the bile acid-binding ability but excreted double the amount of fecal bile acid (Matsumoto et al., 2014). The bile acid-binding compounds are used for the treatment of hypercholesterolemia (Gato et al., 2013). Martínez-Las Heras et al. (2017) mentioned that the passage through the intestine enhanced the persimmon polyphenols and flavonoids solubility from the fruit.

5.3 Persimmon-Based Healthy Ingredients: Application to Dairy- and Muscle-Based Foods

High valued-added Persimmon compounds are of great interest for Mediterranean food industries, and companies and research institutions work together to develop new food products.

5.4 Persimmon-Derived Compounds in Meat Products

Persimmon (*Diospyros kaki* Thunb.) coproducts are good sources of valuable ingredients for the food industry. Thus, the high amount of fiber, sugar, minerals, and bioactive compounds make it a good candidate for use in the meat industry. Lucas-Gonzalez and coworkers (2018) reported that persimmon flours (obtained from co/by-products of the persimmon juice industry) were well accepted by consumers when incorporated in pork liver pate. Technologically, the addition of persimmon flour improves product redness and decreases residual nitrite levels and lipid oxidation. Addition of flour improves the acceptability. Choe et al. (2017) added persimmon peel extracts to raw ground pork and found them a good tool against lipid and protein oxidation of raw meat products during refrigerated storage.

5.5 Persimmon-Derived Compounds in Dairy Products

Carotenoids from persimmon represent a good alternative for obtaining functional dairy products (García-Cayuela et al., 2018). Persimmon leaf extracts have also been used in yogurts, where Young and coworkers (2016) reported they increased the acidification rate and reduced the time to complete fermentation (pH 4.5). They also mentioned that persimmon leaf extracts enhanced the antioxidant activity and generally increasing its acceptability.

6 Date Palm Fruit (*Phoenix Dactylifera L.*)

Although date palm fruit is not considered a "superfruit," evidence concerning its healthy properties has increased in the literature (Pérez-Alvarez, 2016). The date palm, a tropical and subtropical crop, is one of mankind's oldest cultivated plants (Chandrasekaran and Bahkali, 2013). Date palm is the main fruit crop in arid regions of southern Mediterranean countries and the Arabic Peninsula, Iraq, Iran, and Pakistan. The only country along the northern Mediterranean coast with commercial date crops is Spain (Martín-Sánchez et al., 2014a). It is a well-known fruit worldwide, and in Muslim countries, its consumption is frequently guided by Islamic traditions (Karizaki, 2017).

According to Al-Farsi and Lee (2008) and Barreveld (1993), dates have different ripening stages, each with its very different properties; in general, there are four maturity stages, known as Kimri (turgid, astringent, green, and up to 85% moisture), Khalal (hard, juicy, sugar content increases, astringency decreases, color turns yellow or red, and reaches maximum size and weight; some cultivar are edible in this stage), Rutab (semi-ripe, moisture and astringency loss, softer texture, and a darker color), and Tamar (dried stage, fully ripe, 20% moisture, and brownish color). Chandrasekaran and Bahkali (2013) reported that most dates are consumed in the Rutab and Tamar stages, with little or no processing. However, in some soft date cultivars or in adverse climatic conditions, fruits do not reach the Rutab and Tamar stages. Other cultivars are used only for industrial purposes, such as the Spanish Confitera cultivar (Martín-Sánchez et al., 2014a).

Date fruits come from many cultivars in international commerce; Medjool and Deglet-Nour are the most popular. The quantities of processed dates have been rapidly and steadily growing in recent years, which has led to a very big environmental problem. Despite high production, date flesh of low quality (second-grade dates or date by-products) are underutilized (Hasnaoui et al., 2013; Ghnimi et al., 2017) or its by-/coproducts. Martin-Sánchez and coworkers (2014b) reported that underutilized date flesh could be useful as an ingredient in functional foods. The main chemical components of date fruit include carbohydrates, dietary fiber, enzymes, proteins, fats, minerals, vitamins, phenolic acids, and carotenoids (Tang et al., 2013). These authors mentioned that date by-products could be used as raw materials to obtain value-added products, such as organic acids, exopolysaccharides, antibiotics, or food, such as date-flavored probiotic-fermented dairy product, ice cream, or meat products. Borchani et al. (2012) studied their possible use as a potential fiber source in the enrichment of food. Recently, Ghnimi and coworkers (2017) reported that the industrial use of date fruits is based on their distinct phenotypical and biochemical features, which are responsible for their shape, taste, flavor, and nutritional value. Date fiber can be obtained from different sources. For example, Hashim et al. (2009) obtained this fiber from a date syrup by-product. Sánchez-Zapata et al. (2011) obtained a date paste, rich in sugars (53 g/100 g), especially fructose and glucose (reducing sugars), and dietary fiber from fresh dates during processing, mainly obtained during its picking, storage, or conditioning, with significate date losses. Rosa-Alcaraz and coworkers (2017) mentioned that date palm by-products are an important source of phytochemicals. This rich bioactive compound fruit can be useful for obtaining potentially interesting ingredients for the food industry. Martín-Sanchez et al. (2014) reported that Confitera cultivar is a good source of sugars, especially reducing sugars, dietary fiber, minerals, and natural antioxidants. This cultivar is suitable for industrialization, since a high yield (80%) of date paste (IFP) would be an advantage for the industrial processing and it might be successfully used in several food matrixes.

6.1 Date Pits

Date pits represent an average of 10% of the date fruits (Barreveld, 1993). Traditionally, pits are used as a source of feed for small ruminants to overcome the shortage of animal feedstuffs (Dayani et al., 2012). According to Elgasim et al. (1995), date pits contain natural anabolic agents that can act in a similar way to estrogens, since an aqueous extract induced by uterus contraction is not blocked by atropine or cyproheptadine drugs (Amany et al., 2012). The addition of date pits extract (0.50, 0.75, and 1.00%) to ground beef before refrigerated storage (0 °C, 10 days) reduced lipid oxidation, thus improving the shelf-life.

6.2 Technofunctional Properties

Elleuch and coworkers (2008) obtained a dietary fiber concentrate from Tunisian dates (Deglet-Nour and Allig cultivars). The authors in their extraction process obtained a 67% of yield, with high concentration of dietary fiber between 88 and 92.4% (DM). These concentrates had good technofunctional properties, especially their water- (~15.5 g water/g sample) and oil-holding capacities (~9.7 g oil/g sample); rheology of dates fiber concentrates have pseudo-plastic behavior for their food suspensions. Date fiber concentrates were also studied by Borchani and coworkers (2012), who observed interesting functional characteristics in oven-dried date fibers. Water-holding, oil-holding (2.73–4.60 g oil/g dry fiber), and emulsifying capacities (5.93–12.87%) are appropriate for industrial purposes. Hasnaoui et al. (2013) reported that date dietary fiber concentrates could be incorporated in some food products to modify textural properties and stabilize high amounts of fat.

6.3 Application of Date-Based Healthy Ingredients to Dairy and Muscle-Based Foods

6.3.1 Meat products Dates are sparsely used in the Spanish meat industry, and only a few specialized products contain them. The most well-known use is in "datil con bacon" dates with bacon, for oven cooking or deep-frying. Indeed, dates from Confitera cultivar at Khalal maturation stage are ideal for meat products, which need antioxidants but are not sweet in taste (Martín-Sánchez et al., 2014b).

Martín-Sánchez and coworkers (2013) developed a Campagne type pork liver pâté using fresh date by-products at several concentrations (5–15%) to improve shelf-life. This work found that the addition of 10% of dates was enough to avoid lipid oxidation throughout the storage times assayed. Color was the most affected parameter, but overall acceptability was better in products with added date. Sánchez-Zapata et al. (2011) used an IFP date paste in the elaboration of a cooked meat product; unlike other fruit coproducts, date IFP did not affect the nitrite content. The addition of 5% date IFP with dry-cured sausages yielded a product with physicochemical properties similar to the traditional one. Although instrumental color differences were found, these were not detected visually by panelists and the sensory properties were evaluated positively (Martín-Sánchez et al., 2014c).

Therefore, the addition of these potentially functional ingredients obtained from date coproducts could be an economical way of developing healthy meat products, while increasing the eco-efficiency of the food industry.

6.3.2 Dairy products Martín-Sánchez et al. (2014b) described that when an IFP was elaborated from dates in the Confitera Rutab maturation stage, it had higher sugar content but a lower phenolic content, and so it could be used as a natural sweetener in dairy products. Hashim and coworkers (2008) described that yogurt fortified with date fiber did not affect yogurt acidity, although the pH increased. Products had a firmer texture (higher hardness values) and darker color (lower L^* and higher a^*) compared with the control. Fiber fortification affects consumer tests, especially appearance, color, and flavor ratings. Addition of 3% of date fiber showed similar sourness, sweetness, firmness, smoothness, and overall acceptance ratings as the control yogurt. Trigueros et al. (2012) used Confitera date blanching water for reconstituting skim milk powder to produce low-fat yogurt. These authors obtained it in high populations of lactic acid bacteria and the texture was of soft-gel type with a weak structure.

7 Conclusion

Innovation policies of food companies are strongly influenced by the legal frameworks concerning the valorization of fruit coproducts. The high costs of development and the uncertainty of the investment return are obstacles for research, development, and/or innovation in fruit valorization. Companies must plan several strategies for recovering the valued-added compounds and dietary fibers to re-introduce into food products and produce healthier products that satisfy consumer demand. It is necessary to continue the search for locally available natural sources of functional ingredients: for example, fruits traditionally consumed in the Mediterranean area, such as quince, mulberries, jujube fruit, figs, and cactus pear, might be suitable for use in eco-efficient processes to recover valued-added compounds from by-/coproducts.

References

Abid, M., Cheikhrouhou, S., Renard, C.M., Bureau, S., Cuvelier, G., Attia, H. (2017). Characterization of pectins extracted from pomegranate peel and their gelling properties. Food Chemistry, 215:318–325.

Acevedo, C.A., Orellana, N., Avarias, K., Ortiz, R., Benavente, D., Prieto, P. (2018). Micropatterning technology to design an edible film for in vitro meat production. Food and Bioprocess Technology. In press. doi:10.1007/s11947-018-2095-4.

Akter, M.S., Ahmed, M., Eun, J.B. (2010). Dietary fibre components, antioxidant activities and hydration properties of ripe persimmon (Diospyros kaki L. cv. Daebong) peel powders as affected by different washing treatments. International Journal of Food Science and Technology, 45 (7):1464–1471

Aleson-Carbonell, L., Fernández-López, J., Pérez-Alvarez, J.A., Kuri, V. (2005). Characteristics of beef burger as influenced by various types of lemon albedo. Innovative Food Science and Emerging Technologies, 6 (2):247–255

Aleson-Carbonell, L., Fernández-López, J., Sayas-Barberá, E., Sendra, E., Pérez-Alvarez, J.A. (2003). Utilization of lemon albedo in dry-cured sausages. Journal of Food Science, 68 (5):1826–1830

Aleson-Carbonell, L., Fernández-López, J., Sendra, E., Sayas-Barberá, E., Pérez-Alvarez, J.A. (2004). Quality characteristics of a non-fermented dry-cured sausage formulated with lemon albedo. Journal of the Science of Food and Agriculture, 84 (15):2077–2084

Al-Farsi, M.A., Lee, C.Y. (2008). Nutritional and functional properties of dates: A review. Critical Reviews in Food Science and Nutrition, 48:877–887.

Amany, M.M.B., Shaker, M.A., Abeer, A.K. (2012). Antioxidant activities of date pits in a model meat system. International Food Research Journal, 19:223–227.

Amiot, M.J., Coxam, V., Strigler, F. (2012). Les phytomicronutriments. Paris: Lavoisier.
Amiri, S., Shakeri, A., Sohrabib, M.R., Khalajzadeh, S., Ghasemic, E. (2019). Optimization of ultrasonic assisted extraction of fatty acids from Aesculus hippocastanum fruit by response surface methodology. Food Chemistry, 271:762–766.
Andrés, A.I., Petrón, M.J., Adámez, J.D., López, M., Timón, M.L. (2017). Food by-products as potential antioxidant and antimicrobial additives in chill stored raw lamb patties. Meat Science, 129:62–70.
Baiano, A., Del Nobile, M.A. (2016). Antioxidant compounds from vegetable Matrices: Biosynthesis, occurrence, and extraction systems. Critical Reviews in Food Science and Nutrition, 56 (12):2053–2068
Baioumy, A.A., El-Akel, A.T., El-Nikeety, M.M. (2014). Production and evaluation of low fat beef burger using the orange albedo as a partial fat replacer. 8th International Congress of Food Technologists, Biotechnologists and Nutritionists, Proceedings, pp. 272–78.
Banerjee, J., Singh, R., Vijayaraghavan, R., MacFarlane, D., Patti, A.F., Arora, A. (2017). Bioactives from fruit processing wastes: Green approaches to valuable chemicals. Food Chemistry, 225:10–22.
Barba de la Rosa, A.P., Fomsgaard, I.S., Laursen, B., Mortensen, A.G., Olvera-Martínez, L., Silva-Sánchez, C., Mendoza-Herrera, A., González-Castañeda, J., De León-Rodríguez, A. (2009). Amaranth (Amaranthus hypochondriacus) as an alternative crop for sustainable food production: Phenolic acids and flavonoids with potential impact on its nutraceutical quality. Journal of Cereal Science, 49:117–121.
Barbosa-Pereira, L., Bilbao, A., Vilches, P., Angulo, I., LLuis, J., Fité, B., Paseiro-Losada, P., Cruz, J.M. (2014). Brewery waste as a potential source of phenolic compounds: Optimisation of the extraction process and evaluation of antioxidant and antimicrobial activities. Food Chemistry, 145:191–197.
Barreca, D., Gattuso, G., Bellocco, E., Calderaro, A., Trombetta, D., Smeriglio, A., Laganà, G., Daglia, M., Meneghini, S., Nabavi, S.M. (2017). Flavanones: Citrus phytochemical with health-promoting properties. BioFactors, 43 (4):495–506
Barreveld, W.H., 1993. Dates palm products, FAO agricultural services, bulletin No.101
Baugreet, S., Kerry, J.P., Botineştean, C., Allen, P., Hamill, R.M. (2016). Development of novel fortified beef patties with added functional protein ingredients for the elderly. Meat Science, 122 (1):40–47
Beaulieu, J.C., Stein-Chisholm, R.E. (2016). HS-GC-MS volatile compounds recovered in freshly pressed "Wonderful" cultivar and commercial pomegranate juices. Food Chemistry, 190:643–656.
Belc, N., Smeu, I., Macri, A., Vallauri, D., Flynn, K. (2018). Reformulating foods to meet current scientific knowledge about salt, sugar and fats. Trends in Food Science & Technology. In press. doi:10.1016/j.tifs.2018.11.002.
Boff, C.C., Crizel, T.M., de Araujo, R.R., Rios, A.O., Flôres, S.H. (2013). Development of chocolate ice cream using orange peel fiber as fat replacer. Ciencia Rural, 43 (10):1892–1897
Bolger, Z., Brunton, N.P., Monahan, F.J. (2018). Impact of inclusion of flaxseed oil (pre-emulsified or encapsulated) on the physical characteristics of chicken sausages. Journal of Food Engineering, 230:39–48.
Bonet, J.J., Borderías, J., Carbonell, J., Conde M.D., García-Palacios, I., García-Raurich J., Guamis, B., Mestres-Lagarriaga, J., Planas, A., Riera-Vall, E., Sánchez-Soriano, A., Vidal-Caro, C., Viladevall, O., Zahera, M. (2000). Productos Alimentarios Intermedios (PAI). Documentos Cotec sobre oportunidades tecnológicas, 16. Ed. R. Xalabarder-Coca, R. Gráficas Arias Montano S.A. Depósito legal: M. 26.463–2000

Borchani, C., Besbes, S., Masmoudi, M., Bouaziz, M.A., Blecker, C., Attia, H. (2012). Influence of oven-drying temperature on physicochemical and functional properties of date fibre concentrates. Food and Bioprocess Technology, 5 (5):1541–1551

Boukid, F., Folloni, S., Sforza, S., Vittadini, E., Prandi, B. (2018). Current trends in ancient grains-based foodstuffs: Insights into nutritional aspects and technological applications. Comprehensive Reviews in Food Science and Food Safety, 17:123–133.

Brückner, J., Mieth, G., Dabrowski, K., Gwiazda, S., Rutkowski, A. (2010). Functional properties of sunflower protein isolates in a meat model system. Food/Nahrung, 26 (5):457–463

Buss, D. (2019). Closing the loop on food waste. Food Technology, 1:7–14.

Cabanillas, C., Tablada, M., Ferreyra, L., Pérez, A., Sucani, G. (2017). Sustainable management strategies focused on native bio-inputs in Amaranthus cruentus L. in agro-ecological farms in transition. Journal of Cleaner Production., 142 (1):343–350

Canela-Xandri, A., Balcells, M., Villorbina, G., Cubero, M.Á., Canela-Garayoa, R. (2018). Effect of enzymatic treatments on dietary fruit fibre properties. Biocatalysis and Biotransformation, 36 (2):172–179

Chandrasekaran, M., Bahkali, A.H. (2013). Valorization of date palm (Phoenix dactylifera) fruit processing by-products and wastes using bioprocess technology– Review. Saudi Journal of Biological Sciences, 20:105–120.

Choe, J.-H., Kim, H.-Y., Kim, C.-J. (2017). Effect of persimmon peel (Diospyros kaki Thumb.) extracts on lipid and protein oxidation of raw ground pork during refrigerated storage. Korean Journal for Food Science of Animal Resources, 37 (2):254–263

Coksever, E., Saricoban, C. (2010). Effects of bitter orange albedo addition on the quality characteristics of naturally fermented Turkish style sausages (sucuks). Journal of Food, Agriculture and Environment, 8 (2):82–85

Contreras-López, E., Salvá-Ruiz, B.K. (2018). Sensory characterization of llama burger with sanky peel. Journal of High Andean Research, 20 (2):155–168

Damian, C. (2018). Antioxidant activity of citrus peel and seeds extracts. International Multidisciplinary Scientific GeoConference Surveying Geology and Mining Ecology Management, SGEM, 18 (6.2):19–26

Dangles, O. (2012). Antioxidant activity of plant Phenols: Chemical mechanisms and biological significance. Current Organic Chemistry, 16 (6):692–714

de Jong, S.P.L., 2015. Engaging sscientists – organising valorisation in the Netherlands, PhD Thesis. Rathenau Instituut, Den Haag, Netherlands.

Decker, E.A., Park, Y. (2010). Healthier meat products as functional foods. Meat Science., 86 (1):49–55

Delwiche J.F. (2004). The impact of perceptual interactions on perceived flavor. Food Quality and Preference., 15:137–146.

Delwiche, J.F. (2012). You eat with your eyes first. Physiology and Behaviour, 107 (4):502–504

Devatkal, S.K., Narsaiah, K., Borah, A. (2010). Anti-oxidant effect of extracts of kinnow rind, pomegranate rind and seed powders in cooked goat meat patties. Meat Science, 85 (1):155–159

Devatkal, S.K., Narsaiah, K., Borah, A. (2011). The effect of salt, extract of kinnow and pomegranate fruit by-products on colour and oxidative stability of raw chicken patties during refrigerated storage. Journal of Food Science and Technology, 48 (4):472–477

Devatkal, S.K., Thorat, P., Manjunatha, M. (2012). Effect of vacuum packaging and pomegranate peel extract on quality aspects of ground goat meat and nuggets. Journal of Food Science and Technology, doi:10.1007/s13197-012-0753-5.

Directive. (2008). Directive 2008: 98: EC of the European Parliament and of the Council of 19 November 2008 on waste and repealing certain directives, OJ L 312, 22.11.2008. pp. 3–30.

Elgasim, E.A., Alyousef, Y.A., Humeid, A.M. (1995). Possible hormonal activity of date pits and flesh fed to meat animals. Food Chemistry, 52 (2):149–152

Elleuch, M., Besbes, S., Roiseux, O., Blecker, C., Deroanne, C., Drira, N.-E., Attia, H. (2008). Date flesh: Chemical composition and characteristics of the dietary fibre. Food Chemistry, 111 (3):676–682

Evivie, S.E., Huo, G.-C., Igene, J.O., Bian, X. (2017). Some current applications, limitations and future perspectives of lactic acid bacteria as probiotics. Food and Nutrition Research, 61 (1):1318034

FAO. Organización de las Naciones Unidas para la Alimentación y la Agricultura (2013). La contribución de los insectos a la seguridad alimentaria, Los medios de vida y el medio ambiente. www.fao.org/docrep/018/i3264s/i3264s00.pdf.2013. Consulted 22/12/2018

Faucher, M., Serre, E., Langevin, M.E., Mikhaylin, S., Lutin, F., Bazinet, L. (2018). Drastic energy consumption reduction and ecoefficiency improvement of cranberry juice deacidification by electrodialysis with bipolar membranes at semi-industrial scale: Reuse of the recovery solution. Journal of Membrane Science, 555:105–114.

Fayek, N.M., El-Shazly, A.H., Abdel-Monem, A.R., Moussa, M.Y., Abd-Elwahab, S.M., El-Tanbouly, N.D. (2017). Comparative study of the hypocholesterolemic, antidiabetic effects of four agro-waste citrus peels cultivars and their HPLC standardization. Brazilian Journal of Pharmacognosy, 27 (4):488–494

Fernández-Ginés, J.M., Fernández-López, J., Sayas-Barberá, E., Sendra, E., Pérez-Alvarez, J. A. (2003). Effect of storage conditions on quality characteristics of Bologna sausages made with citrus fiber. Journal of Food Science, 68 (2):710–715

Fernández-Ginés, J.M., Fernández-López, J., Sayas-Barberá, E., Sendra, E., Pérez-Álvarez, J. A. (2004a). Lemon albedo as a new source of dietary fiber: Application to bologna sausages. Meat Science, 67 (1):7–13

Fernández-López, J., Fernández-Ginés, J.M., Aleson-Carbonell, L., Sendra, E., Sayas-Barberá, E., Pérez-Alvarez, J.A. (2004b). Application of functional citrus by-products to meat products. Trends in Food Science and Technology, 15 (3-4):176–185

Fernández-López, J., Sendra-Nadal, E., Navarro, C., Sayas, E., Viuda-Martos, M., Pérez-Alvarez, J.A. (2009). Storage stability of a high dietary fibre powder from orange by-products. International Journal of Food Science and Technology, 44 (4):748–756

Fernández-López, J., Viuda-Martos, M., Sendra, E., Sayas-Barberá, E., Navarro, C., Pérez-Alvarez, J.A. (2007). Orange fibre as potential functional ingredient for dry-cured sausages. European Food Research and Technology, 226 (1–2):1–6

Gabriel, A.S., Ninomiya, K., Uneyama, H. (2018). The role of the Japanese traditional diet in healthy and sustainable dietary patterns around the world. Nutrients, 10:173.

Garcia, Y.D., Valles, B.S., Lobo, A.P. (2009). Phenolic and antioxidant composition of by-products from the cider industry: Apple pomace. Food Chemistry, 117 (4):731–738

García-Cayuela, T., Nuño-Escobar, B., Welti-Chanes, J., Cano, M.P. (2018). In vitro bioaccessibility of individual carotenoids from persimmon (Diospyros kaki, cv. Rojo Brillante) used as an ingredient in a model dairy food. Journal of the Science of Food and Agriculture, 98 (9):3246–3254

García-Pérez, F.J., Lario, Y., Fernández-López, J., Sayas, E., Pérez-Alvarez, J.A., Sendra, E. (2005). Effect of orange fiber addition on yogurt color during fermentation and cold storage. Color Research and Application, 30 (6):457–463

García-Pérez, F.J., Sendra, E., Lario, Y., Fernández-López, J., Sayas-Barberá, E., Pérez-Alvarez, J.A. (2006). Rheology of orange fiber enriched yogurt. Milchwissenschaft, 61 (1):55–59

Ghnimi, S., Umer, S., Karim, A., Kamal-Eldin, A. (2017). Date fruit (Phoenix dactylifera L.): An underutilized food seeking industrial valorization. NFS Journal, 6:1–10.

Giordano, S., Clodoveo, M.L., Gennaro, B.D., Corbo, F. (2018). Factors determining neophobia and neophilia with regard to new technologies applied to the food sector: A systematic review. International Journal of Gastronomy and Food Science, 11:1–19.

Goetzke, B., Nitzko, S., Spiller, A. (2014). Consumption of organic and functional food. A matter of well-being and health? Appetite, 77 (1):94–103

Grassino, A.N., Barba, F.J., Brnčića, M., Lorenzo, J.M., Lucinid, L., Brnčić, S.R. (2018). Analytical tools used for the identification and quantification of pectin extracted from plant food matrices, wastes and by-products: A review. Food Chemistry, 266:47–55.

Guo, S., Deng, Q., Xiao, J., Xie, B., Sun, Z. (2007). Evaluation of antioxidant activity and preventing DNA damage effect of pomegranate extracts by chemiluminescence method. Journal of Agricultural and Food Chemistry, 55:3134–3140.

Gupta, S., Kumar, A., Hakeem, H.R., Shafat, S. (2015). Effect of carrageenan on physicochemical and sensory attributes of orange (Citrus sinensis) fortified chicken patties from spent hen meat. Indian Veterinary Journal, 92 (5):13–16

Gutiérrez, T.J. (2018). Plantain flours as potential raw materials for the development of gluten-free functional foods. Carbohydrate Polymers, 202 (15):265–279

Halagarda, M., Kędzior, W., Pyrzyńska, E. (2018). Nutritional value and potential chemical food safety hazards of selected Polish sausages as influenced by their traditionality. Meat Science. 25–34.

Hall, M.B., Kononoff, P.J. (2011). Feed ingredients | feed concentrates: Co-product feeds. In: Encyclopedia of dairy sciences (Second Edition). pp. 342–348.

Hashim, I.B., Khalil, A.H., Afifi, H.S. (2008). Quality characteristics and consumer acceptance of yogurt fortified with date fiber. Journal of Dairy Science, 92 (11):5403–5407

Hasnaoui, A., Elhoumaizi, M.A., Borchani, C., Attia, H., Besbes, S. (2013). Contribution to the valuation of dates production in the region of Figuig (Southeastern Morocco). Acta Horticulturae, 994:111–116.

Ibrahim, H.M., Moawad, R.K., Emam, W.H. (2012). Antioxidant effect of pomegranate rind, seed extracts and pomegranate juice on lipid oxidation and some quality properties of cooked beef patties. Journal of Applied Sciences Research, 8 (8):4023–4032

Jingjing, L., Qipeng, Y. (2008). A new method for ellagic acid production from pomegranate husk. Journal of Food Process Engineering, 31 (4):443–454

Jongaroontaprangsee, S., Chiewchan, N., Devahastin, S. (2014). Composition profiles and functional properties of dietary fiber powder from lime residues: Effects of pretreatment and drying methods. Drying Technology, 32 (4):484–493

Karizaki, V.M. (2017). Iranian dates and ethnic date-based products. Journal of Ethnic Food, 4 (3):204–209

Keşkekoğlu, H., Üren, A. (2014). Inhibitory effects of pomegranate seed extract on the formation of heterocyclic aromatic amines in beef and chicken meatballs after cooking by four different methods. Meat Science, 96 (4):1446–1451

Khwairakpam, A.D., Bordoloi, D., Thakur, K.K., Monisha, J., Arfuso, F., Sethi, G., Mishra, S., Kumar, A.P., Kunnumakkara, A.B. (2018). Possible use of Punica granatum (Pomegranate) in cancer therapy. Pharmacological Research, 133:53–64.

Kim, H.-W., Setyabrata, D., Lee, Y., Jones, O.G., Kim, Y.H.B. (2017). Effect of house cricket (Acheta domesticus) flour addition on physicochemical and textural properties of meat emulsion under various formulations. Journal of Food Science, 82 (12):2787–2793

Kim, H.-W., Setyabrata, D., Lee, Y.J., Jones, O.G., Kim, Y.H.B. (2016). Pre-treated mealworm larvae and silkworm pupae as a novel protein ingredient in emulsion sausages. Innovative Food Science and Emerging Technologies, 38 (1):116–123

King, D.E., Mainous, A.G., Lambourne, C.A. (2012). Trends in dietary fiber intake in the United States, 1999–2008. Journal of the Academy of Nutrition and Dietetics, 112, (5), May:642–648.

Köhler, R., Lambert, C., Biesalski, H.K. (2019). Animal-based food taboos during pregnancy and the postpartum period of Southeast Asian women – A review of literature. Food Research International, 115:480–486.

Kumar, Y., Tyagi, S.K., Vishwakarma, R.K., Kalia, A. (2017). Textural, microstructural, and dynamic rheological properties of low-fat meat emulsion containing aloe gel as potential fat replacer. International Journal of Food Properties, 20:S1132–S1144.

Kursun, E., Karaca, H. (2018). Dried persimmons: Bioactive components, health aspects and current drying techniques. Acta Horticulturae, 1195:169–176.

Lario, Y., Sendra, E., García-Pérez, F.J., Fernández-López, J., Sayas-Barberá, E., Pérez-Alvarez, J.A. (2004a). Evolution of microbial populations of fiber enriched yogurt during commercial shelf-life. Milchwissenschaft, 59 (11-12):646–649

Lario, Y., Sendra, E., García-Pérez, J., Fuentes, C., Sayas-Barberá, E., Fernández-López, J., Pérez-Alvarez, J.A. (2004b). Preparation of high dietary fiber powder from lemon juice by-products. Innovative Food Science and Emerging Technologies, 5 (1):113–117

Latiffi, N.A.A., Mohamed, R.M., Shanmugan, V.A., Pahazri, N.F., Kassim, A.H., Matias-Peralta, H.M., Tajuddin, R.M. (2016). Removal of nutrients from meat food processing industry wastewater by using microalgae botryococcus SP. ARPN Journal of Engineering and Applied Sciences, 11 (16):9863–9867

Laureati, M., Spinelli, S., Monteleone, E., Dinnella, C., Pagliarini, E. (2018). Associations between food neophobia and responsiveness to "warning" chemosensory sensations in food products in a large population sample. Food Quality and Preference, 68:113–124.

Leidi, E.O., Monteros-Altamirano, A., Mercado, G., Rodriguez, J.P., Ramosa, A., Alandia, G., Sørensen, M., Jacobsen, S.E. (2018). Andean roots and tubers crops as sources of functional foods. Journal of Functional Foods., 51:86–93.

Li, C., Schluesener, H. (2017). Health-promoting effects of the citrus flavanone hesperidin. Critical Reviews in Food Science and Nutrition, 57 (3):613–631

López, D.N., Galante, M., Raimundo, G., Spelzini, D., Boeris, V. (2018). Functional properties of amaranth, quinoa and chia proteins and the biological activities of their hydrolyzates. Food Research International, In press.

López-Marcos, M.C., Bailina, C., Viuda-Martos, M., Pérez-Alvarez, J.A., Fernández-López, J. (2015a). Properties of dietary fibers from agroindustrial co-products as source for fiber-enriched foods. Food and Bioprocess Technology, 8 (12):2400–2408

López-Marcos, M.C., Bailina, C., Viuda-Martos, M., Pérez-Alvarez, J.A., Fernández-López, J. (2015b). Effects of various fibre-rich extracts on cholesterol binding capacity during in vitro digestion of pork patties. Food and Function, 6 (11):3473–3478

López-Vargas, J.H., Fernández-López, J., Pérez-Álvarez, J.A., Viuda-Martos, M. (2013). Chemical, physico-chemical, technological, antibacterial and antioxidant properties of dietary fiber powder obtained from yellow passion fruit (Passiflora edulis var. flavicarpa) co-products. Food Research International, 51 (2):756–763

Louati, I., Bahloul, N., Besombes, C., Allaf, K., Kechaou, N. (2019). Instant controlled pressure-drop as texturing pretreatment for intensifying both final drying stage and extraction of phenolic compounds to valorize orange industry by-products (Citrus sinensis L. Food and Bioproducts Processing, 114:85–94.

Lucas-González, R., Pellegrini, M., Viuda-Martos, M., Pérez-Álvarez, J.Á., Fernández-López, J. (2018). Persimmon (Diospyros kaki Thunb.) co-products as a new ingredient in pork liver pâté: Influence on quality properties. International Journal of Food Science and Technology, In Press.

Lucas-González, R., Viuda-Martos, M., Pérez-Álvarez, J.Á., Fernández-López, J. (2017). Evaluation of particle size influence on proximate composition, physicochemical, techno-functional and physio-functional properties of flours obtained from persimmon (Diospyros kaki Trumb). Co-products Plant Foods for Human Nutrition, 72 (1):67–73

Maeda-Yamamoto, M. (2017). Development of functional agricultural products and use of a new health claim system in Japan. Trends in Food Science and Technology, 69:324–332.

Marín, F.R., Soler-Rivas, C., Benavente-García, O., Castillo, J., Pérez-Alvarez, J.A. (2007). By-products from different citrus processes as a source of customized functional fibers. Food Chemistry, 100:736–741.

Martínez-Las Heras, R., Pinazo, A., Heredia, A., Andrés, A. (2017). Evaluation studies of persimmon plant (Diospyros kaki) for physiological benefits and bioaccessibility of antioxidants by in vitro simulated gastrointestinal digestion. Food Chemistry, 214: 478–485

Martins, Z.E., Pinho, O., Ferreira, Isabel M. P. L. V. O. (2017). Food industry by-products used as functional ingredients of bakery products. Trends in Food Science and Technology, 67:106–128.

Martín-Sánchez, A.M. Cherif, S., Vilella-Esplá, J., Ben-Abda, J., Kuri, V., Pérez-Álvarez, J. Á., Sayas-Barberá, E. (2014a). Phytochemicals in date co-products and their antioxidant activity. Food Chemistry, 158:513–520.

Martín-Sánchez, A.M., Cherif, S., Vilella-Esplá, J., Ben-Abda, J., Pérez-Álvarez, J.Á., Sayas-Barberá, E. (2014b). Characterization of novel intermediate food products from Spanish date palm (Phoenix dactylifera L., cv. Confitera) co-products for industrial use. Food Chemistry, 154:269–275.

Martín-Sánchez, A.M., Ciro-Gómez, G., Sayas, E., Vilella-Esplá, J., Ben-Abda, J., Pérez-Álvarez, J.T. (2013). Date palm by-products as a new ingredient for the meat industry: Application to pork liver pâté. Meat Science, 93 (4):880–887

Martín-Sánchez, A.M., Ciro-Gómez, G., Vilella-Esplá, J., Ben-Abda, J., Pérez-Álvarez, J.T., Sayas-Barberá, E. (2014c). Influence of fresh date palm co-products on the ripening of a paprika added dry-cured sausage model system. Meat Science, 97 (2):130–136

Martín-Sánchez, A.M., Navarro, C., Pérez-Álvarez, J.A., Kuri, V. (2009). Alternatives for efficient and sustainable production of surimi: A review. Comprehensive Reviews in Food Science and Food Safety, 8 (4):359–374

Mireles-Arriaga, A.I., Ruiz-Nieto, J.E., Juárez-Abraham, M., Mendoza-Carrillo, M., Martínez-Loperena, R. (2017). Functional restructured meat: Applications of ingredients derived from plants. Vitae, 24 (3):196–204

Mirosa, M., Mangan-Walker, E. (2018). Young Chinese and functional foods for mobility health: Perceptions of importance, trust, and willingness to purchase and pay a premium. Journal of Food Products Marketing, 24 (2):216–234

Narsaiah, K., Jha, S.N., Devatkal, S.K., Borah, A., Singh, D.B., Sahoo, J. (2011). Tenderizing effect of blade tenderizer and pomegranate fruit products in goat meat. Journal of Food Science and Technology, 48 (1):61–68

Naveena, B.M., Sen, A.R., Kingsly, R.P., Singh, D.B., Kondaiah, N. (2008a). Antioxidant activity of pomegranate rind powder extract in cooked chicken patties. International Journal of Food Science and Technology, 43 (10):1807–1812

Naveena, B.M., Sen, A.R., Vaithiyanathan, S., Babji, Y., Kondaiah, N. (2008b). Comparative efficacy of pomegranate juice, pomegranate rind powder extract and BHT as antioxidants in cooked chicken patties. Meat Science, 80 (4):1304–1308

Ningrum, A., Supriyadi, Anggrahini, S., Kusumaningrum, L.D., Hapsari, M.W., Schreiner, M. (2018). Valorization of food by product from selected tropical fruits pomace. IOP Conference Series: Earth and Environmental Science, 205 (1) art. no. 012034.

Nunes, M.C.N. (2015). Correlations between subjective quality and physicochemical attributes of fresh fruits and vegetables. Postharvest Biology and Technology, 107:43–54.

O'Shea, N., Ktenioudaki, A., Smyth, T.P., McLoughlin, P., Doran, L., Auty, M.A.E., Arendt, E., Gallagher, E. (2015). Physicochemical assessment of two fruit by-products as functional ingredients: Apple and orange pomace. Journal of Food Engineering, 153:89–95.

Olsen, N.V. (2015). Design Thinking and food innovation. Trends in Food Science and Technology, 41:182–187.

Omar, H.A., Mohamed, W.R., Arafa, E.-S.A., Shehata, B.A., El Sherbiny, G.A., Arab, H.H., Elgendy, A.N.A.M. (2016). Hesperidin alleviates cisplatin-induced hepatotoxicity in rats without inhibiting its antitumor activity. Pharmacological Reports, 68 (2):349–356

Opazo-Navarrete, M., Freire, D.T., Boom, R.M., Janssen, A.E.M., Schutyser, M.A.I. (2018). Dry fractionation of quinoa sweet varieties Atlas and Riobamba for sustainable production of protein and starch fractions. Journal of Food Composition and Analysis, 74:95–101.

Oreopoulou, V., Tzia, C. (2007). Utilization of plant by-products for the recovery of proteins, dietary fibers, antioxidants, and colorants. In: Utilization of by-products and treatment of waste in the food industry. pp. 209–232.

Ospina-E, J.C., Rojano, B., Ochoa, O., Pérez-Álvarez, J.A., Fernández-López, J. (2015). Development of frankfurter-type sausages with healthy lipid formulation and their nutritional, sensory and stability properties. European Journal of Lipid Science and Technology, 117 (1):122–131

Paraman, I., Sharif, M.K., Supriyadi, S., Rizvi, S.S.H. (2015). Agro-food industry by-products into value-added extruded foods. Food and Bioproducts Processing, 96:78–85.

Pellegrini, M., Lucas-Gonzalez, R., Ricci, A., Fontecha, J., Fernández-López, J., Pérez-Álvarez, J.A., Viuda-Martos, M. (2018). Chemical, fatty acid, polyphenolic profile, techno-functional and antioxidant properties of flours obtained from quinoa (Chenopodium quinoa Willd) seeds. Industrial Crops and Products, 111:38–46.

Perez-Alvarez J.A. (2008). Overwie of meat products as functional food. In: J. Fernandez-Lopez and J.A. Pérez (Eds.). Technological strategies for functional meat products development. 1–7. Kerala, India: Transworld Research Network.

Pérez-Alvarez, J.A. (2016). El dátil, un alimento para la salud y el bienestar en el siglo XXI. En M.E. Sayas-Barberá (Ed.). El dátil en la elaboración de alimentos saludables. pp. 15–25. Universidad Miguel Hernández de Elche.

Pérez-Chabela, M.L., Hernández-Alcántara, A.M. (2018). Agroindustrial co-products as sources of novel functional ingredients. In: A. Mi Grumezescu and A.M. Holban (Eds.). Food processing for increased quality and consumption. Chapter 8. pp. 219–250.

Perussello, C.A., Zhang, Z., Marzocchella, A., Tiwari, B.K. (2017). Valorization of apple pomace by extraction of valuable compounds. Comprehensive Reviews in Food Science and Food Safety, 16 (5):776–796

Pogorzelska-Nowicka, E., Atanasov, A.G., Horbańczuk, J., Wierzbicka, A. (2018). Bioactive compounds in functional meat products. Molecules, 23 (2):307

Predieri, S., Conte, A., Danza, A., Gatti, E., Magli, M., Maria, D.G., Marta, C., Alessandro, D.N.M. (2017). Senior consumer's involvement in developing new fish-based foods through sequential hedonic tests. Current Research in Nutrition and Food Science., 5 (2):66–74

Qin, Y.-Y., Wu, Y., Zhang, Z.-H., Li, B., Liang, X.-B., Cao, J.-X. (2014). Effect of an active film from chitosan and pomegranate rind powder extract on shelf-life extension of pork meat patties. Modern Food Science and Technology, 30 (4):181–188

Rabetafika, H.N., Bchir, B., Blecker, C., Richela, A. (2014). Fractionation of apple by-products as source of new ingredients: Current situation and perspectives. Trends in Food Science & Technology., 40 (1):99–114

Raimondo, M., Caracciolo, F., Cembalo, L., Chinnici, G., Pecorino, B., D'Amico, M. (2018). Making virtue out of necessity: Managing the citrus waste supply chain for bioeconomy applications. Sustainability (Switzerland), 10 (12), art. no. 4821.

Ramandi, N.F., Ghassempour, A., Najafi, N.M. Ghasemi, E. (2012). Optimization of ultrasonic assisted extraction of fatty acids from Borago Officinalis L. flower by central composite design. Arabian Journal of Chemistry, 1066 (S1):1–5

Rashed, K.N., Pan, M.-H. (2014). Antiinflmmatory activity of diospyros kaki peels and bioactive constituents. Pharma Research, 10 (2):7–15

Renard, C.M.G.C. (2018). Extraction of bioactives from fruit and vegetables: State of the art and perspectives. LWT-Food Science and Technology, 93:390–395.

Rice-Evans, C.A., Sampson, J., Bramley, P.M., Holloway, D.E. (1997). Why do we expect carotenoids to be antioxidants in vivo? Free Radical Research, 26 (4):381–398

Rioux, C., Leglaye, L., Lafraire, J. (2018). Inductive reasoning, food neophobia, and domain-specificity in pre-schoolers. Cognitive Development., 47:124–132.

Rocchetti, G., Lucini, L., Lorenzo-Rodriguez, J.M., Barba, F.J., Giuberti, G. (2019). Gluten-free flours from cereals, pseudocereals and legumes: Phenolic fingerprints and in vitro antioxidant properties. Food Chemistry, 271:157–164.

Roselino, M.N., de Almeida, J.F., Canaan, J.M.M., Pinto, R.A., Ract, J.N.R., de Paula, A.V., de Valdez, G.F., Rossi, E.A., Cavallini, D.C.U. (2017). Safety of a low-fat fermented sausage produced with Enterococcus faecium CRL 183 and Lactobacillus acidophilus CRL1014 probiotic strains. International Food Research Journal, 24 (6):2694–2704

Ruiz, K.B., Khakimovb, B., Engelsen, S.B., Bakb, S., Biondia, S., Jacobsenb, S.E. (2017). Quinoa seed coats as an expanding and sustainable source of bioactive compounds: An investigation of genotypic diversity in saponin profiles. Industrial Crops and Products, 104:156–163.

Saeed, M., Naveed, M., Bibi, J., Kamboh, A.A., Arain, M.A., Shah, Q.A., Alagawany, M., El-Hack, M.E.A., Abdel-Latif, M.A., Yatoo, M.I., Tiwari, R., Chakraborty, S., Dhama, K. (2018). The promising pharmacological effects and therapeutic/medicinal applications of Punica Granatum L. (pomegranate) as a functional food in humans and animals. Recent Patents on Inflammation and Allergy Drug Discovery, 12 (1):24–38

Saikia, S., Mahanta, C.L. (2016). In vitro physicochemical, phytochemical and functional properties of fiber rich fractions derived from by-products of six fruits. Journal of Food Science and Technology, 53 (3):1496–1504

Sánchez-Zapata, E., Fernández-López, J., Peñaranda, M., Fuentes-Zaragoza, E., Sendra, E., Sayas, E., Pérez-Alvarez, J.A. (2011). Technological properties of date paste obtained from date by-products and its effect on the quality of a cooked meat product. Food Research International, 44 (7):2401–2407

Satari, B., Karimi, K. (2018). Citrus processing wastes: Environmental impacts, recent advances, and future perspectives in total valorization. Resources, Conservation and Recycling, 129:153–167.

Sau, S., Sarkar, S., Deb, P., Ghosh, B. (2016). Super-fruit: As a potential option to mitigate malnutrition in Indian subcontinent. Asian Journal of Pharmaceutical and Clinical Research, 9 (1):18–22

Sayas-Barberá, E., Viuda-Martos, M., Fernández-López, F., Pérez-Alvarez, J.A., Sendra, E. (2012). Combined use of a probiotic culture and citrus fiber in a traditional sausage "Longaniza de Pascua". Food Control, 27 (2):343–350

Schifferstein, H.N.J., Wehrle, T., Carbon, C.C. (2019). Consumer expectations for vegetables with typical and atypical colors: The case of carrots. Food Quality and Preference, 72:98–108.

Schnettler, B., Bravo, S., Sepúlveda, N., Grunert, K.G., Hueche, C. (2018). Consumer acceptance of a functional processed meat product made with different meat sources. British Food Journal., 120 (2):424–440

Schutyser, M.A.I., Pelgrom, P.J.M., van der Goot, A.J., Boom, R.M. (2015). Dry fractionation for sustainable production of functional legume protein concentrates. Trends in Food Science & Technology, 45 (2):327–335

Sellimi, S., Ksouda, G., Benslima, A., Nasri, R., Rinaudo, M., Nasri, M., Hajji, M. (2017). Enhancing colour and oxidative stabilities of reduced-nitrite turkey meat sausages during refrigerated storage using fucoxanthin purified from the Tunisian seaweed Cystoseira barbata. Food and Chemical Toxicology, 107:620–629.

Sendra, E., Fayos, P., Lario, Y., Fernández-López, J., Sayas-Barberá, E., Pérez-Alvarez, J. (2008b). Incorporation of citrus fibers in fermented milk containing probiotic bacteria. Food Microbiology, 25 (1):13–21

Sendra, E., Lario, Y., Sayas-Barberá, E., Pérez-Alvarez, J.A., Alesón-Carbonell, L., Fernández-Ginés, J.M., Fernández-López, J. (2008a). Obtention of lemon fibre from lemon juice by-products. Agro Food Industry Hi-Tech, 19 (2 Suppl):22–24

Shah, M.A., Bosco, S.J.D., Mir, S.A. (2014). Plant extracts as natural antioxidants in meat and meat products. Meat Science, 98 (1):21–33

Sharif, M.K., Zahid, A., Shah, F-u-H. (2018). Role of food product development in increased food consumption and value addition. In Food processing for increased quality and consumption. Chapter 15. pp. 455–479.

Silva, M.A., Gonçalves-Albuquerque, T., Alves, R.C., Oliveira, M.B.P.P., Costa, H.S. (2018). Melon (Cucumis melo L.) by-products: Potential food ingredients for novel functional foods? Trends in Food Science & Technology, In press.

Singh, B., Singh, J.P., Kaur, A., Singh, N. (2018). Phenolic compounds as beneficial phytochemicals in pomegranate (Punica granatum L.) peel: A review. Food Chemistry, 261:75–86.

Smetana, S., Mathys, A., Knoch, A., Heinz, V. (2015). Meat alternatives: Life cycle assessment of most known meat substitutes. International Journal of Life Cycle Assessment., 20 (9):1254–1267

Sobiecki, J.G., Appleby, P.N., Bradbury, K.E., Key, T.J. (2016). High compliance with dietary recommendations in a cohort of meat eaters, fish eaters, vegetarians, and vegans: Results from the European Prospective Investigation into Cancer and Nutrition–Oxford study. Nutrition Research, 36 (5):464–477

Son, J.E., Hwang, M.K., Lee, E., Seo, S.G., Kim, J.-E., Jung, S.K., Kim, J.R., Ahn, G.-H., Lee, K.W., Lee, H.J. (2013). Persimmon peel extract attenuates PDGF-BB-induced human aortic smooth muscle cell migration and invasion through inhibition of c-Src activity. Food Chemistry, 141 (4):3309–3316

Sonesson, U., Davis, J., Flysjö, A., Gustavsson, J., Witthöft, C. (2017). Protein quality as functional unit – A methodological framework for inclusion in life cycle assessment of food. Journal of Cleaner Production, 140 (1):470–478

Song, J., Pan, T., Wu, J., Ren, F. (2016). The improvement effect and mechanism of citrus fiber on the water-binding ability of low-fat frankfurters. Journal of Food Science and Technology, 53 (12):4197–4204

Soriao, A.B.P., Dale, A.G., Dela Cruz, C.G.M., Indiongco, P.C.K., Manucom, J.L.M., Masangcay, J.N.M., Mañago, J.A.R., Narvaez, M.B. (2015). Dehydration of pomelo (Citrus grandis) albedo and its utilization as a source of dietary fiber in philippine pork sausage. Kasetsart Journal – Natural Science, 49 (4):606–614

Spence, C. (2017). Confort food: A review. International Journal of Gastronomy and Food Science, 9:105–109.

Stephens, N., King, E., Lyall, C. (2018). Blood, meat, and upscaling tissue engineering: Promises, anticipated markets, and performativity in the biomedical and agri-food sectors. BioSocieties. In press. doi:10.1057/s41292-017-0072-1.

Tang, Z.-X., Shi, L.-E., Aleid, S.M. (2013). Date fruit: Chemical composition, nutritional and medicinal values, products. Journal of the Science of Food and Agriculture, 93 (10):2351–2361

Trigueros, L., Sayas-Barberá, E., Pérez-Álvarez, J.A., Sendra, E. (2012). Use of date (Phoenix dactylifera L.) blanching water for reconstituting milk powder: Yogurt manufacture. Food and Bioproducts Processing, 90 (3):506–514

Tulbek, M.C., Lam, R.S.H., Wang, Y.C., Asavajaru, P., Lam, A. (2016). Pea: A sustainable vegetable protein crop. In Sustainable protein sources. pp. 145–164. Saskatoon, SK, Canada: AGT Foods.

van Stokkom, V.L., de Graaf, C., Wang, O., van Kooten., Stieger, M. (2019). Combinations of vegetables can be more accepted than individual vegetables. Food Quality and Preference, 72:147–158.

Vinauskiene, R., Morkunaite, R., Leskauskaite, D. (2015). The influence of pea products on the functional properties of frankfurters. CYTA – Journal of Food, 13 (2):282–292

Viuda-Martos, M., Fernández-López, J., Sayas-Barbera, E., Sendra, E., Navarro, C., Pérez-Álvarez, J.A. (2009a). Citrus co-products as technological strategy to reduce residual nitrite content in meat products. Journal of Food Science, 74 (8):R93–R100

Viuda-Martos, M., Fernandez-Lopez, J., Sayas-Barbera, E., Sendra, E., Perez-Alvarez, J.A. (2011). Physicochemical characterization of the orange juice waste water of a citrus by-product. Journal of Food Processing and Preservation, 35 (2):264–271

Viuda-Martos, M., Ruiz-Navajas, Y., Fernández-López, J., Pérez-Álvarez, J.A. (2009b). Effect of adding citrus waste water, thyme and oregano essential oil on the chemical, physical and sensory characteristics of a bologna sausage. Innovative Food Science & Emerging Technologies, 10 (4):655–660

Viuda-Martos, M., Ruiz-Navajas, Y., Fernández-López, J., Pérez-Álvarez, J.A. (2010a). Effect of orange dietary fibre, oregano essential oil and packaging conditions on shelf-life of bologna sausages. Food Control, 21 (4):436–443

Viuda-Martos, M., Ruiz-Navajas, Y., Fernández-López, J., Pérez-Álvarez, J.A. (2010b). Effect of added citrus fibre and spice essential oils on quality characteristics and shelf-life of mortadella. Meat Science, 85 (3):568–576

Yalinkiliç, B., Kaban, G., Kaya, M. (2012). The effects of different levels of orange fiber and fat on microbiological, physical, chemical and sensorial properties of sucuk. Food Microbiology, 29 (2):255–259

Yang, X., Nisar, T., Hou, Y., Gou, X., Sun, L., Guo, Y. (2018). Pomegranate peel pectin can be used as an effective emulsifier. Food Hydrocolloids, 85:30–38.

Yi, T., Huang, X., Pan, S., Wang, L. (2014a). Physicochemical and functional properties of micronized jincheng orange by-products (Citrus sinensis Osbeck) dietary fiber and its application as a fat replacer in yogurt. International Journal of Food Sciences and Nutrition, 65 (5):565–572

Yi, T., Wang, K., Zhuang, Z., Pan, S., Huang, X. (2014b). Comparative analysis of dietary fibre extract isolated from citrus juice by-products using water extraction, fermentation and enzymatic treatment methods. Advance Journal of Food Science and Technology, 6 (9):1058–1066

Young, J.Y., Lee, J.Y., Ha, Y.S., Shin, Y.K., Kim, Y., Kim, S.H., Oh, N.S. (2016). Enhanced microbial, functional and sensory properties of herbal yogurt fermented with Korean traditional plant extracts. Korean Journal for Food Science of Animal Resources, 36 (1):90–99

Zapata-López G.A. (2016). Efectos de la disminución del ion sodio sobre el sabor en un jamón cocido tipo fiambre elaborado en Medellín, Colombia. Ph.D. Thesis. Universidad Miguel

Hernández de Elche. http://dspace.umh.es/handle/11000/6/browse?type=author&order=ASC&rpp=20&value=Zapata+L%C3%B3pez%2C+Gustavo+Adolfo
Zettel, V., Hitzmann, B. (2018). Applications of chia (Salvia hispanica L.) in food products. Trends in Food Science and Technology, 80:43–50.
Research article Full text access
Żyżelewicz, D., Budryn, G., Oracz, J., Antolak, H., Kręgie, D., Kaczmarska, M. (2018). The effect on bioactive components and characteristics of chocolate by functionalization with raw cocoa beans. Food Research International, 113:234–244.

11

Valuable Compounds in Coffee By-Products

PATRICIA ESQUIVEL

Escuela de Tecnología de Alimentos, Universidad de Costa Rica, 2060 San Pedro, Costa Rica

VÍCTOR M. JIMÉNEZ

CIGRAS/IIA, Universidad de Costa Rica, 2060 San Pedro, Costa Rica
Food Security Center, University of Hohenheim, Stuttgart 70599, Germany

Contents

1 Introduction

Coffee ranks second, after crude oil, among all commodities. It is the most important food product worldwide. The economic importance of coffee is mainly due to the beverage, prepared through the infusion of roasted and ground beans. Coffee beverage is mainly produced from two species, *Coffea arabica* (Arabica) and *Coffea canephora* (Robusta). *C. arabica* is considered to be superior due to its sensory characteristics reaching higher prices in international markets (Alonso-Salces, Serra, Reniero, & Héberger, 2009). The stimulant properties of the coffee brew, mainly due to caffeine, are well known (Wang, Du, & Du, 2018). However, the potential involvement of hundreds of biologically active compounds present in the coffee brew on human health must not be underestimated (O'Keefe, DiNicolantonio, & Lavie, 2018).

Although lower mortality and several health benefits, such as lower risks of Parkinson's and Alzheimer's disease, a favorable effect on liver function, a possible role in weight loss (increased metabolic rate, energy expenditure, lipid oxidation, and lipolytic and thermogenic activities), and a decreased risk of developing certain types of cancer (endometrial, prostatic, colorectal, and liver) have been related to coffee consumption (reviewed by Gonzalez de Mejia & Ramirez-Mares, 2014), there are also reports relating coffee consumption with adverse health effects, such as elevated risk for cardiovascular disease (Ranheim & Halvorsen, 2005). Other beneficial physiological effects of coffee consumption include a stimulating effect on gastrointestinal tract and liver, inhibition of the onset of liver cirrhosis and alcohol-associated pancreatitis, reduction of the probabilities of developing asthma symptoms, and prevention of clinical manifestations of bronchial asthma, reduction in plasma glucose level, protection of lipids, proteins, and DNA against damage, and positive modulation of antioxidant enzymes (Dórea & da Costa, 2005; Martini et al., 2016).

Considering that the coffee beans comprise around 40% w/w of the fruit, it is important to look for possibilities for utilization of the fruit remainings generated during processing to avoid their inadequate disposal, resulting in health and environmental detrimental consequences. In this sense, most processed by-products of the coffee industry are considered to be valuable both as a source of biomass and because of their functional properties (Blinová, Sirotiak, Bartošová, & Soldán, 2017; Esquivel & Jiménez, 2012). The aim of this chapter is to review recent information related to the utilization of coffee processing by-products, focusing mainly on the health beneficial properties, and also on other uses when appropriate.

2 Coffee Processing

The coffee cherry is a fruit composed by a smooth pericarp, green when unripe, but depending on the genotype, it turns yellow, orange, or red at full ripeness. The mesocarp (pulp) is yellowish, fibrous, and sweet, followed by a translucent, colorless, thin, viscous, and highly hydrated layer of mucilage (Figure 11.1). There is also a thin layer of endocarp with a yellowish color, called the "parchment." The coffee bean (endosperm) is also covered by the silverskin (Belitz, Grosch, & Schieberle, 2009; Berbert, Queiroz, & Melo, 2002; Purseglove, 1974).

Coffee is traded as green coffee, which may or may not include the silverskin. To obtain the green coffee, the coffee berries should be processed through either dry, mechanical, or wet processing (Figure 11.2). In the dry process, the mixed fruits (green, ripe, and over-ripe) are generally sun-dried and subsequently mechanically hulled to remove the husks (skin, pulp, and parchment) and as much as possible of the silverskin (Belitz et al., 2009). In the mechanical and wet processes,

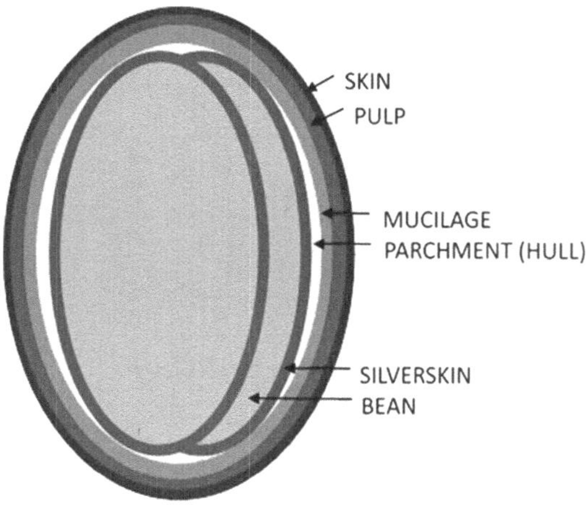

Figure 11.1 Morphology of coffee cherry fruit

fruits are selected during harvesting (only the ripe fruits are picked), and also ripe and unripe cherries are separated by means of density differences in water tanks (Belitz et al., 2009). The skin and the pulp are (partially) separated through pulper sieves (pulping). In the wet processing, pulp remnants and mucilage need to be degraded through controlled fermentation conditions, while in the mechanical processing, the separation is achieved using finer pulping equipment, avoiding the fermentation step. After fermentation, the beans are still covered by parchment, which can be removed after drying and hulling steps, while the silverskin can be removed by a polishing machine (Belitz et al., 2009; Gonzalez-Rios et al., 2007; Joët et al., 2010). In general, wet- and semi-dry-processed coffees are considered to have a superior aroma and higher acceptance (Knopp, Bytof, & Selmar, 2005). Likewise, some authors reported higher contents of chlorogenic acids (CGAs), important bioactive compounds, as will be mentioned later, in wet-processed coffee beans (Duarte, Pereira, & Farah, 2010).

As indicated earlier, several by-products result from the coffee bean processing. In the dry process, there is a single by-product, the coffee husks (Prata & Oliveira, 2007), while the by-product fractions of wet and semi-dry processing are three: skin and pulp in one fraction (43.2% w/w fruit), mucilage and soluble sugars (when not fermented) in a second fraction (11.8% w/w) and, finally, the parchment (6.1% w/w) (Bressani, 1978). Therefore, depending on the processing system, different by-products can be used for further processing and/or utilization.

3 Roasted Coffee Beans and Brews

The coffee beans are subjected to several compositional changes during roasting (Alves, Almeida, Casal, & Oliveira, 2010), resulting in the development of particular

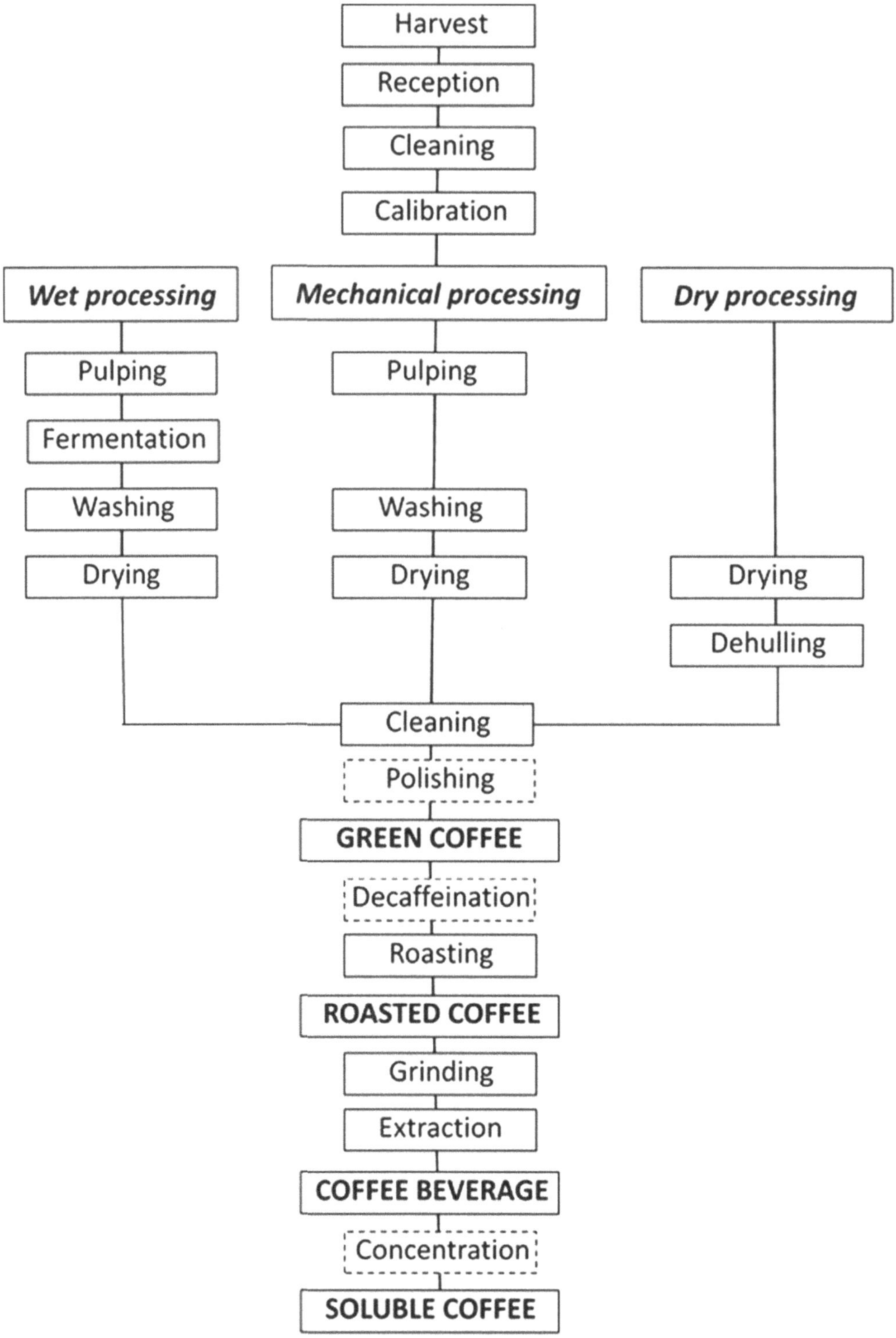

Figure 11.2 Coffee processing scheme

aroma, flavor, and color (Buffo & Cardelli-Freire, 2004). Roasting time and temperature, the latter is usually over 200 °C, affect the physical properties and kinetics of aroma formation (Baggenstoss, Poisson, Kaegi, Perren, & Escher, 2008). The external color of the beans reflects the degree of roasting, affecting flavor properties, among others (Belitz et al., 2009; Franca, Oliveira, Oliveira, Agresti, & Augusti, 2009).

Roasted coffee is composed of carbohydrates (38–42% w/w dry basis), melanoidins (23% w/w), lipids (11–17% w/w), proteins (10% w/w), minerals (4.5–4.7 w/w), CGAs (2.7–3.1% w/w), aliphatic acids (2.4–2.5% w/w), caffeine (1.3–2.4% w/w), and so on. Purine alkaloids found in roasted coffee are theobromine and theophylline (Belitz et al., 2009). From around 850 volatile compounds identified until now in roasted coffee, only around 40 contribute to the aroma (Belitz et al., 2009). During roasting, some polysaccharides are degraded to low molecular weight carbohydrates (Arya & Rao, 2007).

Functional ingredients, such as flavonoids, and caffeic and ferulic acids, are found in the coffee brew (Meletis, 2006). The antioxidant properties of the beverage are mainly attributed to caffeine, CGAs, hydroxycinnamic acids, and Maillard reaction products, such as melanoidins (Patay, Bencsik, & Papp, 2016; Sato et al., 2011; Verzelloni, Tagliazucchi, Del Rio, Calani, & Conte, 2011; Vignoli, Bassoli, & Benassi, 2011). The contents of total CGAs are reduced due to high temperatures commonly used during roasting (Moon, Yoo, & Shibamoto, 2009). However, a reduction in the antioxidant activity of roasted coffee, when compared with green coffee, is not always observed, maybe due to a balance between degradation and formation of antioxidant compounds during the process (Bekedam et al., 2008; Votavová et al., 2009). Antioxidants of the hydroxycinnamic acid group, such as caffeic, chlorogenic, coumaric, ferulic, and sinapic acids, are also found in the coffee beverage (Manach, Scalbert, Morand, Rémésy, & Jiménez, 2004). The highest antioxidant properties were reported for medium-roasted coffee, showing that with higher roasting degrees, the above-mentioned degradation of phenolic compounds may increase (Cho, Park, Kim, Kim, & Han, 2014; del Castillo, Ames, & Gordon, 2002; Sacchetti, Di Mattia, Pittia, & Mastrocola, 2009). In addition, caffeine is considered to have an important contribution to the antioxidant activity of coffee brew (Vignoli et al., 2011). Tocopherols, essential components of vitamin E, are also present in the coffee brew (Alves, Casal, & Oliveira, 2010).

4 Coffee By-Products

Coffee companies annually produce around 2 billion tons of coffee by-products, including husks, skins, pulp, mucilage, parchment, silverskin, and spent coffee. Most of the coffee processing by-products are discarded and not used further (Jiménez-Zamora, Pastoriza, & Rufián-Henares, 2015); however, there is a trend toward valorization of agricultural by-products to reduce waste and also to add value aiming at a profit (Echeverria & Nuti, 2017). Coffee by-products are rich in carbohydrates, proteins, pectins, and bioactive compounds. Efficient recovery of bioactive molecules and the production of value-added products are required for waste management of coffee processing (Murthy & Naidu, 2012).

4.1 Coffee Husks

Coffee husks correspond to the outer skin of the coffee berry, the pulp and the parchment obtained after coffee dry processing. It represents around 12% of the dry coffee bean (Murthy & Naidu, 2012). Coffee husks are composed of 15.0% moisture, 5.4% ash, 7.0% protein, 0.3% lipids, and 72.3% carbohydrates (24.5% cellulose, 29.7% hemicelluloses, 23.7% lignin, and 6.2% ash) (Bekalo & Reinhardt, 2010; Gouvea, Torres, Franca, Oliveira, & Oliveira, 2009), but should not be directly used as animal feed due to the presence of anti-physiological and anti-nutritional factors (e.g., caffeine and tannins) (Brand, Pandey, Roussos, & Soccol, 2000; Orozco et al., 2008). Regarding the direct recovery of valuable compounds, the use of the husks as a source of the anthocyanin cyanidin-3-rutinoside has been suggested (Prata & Oliveira, 2007).

Biotechnological approaches to study the potential of coffee husks have been implemented considering their capacity as a source of organic material. Although uses as fertilizers, livestock feed, and compost have been attempted, they are still not technically efficient. Moreover, fibers of coffee husks were recommended to be used for the production of particle boards, replacing almost 50% of wood (Bekalo & Reinhardt, 2010). The potential of coffee husks as a substrate for bioprocesses was reviewed as well. Coffee husks were reported to be an inexpensive substrate for the production of citric acid by *Aspergillus niger* in a solid-state fermentation system (Shankaranand & Lonsane, 1994). Additional potential uses have been recommended, such as the source of enzymes and aroma compounds, and substrate for vermicomposting. On the other hand, for broader uses, biological detoxification methods have to be considered due to the presence of the above-mentioned anti-physiological and anti-nutritional factors (Brand et al., 2000; Pandey et al., 2000).

4.2 Coffee Skin and Pulp

The coffee fruit skin is considered as an important source of anthocyanins, which may have applications in the food industry as a natural food colorant (Parra-Campos & Ordóñez-Santos, 2019). Cyanidin-3-rutinoside, cyanidin-3-glucoside, and its aglycone were identified as the major anthocyanins in peels and pulp derived from wet-processed fruits, together with important levels of caffeine. Moreover, considerable contents of carotenoids were found in the coffee skin (Esquivel, Kramer, Carle, & Jiménez, 2010). This latter group of pigments has also been obtained after yeast fermentation of coffee husks and pulp (Moreira et al., 2018). This by-product has also been tested for the extraction of reducing sugars by enzymatic hydrolysis of acid cellulose, rendering up to 0.58 mg/mL with 32% glucose of total reducing sugars (Febrianto, 2018b).

In terms of quantity, the coffee pulp is the first by-product resulting from wet processing, and it represents around 29% dry weight of the whole fruit. The pulp is rich in carbohydrates, proteins, and minerals, mostly potassium. The composition of the dry coffee pulp comprises 1.80–8.56% tannins, 6.5% total pectic substances, 12.4% reducing sugars, 2% non-reducing sugars, 1.3% caffeine, 2.6% CGA, and 1.6% caffeic acid (Murthy & Naidu, 2012). While condensed tannins were the main phenolic compounds originally identified in the coffee pulp (Clifford, 1985), in a more recent study, four major classes of polyphenols (namely flavan-3-ols, hydroxycinnamic acids, flavonols, and anthocyanidins) were described in Arabica coffee pulp (Ramirez-Coronel et al., 2004). Recently, coffee pulp extracts showed several biological properties, including inhibition of interleukin-8 release in human gastric epithelial cells, a potential mechanism to treat gastric disorders. Candidates for causing this effect are quinic acid derivatives and procyanidin derivatives (Magoni et al., 2018).

The coffee pulp can also be fermented to obtain bioethanol. Bonilla-Hermosa, Duarte, and Schwan (2014) observed that, from eight yeast strains, *Hanseniaspora uvarum* UFLA CAF76 showed the best fermentation performance. In another work, fermentation of coffee pulp, at laboratory (2 L) and pilot (50 L) scales, using *Bacillus coagulans* allowed obtaining L-(+)-lactic acid with an optical purity of 99.7% (Pleissner et al., 2016).

4.3 Coffee Mucilage

After depulping, coffee mucilage remains adhered to the bean during wet processing. As described earlier, the mucilage can be enzymatically degraded through the action of microorganisms during fermentation or may be mechanically separated. The mucilage is composed of water (84.2%), proteins (8.9%), sugars (4.1%), pectic substances (0.91%), and ash (0.7%) (Belitz et al., 2009). The composition analysis of the alcohol-insoluble residues showed the presence of pectic substances (around 30%), cellulose (around 8%), and neutral non-cellulosic polysaccharides (around 18%). Pectins contained uronic acids (around 60%) with a high degree of methyl esterification and a moderate degree of acetylation (Avallone, Guiraud, Guyot, Olguin, & Brillouet, 2000).

Utilization of coffee mucilage as a substrate to obtain value-added products through biotechnological processes opens the opportunity to expand the number of applications for this by-product. In one example, direct fermentation of coffee mucilage by means of a *Saccharomyces cerevisiae* strain produced bioethanol with very high efficiency (90–94% of theoretically achievable conversion yield) (Orrego, Zapata-Zapata, & Kim, 2018a, 2018b). It was also observed that fermentation of coffee mucilage with yeast extract (as additional N source) by *B. coagulans* produced L-(+)-lactic acid with an optical purity of 99.8% (Neu et al., 2016).

4.4 Coffee Parchment

The parchment is the endocarp covering both halves of the coffee seed. In the dry processing, the parchment is separated from the green coffee beans together with the peel and pulp, in a single step. However, in wet processing, the parchment is removed after drying and hulling in separate steps (Belitz et al., 2009). This coffee residue has been scarcely studied, and its utilization lacks behind that of other coffee by-products. In one of the very few studies conducted, parchment extracts were added to gellan gum-based films, aiming at designing an alternative bioactive food package with antifungal properties (it showed activity against *Fusarium verticillioides, Fusarium* sp., and *Colletotrichum gloeosporioides*) (Mirón-Mérida, Yáñez-Fernández, Montañez-Barragán, & Barragán Huerta, 2019). Care must be taken with direct utilization of this by-product because it can accumulate ochratoxin A, a mycotoxin with nephrotoxic, immunosuppressive, teratogenic, and carcinogenic effects (Suárez-Quiroz et al., 2004).

4.5 Coffee Silverskin

Coffee silverskin is the inner fruit layer surrounding the coffee bean, whose remnants remain attached to the green coffee beans and are only removed during roasting (Belitz et al., 2009; Bresciani, Calani, Bruni, Brighenti, & Del Rio, 2014). They can be easily found as a coffee processing by-product in coffee roasting plants and are presently used as fuel or for composting (Menéndez, Domínguez, Fernández, & Pis, 2007; Saenger, Hartge, Werther, Ogada, & Siagi, 2001). The use of silverskin as a functional ingredient, based on the marked antioxidant activity and the high contents of soluble dietary fiber (86% of total dietary fiber) having cellulose and hemicellulose as main components, has been recommended (Jiménez-Zamora et al., 2015; Napolitano, Fogliano, Tafuri, & Ritieni, 2007). Moreover, monosaccharides, such as glucose, xylose, galactose, mannose, and arabinose, are present in the silverskin, together with diverse proteins and lipids (Bessada, Alves, Costa, Nunes, & Oliveira, 2018; Carneiro, Silva, Musatto, Roberto, & Teixeira, 2009), including linoleic acid and phytosterols (7.0 and 12.1% of lipid fraction, respectively) (Toschi, Cardenia, Bonaga, Mandrioli, & Rodriguez-Estrada, 2014). The most abundant quantified phenolics in coffee silverskin are caffeoylquinic acids, the 5- and 3-isomers being the most relevant ones, amounting 74% of total CGAs. Other important phenolic compounds found in the silverskin are feruloylquinic and coumaroylquinic acids. Probably because of the abundance of these bioactive compounds, silverskin extracts show high antioxidant activity as mentioned previously and polyphenol extraction methods have been developed (Conde & Mussatto, 2016; Costa et al., 2014; Mesías et al., 2014; Narita & Inouye, 2012), which can be imparted to silverskin-reinforced biopolymer-based composites (Sarasini et al.,

2018). Caffeine contents reached 10 mg/g in the silverskin (Bresciani et al., 2014). Notably, in addition, prebiotic and antimicrobial activities of silverskin have been reported (Jiménez-Zamora et al., 2015).

In terms of utilization, Narita and Inouye (2014) in their review proposed the use of coffee silverskin in two ways: one by taking direct advantage of the bioactive properties of its extracts (e.g., inhibiting hyaluronidases, and prebiotic and antioxidant activities), and the other as a substrate for solid-state fermentation for obtaining valuable compounds, such as α-amylase, fructooligosaccharides, and diverse phenolic compounds. In spite of its relevance as a source of bioactive compounds, care should be taken when proposing to include this by-product or its extracts in the human or livestock diet because significant levels of ochratoxin A (like it was observed for the coffee parchment) and phytosterol oxidation products have been reported (Toschi et al., 2014).

4.6 Spent Coffee Grounds

Around 50% of the coffee consumed worldwide is processed as soluble coffee (Ramalakshmi, Rao, Takano-Ishikawa, & Goto, 2009), where 1 ton of green coffee may produce about 650 kg of spent coffee (Murthy & Naidu, 2012), the latter becoming the main coffee industry residue (Mussatto, Carneiro, Silva, Roberto, & Teixeira, 2011). Spent coffee grounds contain large amounts of organic compounds, such as fatty acids, amino acids, polyphenols, minerals, and polysaccharides (Cruz et al., 2012; Panusa, Zuorro, Lavecchia, Marrosu, & Petrucci, 2013; Ramalakshmi et al., 2009; Zuorro & Lavecchia, 2012). This waste could be a valuable resource for the production of hydrophilic bioactive antioxidants (Seo & Park, 2019), with relevant amounts of CGAs, even higher than those of the coffee brew, up to 16 mg gallic acid equivalents/g (Ramalakshmi et al., 2009), with prevalence of dicaffeoylquinic acids (3.31–5.79 mg/g of spent coffee) (Bravo et al., 2012). Melanoidins in spent coffee may exert bacteriostatic activity and may be an antimicrobial alternative for the food industry (Campos-Vega, Loarca-Piña, Vergara-Castañeda, & Oomah, 2015; Jiménez-Zamora et al., 2015). The role of antimutagenic and antimicrobial activities of phenolics, caffeine, and melanoidins present in spent coffee has been discussed as well (Monente et al., 2015). In order to take advantage of these beneficial properties, proper methods for extraction (e.g., microwave-assisted) and delivery (e.g., microencapsulation by using the spray-drying technique) of the bioactive compounds have been developed (Abrahão et al., 2019; Pavlović, Buntić, Šiler-Marinković, & Dimitrijević-Branković, 2013; Ranic et al., 2014).

Spent coffee ground is also rich in sugars, amounting 36.7% (w/w) hemicellulose, composed mainly by mannose, galactose, and arabinose, and cellulose with 8.6% (w/w). Efficiencies of up to 100% were obtained by generating hydrolysates of spent coffee ground for the production of sugars for chemical or fermentation processes, for example, mannose could be used for the production of mannitol, which is highly used in the food industry (Mussatto et al., 2011).

Moreover, the production of short-chain fatty acids from spent coffee ground resulting from fermentation by gut microbiota prevented inflammation, confirming its potential use in the food industry as a dietary fiber source (López-Barrera, Vázquez-Sánchez, Loarca-Piña, & Campos-Vega, 2016). In addition, enzymatic extraction for 2 h with 6% papain allowed extraction of almost 68% of the protein of the spent coffee ground and its hydrolysate possessed relatively high antioxidant activity (Febrianto, 2018a).

Additional extracted fractions have shown interesting properties as well. Spent coffee ground extracts obtained by subcritical water extraction showed antiaging and skin lightening effects by inhibiting elastase and tyrosinase activity (Ribeiro et al., 2018). Dietary fiber extracted from spent coffee grounds, incorporated into biscuits and digested in vitro, releases compounds with the potential to regulate sugar metabolism, and thus antidiabetic effects (Vázquez-Sánchez et al., 2018).

4.7 Green Coffee Press Cake

Another coffee by-product is the defatted green coffee press cake. It is the result of processing green coffee beans for oil extraction and is not part of the process to obtain the coffee beverage, like the other by-products mentioned earlier. This by-product is usually disposed because of the antinutritional potential of its tannins, making it unsuitable for animal feeding. However, a very recent work shows that phenolic antioxidant adjuncts can be obtained from the press cake with potential for the food and pharmaceutical industries (Resende Oliveira, Fonseca da Silva, Ribeiro Santos, & Queiroz, 2019).

5 Conclusion

Considering that coffee by-products represent around 60% of the dry weight of the coffee berry and that coffee is one of the most important food commodities in the world, the relevance of implementing adequate waste-management practices and appropriate utilization of this valuable source of beneficial compounds, by pondering environmental but also economic and health issues, must not be underestimated. Information presented in this review gives an impression of the vast amount of work that has been conducted, aiming at taking advantage of the different coffee by-products that result from the different steps and ways of coffee processing. One option is the direct utilization of some of the by-products as raw material for building purposes and for amendments (not the focus of this chapter), but most relevant and a way to considerably increase the value and cause a deeper impact is the retrieval of bioactive compounds or the manufacture of new products by means of industrial- and biotechnology-based processes. Additional work needs to be conducted for scaling up some of the lab-based methods to the industrial level to be relevant for the manufacturing sector (pharmaceutical, medical, or food) but also to have an impact on waste reduction. Continuous developments, reflected in the sustained number of new publications appearing each year, assure further progress in the near future.

References

Abrahão, F. R., Rocha, L. C. R., Santos, T. A., Carmo, E. L. do, Pereira, L. A. S., Borges, S. V., … Botrel, D. A. (2019). Microencapsulation of bioactive compounds from espresso spent coffee by spray drying. *LWT - Food Science and Technology*, *103*, 116–124. 10.1016/j.lwt.2018.12.061.

Alonso-Salces, R. M., Serra, F., Reniero, F., & Héberger, K. (2009). Botanical and geographical characterization of green coffee (*Coffea arabica* and *Coffea canephora*): Chemometric evaluation of phenolic and methylxanthine contents. *Journal of Agricultural and Food Chemistry*, *57*(10), 4224–4235, 10.1021/jf8037117.

Alves, R. C., Almeida, I. M. C., Casal, S., & Oliveira, M. B. P. P. (2010). Isoflavones in coffee: Influence of species, roast degree, and brewing method. *Journal of Agricultural and Food Chemistry*, *58*(5), 3002–3007, 10.1021/jf9039205.

Alves, R. C., Casal, S., & Oliveira, M. B. P. P. (2010). Tocopherols in coffee brews: Influence of coffee species, roast degree and brewing procedure. *Journal of Food Composition and Analysis*, *23*(8), 802–808. 10.1016/j.jfca.2010.02.009.

Arya, M., & Rao, L. J. M. (2007). An impression of coffee carbohydrates. *Critical Reviews in Food Science and Nutrition*, *47*(1), 51–67. 10.1080/10408390600550315.

Avallone, S., Guiraud, J.-P., Guyot, B., Olguin, E., & Brillouet, J.-M. (2000). Polysaccharide constituents of coffee-bean mucilage. *Journal of Food Science*, *65*(8), 1308–1311. 10.1111/j.1365-2621.2000.tb10602.x.

Baggenstoss, J., Poisson, L., Kaegi, R., Perren, R., & Escher, F. (2008). Coffee roasting and aroma formation: Application of different time-temperature conditions. *Journal of Agricultural and Food Chemistry*, *56*(14), 5836–5846. 10.1021/jf800327j.

Bekalo, S. A., & Reinhardt, H.-W. (2010). Fibers of coffee husk and hulls for the production of particleboard. *Materials and Structures*, *43*(8), 1049–1060. 10.1617/s11527-009 9565–0.

Bekedam, E. K., Schols, H. A., Cämmerer, B., Kroh, L. W., van Boekel, M. A. J. S., & Smit, G. (2008). Electron spin resonance (ESR) studies on the formation of roasting-induced antioxidative structures in coffee brews at different degrees of roast. *Journal of Agricultural and Food Chemistry*, *56*(12), 4597–4604. 10.1021/jf8004004.

Belitz, G., Grosch, W., & Schieberle, P. (2009). *Food Chemistry*. Heidelberg: Springer.

Berbert, P. A., Queiroz, D. M., & Melo, E. C. (2002). PH—postharvest technology: Dielectric properties of common bean. *Biosystems Engineering*, *83*(4), 449–462. 10.1006/bioe.2002.0135.

Bessada, S. M. F., Alves, R. C., Costa, A. S. G., Nunes, M. A., & Oliveira, M. B. P. P. (2018). *Coffea canephora* silverskin from different geographical origins: A comparative study. *Science of The Total Environment*, *645*, 1021–1028. 10.1016/j.scitotenv.2018.07.201.

Blinová, L., Sirotiak, M., Bartošová, A., & Soldán, M. (2017). Review: Utilization of waste from coffee production. *Research Papers Faculty of Materials Science and Technology Slovak University of Technology*, *25*(40), 91–101. Retrieved from https://content.sciendo.com/view/journals/rput/25/40/article-p91.xml.

Bonilla-Hermosa, V. A., Duarte, W. F., & Schwan, R. F. (2014). Utilization of coffee by-products obtained from semi-washed process for production of value-added compounds. *Bioresource Technology*, *166*, 142–150. 10.1016/j.biortech.2014.05.031.

Brand, D., Pandey, A., Roussos, S., & Soccol, C. R. (2000). Biological detoxification of coffee husk by filamentous fungi using a solid state fermentation system. *Enzyme and Microbial Technology*, *27*(1), 127–133. 10.1016/S0141–0229(00)00186–1.

Bravo, J., Juániz, I., Monente, C., Caemmerer, B., Kroh, L. W., De Peña, M. P., & Cid, C. (2012). Evaluation of spent coffee obtained from the most common coffeemakers as a source of hydrophilic bioactive compounds. *Journal of Agricultural and Food Chemistry*, *60*(51), 12565–12573. 10.1021/jf3040594.

Bresciani, L., Calani, L., Bruni, R., Brighenti, F., & Del Rio, D. (2014). Phenolic composition, caffeine content and antioxidant capacity of coffee silverskin. *Food Research International*, *61*, 196–201. 10.1016/j.foodres.2013.10.047.

Bressani, R. (1978). The by-products of coffee berries. In J. E. Braham & R. Bressani (Eds.), *Coffee Pulp - Composition, Technology and Utilization* (pp. 9–17). Canada: INCAP.

Buffo, R. A., & Cardelli-Freire, C. (2004). Coffee flavour: An overview. *Flavour and Fragrance Journal*, *19*(2), 99–104. 10.1002/ffj.1325.

Campos-Vega, R., Loarca-Piña, G., Vergara-Castañeda, H. A., & Oomah, B. D. (2015). Spent coffee grounds: A review on current research and future prospects. *Trends in Food Science & Technology*, *45*(1), 24–36. 10.1016/j.tifs.2015.04.012.

Carneiro, L. M., Silva, J. P. A., Musatto, S. I., Roberto, I. C., & Teixeira, J. A. (2009). Determination of total carbohydrates content in coffee industry residues. In *8th International Meeting of the Portuguese Carbohydrate Group GLUPOR*. (p. 94). Braga.

Cho, A. R., Park, K. W., Kim, K. M., Kim, S. Y., & Han, J. (2014). Influence of roasting conditions on the antioxidant characteristics of Colombian coffee (*Coffea arabica* L.) beans. *Journal of Food Biochemistry*, *38*(3), 271–280. 10.1111/jfbc.12045.

Clifford, M. N. (1985). Chlorogenic acids. In: R. J. Clarke & R. Macrae (Eds.), *Coffee* (Vol. 1, pp. 153–202). London: Elsevier.

Conde, T., & Mussatto, S. I. (2016). Isolation of polyphenols from spent coffee grounds and silverskin by mild hydrothermal pretreatment. *Preparative Biochemistry and Biotechnology*, *46*(4), 406–409. 10.1080/10826068.2015.1084514.

Costa, A. S. G., Alves, R. C., Vinha, A. F., Barreira, S. V. P., Nunes, M. A., Cunha, L. M., & Oliveira, M. B. P. P. (2014). Optimization of antioxidants extraction from coffee silverskin, a roasting by-product, having in view a sustainable process. *Industrial Crops and Products*, *53*, 350–357. 10.1016/j.indcrop.2014.01.006.

Cruz, R., Cardoso, M. M., Fernandes, L., Oliveira, M., Mendes, E., Baptista, P., … Casal, S. (2012). Espresso coffee residues: A valuable source of unextracted compounds. *Journal of Agricultural and Food Chemistry*, *60*(32), 7777–7784. 10.1021/jf3018854.

del Castillo, M. D., Ames, J. M., & Gordon, M. H. (2002). Effect of roasting on the antioxidant activity of coffee brews. *Journal of Agricultural and Food Chemistry*, *50*(13), 3698–3703. 10.1021/jf011702q.

Dórea, J. G., & da Costa, T. H. M. (2005). Is coffee a functional food? *British Journal of Nutrition*, *93*(6), 773–782. 10.1079/BJN20051370.

Duarte, G. S., Pereira, A. A., & Farah, A. (2010). Chlorogenic acids and other relevant compounds in Brazilian coffees processed by semi-dry and wet post-harvesting methods. *Food Chemistry*, *118*(3), 851–855. 10.1016/j.foodchem.2009.05.042.

Echeverria, M. C., & Nuti, M. (2017). Valorisation of the residues of coffee agro-industry: Perspectives and limitations. *The Open Waste Management Journal*, *10*(1). 10.2174/1876400201710010013.

Esquivel, P., & Jiménez, V. M. (2012). Functional properties of coffee and coffee by-products. *Food Research International*, *46*(2), 488–495. 10.1016/j.foodres.2011.05.028.

Esquivel, P., Kramer, M., Carle, R., & Jiménez, V. M. (2010). Anthocyanin profiles and caffeine contents of wet-processed coffee (*Coffea arabica*) husks by HPLC DAD-MS/MS. In

28th International Horticultural Congress. Book of Abstracts. (pp. 129–130). Association for Science and Informaton on Coffee.

Febrianto, N. A. (2018a). Liberation of protein and antioxidative compound from spent coffee ground through protein hydrolysis. *Pelita Perkebunan (a Coffee and Cocoa Research Journal), 34*(2), 95–103.

Febrianto, N. A. (2018b). Utilization of coffee skin fiber as potential source of reducing sugar by means on enzymatic hydrolysis. *Pelita Perkebunan (a Coffee and Cocoa Research Journal), 34*(3), 166–174.

Franca, A. S., Oliveira, L. S., Oliveira, R. C. S., Agresti, P. C. M., & Augusti, R. (2009). A preliminary evaluation of the effect of processing temperature on coffee roasting degree assessment. *Journal of Food Engineering, 92*(3), 345–352. 10.1016/j.jfoodeng.2008.12.012.

Gonzalez de Mejia, E., & Ramirez-Mares, M. V. (2014). Impact of caffeine and coffee on our health. *Trends in Endocrinology & Metabolism, 25*(10), 489–492. 10.1016/j.tem.2014.07.003.

Gonzalez-Rios, O., Suarez-Quiroz, M. L., Boulanger, R., Barel, M., Guyot, B., Guiraud, J.-P., & Schorr-Galindo, S. (2007). Impact of "ecological" post-harvest processing on the volatile fraction of coffee beans: I. Green coffee. *Journal of Food Composition and Analysis, 20* (3), 289–296. 10.1016/j.jfca.2006.07.009.

Gouvea, B. M., Torres, C., Franca, A. S., Oliveira, L. S., & Oliveira, E. S. (2009). Feasibility of ethanol production from coffee husks. *Biotechnology Letters, 31*(9), 1315–1319. 10.1007/s10529-009-0023-4.

Jiménez-Zamora, A., Pastoriza, S., & Rufián-Henares, J. A. (2015). Revalorization of coffee by-products. Prebiotic, antimicrobial and antioxidant properties. *LWT - Food Science and Technology, 61*(1), 12–18. 10.1016/j.lwt.2014.11.031.

Joët, T., Laffargue, A., Descroix, F., Doulbeau, S., Bertrand, B., kochko, A. de, & Dussert, S. (2010). Influence of environmental factors, wet processing and their interactions on the biochemical composition of green Arabica coffee beans. *Food Chemistry, 118*(3), 693–701. 10.1016/j.foodchem.2009.05.048.

Knopp, S., Bytof, G., & Selmar, D. (2005). Influence of processing on the content of sugars in green Arabica coffee beans. *European Food Research and Technology, 223*(2), 195. 10.1007/s00217-005-0172-1.

López-Barrera, D. M., Vázquez-Sánchez, K., Loarca-Piña, M. G. F., & Campos-Vega, R. (2016). Spent coffee grounds, an innovative source of colonic fermentable compounds, inhibit inflammatory mediators in vitro. *Food Chemistry, 212*, 282–290. 10.1016/j.foodchem.2016.05.175.

Magoni, C., Bruni, I., Guzzetti, L., Dell'Agli, M., Sangiovanni, E., Piazza, S., … Labra, M. (2018). Valorizing coffee pulp by-products as anti-inflammatory ingredient of food supplements acting on IL-8 release. *Food Research International, 112*, 129–135. 10.1016/j.foodres.2018.06.026.

Manach, C., Scalbert, A., Morand, C., Rémésy, C., & Jiménez, L. (2004). Polyphenols: Food sources and bioavailability. *The American Journal of Clinical Nutrition, 79*(5), 727–747. 10.1093/ajcn/79.5.727.

Martini, D., Del Bo', C., Tassotti, M., Riso, P., Del Rio, D., Brighenti, F., & Porrini, M. (2016). Coffee consumption and oxidative stress: A review of human intervention studies. *Molecules, 21*(8). 10.3390/molecules21080979.

Meletis, C. D. (2006). Coffee – Functional food and medicinal herb. *Alternative and Complementary Therapies, 12*(1), 7–13. 10.1089/act.2006.12.7.

Menéndez, J. A., Domínguez, A., Fernández, Y., & Pis, J. J. (2007). Evidence of self-gasification during the microwave-induced pyrolysis of coffee hulls. *Energy & Fuels, 21*(1), 373–378. 10.1021/ef060331i.

Mesías, M., Navarro, M., Martínez-Saez, N., Ullate, M., del Castillo, M. D., & Morales, F. J. (2014). Antiglycative and carbonyl trapping properties of the water soluble fraction of coffee silverskin. *Food Research International*, *62*, 1120–1126. 10.1016/j.foodres.2014.05.058.

Mirón-Mérida, V. A., Yáñez-Fernández, J., Montañez-Barragán, B., & Barragán Huerta, B. E. (2019). Valorization of coffee parchment waste (*Coffea arabica*) as a source of caffeine and phenolic compounds in antifungal gellan gum films. *LWT - Food Science and Technology*, *101*, 167–174. 10.1016/j.lwt.2018.11.013.

Monente, C., Bravo, J., Vitas, A. I., Arbillaga, L., De Peña, M. P., & Cid, C. (2015). Coffee and spent coffee extracts protect against cell mutagens and inhibit growth of food-borne pathogen microorganisms. *Journal of Functional Foods*, *12*, 365–374. 10.1016/j.jff.2014.12.006.

Moon, J.-K., Yoo, H. S., & Shibamoto, T. (2009). Role of roasting conditions in the level of chlorogenic acid content in coffee beans: Correlation with coffee acidity. *Journal of Agricultural and Food Chemistry*, *57*(12), 5365–5369. 10.1021/jf900012b.

Moreira, M. D., Melo, M. M., Coimbra, J. M., Reis, K. C. dos, Schwan, R. F., & Silva, C. F. (2018). Solid coffee waste as alternative to produce carotenoids with antioxidant and antimicrobial activities. *Waste Management*, *82*, 93–99. 10.1016/j.wasman.2018.10.017.

Murthy, P. S., & Naidu, M. M. (2012). Sustainable management of coffee industry by-products and value addition—A review. *Resources, Conservation and Recycling*, *66*, 45–58. 10.1016/j.resconrec.2012.06.005.

Mussatto, S. I., Carneiro, L. M., Silva, J. P. A., Roberto, I. C., & Teixeira, J. A. (2011). A study on chemical constituents and sugars extraction from spent coffee grounds. *Carbohydrate Polymers*, *83*(2), 368–374. 10.1016/j.carbpol.2010.07.063.

Napolitano, A., Fogliano, V., Tafuri, A., & Ritieni, A. (2007). Natural occurrence of ochratoxin A and antioxidant activities of green and roasted coffees and corresponding byproducts. *Journal of Agricultural and Food Chemistry*, *55*(25), 10499–10504. 10.1021/jf071959+.

Narita, Y., & Inouye, K. (2012). High antioxidant activity of coffee silverskin extracts obtained by the treatment of coffee silverskin with subcritical water. *Food Chemistry*, *135*(3), 943–949. 10.1016/j.foodchem.2012.05.078.

Narita, Y., & Inouye, K. (2014). Review on utilization and composition of coffee silverskin. *Food Research International*, *61*, 16–22. 10.1016/j.foodres.2014.01.023.

Neu, A.-K., Pleissner, D., Mehlmann, K., Schneider, R., Puerta-Quintero, G. I., & Venus, J. (2016). Fermentative utilization of coffee mucilage using *Bacillus coagulans* and investigation of down-stream processing of fermentation broth for optically pure L(+)-lactic acid production. *Bioresource Technology*, *211*, 398–405. 10.1016/j.biortech.2016.03.122.

O'Keefe, J. H., DiNicolantonio, J. J., & Lavie, C. J. (2018). Coffee for cardioprotection and longevity. *Dietary Prevention of Cardiovascular Diseases*, *61*(1), 38–42. 10.1016/j.pcad.2018.02.002.

Orozco, A. L., Pérez, M. I., Guevara, O., Rodríguez, J., Hernández, M., González-Vila, F. J., … Arias, M. E. (2008). Biotechnological enhancement of coffee pulp residues by solid-state fermentation with *Streptomyces*. Py–GC/MS analysis. *Journal of Analytical and Applied Pyrolysis*, *81*(2), 247–252. 10.1016/j.jaap.2007.12.002.

Orrego, D., Zapata-Zapata, A. D., & Kim, D. (2018a). Ethanol production from coffee mucilage fermentation by *S. cerevisiae* immobilized in calcium-alginate beads. *Bioresource Technology Reports*, *3*, 200–204. 10.1016/j.biteb.2018.08.006.

Orrego, D., Zapata-Zapata, A. D., & Kim, D. (2018b). Optimization and scale-up of coffee mucilage fermentation for ethanol production. *Energies*, *11*(4), 786. 10.3390/en11040786.

Pandey, A., Soccol, C. R., Nigam, P., Brand, D., Mohan, R., & Roussos, S. (2000). Biotechnological potential of coffee pulp and coffee husk for bioprocesses. *Biochemical Engineering Journal*, *6*(2), 153–162. 10.1016/S1369-703X(00)00084–X.

Panusa, A., Zuorro, A., Lavecchia, R., Marrosu, G., & Petrucci, R. (2013). Recovery of natural antioxidants from spent coffee grounds. *Journal of Agricultural and Food Chemistry*, *61*(17), 4162–4168. 10.1021/jf4005719.

Parra-Campos, A., & Ordóñez-Santos, L. E. (2019). Natural pigment extraction optimization from coffee exocarp and its use as a natural dye in French meringue. *Food Chemistry*, *285*, 59–66. 10.1016/j.foodchem.2019.01.158.

Patay, É. B., Bencsik, T., & Papp, N. (2016). Phytochemical overview and medicinal importance of *Coffea* species from the past until now. *Asian Pacific Journal of Tropical Medicine*, *9* (12), 1127–1135. 10.1016/j.apjtm.2016.11.008.

Pavlović, M. D., Buntić, A. V., Šiler-Marinković, S. S., & Dimitrijević-Branković, S. I. (2013). Ethanol influenced fast microwave-assisted extraction for natural antioxidants obtaining from spent filter coffee. *Separation and Purification Technology*, *118*, 503–510. 10.1016/j.seppur.2013.07.035.

Pleissner, D., Neu, A.-K., Mehlmann, K., Schneider, R., Puerta-Quintero, G. I., & Venus, J. (2016). Fermentative lactic acid production from coffee pulp hydrolysate using *Bacillus coagulans* at laboratory and pilot scales. *Bioresource Technology*, *218*, 167–173. 10.1016/j.biortech.2016.06.078.

Prata, E. R. B. A., & Oliveira, L. S. (2007). Fresh coffee husks as potential sources of anthocyanins. *LWT - Food Science and Technology*, *40*(9), 1555–1560. 10.1016/j.lwt.2006.10.003.

Purseglove, J. W. (1974). Rubiaceae. In J. W. Purseglove (Ed.), *Tropical Crops Dicotyledons* (pp. 451–492). London: The English Language Book Society and Longman Group Ltd.

Ramalakshmi, K., Rao, L. J. M., Takano-Ishikawa, Y., & Goto, M. (2009). Bioactivities of low-grade green coffee and spent coffee in different in vitro model systems. *Food Chemistry*, *115*(1), 79–85. 10.1016/j.foodchem.2008.11.063.

Ramirez-Coronel, M. A., Marnet, N., Kolli, V. S. K., Roussos, S., Guyot, S., & Augur, C. (2004). Characterization and estimation of proanthocyanidins and other phenolics in coffee pulp (*Coffea arabica*) by thiolysis-high-performance liquid chromatography. *Journal of Agricultural and Food Chemistry*, *52*(5), 1344–1349. 10.1021/jf035208t.

Ranheim, T., & Halvorsen, B. (2005). Coffee consumption and human health – Beneficial or detrimental? – Mechanisms for effects of coffee consumption on different risk factors for cardiovascular disease and type 2 diabetes mellitus. *Molecular Nutrition & Food Research*, *49*(3), 274–284. 10.1002/mnfr.200400109.

Ranic, M., Nikolic, M., Pavlovic, M., Buntic, A., Siler-Marinkovic, S., & Dimitrijevic-Brankovic, S. (2014). Optimization of microwave-assisted extraction of natural antioxidants from spent espresso coffee grounds by response surface methodology. *Journal of Cleaner Production*, *80*, 69–79. 10.1016/j.jclepro.2014.05.060.

Resende Oliveira, É., Fonseca da Silva, R., Ribeiro Santos, P., & Queiroz, F. (2019). An investigation into green coffee press cake as a renewable source of bioactive compounds. *International Journal of Food Science & Technology*. 10.1111/ijfs.14074.

Ribeiro, H. M., Allegro, M., Marto, J., Pedras, B., Oliveira, N. G., Paiva, A., … Simões, P. (2018). Converting spent coffee grounds into bioactive extracts with potential skin antiaging and lightening effects. *ACS Sustainable Chemistry & Engineering*, *6*(5), 6289–6295. 10.1021/acssuschemeng.8b00108.

Sacchetti, G., Di Mattia, C., Pittia, P., & Mastrocola, D. (2009). Effect of roasting degree, equivalent thermal effect and coffee type on the radical scavenging activity of coffee brews and their phenolic fraction. *Journal of Food Engineering*, *90*(1), 74–80. 10.1016/j.jfoodeng.2008.06.005.

Saenger, M., Hartge, E.-U., Werther, J., Ogada, T., & Siagi, Z. (2001). Combustion of coffee husks. *Renewable Energy*, *23*(1), 103–121. 10.1016/S0960–1481(00)00106–3.

Sarasini, F., Tirillò, J., Zuorro, A., Maffei, G., Lavecchia, R., Puglia, D., … Torre, L. (2018). Recycling coffee silverskin in sustainable composites based on a poly(butylene adipate-co-terephthalate)/poly(3-hydroxybutyrate-co-3-hydroxyvalerate) matrix. *Industrial Crops and Products*, *118*, 311–320. 10.1016/j.indcrop.2018.03.070.

Sato, Y., Itagaki, S., Kurokawa, T., Ogura, J., Kobayashi, M., Hirano, T., … Iseki, K. (2011). In vitro and in vivo antioxidant properties of chlorogenic acid and caffeic acid. *International Journal of Pharmaceutics*, *403*(1), 136–138. 10.1016/j.ijpharm.2010.09.035.

Seo, H. S., & Park, B. H. (2019). Phenolic compound extraction from spent coffee grounds for antioxidant recovery. *Korean Journal of Chemical Engineering*, *36*(2), 186–190. 10.1007/s11814-018-0208–4.

Shankaranand, V. S., & Lonsane, B. K. (1994). Coffee husk: An inexpensive substrate for production of citric acid by *Aspergillus niger* in a solid-state fermentation system. *World Journal of Microbiology and Biotechnology*, *10*(2), 165–168. 10.1007/BF00360879.

Suárez-Quiroz, M., González-Rios, O., Barel, M., Guyot, B., Schorr-Galindo, S., & Guiraud, J.-P. (2004). Study of ochratoxin A-producing strains in coffee processing. *International Journal of Food Science & Technology*, *39*(5), 501–507. 10.1111/j.1365-2621.2004.00810.x.

Toschi, T. G., Cardenia, V., Bonaga, G., Mandrioli, M., & Rodriguez-Estrada, M. T. (2014). Coffee silverskin: Characterization, possible uses, and safety aspects. *Journal of Agricultural and Food Chemistry*, *62*(44), 10836–10844. 10.1021/jf503200z.

Vázquez-Sánchez, K., Martinez-Saez, N., Rebollo-Hernanz, M., del Castillo, M. D., Gaytán-Martínez, M., & Campos-Vega, R. (2018). In vitro health promoting properties of antioxidant dietary fiber extracted from spent coffee (*Coffea arabica* L.) grounds. *Food Chemistry*, *261*, 253–259. 10.1016/j.foodchem.2018.04.064.

Verzelloni, E., Tagliazucchi, D., Del Rio, D., Calani, L., & Conte, A. (2011). Antiglycative and antioxidative properties of coffee fractions. *Food Chemistry*, *124*(4), 1430–1435. 10.1016/j.foodchem.2010.07.103.

Vignoli, J. A., Bassoli, D. G., & Benassi, M. T. (2011). Antioxidant activity, polyphenols, caffeine and melanoidins in soluble coffee: The influence of processing conditions and raw material. *Food Chemistry*, *124*(3), 863–868. 10.1016/j.foodchem.2010.07.008.

Votavová, L., Voldřich, M., Ševčík, R., Čížková, H., Mlejnecká, J., Stolar, M., & Fleišman, T. (2009). Changes of antioxidant capacity of Robusta coffee during roasting. *Czech Journal of Food Science*, *27*, S49–S52.

Wang, L., Du, L.-D., & Du, G.-H. (2018). Caffeine. In G.-H. Du (Ed.), *Natural Small Molecule Drugs from Plants* (pp. 199–204). Singapore: Springer. 10.1007/978-981-10-8022-7_32.

Zuorro, A., & Lavecchia, R. (2012). Spent coffee grounds as a valuable source of phenolic compounds and bioenergy. *Journal of Cleaner Production*, *34*, 49–56. 10.1016/j.jclepro.2011.12.003.

12

Extraction of Valuable Compounds from Leaf Vegetables By-Products

JOÃO CARLOS MARTINS BARREIRA AND ISABEL CRISTINA FERNANDES RODRIGUES FERREIRA

Centro de Investigação de Montanha, Instituto Politécnico de Bragança, Campus de Santa Apolónia, 5300-253 Bragança, Portugal

Contents

1 Introduction

Fruits and vegetables represent an essential part of human diet, and their consumption has been stimulated by the medical community, demanding increasingly higher production levels to fulfil these challenging dietary habits (Vilariño et al., 2017). Data from 2016 indicate primary vegetables as the fourth commodity produced worldwide (1,075,203,877 tons), only surpassed by cereals, sugarcane, and coarse grains. Some of the most representative vegetables include potato: 376,826,967 tons; tomatoes: 177,042,359 tons; cabbages and other brassicas: 71,259,199 tons; carrots and turnips: 42,711,830 tons; lettuce and chicory: 26,779,564 tons; or spinach: 26,684,493 tons ("FAOSTAT," 2018).

This amazing increase in production level was not properly followed by the adoption of proper handling methods, nor the development of suitable infrastructures, conducting to massive losses and generating huge quantities of wastes and by-products. In fact, it has been estimated that the global volume of food wastage (i.e., the sum of food losses and food waste) is more than 25% (1.6 gigatons) of total agricultural production (for food and nonfood uses), which is about 6 gigatons per year (Food and Agriculture Organization of the United Nations, 2013). To aggravate this scenario, losses and waste of horticultural products are generated in higher percentages (up to 60%) than in other cases (Gustavsson et al., 2011). A possible explanation for this fact is related with the high number of stages that these products undergo: harvesting, transportation, grading, storage, industrial processing, market distribution, handling by consumers, among others (Parfitt et al., 2010). Fruit and vegetable processing units are responsible for producing the highest volume of wastes, immediately after household garbage (Gowe, 2015).

In addition to their direct disadvantages, the huge amounts of wastes also contribute to major environmental problems, mainly due to the high emission of harmful greenhouse gases throughout their decomposition in landfills (Vilariño et al., 2017; Wijngaard et al., 2012), thereby demanding effective management strategies to prevent serious environmental problems. Such strategies could be highly facilitated if associated with suitable processes of biomass valorization, or as currently designated, biorefineries. This is particularly interesting if we consider the colossal volume of horticultural wastes and by-products, as well as their practical low cost when compared to other primary sources (Lin et al., 2014).

At present, the most well-known examples of agri-food wastes valorization include animal feed, compost, biomass-based production of energy, and a wide variety of industrial products, such as wood-based panels, bio-fertilizers, or biofibers (Santana-Méridas et al., 2012).

However, these agri-food matrices have also proved to be rich sources of valuable bioactive and functional compounds (vitamins, proteins, alkaloids, phenolic compounds, steroids, triterpenoids, dietary fibers, volatile compounds, among others) (Lin et al., 2014) associated with different functionalities. Accordingly, a suitable valorization of bioresources should start by an adequate extraction of these valuable biomolecules, followed by a multistage valorization process aligned with the maximization of added value associated with a given resource (Rombaut et al., 2014). This concept might also be analyzed from a qualitative point of view; that is, more than one bioactive component or functional component could be obtained from a given horticultural waste or by-product, by applying sequential processes suitable to recover different types of valuable biomolecules (Galanakis, 2012).

Actually, the biorefinery concept has been gaining relevance, especially concerning the conversion of biomass to produce fuels, power, heat, or value-added chemicals (Zuin and Ramin, 2018). Independently of the selected approach, the identification and evaluation of potentially high-value compounds (or enriched extracts) that may be recovered from horticultural wastes and by-products must be achieved through efficient and reproducible methodologies, fulfilling green chemistry principles. Hence, the extraction and purification of bioactive compounds from that remaining material represent an attractive approach, both from the socioenvironmental or economic point of view. In either case, the overall process demands a continuous effort to develop groundbreaking research to achieve the most sustainable and efficient extraction methodologies (Zuin and Ramin, 2018).

This whole paradigm is furthermore associated with some currently relevant concepts, such as bioeconomy and circular economy. Circular economy represents a quite appealing approach, as it might be characterized as a system that is restorative by intention and design. Fundamentally, it intends to replace the orthodox end-of-life idea by the regeneration concept, to increase the use of renewable energy sources, minimizing the use of toxic chemicals, and generally eliminating the existence of waste, all supported in an optimized design of materials, products, and systems (The Ellen MacArthur Foundation, 2012).

In the next sections, selected examples of the most valuable bioactive compounds that might be found in leaf vegetables by-products and wastes, as well as associated extraction technologies and their most prominent bioactivities, will be described.

2 Extraction of Bioactive/Functional Compounds from Leaf Vegetable By-Products

Leaf vegetable by-products are frequently subjected to conventional extraction processes (e.g., maceration, decoction, infusion, Soxhlet extraction, among others). Nevertheless, these processes require high energy levels and implicate using high volumes of solvent, which is being tentatively overcome by applying alternative methods, such as ultrasound-assisted extraction (UAE), microwave-assisted extraction (MAE), pressurized liquid extraction (PLE), supercritical fluid extraction (SFE), or enzyme-assisted extraction (EAE). One of the main advantages of these processes is their higher proximity to the principles of Green Extraction of Natural Products (Chemat et al., 2012):

i. Innovation by **selection of varieties** and use of **renewable plant resources**.
ii. Use of **alternative solvents**, mainly **water** or **agro-solvents**.
iii. **Reduce energy consumption** through **recovery processes** and **innovative technologies**.
iv. Production of **by-products instead of waste** to be aligned with the **bio- and agro-refining premises**.

v. **Reduce unit operations** and favor **safe, robust, and controlled processes**.
vi. Aim for a **nondenatured and biodegradable extract without contaminants**.

The fulfilling of the fourth principle is mandatory in the extraction of valuable bioactive molecules, but this process should also consider the great variability of natural matrices, as well as technological aspects, such as the extraction process, solvent type, or operating conditions, more adapted to a given matrix. A series of works has detailed the extraction conditions that are more appropriate to different kinds of natural products, including the book published by Galanakis (2015), the book chapter by Carciochi et al. (2017), and different review papers (Araújo et al., 2015; Baaka et al., 2017; Kammerer et al., 2014; Roselló-Soto et al., 2015; Teixeira et al., 2014).

In general, the extraction process has great influence over downstream processes, such as the identification and characterization of bioactive compounds. Generally, and regardless of bioactive compounds type, the extraction techniques are designed to:

i. extract specific target compounds;
ii. increase the overall selectivity of a given analytical method;
iii. increase the sensitivity of bioassays performed afterwards;
iv. convert the target compounds into chemical derivatives more easy to be detected and separated; and
v. provide a generally strong and reproducible method (Vieira da Silva et al., 2016).

Some operational characteristics of different extraction processes will be described in next sections.

2.1 Ultrasound-Assisted Extraction

UAE is based on the application of ultrasonic vibrations to the sample to be extracted. The major drawback is related to its high dependency on the solvent type (usually methanol, acetone, water, or ethyl acetate), sample size, pH, temperature, and pressure. In what concerns solvent type, using water or water–ethanol mixtures is generally preferred, as it fulfils the second principle of Green Extraction of Natural Products. On the advantages side, UAE is a relatively simple process and allows extracting several samples simultaneously. Likewise, UAE takes lesser time and is more effective than related processes (allowing a reduction in energy consumption), besides using moderate temperatures (preserving heat-sensitive compounds) (Carciochi et al., 2017).

In solid–liquid UAE, the mechanical waves (typically varying from 10 to 20 MHz) transmitted through the liquid medium dislocate particles longitudinally, while the sound source creates successive compressions and distensions (like a piston) on the medium (Chemat et al., 2011). The distensions move away the liquid molecules, sometimes beyond the critical molecular distance, creating voids (cavitation bubbles). These bubbles increase in size during distension and decrease throughout compression. When the bubble size reaches a critical size (at maximum distension), they end up by collapsing during the next compression step, causing a violent implosion that releases a large amount of energy. This collapse induces mechanical effects (e.g., shockwave damages or microjet impacts) in the solid–liquid interface (Esclapez et al., 2011), which might disrupt the cell walls of plant cells, increasing the solvent penetration and improving the extraction efficiency (Toma et al., 2001).

Considering the information mentioned above, it might become surprising that UAE preserves heat-sensitive compounds, as it could be expected to reach very high temperatures. However, this is not the case, since the described reactions are extremely fast, occurring without changing macroscopic properties (Roselló-Soto et al., 2015).

In the specific field of vegetable leaves, UAE has been successfully applied to extract phenolic compounds (Galván D'Alessandro et al., 2014; Pingret et al., 2012; Virot et al., 2010), polysaccharides, proteins, aromatic compounds, carotenoids, or sterols (Vilkhu et al., 2008; Villares et al., 2012; Wu et al., 2015).

In general, UAE allows shorter and more efficient extraction times, simplifying the handling and work-up conditions; it is a versatile and low-cost technology, and it is suitable to be performed with green solvents and to be used in upscaling processes (Barba et al., 2016).

However, to be used in its full potential, UAE still needs important analytical studies, namely to optimize the most adequate ultrasonic power/frequency, temperature, solvent type, particle size, or sample-to-solvent ratio, that is, to improve the kinetics of the whole process (Carciochi and Dimitrov, 2014).

2.2 Microwave-Assisted Extraction

Microwaves have two orthogonal components—magnetic field and electric field (Letellier and Budzinski, 1999). The basic working principle of MAE is related to the direct effect of microwaves through ionic conduction and dipole rotation. Hence, the heating effect of microwaves is particularly verified in polar solvents (specifically those with a permanent dipole moment, such as water or methanol) and ionic solutions (usually acids). In turn, nonpolar solvents, such as hexane, will not heat with microwaves (Wu et al., 2012). MAE process might be divided into three main steps (Alupului et al., 2012):

i. **Splitting and separation of solute molecules** due to the increase in pressure and temperature.
ii. **Solvent diffusion** through the sample matrix.
iii. **Liberation of different solutes** into the solvent.

When used for assisting the extraction of bioactive compounds from natural matrices (typically by applying electromagnetic fields at frequencies ranging from 0.3 to 300 GHz), MAE presents several advantages, such as the possibility of using temperature gradients, requiring small-size equipment, allowing high extraction yields in shorter periods when compared with conventional techniques, or working with "green" solvents (Alupului et al., 2012; Cravotto et al., 2008). Actually, its typically short duration makes MAE highly suitable for extracting heat-sensitive bioactive compounds (Chumnanpaisont et al., 2014). Another important advantage is related to the possibility of working without solvents (a process designated solvent-free MAE), but this is mainly appropriate for extracting essential oils (Chemat et al., 2011). The described advantages, together with the fact of being a nondestructive technique, as well as working with low solvent volumes (or no solvents at all), make MAE a very interesting alternative to food industry, particularly because it also achieves high quality extracts (Baiano et al., 2014; Dahmoune et al., 2015). In fact, this technique has been extensively used in analytical chemistry to extract different biomolecules (Carciochi et al., 2017).

Despite its wide utilization, MAE processes are still far from being fully optimized. Generally, the optimization of any given MAE process should include (Madej, 2009; Wu et al., 2012):

i. **Solvent type and composition**: specifically its capacity to absorb microwaves and its affinity toward the molecule to be extracted.
ii. **Solvent volume**: despite depending on sample size and type, it should be around 10 times less than the volume used in conventional extraction methods.
iii. **Solid-to-solvent ratio**
iv. **Microwave power**: in general, high-power microwaves are preferred to reduce the extraction time; nevertheless, above some critical values this might degrade the target compounds or induce solvent boiling, hindering the contact with the solute (in open vessels).
v. **Exposure time**: should be the shortest possible, inclusively because extending it beyond specific values will not necessarily increase the extractability (in the case of thermolabile compounds it is more prone to decrease).
vi. **Temperature**: this will depend, in any case, on microwave power and extraction time.

Owing to the described properties, MAE has an important track of success in the recovery of valuable substances from agri-food by-products (Carciochi et al., 2017).

2.3 Pressurized Liquid Extraction

PLE technique (also known as pressurized solvent extraction, accelerated solvent extraction, or enhanced solvent extraction) employs liquid solvents at high temperature and pressure (but without, obviously, reaching the critical point values) allowing to achieve higher solubility and mass transfer than those obtained at NTP (normal temperature and pressure) (Pronyk and Mazza, 2009). The reached temperatures decrease the viscosity and surface tension of the solvent, improving its diffusion and capacity to break the solute–matrix bonds, thereby increasing the extraction effectiveness (Ramos et al., 2002).

PLE fulfils the second principle defined in the Green Extraction of Natural Products, especially when performed with water or ethanol. In general terms, PLE might be operated in two different modes (Carabias-Martínez et al., 2005; Nieto et al., 2010):

i. **Static**: the solvent is pressurized in the extractor in a close system (batch process). After extraction, the extract is liberated (through a valve), and further extraction cycles are carried out by adding "fresh" solvent. This mode is usually considered as being more efficient since the solvent is able to cross the pores of vegetal material more efficiently.
ii. **Dynamic**: in this case, the solvent is added continuously (inlet and outlet nozzles are maintained open).

There is a high diversity of solvents already used in PLE to extract compounds from leaf vegetables. However, it should be taken into account that the applied conditions (high temperature and pressure) change the solvent properties (as previously indicated: polarity, viscosity, or surface tension) in a great extent. Therefore, the same solvent might be used to extract compounds of varying polarity, just by regulating temperature and pressure (Alvarez et al., 2014). The pressure value (usually varying between 4 and 20 MPa) should be high enough to maintain the solvent in the liquid state, enhancing the solubility of compounds (Mustafa and Turner, 2011).

In comparison to conventional methodologies, PLE presents some advantages, such as using equipment that protects the vegetal material from oxygen and light, being performed in short time, using low volumes of solvent, or allowing achieving high yields (Mustafa and Turner, 2011). In turn, PLE does not represent a feasible solution to extract thermolabile compounds, owing to the high temperatures reached throughout the process (Piasek et al., 2011).

2.4 Supercritical Fluid Extraction

A fluid is considered to be in supercritical condition if temperature and pressure are both above the corresponding critical point (a combination of temperature and pressure that makes the liquid indistinguishable from its gas). There is a variety of substances that reach excellent solubilizing capacities when maintained in supercritical conditions (Carciochi et al., 2017).

The top choice of fluid in SFE is carbon dioxide, a nontoxic and nonflammable gas that allows working with low temperatures, reduced volumes of solvent and small size samples, besides reaching high selectivity. Furthermore, it is an environmentally friendly solvent and generally recognized as safe for use in the food industry (Arceusz et al., 2013). In either case, based on the individuality of the selected fluid, SFE might be divided into four stages:

i. **Pressurization step**: pressure is raised above the critical point.
ii. **Temperature adjustment**: the temperature of the compressed fluid will be elevated above the critical point.
iii. **Extraction**: the supercritical fluid and the solid matrix are maintained in continuous contact.
iv. **Separation**: the solute is precipitated by reducing the pressure below the critical point.

In terms of vegetal material, SFE was first applied (the early 1970s) in the decaffeination of coffee beans. Since then, it has been applied for extracting several compounds, including carotenoids, sterols, tocopherols, or polyphenolic compounds (Srinivas and King, 2010).

Its exponential application might be related to the fact of being an environmentally advantageous alternative to conventional methodologies. Furthermore, a supercritical fluid presents low viscosity and high diffusivity, a combination of factors that increases its penetrating power and mass-transfer capacity. Another interesting fact is represented by the possibility of producing great variation in the supercritical fluid properties (e.g., its density) through minor changes in temperature or pressure, without phase change (as the fluid is in supercritical condition). In comparison to similar technologies, SFE has a major advantage in the solvent elimination step, as it is simply necessary to regulate temperature and pressure to values in which the fluid becomes a gas. However, as described in the case of PLE, this methodology is generally unsuitable for extracting thermolabile compounds, due to the need of raising the temperature above the critical value (Carciochi et al., 2017).

Nevertheless, this limitation might be overcome when using carbon dioxide, owing to its low critical temperature (31.05 °C) and mild critical pressure (7.38 MPa). However, supercritical CO_2 is better for nonpolar compounds, such as sterols (Ekinci and Gürü, 2014; Sajfrtová et al., 2010), fatty acids (Casas et al., 2012; Pradhan et al., 2010), carotenoids (Filho et al., 2008; Goto et al., 2015; MacHmudah et al., 2012), and tocopherols (Esquivel-Hernández et al., 2016; Kraujalis and Venskutonis, 2013). Its lower effectiveness toward polar compounds can, nonetheless, be improved by using polar modifiers or cosolvents (e.g., water, ethanol, methanol, acetic acid, and ethylene glycol) (Goyeneche et al., 2018; He et al., 2012; Massias et al., 2015; Seabra et al., 2010).

2.5 *Enzyme-Assisted Extraction*

EAE is a nonconventional technology that might be divided in **EA cold pressing** and **EA aqueous extraction** (Latif and Anwar, 2009; Mat Yusoff et al., 2015).

In either case, EAE is based on an enzymatic treatment of the plant material prior to the solvent extraction process. The selected enzymes have the capacity to degrade cell walls and membranes, facilitating the solvent access to the target compounds. Considering the cell wall composition (mainly interconnecting polysaccharides, such as cellulose, hemicellulose, lignin, and pectin), the enzymes (food-grade) of choice are usually cellulases, pectinases, or hemicellulases. In addition to the cell wall composition, it is equally important to know the enzyme characteristics, for example, its catalytic properties, mode of action, and optimum operational conditions (including enzyme concentration, incubation time, temperature, pH, and enzyme-to-substrate ratio). When properly defined, the enzymatic pretreatment will provide a reduction in the extraction time, besides allowing performing the process at lower temperatures (Puri et al., 2012).

EAE is considered an eco-friendly (water is used instead of organic solvents) method particularly suitable for extracting oils and bioactive compounds. In general, the factors that exert the highest influence over EAE effectiveness are the enzyme itself, the molecular size of plant materials (the lowest, the better), and the catalytic time (Pinelo et al., 2006; Puri et al., 2012).

Furthermore, it should be taken into account that enzymes behave differently according to the system conditions (oxygen diffusivity, temperature, pH, or nutrient availability), particularly when EAE is applied in upscaling processes (Baiano, 2014).

In turn, EAE might be especially important to ease the extraction of bound phenolic compounds, to which the access is often hindered by lignin or pectin materials. In this way, EAE increases the extraction yield when compared with conventional solvent extraction, which can only recover the most accessible and weakly bonded phenols (Pinelo et al., 2008). However, the complexity of the structure of phenolic compounds (briefly described in the next chapter) will certainly be associated with differences in their solubility, thermal and chemical stability, and affinity to bond to other compounds, demanding specific studies to anticipate their interaction with enzymes (Arnous and Meyer, 2010).

3 Main Classes of Extracted Bioactive Compounds

Compounds produced by plants might be divided into **primary metabolites** (e.g., including carbohydrates, amino acids, proteins, or lipids), which represent substances essential for cell maintenance, growth and development, and **secondary metabolites** (e.g., phenolic acids, sterols, or terpenoids), which are typically low-molecular-weight substances with different functions mostly associated with plant protection from surroundings threats (Azmir et al., 2013).

These secondary metabolites represent the major part of plant bioactive compounds, which can be extracted from different botanical parts, such as roots, stem, bark, leaves, flowers, fruits and seeds, and those which are usually associated with pharmacological or toxicological effects in human and animals (Aksel, 2010).

Owing to the complexity of different bioactive compounds classes, next sections will be dedicated only to selected groups; namely, phenolic compounds, triterpenoids, and sterols.

3.1 Phenolic Compounds

Phenolics are widely distributed in Plant kingdom, being particularly abundant in berries, apples, citrus fruits, cocoa, grapes, onions, olives, tea, coffee beans, grains, cereals, soybeans, tomatoes, broccoli, and lettuce (Birt, Hendrich, & Wang, 2001). Besides their wide distribution, phenolic compounds are also very heterogenous, having been reported more than 8,000 phenolic structures (Arceusz et al., 2013). Nevertheless, all phenolic compounds present an aromatic ring, with one or more hydroxyl substituents, ranging from simple molecules, such as phenolic acids, to highly polymerized compounds, such as tannins. According to their carbon chain, phenolic compounds can be divided into 16 major classes (Table 12.1) (Bravo, 2009). However, another common division criterion separates phenolic compounds in two main groups, non-flavonoids and flavonoids. Non-flavonoids, despite being structurally heterogeneous, are aggregated into three main subgroups: **phenolic acids, lignans** $(C_6–C_3)_2$, and **stilbenes** $C_6–C_2–C_6$. Two additional subclasses of non-flavonoid compounds are composed of tannins and lignins, which occur naturally as complicated biopolymers (Del Rio et al., 2013).

Phenolic acids might be divided into **benzoic acids** (the most common in vegetables), with a seven carbon atoms ($C_6–C_1$) structure, and hydroxycinnamic acids, with nine carbon atoms ($C_6–C_3$), both usually found in plants esterified with other compounds (Lafay and Gil-Izquierdo, 2008). Besides their protecting role, phenolic compounds have some relevant functions, such as contributing to the sensory properties (e.g., hop), providing attractive colors to leaves, flowers, and fruits, or controlling the dynamics of germination, growth, and reproduction (Altemimi et al., 2017).

Together with phenolic acids, flavonoids constitute the most abundant group of phenolic compounds in plants. From a structural point of view, flavonoids are phenylpropanoid derivatives, presenting a 15-carbon skeleton with two benzene rings linked via a heterocyclic pyran ring (Figure 12.1). Their basic structure consists of two aromatic rings linked by a three-carbon bridge ($C_6–C_3–C_6$), but usually they occur in association with sugar molecules (glycoside derivatives) (Del Rio et al., 2013).

Flavonoids are subdivided into several subclasses, which are mainly determined by the carbon of C ring on which the B ring is attached and the degree of

Table 12.1 Basic structures of the 16 major subclasses of phenolic compounds

CLASS	BASIC SKELETON	BASIC STRUCTURE
Simple phenols	C_6	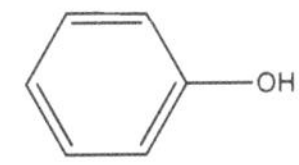
Benzoquinones	C_6	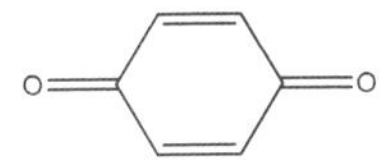
Phenolic acids	C_6–C_1	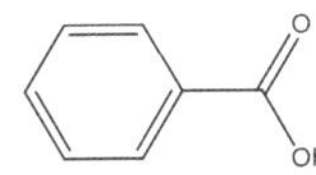
Acetophenones	C_6–C_2	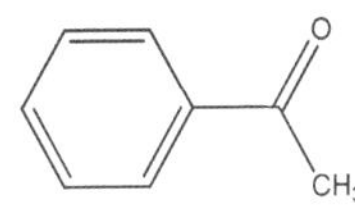
Phenylacetic acids	C_6–C_2	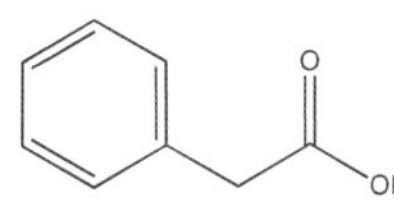
Hydroxycinnamic acids	C_6–C_3	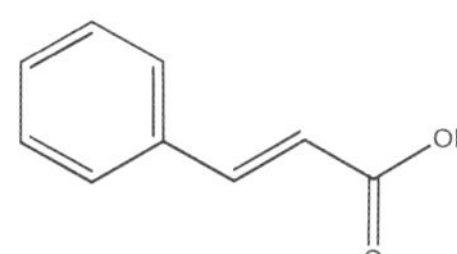
Phenylpropenes	C_6–C_3	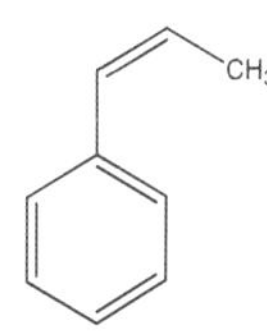
Coumarins	C_6–C_3	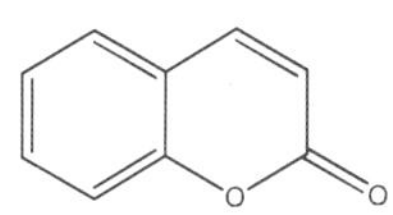
Chromones	C_6–C_3	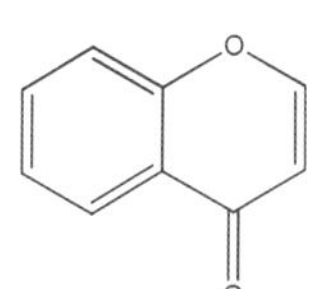
Naphthoquinones	C_6–C_4	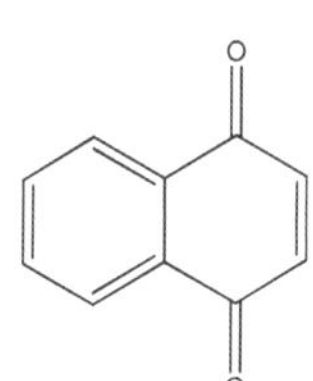

(Continued)

Table 12.1 (Cont.)

CLASS	BASIC SKELETON	BASIC STRUCTURE
Xanthones	C_6–C_1–C_6	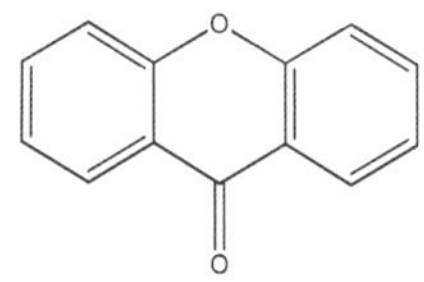
Stilbenes	C_6–C_2–C_6	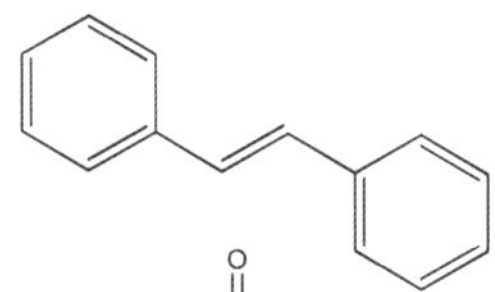
Anthraquinones	C_6–C_2–C_6	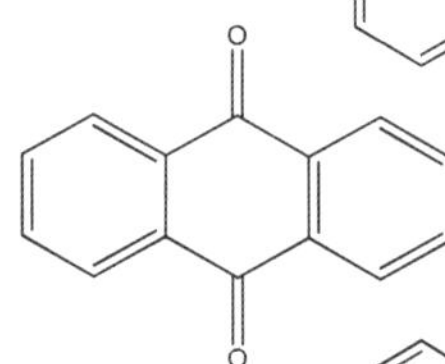
Flavonoids	C_6–C_3–C_6	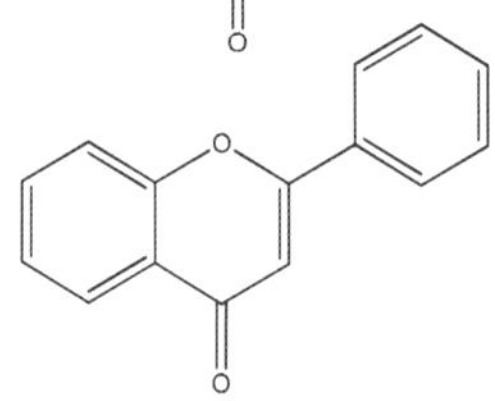
Lignans	$(C_6$–$C_3)_2$	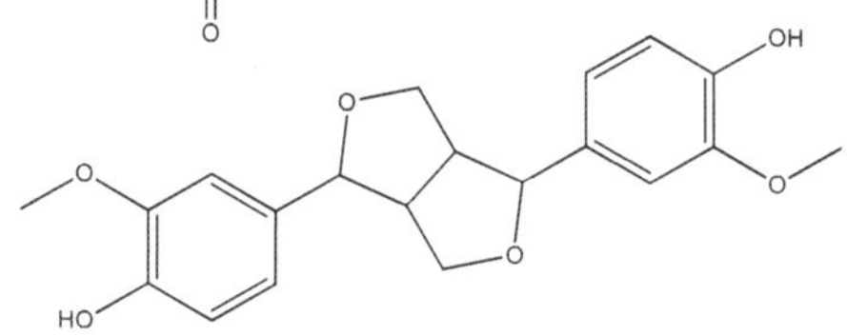
Ellagitannins	$(C_6$–$C_3)_n$	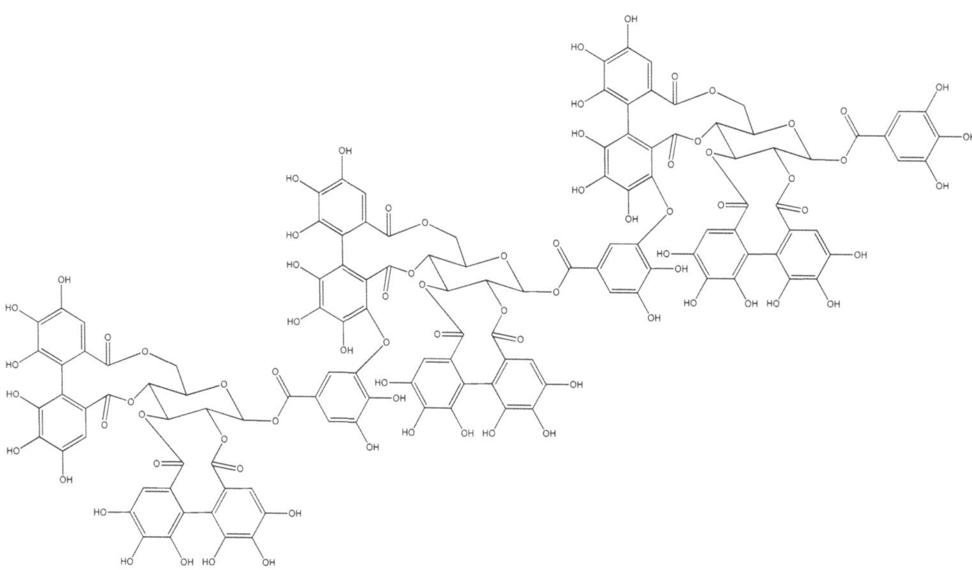

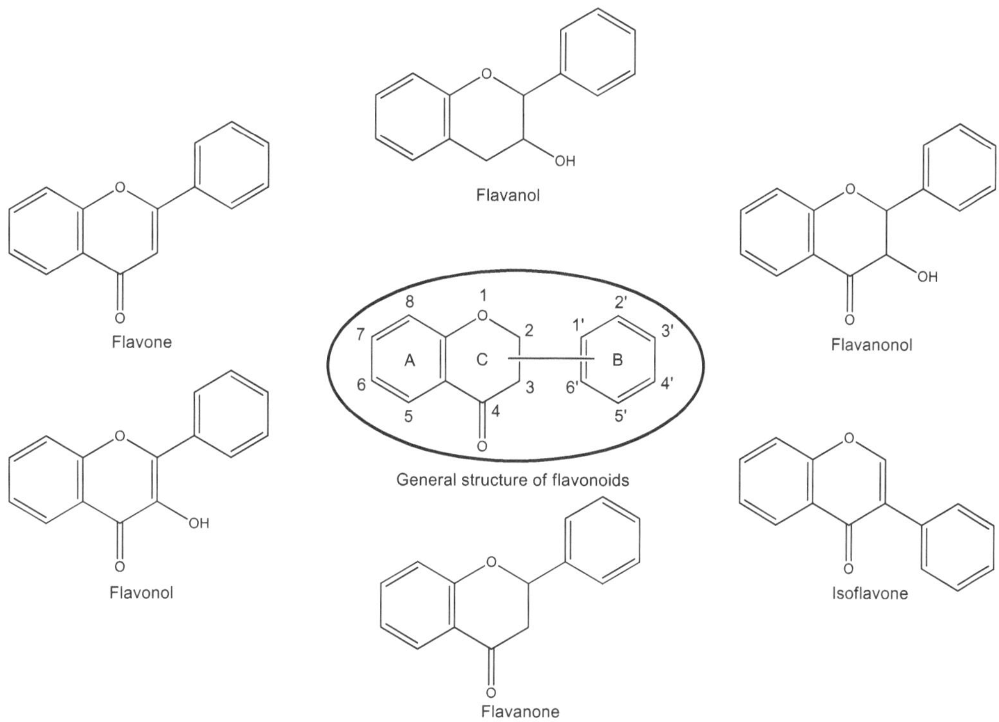

Figure 12.1 Primary structure of flavonoids and some of their main derivatives.

unsaturation and oxidation of C ring. Flavonoids in which the B ring is linked in position 3 of C ring are called isoflavones. In turn, those in which the B ring is linked in position 4 are called neoflavonoids, while those with the B ring in position 2 can be further subdivided into several subgroups (e.g., flavones, flavonols, flavanones, flavanonols, flavanols or catechins, anthocyanins, and chalcones) (Panche et al., 2016).

Independent of their subclass, flavonoids, as well as phenolic acids, demonstrate important biological activities, including antioxidant, hepatoprotective, antibacterial, anti-inflammatory, anticancer, and antiviral effects (Heleno et al., 2015; Kaushik et al., 2015; Viskupičová et al., 2008).

This bioactivity is modulated by specific structural properties. The antioxidant activity, for instance, is explained by their ability to neutralize free radicals through the interaction of hydroxyl groups with the *p*-electrons of the benzene ring. Therefore, phenolics with *ortho*-positioned hydroxyl groups are strong antioxidants (Parr and Bolwell, 2000). Likewise, their capacity to chelate metal ions, to which the presence of vicinal hydroxyl groups is essential, is an important source of antioxidant activity, particularly because this metal-binding capacity is, once again, relevant to provide antioxidant activity, particularly because free transition metal ions are pro-oxidants capable of producing free-radicals in the presence of hydrogen peroxide (Croft, 1998; Psotová et al., 2003).

3.2 Triterpenoids

Triterpenoids are a particular class of secondary metabolites produced from isopentenyl pyrophosphate oligomers that represent one of the largest groups of phytochemicals, and their production seems to be related with mechanisms developed by plants to fight thermal and oxidative stress (Bishayee et al., 2011). From a structural point of view, terpenoids are composed of a varying number of isoprene units (C_5H_8). Commonly, isoprene-derived compounds undergo chemical modifications, such as oxidation or rearrangement of carbon backbones. Triterpenoids, which result from the rearrangement of six isoprene units (C_{30}), are being generally regarded as promising secondary plant metabolites (Domingues et al., 2014; Mueller et al., 2013).

Some of the leaf vegetables with the highest contents in triterpenoids include medicinal herbs, mistletoe, lavender, oregano, rosemary, and thyme (Bishayee et al., 2011).

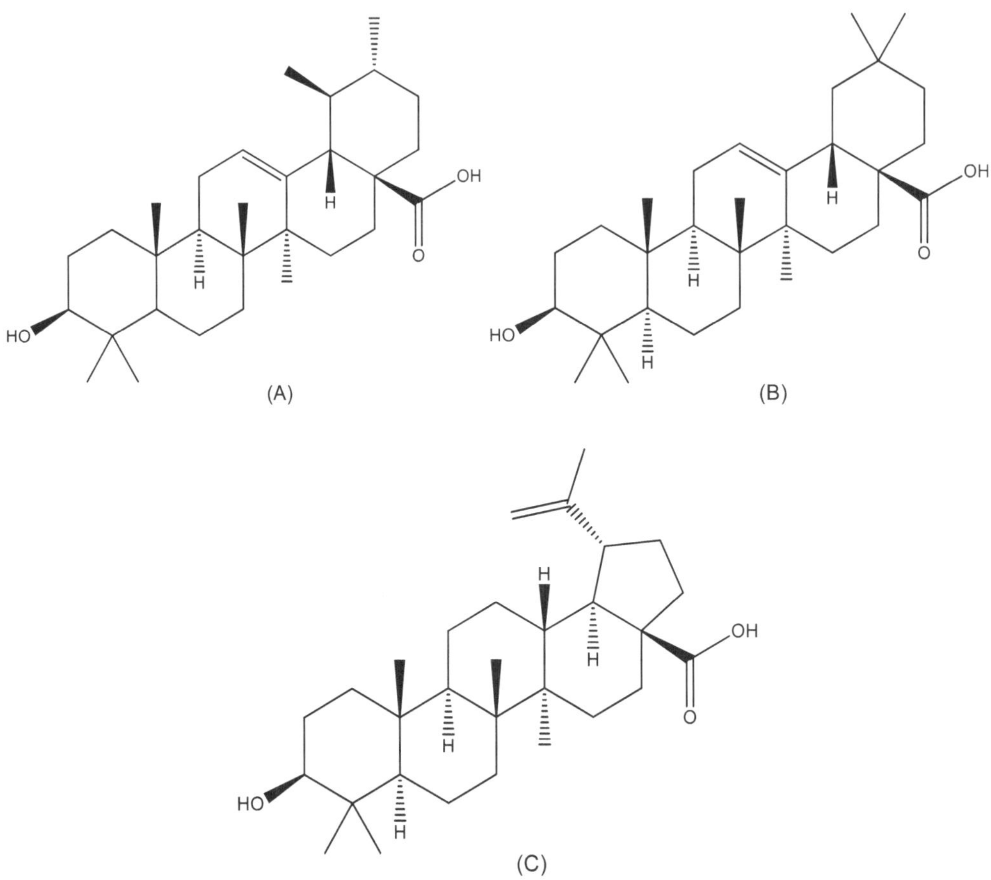

Figure 12.2 Chemical structures of ursolic (A), oleanolic (B), and betulinic (C) acids.

In terms of bioactivity, the subgroup of triterpenoids with the highest potential is represented by pentacyclic compounds, particularly those bearing lupane, ursane, or oleanane backbone structures. Ursane-type triterpenoids, such as ursolic acid and its isomers oleanolic and betulinic acids (Figure 12.2), are among the most studied (Babalola and Shode, 2013), particularly for their antibacterial activity, namely against Gram-positive bacteria (Huang et al., 2015), and capacity to modulate the inflammatory response (Waldbauer et al., 2016).

In addition, ursolic and oleanolic acids have further shown proapoptotic, antiangiogenic, and antitumorigenic activities (Bishayee et al., 2011; Cargnin and Gnoatto, 2017; Yan et al., 2014), and hepatoprotective properties (Xu et al., 2018).

3.2.1 Sterols Sterols obtained from plants are designed as phytosterols and include plant sterols and stanols (Jones and AbuMweis, 2009). These compounds are included in the triterpene family, representing important structural components of plant membranes, specifically to stabilize phospholipid bilayers (similarly to cholesterol in animal cells). Phytosterols might occur in free or esterified forms (Smet et al., 2012), included in the nonsaponifiable fraction of plant oil (Jones and AbuMweis, 2009).

Structurally, phytosterols resemble cholesterol, containing 28 or 29 carbons and one or two double C–C bonds, characteristically one in the sterol nucleus and

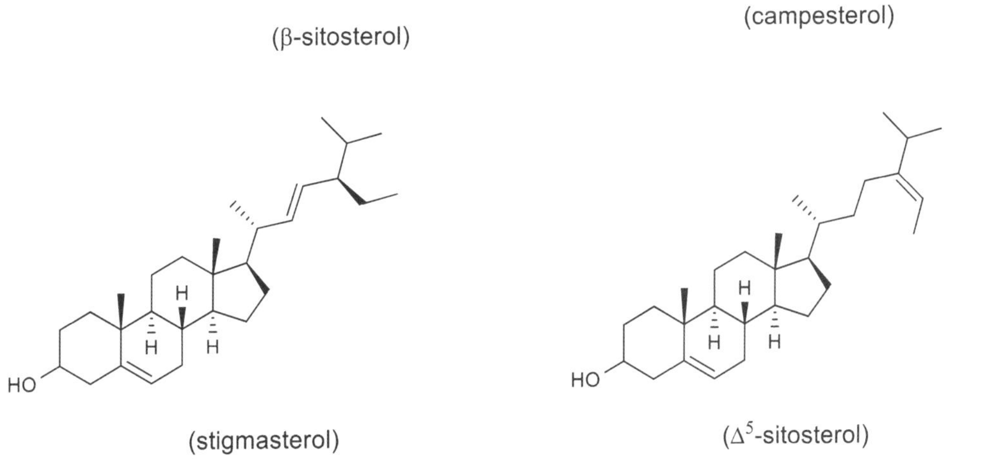

Figure 12.3 Chemical structures of some of the most relevant phytosterols.

often a second in the alkyl side chain. Sterols derive from hydroxylated polycyclic isopentenoids having a 1,2-cyclopentanophenanthrene structure and comprise the largest proportion of the unsaponifiable fraction of lipids (Valitova et al., 2016). More than 200 sterols have been reported, among which the 4-desmethylsterols, such as β-sitosterol, campesterol, stigmasterol, and Δ^5-avenasterol (Figure 12.3). In plants, β-sitosterol is clearly the predominant sterol (Barreira and Ferreira, 2015).

Phytosterols occur in five common forms: free, esterified with fatty acids or hydroxycinnamic acids, steryl glycosides, and acylated steryl glycosides; the last four forms generically called phytosterol conjugates (Moreau et al., 2002). Nuts (walnuts, almonds, peanuts, hazelnuts, and the macadamia nuts) and cereal products (rye, wheat, barley, and oat) are typically considered good sources of sterols (free, esterified, and steryl glycosides), but there are some leaf vegetables, particularly from *Brassica* genus, which also contain relevant quantities of these bioactive compounds (Han et al., 2008).

4 Conclusion

In general, the methodologies applied to extract valuable compounds from leaf vegetables should be supported in an adequate extraction technique. However, the extraction efficiency might be highly improved by applying selected pretreatment processes (e.g., milling, homogenizing, extrusion, dehydration, drying, among others).

Concerning the extraction methodology itself, there is no perfect solution, so it should be selected according to specific purposes or matrix features. For instance, the enzymatic treatment of the solid matrix should ease the access of the solvent, in that way promoting the diffusion of extractable molecules. Likewise, EAE might be the best choice in extraction processes dealing with heat-sensitive compounds. SFE, in turn, has an unparalleled advantage in what concerns the elimination of the solvent, but it is not equally effective for polar and nonpolar compounds. All other extraction techniques will obviously present specific advantages and weak points, so it is not possible to define the best choice for all single occasions.

The extract purity, associated with the solvent specificity, is a key point; its quality, particularly the maintenance of bioactive properties, is also of paramount importance, as well as the stability of the extract under different conditions. The quality and the stability of a given extract can be enhanced by techniques, such as freeze drying, spray drying, or encapsulation.

In addition to the extraction techniques, this chapter also addressed some examples of the most important bioactive compounds (phenolic compounds, triterpenoids, and steroids) that are usually targeted for extraction. It is obvious that leaf vegetables might be good sources of these biomolecules, which could be used in different applications, such as the food, pharmaceutical, or cosmetics industries. Likewise, losses and wastes of horticultural products could be used in the same way, contributing to sustainable development and mitigating environmental problems.

Acknowledgments

The authors would like to thank.

References

Aksel, B., 2010. A brief review on bioactive compounds in plants, in: Bioactive Compounds in Plants – Benefits and Risks for Man and Animals. doi:10.1016/S2005-2901(11)60010-1

Altemimi, A., Lakhssassi, N., Baharlouei, A., Watson, D., Lightfoot, D., 2017. Phytochemicals: Extraction, isolation, and identification of bioactive compounds from plant extracts. Plants. doi:10.3390/plants6040042

Alupului, A., Călinescu, I., Lavric, V., 2012. Microwave Extraction of active principles from medicinal plants. UPB Sci. Bull. Ser. B Chem. Mater. Sci. doi:10.3303/CET0917171

Alvarez, V.H., Cahyadi, J., Xu, D., Saldaña, M.D.A., 2014. Optimization of phytochemicals production from potato peel using subcritical water: Experimental and dynamic modeling. J. Supercrit. Fluids. doi:10.1016/j.supflu.2014.02.013

Araújo, M., Pimentel, F.B., Alves, R.C., Oliveira, M.B.P.P., 2015. Phenolic compounds from olive mill wastes: Health effects, analytical approach and application as food antioxidants. Trends Food Sci. Technol. doi:10.1016/j.tifs.2015.06.010

Arceusz, A., Wesolowski, M., Konieczynski, P., 2013. Methods for extraction and determination of phenolic acids in medicinal plants: A review. Nat Prod. Commun. 8, 1821–1829.

Arnous, A., Meyer, A.S., 2010. Discriminated release of phenolic substances from red wine grape skins (Vitis vinifera L.) by multicomponent enzymes treatment. Biochem. Eng. J. doi:10.1016/j.bej.2009.11.012

Azmir, J., Zaidul, I.S.M., Rahman, M.M., Sharif, K.M., Mohamed, A., Sahena, F., Jahurul, M.H.A., Ghafoor, K., Norulaini, N.A.N., Omar, A.K.M., 2013. Techniques for extraction of bioactive compounds from plant materials: A review. J. Food Eng. doi:10.1016/j.jfoodeng.2013.01.014

Baaka, N., El Ksibi, I., Mhenni, M.F., 2017. Optimisation of the recovery of carotenoids from tomato processing wastes: Application on textile dyeing and assessment of its antioxidant activity. Nat Prod. Res. doi:10.1080/14786419.2016.1226828

Babalola, I., Shode, F., 2013. Ubiquitous ursolic acid : A potential pentacyclic triterpene natural product. J. Pharmacogn. Phytochem. Ubiquitous. doi:10.1029/2004GL019972

Baiano, A., 2014. Recovery of biomolecules from food wastes – A review. Molecules. doi:10.3390/molecules190914821

Baiano, A., Bevilacqua, L., Terracone, C., Contò, F., Del Nobile, M.A., 2014. Single and interactive effects of process variables on microwave-assisted and conventional extractions of antioxidants from vegetable solid wastes. J. Food Eng. doi:10.1016/j.jfoodeng.2013.07.010

Barba, F.J., Zhu, Z., Koubaa, M., Sant'Ana, A.S., Orlien, V., 2016. Green alternative methods for the extraction of antioxidant bioactive compounds from winery wastes and by-products: A review. Trends Food Sci. Technol. doi:10.1016/j.tifs.2016.01.006

Barreira, J.C.M., Ferreira, I.C.F.R., 2015. Steroids in natural matrices: Chemical features and bioactive properties, in: Biotechnology of Bioactive Compounds: Sources and Applications. doi:10.1002/9781118733103.ch16

Bishayee, A., Ahmed, S., Brankov, N., Perloff, M., 2011. Triterpenoids as potential agents for the chemoprevention and therapy of breast cancer. Front. Biosci-Landmark. 16, 980–996. doi:10.2741/3730

Bravo, L., 2009. Polyphenols: Chemistry, dietary sources, metabolism, and nutritional significance. Nutr. Rev. 56, 317–333. doi:10.1111/j.1753-4887.1998.tb01670.x

Carabias-Martínez, R., Rodríguez-Gonzalo, E., Revilla-Ruiz, P., Hernández-Méndez, J., 2005. Pressurized liquid extraction in the analysis of food and biological samples. J. Chromatogr. A. doi:10.1016/j.chroma.2005.06.072

Carciochi, R.A., D'Alessandro, L.G., Vauchel, P., Rodriguez, M.M., Nolasco, S.M., Dimitrov, K., 2017. Valorization of agrifood by-products by extracting valuable bioactive compounds using green processes, in: Ingredients Extraction by Physicochemical Methods in Food. doi:10.1016/B978-0-12-811521-3.00004-1

Carciochi, R.A., Dimitrov, K., 2014. Optimization of antioxidant phenolic compounds extraction from quinoa (Chenopodium quinoa) seeds. J. Food Sci. Technol. doi:10.1007/s13197-014-1514-4

Cargnin, S.T., Gnoatto, S.B., 2017. Ursolic acid from apple pomace and traditional plants: A valuable triterpenoid with functional properties. Food Chem. doi:10.1016/j.foodchem.2016.10.029

Casas, L., Serrano, C.M., Rodríguez, M.R., Martinez de la Ossa, E.J., Lubian, L.M., 2012. Extraction of carotenoids and fatty acids from microalgae using supercritical technology. Am. J. Anal. Chem. doi:10.4236/ajac.2012.312A116

Chemat, F., Vian, M.A., Cravotto, G., 2012. Green extraction of natural products: Concept and principles. Int. J. Mol. Sci. doi:10.3390/ijms13078615

Chemat, F., Zill-E-Huma, Khan, M.K., 2011. Applications of ultrasound in food technology: Processing, preservation and extraction. Ultrason. Sonochem. doi:10.1016/j.ultsonch.2010.11.023

Chumnanpaisont, N., Niamnuy, C., Devahastin, S., 2014. Mathematical model for continuous and intermittent microwave-assisted extraction of bioactive compound from plant material: Extraction of β-carotene from carrot peels. Chem. Eng. Sci. doi:10.1016/j.ces.2014.05.010

Cravotto, G., Boffa, L., Mantegna, S., Perego, P., Avogadro, M., Cintas, P., 2008. Improved extraction of vegetable oils under high-intensity ultrasound and/or microwaves. Ultrason. Sonochem. doi:10.1016/j.ultsonch.2007.10.009

Croft, K.D., 1998. The chemistry and biological effects of flavonoids and phenolic acids. Ann. N. Y. Acad. Sci. 435–442. doi:10.1111/j.1749-6632.1998.tb09922.x

Dahmoune, F., Nayak, B., Moussi, K., Remini, H., Madani, K., 2015. Optimization of microwave-assisted extraction of polyphenols from *Myrtus communis* L. leaves. Food Chem. doi:10.1016/j.foodchem.2014.06.066

Del Rio, D., Rodriguez-Mateos, A., Spencer, J.P.E., Tognolini, M., Borges, G., Crozier, A., 2013. Dietary (Poly)phenolics in human health: Structures, bioavailability, and evidence of protective effects against chronic diseases. Antioxid. Redox Signal. 18, 1818–1892. doi:10.1089/ars.2012.4581

Domingues, R., Guerra, A.R., Duarte, M.F., Freire, C.S.R., Neto, C.S., Silvestre, A.J.D., 2014. Bioactive triterpenic acids: From agroforestry biomass residues to promising therapeutic tools. Mini. Rev. Org. Chem. doi:10.2174/1570193X113106660001

Ekinci, M.S., Gürü, M., 2014. Extraction of oil and β-sitosterol from peach (Prunus persica) seeds using supercritical carbon dioxide. J. Supercrit. Fluids. doi:10.1016/j.supflu.2014.06.004

Esclapez, M.D., García-Pérez, J. V., Mulet, A., Cárcel, J.A., 2011. Ultrasound-assisted extraction of natural products. Food Eng. Rev. doi:10.1007/s12393-011-9036-6

Esquivel-Hernández, D.A., López, V.H., Rodríguez-Rodríguez, J., Alemán-Nava, G.S., Cuéllar-Bermúdez, S.P., Rostro-Alanis, M., Parra-Saldívar, R., 2016. Supercritical carbon dioxide and microwave-assisted extraction of functional lipophilic compounds from Arthrospira platensis. Int. J. Mol. Sci. doi:10.3390/ijms17050658

FAOSTAT [WWW Document], 2018. URL www.fao.org/faostat/en/#data/QC/visualize

Filho, G.L., De Rosso, V.V., Meireles, M.A.A., Rosa, P.T.V., Oliveira, A.L., Mercadante, A. Z., Cabral, F.A., 2008. Supercritical CO2 extraction of carotenoids from pitanga fruits (Eugenia uniflora L.). J. Supercrit. Fluids. doi:10.1016/j.supflu.2008.02.014

Food and Agriculture Organization of the United Nations., 2013. Food Wastage Footprint: Impacts on Natural Resources: Summary Report. doi:ISBN 978-92-5-107752-8

Galanakis, C.M., 2012. Recovery of high added-value components from food wastes: Conventional, emerging technologies and commercialized applications. Trends Food Sci. Technol. doi:10.1016/j.tifs.2012.03.003

Galanakis, C.M., 2015. Food Waste Recovery: Processing Technologies and Industrial Techniques. doi:10.1016/C2013-0-16046-1

Galván D'Alessandro, L., Dimitrov, K., Vauchel, P., Nikov, I., 2014. Kinetics of ultrasound assisted extraction of anthocyanins from Aronia melanocarpa (black chokeberry) wastes. Chem. Eng. Res. Des. doi:10.1016/j.cherd.2013.11.020

Goto, M., Kanda, H., Wahyudiono, Machmudah, S., 2015. Extraction of carotenoids and lipids from algae by supercritical CO2and subcritical dimethyl ether. J. Supercrit. Fluids. doi:10.1016/j.supflu.2014.10.003

Gowe, C., 2015. Review on potential use of fruit and vegetables by – Products as a valuable source of natural food additives. ISSN. doi:10.1109/FPT.2016.7929515

Goyeneche, R., Fanovich, A., Rodriguez Rodrigues, C., Nicolao, M.C., Di Scala, K., 2018. Supercritical CO2extraction of bioactive compounds from radish leaves: Yield, antioxidant capacity and cytotoxicity. J. Supercrit. Fluids. doi:10.1016/j.supflu.2018.01.004

Gustavsson, J., Cederberg, C., Sonesson, U., van Otterdijk, R., Meybeck, A., 2011. Global Food Losses and Food Waste. Food and Agriculture Organization of the United Nations. doi:doi:10.1098/rstb.2010.0126

Han, J.H., Yang, Y.X., Feng, M.Y., 2008. Contents of phytosterols in vegetables and fruits commonly consumed in China. Biomed. Environ. Sci. doi:10.1016/S0895-3988(09)60001-5

He, J.Z., Shao, P., Liu, J.H., Ru, Q.M., 2012. Supercritical carbon dioxide extraction of flavonoids from pomelo (citrus grandis (L.) osbeck) peel and their antioxidant activity. Int. J. Mol. Sci. doi:10.3390/ijms131013065

Heleno, S.A., Martins, A., Queiroz, M.J.R.P., Ferreira, I.C.F.R., 2015. Bioactivity of phenolic acids: Metabolites versus parent compounds: A review. Food Chem. doi:10.1016/j.foodchem.2014.10.057

Huang, L., Luo, H., Li, Q., Wang, D., Zhang, J., Hao, X., Yang, X., 2015. Pentacyclic triterpene derivatives possessing polyhydroxyl ring A inhibit Gram-positive bacteria growth by regulating metabolism and virulence genes expression. Eur. J. Med. Chem. doi:10.1016/j.ejmech.2015.01.015

Jones, P.J.H., AbuMweis, S.S., 2009. Phytosterols as functional food ingredients: Linkages to cardiovascular disease and cancer. Curr. Opin. Clin. Nutr. Metab. Care. 12, 147–151. doi:10.1097/MCO.0b013e328326770f

Kammerer, D.R., Kammerer, J., Valet, R., Carle, R., 2014. Recovery of polyphenols from the by-products of plant food processing and application as valuable food ingredients. Food Res. Int. doi:10.1016/j.foodres.2014.06.012

Kaushik, P., Andújar, I., Vilanova, S., Plazas, M., Gramazio, P., Herraiz, F.J., Brar, N.S., Prohens, J., 2015. Breeding vegetables with increased content in bioactive phenolic acids. Molecules. doi:10.3390/molecules201018464

Kraujalis, P., Venskutonis, P.R., 2013. Supercritical carbon dioxide extraction of squalene and tocopherols from amaranth and assessment of extracts antioxidant activity. J. Supercrit. Fluids. doi:10.1016/j.supflu.2013.04.005

Lafay, S., Gil-Izquierdo, A., 2008. Bioavailability of phenolic acids. Phytochem. Rev. doi:10.1007/s11101-007-9077-x

Latif, S., Anwar, F., 2009. Physicochemical studies of hemp (Cannabis sativa) seed oil using enzyme-assisted cold-pressing. Eur. J. Lipid Sci. Technol. doi:10.1002/ejlt.200900008

Letellier, M., Budzinski, H., 1999. Microwave assisted extraction of organic compounds. Analusis. doi:10.1051/analusis:1999116

Lin, C.S.K., Koutinas, A.A., Stamatelatou, K., Mubofu, E.B., Matharu, A.S., Kopsahelis, N., Pfaltzgraff, L.A., Clark, J.H., Papanikolaou, S., Kwan, T.H., Luque, R., 2014. Current and future trends in food waste valorization for the production of chemicals, materials and fuels: A global perspective. Biofuels, Bioprod. Biorefining. 8, 686–715. doi:10.1002/bbb.1506

MacHmudah, S., Zakaria, Winardi, S., Sasaki, M., Goto, M., Kusumoto, N., Hayakawa, K., 2012. Lycopene extraction from tomato peel by-product containing tomato seed using supercritical carbon dioxide. J. Food Eng. doi:10.1016/j.jfoodeng.2011.08.012

Madej, K., 2009. Microwave-assisted and cloud-point extraction in determination of drugs and other bioactive compounds. TrAC – Trends Anal. Chem. doi:10.1016/j.trac.2009.02.002

Massias, A., Boisard, S., Baccaunaud, M., Leal Calderon, F., Subra-Paternault, P., 2015. Recovery of phenolics from apple peels using CO2 + ethanol extraction: Kinetics and antioxidant activity of extracts. J. Supercrit. Fluids. doi:10.1016/j.supflu.2014.12.007

Mat Yusoff, M., Gordon, M.H., Niranjan, K., 2015. Aqueous enzyme assisted oil extraction from oilseeds andemulsion de-emulsifying methods: A review. Trends Food Sci. Technol. doi:10.1016/j.tifs.2014.09.003

Moreau, R.A., Whitaker, B.D., Hicks, K.B., 2002. Phytosterols, phytostanols, and their conjugates in foods: Structural diversity, quantitative analysis, and health-promoting uses. Prog. Lipid Res. doi:10.1016/S0163-7827(02)00006-1

Mueller, D., Triebel, S., Rudakovski, O., Richling, E., 2013. Influence of triterpenoids present in apple peel on inflammatory gene expression associated with inflammatory bowel disease (IBD). Food Chem. doi:10.1016/j.foodchem.2013.01.101

Mustafa, A., Turner, C., 2011. Pressurized liquid extraction as a green approach in food and herbal plants extraction: A review. Anal. Chim. Acta. doi:10.1016/j.aca.2011.07.018

Nieto, A., Borrull, F., Pocurull, E., Marcé, R.M., 2010. Pressurized liquid extraction: A useful technique to extract pharmaceuticals and personal-care products from sewage sludge. TrAC – Trends Anal. Chem. doi:10.1016/j.trac.2010.03.014

Panche, A.N., Diwan, A.D., Chandra, S.R., 2016. Flavonoids: An overview. J. Nutr. Sci. doi:10.1017/jns.2016.41

Parfitt, J., Barthel, M., MacNaughton, S., 2010. Food waste within food supply chains: Quantification and potential for change to 2050. Philos. Trans. R. Soc. B Biol. Sci. doi:10.1098/rstb.2010.0126

Parr, A.J., Bolwell, G.P., 2000. Phenols in the plant and in man. The potential for possible nutritional enhancement of the diet by modifying the phenols content or profile. J. Sci. Food Agric. doi:10.1002/%28SICI%291097%2D0010%2820000515%2980:7%3C;985::AID%2DJSFA572%3E;3.0.CO;2%2D7

Piasek, A., Kusznierewicz, B., Grzybowska, I., Malinowska-Pańczyk, E., Piekarska, A., Azqueta, A., Collins, A.R., Namieśnik, J., Bartoszek, A., 2011. The influence of sterilization with EnbioJet® Microwave Flow Pasteurizer on composition and bioactivity of aronia and blue-berried honeysuckle juices. J. Food Compos. Anal. doi:10.1016/j.jfca.2011.04.005

Pinelo, M., Arnous, A., Meyer, A.S., 2006. Upgrading of grape skins: Significance of plant cell-wall structural components and extraction techniques for phenol release. Trends Food Sci. Technol. doi:10.1016/j.tifs.2006.05.003

Pinelo, M., Zornoza, B., Meyer, A.S., 2008. Selective release of phenols from apple skin: Mass transfer kinetics during solvent and enzyme-assisted extraction. Sep. Purif. Technol. doi:10.1016/j.seppur.2008.07.007

Pingret, D., Fabiano-Tixier, A.S., Bourvellec, C. Le, Renard, C.M.G.C., Chemat, F., 2012. Lab and pilot-scale ultrasound-assisted water extraction of polyphenols from apple pomace. J. Food Eng. doi:10.1016/j.jfoodeng.2012.01.026

Pradhan, R.C., Meda, V., Rout, P.K., Naik, S., Dalai, A.K., 2010. Supercritical CO2 extraction of fatty oil from flaxseed and comparison with screw press expression and solvent extraction processes. J. Food Eng. doi:10.1016/j.jfoodeng.2009.11.021

Pronyk, C., Mazza, G., 2009. Design and scale-up of pressurized fluid extractors for food and bioproducts. J. Food Eng. doi:10.1016/j.jfoodeng.2009.06.002

Psotová, J., Lasovský, J., Vicar, J., 2003. Metal-chelating properties, electrochemical behavior, scavenging and cytoprotective activities of six natural phenolics. Biomed. Pap. Med. Fac. Univ. Palacky. Olomouc. Czech. Repub. 147, 147–153. doi:10.5507/bp.2003.020

Puri, M., Sharma, D., Barrow, C.J., 2012. Enzyme-assisted extraction of bioactives from plants. Trends Biotechnol. doi:10.1016/j.tibtech.2011.06.014

Ramos, L., Kristenson, E.M., Brinkman, U.A.T., 2002. Current use of pressurised liquid extraction and subcritical water extraction in environmental analysis. J. Chromatogr. A. doi:10.1016/S0021-9673(02)01336-5

Rombaut, N., Tixier, A.S., Bily, A., Chemat, F., 2014. Green extraction processes of natural products as tools for biorefinery. Biofuels, Bioprod. Biorefining. doi:10.1002/bbb.1486

Roselló-Soto, E., Barba, F.J., Parniakov, O., Galanakis, C.M., Lebovka, N., Grimi, N., Vorobiev, E., 2015. High voltage electrical discharges, pulsed electric field, and ultrasound assisted extraction of protein and phenolic compounds from Olive Kernel. Food Bioprocess Technol. doi:10.1007/s11947-014-1456-x

Sajfrtová, M., Ličková, I., Wimmerová, M., Sovová, H., Wimmer, Z., 2010. β-sitosterol: Supercritical carbon dioxide extraction from sea buckthorn (Hippophae rhamnoides L.) seeds. Int. J. Mol. Sci. doi:10.3390/ijms11041842

Santana-Méridas, O., González-Coloma, A., Sánchez-Vioque, R., 2012. Agricultural residues as a source of bioactive natural products. Phytochem. Rev. doi:10.1007/s11101-012-9266-0

Seabra, I.J., Braga, M.E.M., Batista, M.T., De Sousa, H.C., 2010. Effect of solvent (CO2/ethanol/H2O) on the fractionated enhanced solvent extraction of anthocyanins from elderberry pomace. J. Supercrit. Fluids. doi:10.1016/j.supflu.2010.05.001

Smet, E. De, Mensink, R.P., Plat, J., 2012. Effects of plant sterols and stanols on intestinal cholesterol metabolism: Suggested mechanisms from past to present. Mol. Nutr. Food Res. doi:10.1002/mnfr.201100722

Srinivas, K., King, J.W., 2010. Supercritical carbon dioxide and subcritical water: Complementary agents in the processing of functional foods, in: Functional Food Product Development. doi:10.1002/9781444323351.ch3

Teixeira, A., Baenas, N., Dominguez-Perles, R., Barros, A., Rosa, E., 2014. Natural bioactive compounds from winery by-products as health promoters: A review. Int. J. Mol. Sci. doi:10.3390/ijms150915638

The Ellen MacArthur Foundation, 2012. Towards a circular economy – Economic and business rationale for an accelerated transition. Greener Manag. Int. 97. doi:2012-04-03

Toma, M., Vinatoru, M., Paniwnyk, L., Mason, T.J., 2001. Investigation of the effects of ultrasound on vegetal tissues during solvent extraction. Ultrason. Sonochem. doi:10.1016/S1350-4177(00)00033-X

Valitova, J.N., Sulkarnayeva, A.G., Minibayeva, F.V., 2016. Plant sterols: Diversity, biosynthesis, and physiological functions. Biochem. 81, 819–834. doi:10.1134/S0006297916080046

Vieira da Silva, B., Barreira, J.C.M., Oliveira, M.B.P.P., 2016. Natural phytochemicals and probiotics as bioactive ingredients for functional foods: Extraction, biochemistry and protected-delivery technologies. Trends Food Sci. Technol. doi:10.1016/j.tifs.2015.12.007

Vilariño, M.V., Franco, C., Quarrington, C., 2017. Food loss and waste reduction as an integral part of a circular economy. Front. Environ. Sci. doi:10.3389/fenvs.2017.00021

Vilkhu, K., Mawson, R., Simons, L., Bates, D., 2008. Applications and opportunities for ultrasound assisted extraction in the food industry – A review. Innov. Food Sci. Emerg. Technol. doi:10.1016/j.ifset.2007.04.014

Villares, A., García-Lafuente, A., Guillamón, E., Ramos, Á., 2012. Identification and quantification of ergosterol and phenolic compounds occurring in Tuber spp. truffles. J. Food Compos. Anal. doi:10.1016/j.jfca.2011.12.003

Virot, M., Tomao, V., Le Bourvellec, C., Renard, C.M.C.G., Chemat, F., 2010. Towards the industrial production of antioxidants from food processing by-products with ultrasound-assisted extraction. Ultrason. Sonochem. doi:10.1016/j.ultsonch.2009.10.015

Viskupičová, J., Ondrejovič, M., Šturdík, E., 2008. Bioavailability and metabolism of flavonoids. J. Food Nutr. Res. doi:10.1080/13880200490893492

Waldbauer, K., Seiringer, G., Nguyen, D.L., Winkler, J., Blaschke, M., McKinnon, R., Urban, E., Ladurner, A., Dirsch, V.M., Zehl, M., Kopp, B., 2016. Triterpenoic acids from apple pomace enhance the activity of the endothelial nitric oxide synthase (eNOS). J. Agric. Food Chem. 64, 185–194. doi:10.1021/acs.jafc.5b05061

Wijngaard, H., Hossain, M.B., Rai, D.K., Brunton, N., 2012. Techniques to extract bioactive compounds from food by-products of plant origin. Food Res. Int. doi:10.1016/j.foodres.2011.09.027

Wu, C., Wang, F., Liu, J., Zou, Y., Chen, X., 2015. A comparison of volatile fraction obtained from Lonicera macranthoides via different extraction processes: Ultrasound, microwave, soxhlet extraction, hydrodistillation and cold-maceration. Integr. Med. Res. doi:10.1016/j.imr.2015.06.001

Wu, T., Yan, J., Liu, R., Marcone, M.F., Aisa, H.A., Tsao, R., 2012. Optimization of microwave-assisted extraction of phenolics from potato and its downstream waste using orthogonal array design. Food Chem. doi:10.1016/j.foodchem.2011.08.002

Xu, G.-B., Xiao, Y.-H., Zhang, Q.-Y., Zhou, M., Liao, S.-G., 2018. Hepatoprotective natural triterpenoids. Eur. J. Med. Chem. 145, 691–716. doi:10.1016/j.ejmech.2018.01.011

Yan, X.J., Gong, L.H., Zheng, F.Y., Cheng, K.J., Chen, Z.S., Shi, Z., 2014. Triterpenoids as reversal agents for anticancer drug resistance treatment. Drug Discov. Today. doi:10.1016/j.drudis.2013.07.018

Zuin, V.G., Ramin, L.Z., 2018. Green and sustainable separation of natural products from agro-industrial waste: Challenges, potentialities, and perspectives on emerging approaches. Top. Curr. Chem. 376, 3. doi:10.1007/s41061-017-0182-z

13
Valuable Compounds in Algae

LARS LEONHARDT

Chair of Food Engineering, Technische Universität Dresden, 01062 Dresden, Germany
Elea Vertriebs- und Vermarktungsgesellschaft mbH, Prof.-von-Klitzing Str. 9, D-49610 Quakenbrück, Germany

JULIAN WITT AND STEFAN TOEPFL

Elea Vertriebs- und Vermarktungsgesellschaft mbH, Prof.-von-Klitzing Str. 9, D-49610 Quakenbrück, Germany

HARALD ROHM

Chair of Food Engineering, Technische Universität Dresden, 01062 Dresden, Germany

OLEKSII PARNIAKOV

Elea Vertriebs- und Vermarktungsgesellschaft mbH, Prof.-von-Klitzing Str. 9, D-49610 Quakenbrück, Germany

Contents

1 Introduction

In recent years, the interest in algae as production organisms has increased. This results from rising population growth, with serious effects in terms of land and water consumption, as well as greenhouse gases. To be able to meet the global protein requirements in future as well, alternative sources of proteins are urgently needed (Kinkel, 2018; Wolkers et al., 2011). The cultivation of microalgae, initially in the focus for the production of biodiesel and bioethanol (Enzing, Ploeg, Barbosa, & Sijtsma, 2014; Eppink et al., 2017; Wahlen, Willis, & Seefeldt, 2011), can make an important contribution to this. Compared with agricultural grown biomass, there are three main advantages. First, the high content of valuable compounds (>50 g/100 g_{dw}), such as lipids, proteins, polysaccharides, antioxidants, and pigments (Eing, Goettel, Straessner, Gusbeth, & Frey, 2013). Second, their high biomass growth rate and productivity, which is three to five times higher compared with cultivated plants, such as corn, raps, or soya. Third, the possibility to cultivate them on infertile land. As an aquatic biomass, they do not compete with food crops for land. Thus, microalgae are often referred to as "sustainable all-rounders." They are not only used in the food and feed industry but also in the pharmaceutical industry and the waste management industry for the purification of wastewater (Kinkel, 2018). While most of the extraction applications are focused on specific components, such as oil or pigments, conscious biorefinery of a broader spectrum of microalgal constituents would enable an increased economy. However, the technology for the production of microalgae is still within its early stages. Cultivation and downstreaming processes are highly energy- and cost-intensive, so the use of algae is limited to the production of high-value and low-volume products (Goettel, Eing, Gusbeth, Straessner, & Frey, 2013). The largest proportion of algae biomass is currently produced in Asia, where it is harvested, dried, and shipped to Europe. In Europe, it is resuspended and prepared for extraction. Next to large-distance postage expenses, both processing steps, drying, as well as the extraction, consume a large amount of energy. This is referred to the strong cell walls of algae, making cell disintegration and extraction rather difficult (Goettel et al., 2013). As conventional processing of algae is commonly based on heat extraction with organic solvents holding risks of environmental pollution and health effects, progression of innovative techniques is needed to realize an eco-friendly and economical usage of microalgae as a sustainable resource of the future.

In this chapter, the general description of microalgae, as well as their composition are presented. Therefore, the origin and location of valuable compounds inside microalgae are discussed. Moreover, different conventional and novel (thermal and nonthermal) extraction techniques are reviewed. This implies a detailed description of some nonconventional methods, including pulsed electric fields (PEF), high-voltage electric discharges (HVED), ultrasound (US), microwave (MW), high-pressure homogenization (HPH), pressure change technology (PCT), sub- and supercritical fluid, and ionic liquid (IL) extraction for recovering of nutritionally valuable compounds from microalgae.

2 General Information about Algae

Algae, in general, can be defined as plants without leaves, stems, and roots (thallophytes). They utilize the pigment chlorophyll *a* to perform oxygenic photosynthesis and have a lack of cells covering the reproductive cells (R. E. Lee, 2008). On the basis of their size, algae can be divided into macroalgae and microalgae. While macroalgae (also called seaweed) are larger multicellular plants, microalgae are unicellular species ranging between a few micrometers and a few hundreds of micrometers (Suganya, Varman, Masjuki, & Renganathan, 2016). Many strains of microalgae (traditionally covering eukaryotic microalgae and prokaryotic cyanobacteria) hold high quantities of lipids and proteins. Additionally, with their

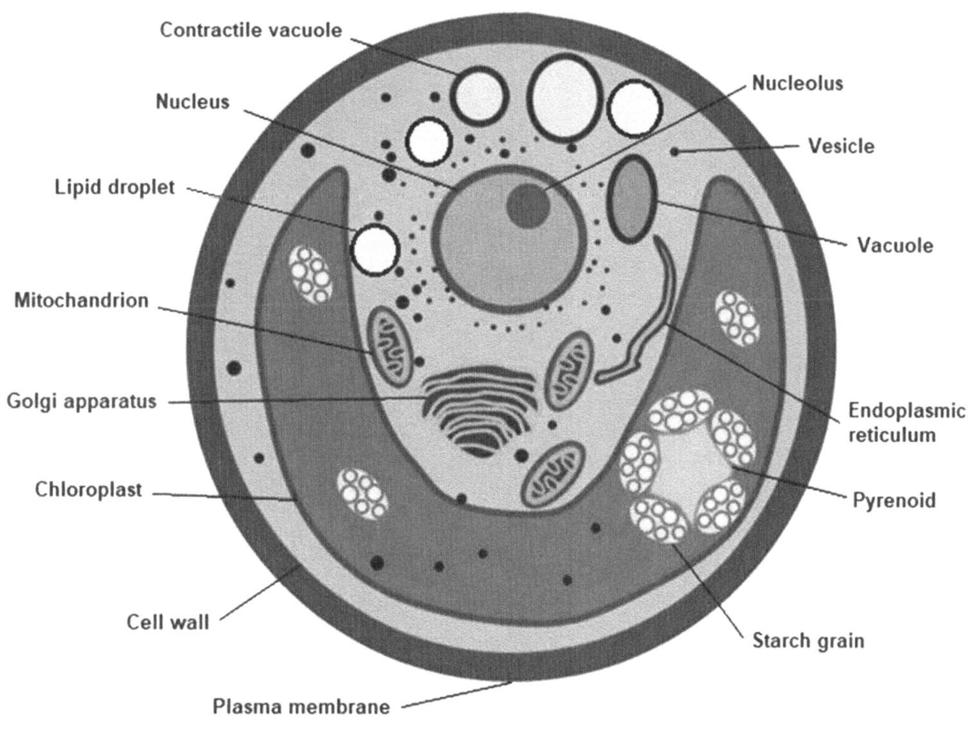

Figure 13.1 Schematic illustration of microalgal cell structure; adopted with modification from Poojary et al. (2016).

simple cellular system (Figure 13.1) and big surface-to-quantity ratio, microalgae are able to utilize high quantity of nutrients and achieve high growth rates (Rajkumar, Yaakob, & Takriff, 2013), which makes them more attractive for industrial applications.

Microalgae are highly diverse. However, among thousands of algae species [over 50,000 described (Cesário, Da Fonseca, Marques, & de Almeida, 2018)], commercial production of microalgae is mainly focused on a few: *Spirulina* sp., *Chlorella* sp., *Dunaliella* sp., *Haematococcus* sp., *Tetraselmis* sp., and *Nannochloropsis* sp. (Eppink et al., 2017). Especially considering the European food market, *Spirulina* sp. and *Chlorella* sp. do not apply as novel foods as they were consumed to a significant degree by humans in the EU prior to May 15, 1997 (Hayes, 2018; Regulation (EU) 2283, 2015). Nutritional profiles of some microalgae strains, which are mainly used as a source of valuable compounds, are presented in Table 13.1. In general, valuable compounds and recovered products of microalgae can be classified into lipids, carbohydrates, proteins, and pigments.

Table 13.1 Chemical composition of some microalgae species.

ALGAE		MACRONUTRIENT CONTENT [G/100 G_{DW}]		
CLASS	SPECIES	PROTEIN	LIPID	CARBOHYDRATE
Bacillariophyceae	*Phaeodactylum tricornutum*	30	14	8.4
	Nitzschia sp.	26	13	9.8
Chlorodendrophyceae	*Tetraselmis maculata*	52	3	15
Chlorophyceae	*Chlorella vulgaris*	51–58	14–22	12–17
	Dunaliella sp.	20–57	6–15	4–32
	Haematococcus sp.	17–27	25	37–40
	Scenedesmus obliquus	50–56	12–14	10–17
	Chlamydomonas reinhardtii	48	21	17
Coscinodiscophyceae	*Chaetoceros* sp.	12–34	7.2–16	4.7–6.0
	Thalassiosira pseudonana	34	19	8.8
Cyanophyceae	*Spirulina platensis*	46–63	4–9	8–14
	Anabaena cylindrica	43–56	4–7	25–30
	Aphanizomenon flos-aquae	62	3	23
Euglenophyceae	*Euglena gracilis*	39–61	14–20	14–18
Eustigmatophyceae	*Nannochloropsis* sp.	35	18	7.8
Pavlovophyceae	*Pavlova* sp.	29	12	9
Porphyridiophyceae	*Porphyridium cruentum*	28–39	9–14	40–57
Prymnesiophyceae	*Isochrysis* sp.	23–29	20–23	6.0–12.9
	Prymnesium parvum	28–45	22–38	25–33
Trebouxiophyceae	*Nannochloris atomus*	30	21	23
Zygnemophyceae	*Spirogyra* sp.	6–20	11–21	33–64

Adapted from Barba, Grimi, & Vorobiev, 2014; Becker, 2007; Demirbas & Fatih Demirbas, 2011; Raja, Vipin, & Aiyappan (2013).

2.1 *Proteins*

Microalgal proteins mainly consist of structural glycoprotein within the cell wall and membranes and water-soluble enzymatic proteins, such as RuBisCO, located in the cytoplasm and chloroplast and involved in carbon fixation. In general, microalgae are recognized as a viable source of protein, as some strains are characterized by high protein contents (~50 g/100 g_{dw}). For example, *Spirulina* and *Chlorella* have high protein content up to 70 g/100 g_{dw}, and they are produced in large-scale all over the world (Wells et al., 2017). Regarding the protein quality, in vivo studies of microalgal proteins showed absorption, digestibility, and mass gain properties, which are close to those of casein and egg (Becker, 2007). By taking into account the amino acid composition, profiles of various microalgae species compare well to conventional protein sources. The essential amino acid content is even far superior to plant sources, such as soybean (Barka & Blecker, 2016). Thus, protein recovered from microalgae can be considered not only as a dietary supplement but also as substitution of conventional protein sources.

2.2 *Lipids*

Holding a high lipid content and productivity, microalgae are particularly in focus for biofuel production and food applications. On the basis of different functionalities, lipids of microalgae can be divided into natural nonpolar storage lipids and structural polar membrane lipids (Lang, 2007). While storage lipids mainly consist of triacylglycerols (TAGs), structural lipids (phospholipids, glycolipids and sphingolipids) are amphiphilic molecules containing a polar group to build up cell and organelle membranes (Klok, Lamers, Martens, Draaisma, & Wijffels, 2014). Under specific stress conditions, for example, nitrogen starvation, some microalgae species are capable of accumulating larger quantities of natural lipids (>20 g/100g_{dw}). Lipid content of these so-called oleaginous algae genera, for example, *Porphyridium, Dunaliella, Isochrysis, Nannochloropsis, Tetraselmis, Phaeodactylum, Chlorella*, and *Schizochytrium*, commonly varies between 20 and 50 g/100 g_{dw} (Bellou et al., 2014) and even up to 80 g/100 g_{dw} can be reached (Chisti, 2007). Especially considering the amount of polyunsaturated fatty acids (PUFAs), which for instance can be up to 78% of total FAs for *Dunaliella* (Barba et al., 2014), microalgal lipids outline as an important nutraceutical. Large amounts of essential FAs, such as linoleic acid (C18:2), α-linolenic acid (C18:3), eicosapentaenoic acid (C20:5), and docosahexaenoic acid (C22:6), have several benefits for human health, inter alia, prevention of cardiovascular disease, cancer, asthma, and psychiatric disorders (Hibbeln, Nieminen, Blasbalg, Riggs, & Lands, 2006).

2.3 *Carbohydrates*

Algal carbohydrates can be classified into three main groups: (i) energy providers within the cell wall and intracellular vacuoles, whereby starch and glycogen are the main carbohydrate energy storage compounds; (ii) structural carbohydrates, such as peptidoglycans (cyanobacteria), (hemi-)cellulose, pectin (microalgae), agar, carrageenan, alginate and fucoidan (macroalgae), which make up the cell wall; and (iii) exopolysaccharides (EPSs) exuding into the extracellular environment and protecting against biotic stress. Agar, carrageenan, alginate, and various EPSs are hydrocolloids and can be added to food as thickeners, stabilizers, and gelling agents (Atobe et al., 2015; Donot, Fontana, Baccou, & Schorr-Galindo, 2012). Additionally, bioactive properties of some EPS are attractive for cosmetic or medical applications (de Jesus Raposo, de Morais, & de Morais, 2015). In general, content and type of present carbohydrates (mainly starch and cellulose for microalgae) is dependent on the particular species. Manipulation of growth conditions can increase the amount of carbohydrate formed by microalgae. Similar to the production of lipids, many algae species stop reproduction and accumulate storage carbohydrates during nutrient deficiency, such as nitrogen depletion (Hanifzadeh, Garcia, & Viamajala, 2018; Zhu et al., 2014).

2.4 Pigments

Microalgal pigments consist of three major groups: (i) chlorophyll—green, (ii) carotenoids (carotenes—orange and xanthophylls—yellow), and (iii) phycobilins—red and blue (Koller, Muhr, & Braunegg, 2014). They are responsible for coloration of the plant and hold several functions, such as light capturing during photosynthesis and protection against illumination damage and oxygen radicals. Therefore, most of the carotenoids are produced during autotrophic cultivation. However, some pigments, such as xanthophylls (lutein and astaxanthin), can also be produced heterotrophically (Venkata Subhash et al., 2017). Compared with the synthetic production, natural microalgal pigments are in interest of the food, pharmaceutical, and cosmetic industry. Other than their use as colorants, they have high contents of pro-vitamin A and showed benefits, such as antioxidant activity, anti-inflammatory effects, neuroprotection, and cancer prevention (Pangestuti & Kim, 2011).

3 Cultivation and Processing of Microalgae

Placed at the beginning of the food chain, microalgae only need sunlight, water, mineral nutriments, such as carbon, nitrate and phosphorus, and carbon dioxide for reproduction. The ability to use carbon dioxide-rich flue gas and mineral nutriment-rich residual water to produce algal biomass enables bio-fixation of greenhouse gas and recycling of nutrients/pollutants (Medeiros, Sales, & Kiperstok, 2015). In addition to the autotrophic cultivation named earlier, three further cultivation modes of microalgae are possible: heterotrophy, mixotrophy, and photoheterotrophy. Depending on the capability of particular species, cultivation mode can enhance the production of several compounds (Hu, Nagarajan, Zhang, Chang, & Lee, 2018; Zhan, Rong, & Wang, 2017).

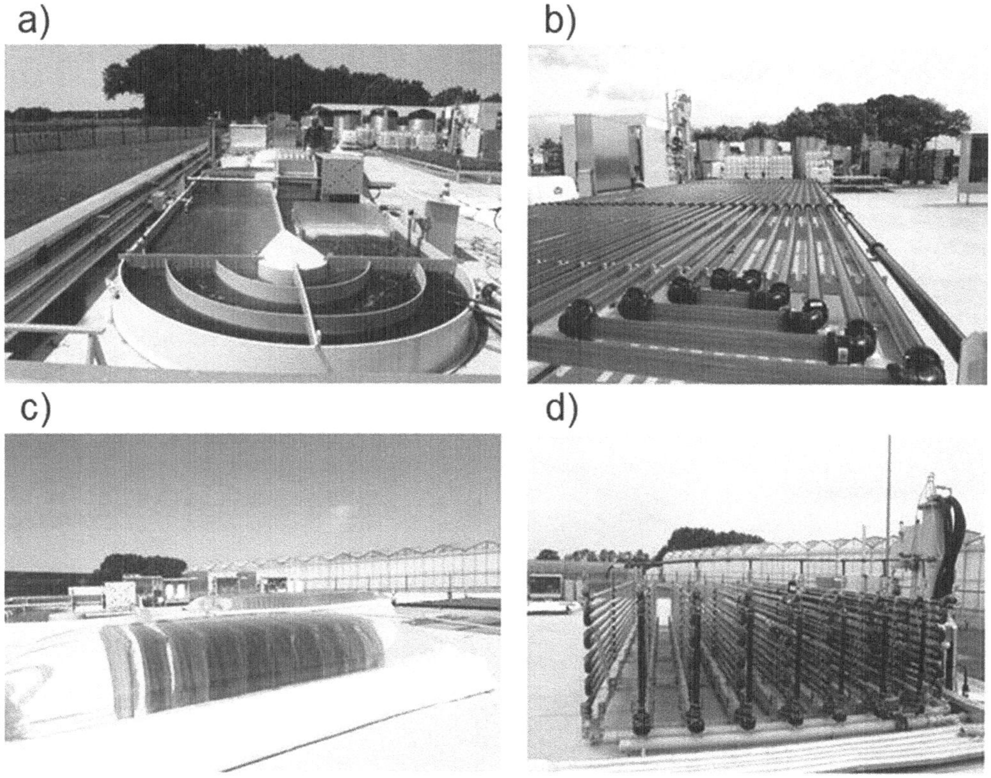

Figure 13.2 Pilot photobioreactors (PBR): (a) raceway pond, (b) horizontal tubular PBR, (c) flat panel PBR, and (d) vertical tubular PBR at AlgaePARC, Wageningen UR, the Netherlands; adapted from de Vree, Bosma, Janssen, Barbosa, & Wijffels (2015).

Cultivation of microalgae can be realized in open ponds or photobioreactors (PBR), which are mainly closed systems (Figure 13.2). Although operating costs of a PBR are generally higher, they offer higher flexibility and controlled environment, which enable optimized growth conditions. In addition to the microalgae strain and culture purity, influencing parameters on biomass productivity are temperature, pH, nutrients, light, and mixing (Rizwan, Mujtaba, Memon, Lee, & Rashid, 2018).

After cultivation, traditional microalgae processing is carried out by harvesting, concentration and drying to preserve biomass, and preparation for further processing, especially lipid extraction. Subsequently, cell disruption, extraction, and further separation of target products take place (Figure 13.3).

4 Conventional Extraction

Conventional extraction of proteins, lipids, carbohydrates, and pigments is usually done by maceration, percolation, or countercurrent extraction with aqueous or

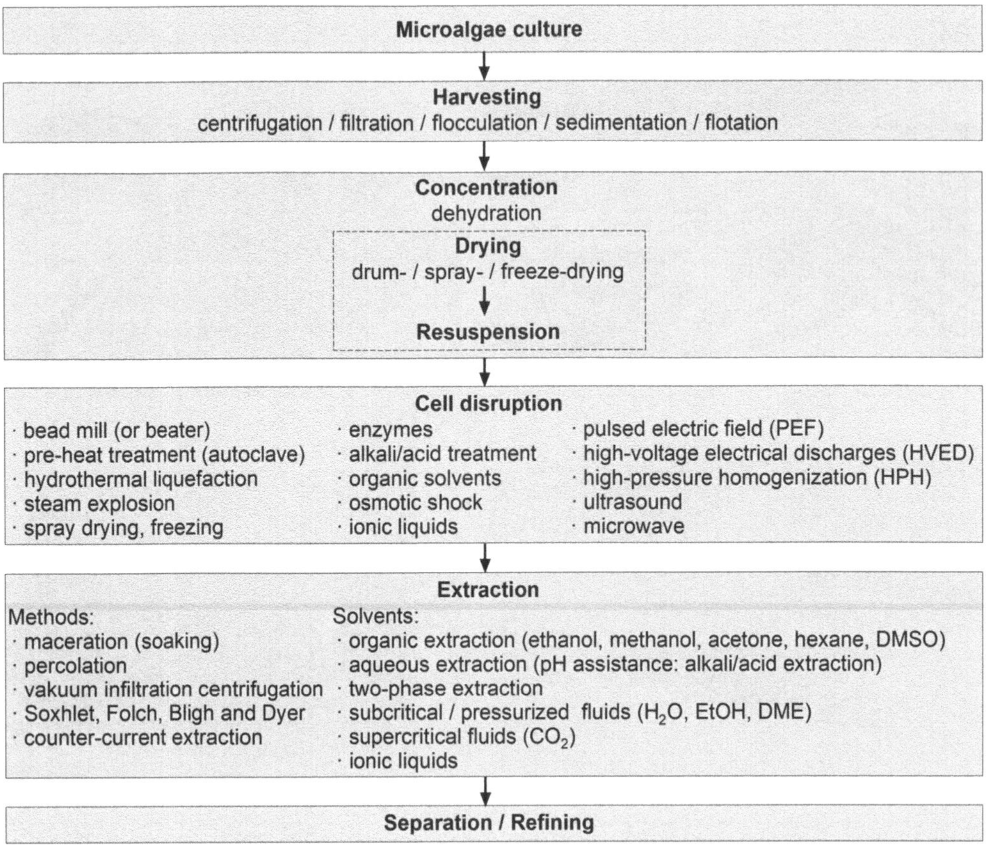

Figure 13.3 Traditional downstream processing of microalgae cultures; steps within the dashed box represent optional processing usually implemented for preservation and subsequent hydrophobic extraction.

organic solvents at different temperatures and pressures. Because of high microalgal diversity and variations of cell structures, release of valuable compounds from their intracellular location is a difficult task. Hence, cell wall disruption is the most determinant step for recovery of valuable compounds to ensure intensified mass transfer and an efficient extraction (Safi, Zebib, Merah, Pontalier, & Vaca-Garcia, 2014). Conventional disintegration methods are mechanical shearing (bead mill), preheat treatment, and chemical or enzymatic hydrolysis. Subsequent recovery of the solvents is usually done by distillation. In general, a rapid extraction of microalgal components is possible with most of the solvents. However, large amounts of solvents are needed for these extraction processes, which involve risks of environmental pollution and health hazards. Additionally, heat extraction can lead to thermal spoilage effects, such as denaturation, transformation of molecules, or degradation of valuable compounds (Manzan, Toniolo, Bredow, & Povh, 2003; Peshin, Lall, & Gupta, 2002; L. Wang & Weller, 2006).

4.1 Protein Extraction

After cell disruption, conventional extraction process of microalgal protein is carried out by alkali solubilization (pH 11–12) and separation via isoelectric precipitation at pH 3–5. In addition to the isoelectric point, fractionation can also be achieved by salting out, solvent precipitation (trichloroacetic acid/acetone), thermal treatment (cold precipitation at 4 °C, coagulation at 60–95 °C), ultrafiltration, or adsorption (Muñoz et al., 2015; Safi et al., 2014; Slocombe, Ross, Thomas, McNeill, & Stanley, 2013). Further concentration of larger quantities can be realized by centrifugation, membrane filtration (crossflow), and drying of the extracted protein. Higher purities of recovered protein for laboratory application can be obtained by chromatography, such as ion exchange, size exclusion, and affinity and hydrophobic interaction chromatography. Most of the existing protein extraction methods are performed in small-scale for scientific research. Upscaling and protein dissociation from carbohydrates and pigments are challenges that still need to be overcome (Khanra et al., 2018; Tamayo Tenorio, Kyriakopoulou, Suarez-Garcia, van Den Berg, & van der Goot, 2018).

4.2 Lipid Extraction

As microalgae production takes place by aquatic cultivation, the water present in the harvested biomass has a great influence on lipid extraction. Therefore, conventional processing for recovering microalgal oil for transformation into biodiesel (fatty acid methyl esters) is based on dehydration and resuspension of dried biomass in organic solvents. Quantification of the whole lipid content is performed by standardized methods of Soxhlet, Folch, or Bligh and Dyer method. Here, the polar and nonpolar characteristic of a hexane–acetone or chloroform–methanol mixture leads to the dissolution of nonpolar storage and polar structural lipids. However, due to solvent toxicity and the extraction of polar lipids, such as phospholipids, which negatively affect oil quality and stability (Chaiyasit, Elias, McClements, & Decker, 2007), these methods are not suitable for industrial production of edible oil where TAGs are preferred (Vanthoor-Koopmans, Wijffels, Barbosa, & Eppink, 2013). Within the food sector, hexane is already applied for large-scale lipid extraction (Sicaire et al., 2015). Despite its lower efficiency, hexane is a cost-effective nonpolar solvent holding a high selectivity toward the targeted TAGs. Nevertheless, it should be noted that this solvent is registered under the REACH Regulation as a category 2 reprotoxic and aquatic chronic toxic. In addition to environmental issues, hexane shows neurotoxic effects during occupational exposure, mainly inhalation by workers (ECHA, 2017).

4.3 Carbohydrate Extraction

Apart from the commercial production of macroalgal carbohydrates, such as agar, alginate, and carrageenan, which is performed by boiling water extraction (agar) and alkali extraction (alginate and carrageenan), significant amounts of starch and cellulose are either processed by saccharification of remaining carbohydrate-rich biomass for producing biofuel or discarded as debris. Usually, the aqueous extraction of hydrophilic components, such as proteins, also leads to higher amounts of carbohydrates washed out of the cells that can be separated by filtration (Eppink et al., 2017; Postma et al., 2016).

4.4 Pigment Extraction

Conventional extraction of pigments is based on organic solvent extraction. By application of solvents, such as methanol, ethanol, diethyl formamide, acetone, and dimethyl sulfoxide (DMSO), cell membranes of the chloroplasts get penetrated (dissolution of lipoproteins) and chlorophylls can be extracted (Cuellar-Bermudez et al., 2015). As most of the carotenoids (carotenes and esterified xanthophylls) are holding a hydrophobic character, nonpolar solvents, such as hexane, tetrahydrofuran, or petroleum, are conventionally used for extraction. Polar solvents, on the other side, such as ethanol, acetone, or ethyl acetate, are applied for polar carotenoids, such as astaxanthin (Saini & Keum, 2018). Lipids, which also get dissolved, can be separated, for example, by adsorption or membrane diafiltrating method (Kale, 2012). Nevertheless, efficient downstream processing of pigments is still a challenge to face. Several purification steps and pigment degradation lead to low final recoveries (Cuellar-Bermudez et al., 2015).

5 Green Extraction and Valorization

Over the past decades, technologies, such as extraction assisted by PEF, HVED, HPH, PCT, US, MW, sub- or supercritical pressurized fluids, and IL, emerged and gained interest in the industries. These so-called green extraction technologies have several advantages compared with conventional cell disruption and extraction methods. They are eco-friendly due to the reduced usage or elimination of toxic solvents and cost-efficient by less energy consumption. Furthermore, no additional (chemical) substances or means, such as beads, need to be added and separated afterward. Finally, higher yields and reduced treatment times can be reached within shorter operating times.

5.1 Pulsed Electric Fields

PEF technology is based on the phenomenon called electroporation. Exceeding the transmembrane potential, a critical voltage is needed to cause electro-compression of the membrane and occurrence of dielectric breaks that pierce the cell membrane at high electric fields. Main parameters characterizing PEF treatments are the specific energy input and electric field strength (Pataro, De Lisi, Donsì, & Ferrari,

2014). Mainly depending on the energy input, the formation of pores inside the membranes can be permanent or temporary in case of pore resealing and membrane recovery. Applied electric field strengths can range from 100 V/cm to 300 kV/cm. Because of the usage of short duration pulses (several nanoseconds to milliseconds), energy efficiency is enhanced (Kotnik, Kramar, Pucihar, Miklavcic, & Tarek, 2012) and excessive joule heating can be avoided, which makes PEF a nonthermal mild treatment method. Additionally, mild temperatures (<50 °C) are causing a synergetic effect increasing electroporation. Nowadays, PEF systems can be realized in continuous mode and are used within a wide range of food applications, such as PEF-assisted extraction, drying, freezing, cutting, and cold pasteurization (Barba et al., 2015). Only considering microalgae processing, PEF can be used for the control of contaminants during cultivation (Buchmann et al., 2018; Rego et al., 2015) to stimulate growth (Bai, Gusbeth, Frey, & Nick, 2017; Buchmann, Frey, Gusbeth, Ravaynia, & Mathys, 2019; Gusbeth, Eing, Göttel, & Frey, 2013) up to selective extraction of valuable compounds (Pereira, Rego, & Redondo, 2016). However, some limitations have to be considered. Cell suspensions have to be pumpable for continuous operation and conductivity of the transmission medium needs to be considered for ensuring an efficient flow of electricity leading to cell electroporation. Additionally, enclosed air inside the treatment chamber can lead to adverse flashovers.

In contrast to conventional cell disruption methods, where the cell membrane gets disintegrated completely and all components are released, electroporation enables a highly selective extraction of intracellular products (Figure 4), thus avoiding quality loss and subsequent fractionation of non-pure extracts (Carullo et al., 2018; Coustets, Al-Karablieh, Thomsen, & Teissié, 2013; Grimi et al., 2014; Luengo, Martínez, Bordetas, Álvarez, & Raso, 2015; Parniakov et al., 2015a; Postma et al., 2018). As induction of electropermeabilization within smaller organelles requires higher field strength than for the cell membrane (Esser, Smith, Gowrishankar, Vasilkoski, & Weaver, 2010), treatment of microalgae by PEF, for instance, allows perforation of the cell membrane and release of hydrophilic components (protein and carbohydrates) without disintegration of intracellular vacuoles-containing proteases (Coustets et al., 2013).

For the extraction of hydrophobic molecules, PEF is rather ineffective (Parniakov et al., 2015a). However, designing a multistep extraction process can lead to an effective extraction where larger specific energy inputs result in bigger pores. Here, reduced organic solvent usage might be realized by solvent exchange to organic solvents or binary mixtures (Parniakov et al., 2015b). In case of pigment extraction, for instance, a reduction of DMSO from 70 to 50% within a DMSO–water mixture was realized. Here, yield of the pigments was at the same level as the higher concentrated solvent mixture (Siemer & Parniakov, 2018).

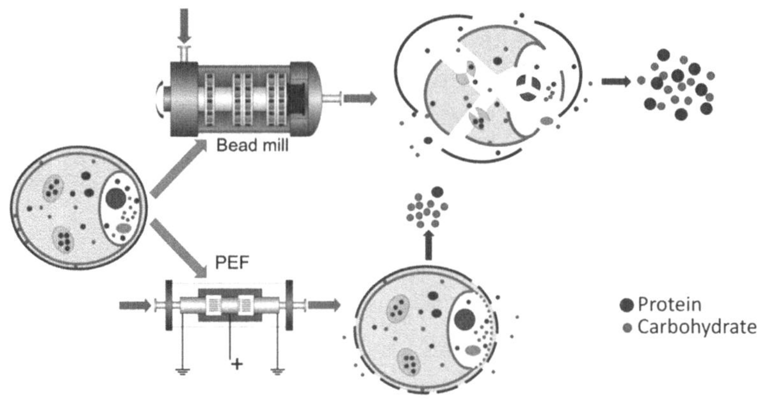

Figure 13.4 Illustration of cell disruption differences between conventional bead milling and pulsed electric field treatment; adopted from Postma et al. (2016).

5.2 High-Voltage Electrical Discharges

In addition to high electrostatic fields (HEF), HVED belongs to the high-voltage electric field technologies. While HEF is working without any flow of electricity, the application of HVED within aqueous solutions leads to fluid ionization and the formation of a high current plasma channel between two electrodes. For extraction purposes, these electrical discharges are commonly characterized by field strengths of ~40 kV/cm and current intensities ~10 kA (Boussetta & Vorobiev, 2014). Because of no significant temperature change during application, HVED is a nonthermal processing technology. Compared with PEF, HVED is a nonselective method. The more powerful disintegration enables the release of high molecular weight compounds, but is also accompanied by larger quantities of cell debris and contaminants inside the extract (Liu, Lebovka, & Vorobiev, 2013). By now, there is not much scientific literature published about HVED-assisted downstream processing of microalgae. However, initial studies have been done (Grimi et al., 2014).

5.3 High-Pressure Homogenization

HPH, originally employed for emulsification applications, belongs to the most commonly used large-scale disruption methods for microbial cells, especially bacteria and yeast (Norton & Sun, 2008). By striking a stationary impact ring inside the homogenization valve, cell disintegration is achieved by shear forces at high pressures. Additionally, cavitation induces shear stress to the cells (Chisti & Moo-Young, 1986). These extremely high impacts on the membranes lead to high rupturing and extraction efficiencies of high molecular mass compounds (Carullo et al., 2018). For example, high yields of water-soluble protein extracted from *Nannochloropsis* sp. can be achieved by the application of HPH (Figure 13.5).

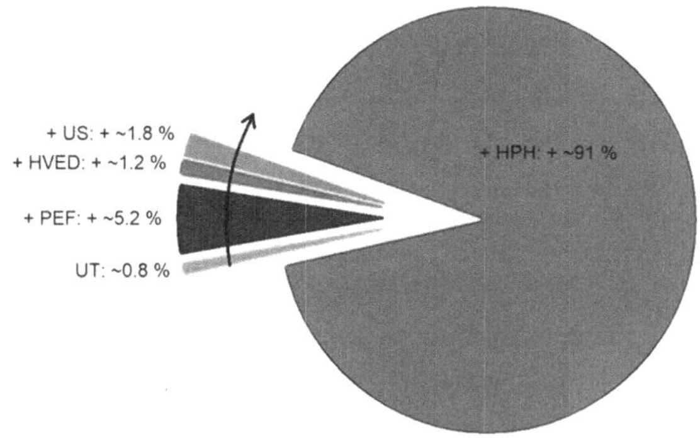

Figure 13.5 Extraction yields of water-soluble protein from *Nannochloropsis* sp. untreated (UT) and after sequential treatment with pulsed electric field (PEF), high-voltage electrical discharges (HVED), ultrasound (US), and high-pressure homogenization (HPH); adapted from Grimi et al. (2014).

Nevertheless, HPH is characterized by the nonselective release of intracellular components and appearance of a very fine dispersion of cell debris complicating downstream processing (Grimi et al., 2014), thus being more suitable at the end of a modern multistage extraction process. Because of high energy amounts needed for the treatment (A. K. Lee, Lewis, & Ashman, 2013) and impacts on the quality of valuable compounds, such as formation of free FAs (FFAs) (Balduyck, Bruneel, Goiris, Dejonghe, & Foubert, 2018), HPH cannot be identified as a mild disruption method, even its potential as a green extraction technique might be doubted.

5.4 Pressure Change Technology

PCT represents an innovative, nonthermal technology that was developed for the inactivation of microbial loads in liquid food and pharmaceuticals (Bönsch, Wecks, Ondruschka, & Staudt, 2007). By applying pressures up to 50 MPa, the liquid gets saturated with gas, preferably inert gases, such as N_2, CO_2, or Ar. After diffusion into the cells, pressure release and a rapid expansion lead to their disruption. PCT can be run in continuous operation and already showed its potential to extend the shelf-life of fresh orange juice (Aschoff et al., 2016). By now, PCT has not been applied for the disintegration of microalgal cells.

5.5 Ultrasound

Ultrasonic treatment of cell suspensions is based on the cavitation effect. The creation and collapse of small vacuum bubbles lead to shock waves and shearing forces that rupture the ridged cell walls of microalgae. Additionally, acoustic streaming enhances dispensation by mixing the treated medium (Gerde, Montalbo-Lomboy,

Yao, Grewell, & Wang, 2012). Past studies on ultrasound-assisted extraction (UAE) demonstrated its potential for effective disintegration of various algal species and increased yields of compounds, such as lipids and pigments (Barba et al., 2014). In general, most of the applications use non-focused US at low frequencies of 20–40 kHz (Ido, de Luna, Capareda, Maglinao, & Nam, 2018; Phong et al., 2018; Wang, Yuan, Jiang, Jing, & Wang, 2014). The treatment of microalgae leads to the disruption of cells into smaller particles (Wang, Yuan, & Hale, 2016); thus ultrasonic disintegration represents a nonselective disruption technique. The treatment process needs to be controlled carefully to reduce the formation of free radicals, which can influence the product quality by inducing lipid oxidation (Gerde et al., 2012). Furthermore, lipids are degraded into FFAs (Meullemiestre, Breil, Abert-Vian, & Chemat, 2016) and, in case of increased energy inputs, degradation of large proteins can also be possible (Pchelintsev, Adams, & Nelson, 2016). Compared to microwave-assisted extractions (MAE) (100 °C) and autoclaving (121–125 °C), the UAE can be realized at relatively low temperatures. Nevertheless, counter cooling is often applied to avoid, for instance, possible chlorophyll and carotenoid degradation (Gerde et al., 2012; Parniakov et al., 2015; Rahimi, Mohamadian, Dadari, Arbab, & Karimi, 2017; Zhao et al., 2006).

5.6 Microwaves

MW treatment is an electromagnetic radiation method inducing a thermal impact, which causes moisture evaporation. As a consequence, formed bubbles holding high pressure are able to rupture cells. Increased temperatures and disintegrated membranes accelerate the mass transfer of intracellular compounds. MAE can be conducted in open or closed systems. Here, MW is directly applied on the microalgae–solvent suspension used for extraction. To enable an efficient process, solvents need to hold a significant dipole moment (Chemat, Zill-e-Huma, & Khan, 2011). As drying is consuming major energy during the processing of microalgae MAE of microalgal lipids, especially based on wet biomass (Moura et al., 2018), it is currently in scientific focus for the third generation of biofuel. However, regarding food applications, further aspects need to be considered. An intense MW treatment, which is needed due to limited solvent access through rigid cell walls or thick EPS envelopes of some microalgae, can lead to the degradation of PUFAs (Moura et al., 2018) and chlorophyll a (Pasquet et al., 2011). Furthermore, the oxidative formation of FFAs, similar to UAE, was shown (Meullemiestre et al., 2016).

5.7 Sub- and Supercritical Fluid Extraction

Subcritical fluid extraction, also known as pressurized liquid extraction (PLE) or subcritical water extraction (SWE), and supercritical fluid extraction (SFE) are based on the phenomenon that fluids maintain their state of aggregation at increased pressure and temperature. Within that state, mass transfer is increased, and solvents that are unsuitable at ambient conditions can be used. SWE is applied by heating water within the range of 100–374 °C. At 374 °C and 22.1 MPa, water is at its critical temperature, where liquid and gaseous properties can be seen. Higher temperatures involve a decrease of the dielectric constant, ε_r, whereby lower constants are related to a better solubility of nonpolar compounds and, on the other hand, higher constants to polar compounds. For example, water, a polar solvent at room temperature ($\varepsilon_r \sim 80$), has a dielectric constant of ~35 at 200 °C, which is close to that of ethanol (~33) at ambient conditions (Miller & Hawthorne, 2000). SWE is usually applied at 1–6 MPa (Herrero, Cifuentes, & Ibañez, 2006). In addition to water, other solvents, such as ethanol, hexane, subcritical compressed gasses (butane and dimethylether), and additional cosolvents to enhance solubility can be used to extract lipids, carotenoids, proteins, and carbohydrates (Awaluddin et al., 2016; Bian et al., 2018; Chen et al., 2012; Denery, Dragull, Tang, & Li, 2004; Feller et al., 2018; Goto, Kanda, Wahyudiono, & Machmudah, 2015; Sánchez-Camargo et al., 2018). One of the biggest drawbacks of water-based PLE is high-process temperatures, which lead to degradation of many valuable microalgal compounds, such as hydrolysation of lipids into FFAs (Saber, Golzary, Wu, Takahashi, & Yoshikawa, 2018), denaturation of proteins, and Maillard reaction (Rodríguez-Meizoso et al., 2010).

SFE, on the other hand, is characterized by better mass transfer than liquids due to the high diffusivity of supercritical fluids. Furthermore, fluid density can be modified by pressure and temperature. CO_2 is most commonly used for SFE, as it is cheap, nonflammable, holding a mild critical temperature of 31.3 °C at ~7.4 MPa, generally recognized as safe, and elimination can be done by evaporation without residues. Although CO_2 has higher efficiencies for the extraction of nonpolar components, cosolvents (mainly up to 10% ethanol) can be added to enhance the solubility of polar components. Generally, operating conditions for CO_2 are 40–60 °C and 20–50 MPa. Consequently, mild extraction and higher recovery of valuable compounds, such as lipids, especially PUFAs and carotenoids, can be achieved compared with conventional extractions (Abrahamsson, Cunico, Andersson, Nilsson, & Turner, 2018; Cardoso, Serrano, Rodríguez, Ossa, & Lubián, 2012; Mendes, Nobre, Cardoso, Pereira, & Palavra, 2003; Mendiola et al., 2008). However, SFE is strongly limited by residual moisture of the material. As surface water leads to a sharp drop of extraction yield, SFE–CO_2 is not suitable for wet biomass extraction intends (Viguera, Marti, Masca, Prieto, & Calvo, 2016).

5.8 Ionic Liquids

Next to pressurized fluids, IL showed their potential as solvents for microalgae extraction. IL are salts that hold a large asymmetrical organic cation, making them liquid at ambient conditions and up to 100 °C (Orr & Rehmann, 2016). By disrupting hydrogen bonds of cellulose networks at increased temperatures (Pinkert, Marsh, Pang, & Staiger, 2009), they can also facilitate the extraction of intracellular compounds, even of wet microalgal biomass. On one side, their low volatilities, as well as chemical and thermal stabilities, are favorable compared with conventional organic solvents. However, on the other side, this makes it more difficult to separate the ILs from the extracted components, which is why an immiscible laver of organic solvent, mainly hexane, is often used for facilitating the handling of lipids (Orr & Rehmann, 2016). While lipid extractions are the field of main application, some studies also demonstrated successful separation and refinery of proteins, carbohydrates, and pigments by using aqueous two-phase systems (Desai, Streefland, Wijffels, & Eppink, 2016; Dreyer & Kragl, 2008; Suarez Garcia et al., 2018; Suarez Ruiz, Emmery, Wijffels, Eppink, & van Den Berg, 2018).

6 Conclusions

Green extraction and downstream processing of various valuable microalgae compounds have attracted the interest of industries over the past years. Most of them are by-products of conventional production lines, such as for biofuel, and could be recovered by sequential extraction steps. Although many efforts toward multiproduct extractions have been done within recent years, there is still further research needed to implement an efficient combined commercial application for food processing and enabling competitive production. Furthermore, harmful conventional organic solvents, such as hexane and chloroform, pose a risk to human health and should be replaced to reduce hazard potential. Innovative technologies, such as PEF and supercritical CO_2 extraction, enable selective and mild extraction of valuable compounds by the usage of eco-friendly solvents paving the way for less harmful food applications.

References

Abrahamsson, V., Cunico, L.P., Andersson, N., Nilsson, B., & Turner, C. (2018). Multicomponent inverse modeling of supercritical fluid extraction of carotenoids, chlorophyll *a*, ergosterol and lipids from microalgae. *The Journal of Supercritical Fluids*, *139*, 53–61. https://doi.org/10.1016/j.supflu.2018.05.007

Aschoff, J.K., Knoblauch, K., Hüttner, C., Vásquez-Caicedo, A.L., Carle, R., & Schweiggert, R.M. (2016). Non-thermal pasteurization of orange (*Citrus sinensis* (L.) Osbeck) juices using continuous pressure change technology (PCT): A proof-of-concept. *Food and Bioprocess Technology*, *9*(10), 1681–1691. https://doi.org/10.1007/s11947-016-1754-6

Atobe, S., Saga, K., Hasegawa, F., Furuhashi, K., Tashiro, Y., Suzuki, T., … Imou, K. (2015). Effect of amphiphilic polysaccharides released from *Botryococcus braunii* Showa on hydrocarbon recovery. *Algal Research*, *10*, 172–176. https://doi.org/10.1016/j.algal.2015.05.007

Awaluddin, S.A., Thiruvenkadam, S., Izhar, S., Hiroyuki, Y., Danquah, M.K., & Harun, R. (2016). Subcritical water technology for enhanced extraction of biochemical compounds from *Chlorella vulgaris*. *BioMed Research International*. https://doi.org/10.1155/2016/5816974

Bai, F., Gusbeth, C., Frey, W., & Nick, P. (2017). Nanosecond pulsed electric fields trigger cell differentiation in *Chlamydomonas reinhardtii*. *Biochimica Et Biophysica Acta (BBA) - Biomembranes*, *1859*(5), 651–661. https://doi.org/10.1016/j.bbamem.2017.01.007

Balduyck, L., Bruneel, C., Goiris, K., Dejonghe, C., & Foubert, I. (2018). Influence of high pressure homogenization on free fatty acid formation in *Nannochloropsis* sp. *European Journal of Lipid Science and Technology*, *120*(4), 1700436. https://doi.org/10.1002/ejlt.201700436

Barba, F.J., Grimi, N., & Vorobiev, E. (2014). New approaches for the use of non-conventional cell disruption technologies to extract potential food additives and nutraceuticals from microalgae. *Food Engineering Reviews*, *7*(1), 45–62.

Barba, F.J., Parniakov, O., Pereira, S.A., Wiktor, A., Grimi, N., Boussetta, N., … Vorobiev, E. (2015). Current applications and new opportunities for the use of pulsed electric fields in food science and industry. *Food Research International*, *77*, 773–798.

Barka, A., & Blecker, C. (2016). Microalgae as a potential source of single-cell proteins. a review. *BASE*. Retrieved from https://popups.uliege.be:443/1780-4507/index.php?id=13132

Becker, E.W. (2007). Micro-algae as a source of protein. *Biotechnology Advances*, *25*(2), 207–210. https://doi.org/10.1016/j.biotechadv.2006.11.002

Bellou, S., Baeshen, M.N., Elazzazy, A.M., Aggeli, D., Sayegh, F., & Aggelis, G. (2014). Microalgal lipids biochemistry and biotechnological perspectives. *Biotechnology Advances*, *32*(8), 1476–1493. https://doi.org/10.1016/j.biotechadv.2014.10.003

Bian, X., Jin, W., Gu, Q., Zhou, X., Xi, Y., Tu, R., … Wang, Q. (2018). Subcritical n-hexane/isopropanol extraction of lipid from wet microalgal pastes of *Scenedesmus obliquus*. *World Journal of Microbiology and Biotechnology*, *34*(3), 39. https://doi.org/10.1007/s11274-018-2421-z

Bönsch, K., Wecks, M., Ondruschka, J., & Staudt, R. (2007). The effect of a new pressure change technology (PCT) on microorganisms: An innovate concept for food safety. *Chemical Engineering & Technology*, *30*(6), 755–757. https://doi.org/10.1002/ceat.200600402

Boussetta, N., & Vorobiev, E. (2014). Extraction of valuable biocompounds assisted by high voltage electrical discharges: A review. *Comptes Rendus Chimie*, *17*(3), 197–203. https://doi.org/10.1016/j.crci.2013.11.011

Buchmann, L., Böcker, L., Frey, W., Haberkorn, I., Nyffeler, M., & Mathys, A. (2018). Energy input assessment for nanosecond pulsed electric field processing and its application in a case study with *Chlorella vulgaris*. *Innovative Food Science & Emerging Technologies*, *47*, 445–453. https://doi.org/10.1016/j.ifset.2018.04.013

Buchmann, L., Frey, W., Gusbeth, C., Ravaynia, P.S., & Mathys, A. (2019). Effect of nanosecond pulsed electric field treatment on cell proliferation of microalgae. *Bioresource Technology*, *271*, 402–408. https://doi.org/10.1016/j.biortech.2018.09.124

Cardoso, L.C., Serrano, C.M., Rodríguez, M.R., Ossa, E.J.M.D.L., & Lubián, L.M. (2012). Extraction of carotenoids and fatty acids from microalgae using supercritical technology. *American Journal of Analytical Chemistry*, *03*, 877. https://doi.org/10.4236/ajac.2012.312A116

Carullo, D., Abera, B.D., Casazza, A.A., Donsì, F., Perego, P., Ferrari, G., & Pataro, G. (2018). Effect of pulsed electric fields and high pressure homogenization on the aqueous extraction of intracellular compounds from the microalgae *Chlorella vulgaris*. *Algal Research*, *31*, 60–69. https://doi.org/10.1016/j.algal.2018.01.017

Cesário, M.T., Da Fonseca, M.M.R., Marques, M.M., & de Almeida, M.C.M.D. (2018). Marine algal carbohydrates as carbon sources for the production of biochemicals and biomaterials. *Biotechnology Advances*, *36*(3), 798–817. https://doi.org/10.1016/j.biotechadv.2018.02.006

Chaiyasit, W., Elias, R.J., McClements, D.J., & Decker, E.A. (2007). Role of physical structures in bulk oils on lipid oxidation. *Critical Reviews in Food Science and Nutrition*, *47*(3), 299–317. https://doi.org/10.1080/10408390600754248

Chemat, F., Zill-e-Huma, & Khan, M.K. (2011). Applications of ultrasound in food technology: Processing, preservation and extraction. *Ultrasonics Sonochemistry*, *18*(4), 813–835. https://doi.org/10.1016/j.ultsonch.2010.11.023

Chen, M., Liu, T., Chen, X., Chen, L., Zhang, W., Wang, J., ... Peng, X. (2012). Subcritical co-solvents extraction of lipid from wet microalgae pastes of *Nannochloropsis* sp. *European Journal of Lipid Science and Technology*, *114*(2), 205–212. https://doi.org/10.1002/ejlt.201100120

Chisti, Y. (2007). Biodiesel from microalgae. *Biotechnology Advances*, *25*(3), 294–306. https://doi.org/10.1016/j.biotechadv.2007.02.001

Chisti, Y., & Moo-Young, M. (1986). Disruption of microbial cells for intracellular products. *Enzyme and Microbial Technology*, *8*(4), 194–204. https://doi.org/10.1016/0141-0229(86)90087-6

Coustets, M., Al-Karablieh, N., Thomsen, C., & Teissié, J. (2013). Flow process for electroextraction of total proteins from microalgae. *Journal of Membrane Biology*, *246*(10), 751–760.

Cuellar-Bermudez, S.P., Aguilar-Hernandez, I., Cardenas-Chavez, D.L., Ornelas-Soto, N., Romero-Ogawa, M.A., & Parra-Saldivar, R. (2015). Extraction and purification of high-value metabolites from microalgae: Essential lipids, astaxanthin and phycobiliproteins. *Microbial Biotechnology*, *8*(2), 190–209. https://doi.org/10.1111/1751-7915.12167

de Jesus Raposo, M.F., de Morais, A.M.B., & de Morais, R.M.S.C. (2015). Marine polysaccharides from algae with potential biomedical applications. *Marine Drugs*, *13*(5), 2967–3028. https://doi.org/10.3390/md13052967

de Vree, J.H., Bosma, R., Janssen, M., Barbosa, M.J., & Wijffels, R.H. (2015). Comparison of four outdoor pilot-scale photobioreactors. *Biotechnology for Biofuels*, *8*. https://doi.org/10.1186/s13068-015-0400-2

Demirbas, A., & Fatih Demirbas, M. (2011). Importance of algae oil as a source of biodiesel. *Energy Conversion and Management*, *52*(1), 163–170. https://doi.org/10.1016/j.enconman.2010.06.055

Denery, J.R., Dragull, K., Tang, C.S., & Li, Q.X. (2004). Pressurized fluid extraction of carotenoids from *Haematococcus pluvialis* and *Dunaliella salina* and kavalactones from *Piper methysticum*. *Analytica Chimica Acta*, *501*(2), 175–181. https://doi.org/10.1016/j.aca.2003.09.026

Desai, R.K., Streefland, M., Wijffels, R.H., & Eppink, M.H.M. (2016). Extraction of proteins with ABS. In Freire M.G. (Ed.), *Ionic-liquid-based aqueous biphasic systems: Fundamentals and applications* (pp. 123–134). Green Chemistry and Sustainable Technology: Springer, Berlin, Heidelberg. https://doi.org/10.1007/978-3-662-52875-4_6

Donot, F., Fontana, A., Baccou, J.C., & Schorr-Galindo, S. (2012). Microbial exopolysaccharides: Main examples of synthesis, excretion, genetics and extraction. *Carbohydrate Polymers*, *87*(2), 951–962. https://doi.org/10.1016/j.carbpol.2011.08.083

Dreyer, S., & Kragl, U. (2008). Ionic liquids for aqueous two-phase extraction and stabilization of enzymes. *Biotechnology and Bioengineering*, *99*(6), 1416–1424. https://doi.org/10.1002/bit.21720

ECHA. (2017). *Substance evaluation conclusion as required by REACH article 48 and evaluation report for n-hexane*. 1–55. Environmental Protection Agency, Vilnius, Lithuania.

Eing, C., Goettel, M., Straessner, R., Gusbeth, C., & Frey, W. (2013). Pulsed electric field treatment of microalgae - benefits for microalgae biomass processing. *IEEE Transactions on Plasma Science*, *41*(10), 2901–2907. https://doi.org/10.1109/TPS.2013.2274805

Enzing, C., Ploeg, M., Barbosa, M.J., & Sijtsma, L. (2014). Microalgae-based products for food and feed sector: An outlook for Europe. In: Vigani M., Parisi C., Rodríguez-Cerezo E. (Ed), *JRC Scientific and Policy Reports*. European Comission. Publications office of the European Union, Luxembourg. https://doi.org/10.2791/3339

Eppink M.H.M., Olivieri G., Reith H., van den Berg C., Barbosa M.J., Wijffels R.H. (2017) From Current Algae Products to Future Biorefinery Practices: A Review. In Wagemann K., Tippkötter N. (Ed), *Biorefineries. Advances in Biochemical Engineering/Biotechnology*, vol 166. Springer, Cham. https://doi.org/10.1007/10_2016_64

Esser, A.T., Smith, K.C., Gowrishankar, T.R., Vasilkoski, Z., & Weaver, J.C. (2010). Mechanisms for the intracellular manipulation of organelles by conventional electroporation. *Biophysical Journal*, *98*(11), 2506–2514. https://doi.org/10.1016/j.bpj.2010.02.035

Feller, R., Matos, Â.P., Mazzutti, S., Moecke, E.H.S., Tres, M.V., Derner, R.B., … Junior, A.F. (2018). Polyunsaturated ω-3 and ω-6 fatty acids, total carotenoids and antioxidant activity of three marine microalgae extracts obtained by supercritical CO2 and subcritical n-butane. *The Journal of Supercritical Fluids*, *133*, 437–443. https://doi.org/10.1016/j.supflu.2017.11.015

Gerde, J.A., Montalbo-Lomboy, M., Yao, L., Grewell, D., & Wang, T. (2012). Evaluation of microalgae cell disruption by ultrasonic treatment. *Bioresource Technology*, *125*, 175–181. https://doi.org/10.1016/j.biortech.2012.08.110

Goettel, M., Eing, C., Gusbeth, C., Straessner, R., & Frey, W. (2013). Pulsed electric field assisted extraction of intracellular valuables from microalgae. *Algal Research*, *2*(4), 401–408. https://doi.org/10.1016/j.algal.2013.07.004

Goto, M., Kanda, H., Wahyudiono, & Machmudah, S. (2015). Extraction of carotenoids and lipids from algae by supercritical CO2 and subcritical dimethyl ether. *The Journal of Supercritical Fluids*, *96*, 245–251. https://doi.org/10.1016/j.supflu.2014.10.003

Grimi, N., Dubois, A., Marchal, L., Jubeau, S., Lebovka, N.I., & Vorobiev, E. (2014). Selective extraction from microalgae Nannochloropsis sp. using different methods of cell disruption. *Bioresource Technology*, *153*, 254–259.

Gusbeth, C.A., Eing, C., Göttel, M., & Frey, W. (2013). Boost of algae growth by ultra short pulsed electric field treatment. *2013 Abstracts IEEE International Conference on Plasma Science (ICOPS)*, 1. https://doi.org/10.1109/PLASMA.2013.6633325

Hanifzadeh, M., Garcia, E.C., & Viamajala, S. (2018). Production of lipid and carbohydrate from microalgae without compromising biomass productivities: Role of Ca and Mg. *Renewable Energy*, *127*, 989–997. https://doi.org/10.1016/j.renene.2018.05.012

Hayes, M. (2018). *Novel proteins for food, pharmaceuticals, and agriculture: Sources, applications, and advances*. Wiley-Blackwell, Hoboken, NJ, USA.

Herrero, M., Cifuentes, A., & Ibañez, E. (2006). Sub- and supercritical fluid extraction of functional ingredients from different natural sources: Plants, food-by-products, algae and microalgae: A review. *Food Chemistry*, *98*(1), 136–148. https://doi.org/10.1016/j.foodchem.2005.05.058

Hibbeln, J.R., Nieminen, L.R., Blasbalg, T.L., Riggs, J.A., & Lands, W.E. (2006). Healthy intakes of n-3 and n–6 fatty acids: Estimations considering worldwide diversity. *The American Journal of Clinical Nutrition*, *83*(6), 1483S-1493S. https://doi.org/10.1093/ajcn/83.6.1483S

Hu, J., Nagarajan, D., Zhang, Q., Chang, J.-S., & Lee, D.-J. (2018). Heterotrophic cultivation of microalgae for pigment production: A review. *Biotechnology Advances*, *36*(1), 54–67. https://doi.org/10.1016/j.biotechadv.2017.09.009

Ido, A.L., de Luna, M.D.G., Capareda, S.C., Maglinao, A.L., & Nam, H. (2018). Application of central composite design in the optimization of lipid yield from *Scenedesmus obliquus* microalgae by ultrasound-assisted solvent extraction. *Energy*, *157*, 949–956. https://doi.org/10.1016/j.energy.2018.04.171

Kale, A. (2012). *United States Patent No. US8115022B2*. Retrieved from https://patents.google.com/patent/US8115022/en

Khanra, S., Mondal, M., Halder, G., Tiwari, O.N., Gayen, K., & Bhowmick, T.K. (2018). Downstream processing of microalgae for pigments, protein and carbohydrate in industrial application: A review. *Food and Bioproducts Processing*, *110*, 60–84. https://doi.org/10.1016/j.fbp.2018.02.002

Kinkel, D. (2018). *Mikroalgen – Eine Rohstoffquelle der Zukunft!* Retrieved from Forschungskreis der Ernährungsindustrie E.V. website: www.fei-bonn.de/gefoerderte-projekte/best-practice-projekte/bpp-2018-06-mikroalgen

Klok, A.J., Lamers, P.P., Martens, D.E., Draaisma, R.B., & Wijffels, R.H. (2014). Edible oils from microalgae: Insights in TAG accumulation. *Trends in Biotechnology*, *32*(10), 521–528. https://doi.org/10.1016/j.tibtech.2014.07.004

Koller, M., Muhr, A., & Braunegg, G. (2014). Microalgae as versatile cellular factories for valued products. *Algal Research*, *6*, 52–63. https://doi.org/10.1016/j.algal.2014.09.002

Kotnik, T., Kramar, P., Pucihar, G., Miklavcic, D., & Tarek, M. (2012). Cell membrane electroporation- part 1: The phenomenon. *IEEE Electrical Insulation Magazine*, *28*(5), 14–23. https://doi.org/10.1109/MEI.2012.6268438

Lang, I. (2007). *New fatty acids, oxylipins and volatiles in microalgae*. Georg-August-Universität zu Göttingen, Göttingen, Germany.

Lee, A.K., Lewis, D.M., & Ashman, P.J. (2013). Force and energy requirement for microalgal cell disruption: An atomic force microscope evaluation. *Bioresource Technology*, *128*, 199–206. https://doi.org/10.1016/j.biortech.2012.10.032

Lee, R.E. (2008). *Phycology*. Cambridge University Press, New York.

Liu, D., Lebovka, N.I., & Vorobiev, E. (2013). Impact of electric pulse treatment on selective extraction of intracellular compounds from *Saccharomyces cerevisiae* yeasts. *Food and Bioprocess Technology*, *6*(2), 576–584. https://doi.org/10.1007/s11947-011-0703-7

Luengo, E., Martínez, J.M., Bordetas, A., Álvarez, I., & Raso, J. (2015). Influence of the treatment medium temperature on lutein extraction assisted by pulsed electric fields from *Chlorella vulgaris*. *Innovative Food Science & Emerging Technologies*, *29*, 15–22. https://doi.org/10.1016/j.ifset.2015.02.012

Manzan, A.C.C.M., Toniolo, F.S., Bredow, E., & Povh, N.P. (2003). Extraction of essential oil and pigments from *Curcuma longa* [L] by steam distillation and extraction with volatile solvents. *Journal of Agricultural and Food Chemistry*, *51*(23), 6802–6807. https://doi.org/10.1021/jf030161x

Medeiros, D.L., Sales, E.A., & Kiperstok, A. (2015). Energy production from microalgae biomass: Carbon footprint and energy balance. *Journal of Cleaner Production*, *96*, 493–500. https://doi.org/10.1016/j.jclepro.2014.07.038

Mendes, R.L., Nobre, B.P., Cardoso, M.T., Pereira, A.P., & Palavra, A.F. (2003). Supercritical carbon dioxide extraction of compounds with pharmaceutical importance from microalgae. *Inorganica Chimica Acta*, *356*, 328–334. https://doi.org/10.1016/S0020-1693(03)00363-3

Mendiola, J.A., Santoyo, S., Cifuentes, A., Reglero, G., Ibáñez, E., & Señoráns, F.J. (2008). Antimicrobial activity of sub- and supercritical CO2 extracts of the green alga. *Dunaliella Salina. Journal of Food Protection*, *71*(10), 2138–2143.

Meullemiestre, A., Breil, C., Abert-Vian, M., & Chemat, F. (2016). Microwave, ultrasound, thermal treatments, and bead milling as intensification techniques for extraction of lipids from oleaginous *Yarrowia lipolytica* yeast for a biojetfuel application. *Bioresource Technology*, *211*, 190–199. https://doi.org/10.1016/j.biortech.2016.03.040

Miller, D.J., & Hawthorne, S.B. (2000). Solubility of liquid organic flavor and fragrance compounds in subcritical (hot/liquid) water from 298 K to 473 K. *Journal of Chemical & Engineering Data*, *45*(2), 315–318. https://doi.org/10.1021/je990278a

Moura, R.R.D., Etges, B.J., Santos, E.O.D., Martins, T.G., Roselet, F., Abreu, P.C., ... D'Oca, M.G.M. (2018). Microwave-assisted extraction of lipids from wet microalgae paste: A quick and efficient method. *European Journal of Lipid Science and Technology*, *120* (7), 1700419. https://doi.org/10.1002/ejlt.201700419

Muñoz, R., Navia, R., Ciudad, G., Tessini, C., Jeison, D., Mella, R., ... Azócar, L. (2015). Preliminary biorefinery process proposal for protein and biofuels recovery from microalgae. *Fuel*, *150*, 425–433. https://doi.org/10.1016/j.fuel.2015.02.004

Norton, T., & Sun, D.-W. (2008). Recent advances in the use of high pressure as an effective processing technique in the food industry. *Food and Bioprocess Technology*, *1*(1), 2–34. https://doi.org/10.1007/s11947-007-0007-0

Orr, V.C.A., & Rehmann, L. (2016). Ionic liquids for the fractionation of microalgae biomass. *Current Opinion in Green and Sustainable Chemistry*, *2*, 22–27. https://doi.org/10.1016/j.cogsc.2016.09.006

Pangestuti, R., & Kim, S.-K. (2011). Biological activities and health benefit effects of natural pigments derived from marine algae. *Journal of Functional Foods*, *3*(4), 255–266. https://doi.org/10.1016/j.jff.2011.07.001

Parniakov, O., Apicella, E., Koubaa, M., Barba, F.J., Grimi, N., Lebovka, N., ... Vorobiev, E. (2015). Ultrasound-assisted green solvent extraction of high-added value compounds from microalgae Nannochloropsis spp. *Bioresource Technology*, *198*, 262–267. https://doi.org/10.1016/j.biortech.2015.09.020

Parniakov, O., Barba, F.J., Grimi, N., Marchal, L., Jubeau, S., Lebovka, N., & Vorobiev, E. (2015a). Pulsed electric field and pH assisted selective extraction of intracellular components from microalgae nannochloropsis. *Algal Research*, *8*, 128–134. https://doi.org/10.1016/j.algal.2015.01.014

Parniakov, O., Barba, F.J., Grimi, N., Marchal, L., Jubeau, S., Lebovka, N., & Vorobiev, E. (2015b). Pulsed electric field assisted extraction of nutritionally valuable compounds from microalgae Nannochloropsis spp. using the binary mixture of organic solvents and water. *Innovative Food Science and Emerging Technologies*, *27*, 79–85. https://doi.org/10.1016/j.ifset.2014.11.002

Pasquet, V., Chérouvrier, J.-R., Farhat, F., Thiéry, V., Piot, J.-M., Bérard, J.-B., ... Picot, L. (2011). Study on the microalgal pigments extraction process: Performance of microwave assisted extraction. *Process Biochemistry*, *46*(1), 59–67. https://doi.org/10.1016/j.procbio.2010.07.009

Pataro, G., De Lisi, M., Donsì, G., & Ferrari, G. (2014). Microbial inactivation of *E. coli* cells by a combined PEF–HPCD treatment in a continuous flow system. *Innovative Food Science & Emerging Technologies*, *22*, 102–109. https://doi.org/10.1016/j.ifset.2013.12.009

Pchelintsev, N.A., Adams, P.D., & Nelson, D.M. (2016). Critical parameters for efficient sonication and improved chromatin immunoprecipitation of high molecular weight proteins. *PLoS ONE*, *11*, 1. https://doi.org/10.1371/journal.pone.0148023

Pereira, M.T., Rego, D.S., & Redondo, L.M.S. (2016). Advantages of pulsed electric field use for treatment of algae. In D. Miklavcic (Ed.), *Handbook of electroporation* (1–14). Springer, Cham. https://doi.org/10.1007/978-3-319-26779-1_171-1

Peshin, S.S., Lall, S.B., & Gupta, S.K. (2002). Potential food contaminants and associated health risks. *Acta Pharmacologica Sinica*, *23*(3), Retrieved from Scopus, 193–202.

Phong, W.N., Show, P.L., Le, C.F., Tao, Y., Chang, J.-S., & Ling, T.C. (2018). Improving cell disruption efficiency to facilitate protein release from microalgae using chemical and mechanical integrated method. *Biochemical Engineering Journal*, *135*, 83–90. https://doi.org/10.1016/j.bej.2018.04.002

Pinkert, A., Marsh, K.N., Pang, S., & Staiger, M.P. (2009). Ionic liquids and their interaction with cellulose. *Chemical Reviews*, *109*(12), 6712–6728. https://doi.org/10.1021/cr9001947

Poojary, M.M., Barba, F.J., Aliakbarian, B., Donsì, F., Pataro, G., Dias, D.A., & Juliano, P. (2016). Innovative alternative technologies to extract carotenoids from microalgae and seaweeds. *Marine Drugs*, *14*, 11. https://doi.org/10.3390/md14110214

Postma, P.R., Cerezo-Chinarro, O., Akkerman, R.J., Olivieri, G., Wijffels, R.H., Brandenburg, W.A., & Eppink, M.H.M. (2018). Biorefinery of the macroalgae *Ulva lactuca*: Extraction of proteins and carbohydrates by mild disintegration. *Journal of Applied Phycology*, *30*(2), 1281–1293. https://doi.org/10.1007/s10811-017-1319-8

Postma, P.R., Pataro, G., Capitoli, M., Barbosa, M.J., Wijffels, R.H., Eppink, M.H.M., … Ferrari, G. (2016). Selective extraction of intracellular components from the microalga *Chlorella vulgaris* by combined pulsed electric field–Temperature treatment. *Bioresource Technology*, *203*, 80–88. https://doi.org/10.1016/j.biortech.2015.12.012

Rahimi, M., Mohamadian, E., Dadari, S., Arbab, M.M., & Karimi, N. (2017). Application of high frequency ultrasound in different irradiation systems for photosynthesis pigment extraction from *Chlorella* microalgae. *Korean Journal of Chemical Engineering*, *34*(4), 1100–1108. https://doi.org/10.1007/s11814-016-0336-7

Raja, A., Vipin, C., & Aiyappan, A. (2013). Biological importance of marine algae- an overview. *International Journal of Current Microbiology and Applied Science*, *2*(5), 222–227.

Rajkumar, R., Yaakob, Z., & Takriff, M.S. (2013). Potential of micro and macro algae for biofuel production: A brief review. *BioResources*, *9*(1), 1606–1633. https://doi.org/10.15376/biores.9.1.1606-1633

Rego, D., Redondo, L.M., Geraldes, V., Costa, L., Navalho, J., & Pereira, M.T. (2015). Control of predators in industrial scale microalgae cultures with pulsed electric fields. *Bioelectrochemistry*, *103*, 60–64. https://doi.org/10.1016/j.bioelechem.2014.08.004

Regulation (EU) 2283. (2015). Regulation (EU) 2015/2283 of the European Parliament and of the Council of 25 November 2015 on novel foods, amending Regulation (EU) No 1169/2011 of the European Parliament and of the Council and repealing Regulation (EC) No 258/97 of the European Parliament and of the Council and Commission Regulation (EC) No 1852/2001., Pub. L. No. 32015R2283, 327 OJ L.

Rizwan, M., Mujtaba, G., Memon, S.A., Lee, K., & Rashid, N. (2018). Exploring the potential of microalgae for new biotechnology applications and beyond: A review. *Renewable and Sustainable Energy Reviews*, *92*, 394–404. https://doi.org/10.1016/j.rser.2018.04.034

Rodríguez-Meizoso, I., Jaime, L., Santoyo, S., Señoráns, F.J., Cifuentes, A., & Ibáñez, E. (2010). Subcritical water extraction and characterization of bioactive compounds from *Haematococcus pluvialis* microalga. *Journal of Pharmaceutical and Biomedical Analysis*, *51*(2), 456–463. https://doi.org/10.1016/j.jpba.2009.03.014

Saber, M., Golzary, A., Wu, H., Takahashi, F., & Yoshikawa, K. (2018). Ultrasonic pretreatment for low-temperature hydrothermal liquefaction of microalgae: Enhancing the bio-oil yield and heating value. *Biomass Conversion and Biorefinery*, *8*(3), 509–519. https://doi.org/10.1007/s13399-017-0300-8

Safi, C., Zebib, B., Merah, O., Pontalier, P.-Y., & Vaca-Garcia, C. (2014). Morphology, composition, production, processing and applications of *Chlorella vulgaris*: A review. *Renewable and Sustainable Energy Reviews*, *35*, 265–278. https://doi.org/10.1016/j.rser.2014.04.007

Saini, R.K., & Keum, Y.-S. (2018). Carotenoid extraction methods: A review of recent developments. *Food Chemistry*, *240*, 90–103. https://doi.org/10.1016/j.foodchem.2017.07.099

Sánchez-Camargo, A.D.P., Pleite, N., Mendiola, J.A., Cifuentes, A., Herrero, M., Gilbert-López, B., & Ibáñez, E. (2018). Development of green extraction processes for *Nannochloropsis gaditana* biomass valorization. *Electrophoresis*, *39*(15), 1875–1883. https://doi.org/10.1002/elps.201800122

Sicaire, A.-G., Vian, M., Fine, F., Joffre, F., Carré, P., Tostain, S., … Chemat, F. (2015). Alternative bio-based solvents for extraction of fat and oils: Solubility prediction, global yield, extraction kinetics, chemical composition and cost of manufacturing. *International Journal of Molecular Sciences*, *16*(4), 8430–8453. https://doi.org/10.3390/ijms16048430

Siemer, C., & Parniakov, O. (2018). *PEF treatment of algae – eAlgae*. Quakenbrück: R&D Department, Elea Vertriebs- und Vermarktungsgesellschaft mbH, 1–8.

Slocombe, S.P., Ross, M., Thomas, N., McNeill, S., & Stanley, M.S. (2013). A rapid and general method for measurement of protein in micro-algal biomass. *Bioresource Technology*, *129*, 51–57. https://doi.org/10.1016/j.biortech.2012.10.163

Suarez Garcia, E., Suarez Ruiz, C.A., Tilaye, T., Eppink, M.H.M., Wijffels, R.H., & van Den Berg, C. (2018). Fractionation of proteins and carbohydrates from crude microalgae extracts using an ionic liquid based-aqueous two phase system. *Separation and Purification Technology*, *204*, 56–65. https://doi.org/10.1016/j.seppur.2018.04.043

Suarez Ruiz, C.A., Emmery, D.P., Wijffels, R.H., Eppink, M.H., & van Den Berg, C. (2018). Selective and mild fractionation of microalgal proteins and pigments using aqueous two-phase systems. *Journal of Chemical Technology and Biotechnology*, *93*(9), 2774–2783. https://doi.org/10.1002/jctb.5711

Suganya, T., Varman, M., Masjuki, H.H., & Renganathan, S. (2016). Macroalgae and microalgae as a potential source for commercial applications along with biofuels production: A biorefinery approach. *Renewable and Sustainable Energy Reviews*, *55*, 909–941. https://doi.org/10.1016/j.rser.2015.11.026

Tamayo Tenorio, A., Kyriakopoulou, K.E., Suarez-Garcia, E., van Den Berg, C., & van der Goot, A.J. (2018). Understanding differences in protein fractionation from conventional crops, and herbaceous and aquatic biomass - consequences for industrial use. *Trends in Food Science & Technology*, *71*, 235–245. https://doi.org/10.1016/j.tifs.2017.11.010

Vanthoor-Koopmans, M., Wijffels, R.H., Barbosa, M.J., & Eppink, M.H.M. (2013). Biorefinery of microalgae for food and fuel. *Bioresource Technology*, *135*, 142–149. https://doi.org/10.1016/j.biortech.2012.10.135

Venkata Subhash, G., Rajvanshi, M., Navish Kumar, B., Govindachary, S., Prasad, V., & Dasgupta, S. (2017). Carbon streaming in microalgae: Extraction and analysis methods for high value compounds. *Bioresource Technology*, *244*, 1304–1316. https://doi.org/10.1016/j.biortech.2017.07.024

Viguera, M., Marti, A., Masca, F., Prieto, C., & Calvo, L. (2016). The process parameters and solid conditions that affect the supercritical CO2 extraction of the lipids produced by microalgae. *The Journal of Supercritical Fluids*, *113*, 16–22. https://doi.org/10.1016/j.supflu.2016.03.001

Wahlen, B.D., Willis, R.M., & Seefeldt, L.C. (2011). Biodiesel production by simultaneous extraction and conversion of total lipids from microalgae, cyanobacteria, and wild mixed-cultures. *Bioresource Technology*, *102*(3), 2724–2730. https://doi.org/10.1016/j.biortech.2010.11.026

Wang, L., & Weller, C.L. (2006). Recent advances in extraction of nutraceuticals from plants. *Trends in Food Science & Technology*, *17*(6), 300–312. https://doi.org/10.1016/j.tifs.2005.12.004

Wang, M., Yuan, W., & Hale, A. (2016). Three-dimensional simulation of ultrasound-induced microalgal cell disruption. *Applied Biochemistry and Biotechnology*, *178*(6), 1184–1195. https://doi.org/10.1007/s12010-015-1937-z

Wang, M., Yuan, W., Jiang, X., Jing, Y., & Wang, Z. (2014). Disruption of microalgal cells using high-frequency focused ultrasound. *Bioresource Technology*, *153*, 315–321. https://doi.org/10.1016/j.biortech.2013.11.054

Wells, M.L., Potin, P., Craigie, J.S., Raven, J.A., Merchant, S.S., Helliwell, K.E., ... Brawley, S.H. (2017). Algae as nutritional and functional food sources: Revisiting our understanding. *Journal of Applied Phycology*, *29*(2), 949–982. https://doi.org/10.1007/s10811-016-0974-5

Wolkers, H., Barbosa, M.J., Kleinegris, D.M.M., Bosma, R., Wijffels, R.H., & Harmsen, P.F.H. (2011). *Microalgae: The green gold of the future?: Large-scale sustainable cultivation of microalgae for the production of bulk commodities*. Retrieved from http://agris.fao.org/agris-search/search.do?recordID=NL2012006570

Zhan, J., Rong, J., & Wang, Q. (2017). Mixotrophic cultivation, a preferable microalgae cultivation mode for biomass/bioenergy production, and bioremediation, advances and prospect. *International Journal of Hydrogen Energy*, *42*(12), 8505–8517. https://doi.org/10.1016/j.ijhydene.2016.12.021

Zhao, L., Zhao, G., Chen, F., Wang, Z., Wu, J., & Hu, X. (2006). Different effects of microwave and ultrasound on the stability of (all-E)-astaxanthin. *Journal of Agricultural and Food Chemistry*, *54*(21), 8346–8351. https://doi.org/10.1021/jf061876d

Zhu, S., Wang, Y., Huang, W., Xu, J., Wang, Z., Xu, J., & Yuan, Z. (2014). Enhanced accumulation of carbohydrate and starch in *Chlorella zofingiensis* induced by nitrogen starvation. *Applied Biochemistry and Biotechnology*, *174*(7), 2435–2445. https://doi.org/10.1007/s12010-014-1183-9

14

Environmental Life Cycle Assessment and Regulatory Issues of Innovative Green Extraction Procedures

ERASMO CADENA

Vertech Group SAS, 11 Rue Defly, 06000 Nice, France
Université Nice Sophia Antipolis, IMREDD, Nice, France

MATHILDE FIORLETTA

Vertech Group SAS, 11 Rue Defly, 06000 Nice, France

Contents

1 Introduction

The food industry, one of the largest industrial sectors in the world, implies a high consumption of energy and a large production of waste (Roy et al., 2009). According to the Food and Agriculture Organization of the United Nations, the global food waste is calculated to 1.3 billion tons per year, from which 53% occurs upstream (during production, post-harvest handling, and storage) and 46% downstream (during processing, distribution, and consumption steps) (FAO, 2011). Considering the potential impacts at both the environmental and security levels, several policies and initiatives have been developed to address this problem effectively. This is the case of the United Nations Sustainable Development Goals, which focuses on the food supply chain trying to ensure sustainable consumption and production patterns (UN, 2015); in the same line, the European Commission (EC) has included the food waste among the priority areas of the Circular Economy Action Plan (European Commission, 2015a). Part of the proposed and promising technical solutions are the valorization of this sort of residues into valuable materials; for example, the fruit and vegetable industries provide large streams with high purity from which complex organic by-products can be extracted (Poltronieri & D'Urso, 2016). The recovery of targeted compounds from food waste has been proved to be possible by many researchers; however, conventional extraction methods face different major obstacles. First, the energy efficiency of these techniques used to be low and the cost involvement is high. Then, the solvent used is mostly non-food grade and processes, such as spray drying or thermal processing, involved in conventional extraction might damage the quality of the bioactive compounds. Finally, the organic nature of food waste implies high moisture content and a potential source of pathogens and on top of that, consumers are often skeptical with the safety of products from waste (Anal, 2017).

Including environmental considerations in processes design, control, and optimization is nowadays a real topic of interest, as well as a necessity. Most of the extraction technologies that have emerged during the last few years are known as green extraction processes. Microwave-, ultrasound-, pulsed electric field (PEF)-assisted extractions, and extractions using pressurized fluids can be found among them. The concept of Green Chemistry promotes six specific principles for the extraction of natural products (Herrero & Ibañez, 2018): (i) selection of renewable plant resources; (ii) use of alternative solvents and water or agro-solvents; (iii) reduction of energy consumption by energy recovery and using innovative technologies; (iv) production of coproducts instead of waste to include the bio- and agro-refining industry; (v) reduction of unit operations and favor safe, robust, and controlled processes; and (vi) aim for a non-denatured and biodegradable extract without contaminants.

An option for greener extraction techniques is not to use any solvent. Some processes can completely avoid solvent use, but it is not always achievable. Indeed, the absence of solvent can cause dangerous overheating or higher energy consumption and counterbalance its positive impact (Häckl & Kunz, 2018). According to Herrero and Ibañez (2018), the properties expected from a green solvent are the following: (i) low toxicity to humans and other organisms; (ii) easily biodegradable in the environment without adverse effects; (iii) naturally occurring; (iv) produced from renewable sources; (v) produced as a by-product; (vi) low vapor pressure; and (vii) no need of a traditional evaporation step after extraction. Nevertheless, solvents are rarely matching all these properties and even if the solvent seems totally green, a poor extraction rate can discard it. For instance, water is the greenest solvent and complies with all the requirements. The advantages are plentiful: low hazard potential, high availability, low cost, uncommon selectivity, and adjustable pH values, among others. Nonetheless, the extraction process can be very energy consuming because of the small range of temperature allowing it to stay in the liquid state, as well as its high thermal capacity. It is important to note that the natural origin of a solvent is not sufficient to justify its greenness. For example, in the case of bio-based furfural, which is a renewable and sustainable chemical but also toxic and carcinogenic. By contrast, some petrol-based solvents, such as 3-methoxy-3-methyl-butane-1-ol, have a good human safety environment profile. For food extraction, the most common solvent is hexane and it is also particularly problematic. Indeed, hexane is about to be banned in Europe because of the serious health concerns that it can cause. However, no adequate substitute has been found yet and its prohibition is still under negotiation. Some solutions exist, such as using supercritical CO_2 or 2-methyltetrahydrofuran, but these solvents also present disadvantages (Häckl & Kunz, 2018).

By-products coming from the food industry are a real source of biomass. The use of lignocellulosic biomass residues and waste from the agri-food industry is a potential alternative to produce value-added compounds, avoiding the cultivation of dedicated crops. This utilization of by-products allows managing the residual streams without entering in confrontation with the food and feed availability. For example, by-products from food industry represent billions of tons per year and are, consequently, a renewable resource. Nevertheless, the recovery of biomass residues from croplands can have negative consequences on cultivation due to the loss of soil organic carbon content, which could lead to a reduction of crops yields (Vauchel et al., 2018). From the dairy industry, whey by-products have critical impacts on the environment. For instance, lactose, which represents 75% of the total whey solids, is highly polluting because it affects the chemical and physical structure of soil. It is besides considered as the most polluting food by-product stream. In this sense, its valorization by green extraction processes is very interesting since it is a potential raw material for value-added products in the pharmaceutical, agri-food, and biotechnology industries (Banaszewska et al., 2014).

In summary, even if the green extraction procedures saving time, water, energy, and organic solvent are achieved, the environmental performances of these technologies are not always proved to be completely free of impacts compared with conventional methods (Vauchel et al., 2018). In this regard, the life cycle assessment (LCA) appears as a useful tool to prove if these extraction processes are sustainable from an environmental perspective. In the same line, the regulatory foresight and identification of legal barriers with the uptake of the product from these sorts of procedures are essential.

This chapter focuses on the critical analysis of the potential environmental impacts of various state-of-the-art green extraction methods applied to valorize food processing. In addition, the regulatory and legal constraints of these procedures will be addressed.

2 Life Cycle Assessment Methodology

LCA is a management tool for evaluating the environmental performance of products (goods and services). LCA takes into account a product's full life cycle, from the extraction of resources and the processing of raw materials, through production, use, possible recycling, to the final disposal of remaining waste (ISO, 2006a). In brief, LCA is a material and energy balance applied to the product's system, combined with an assessment of the environmental impacts related to the inputs and outputs to and from the product system. In this sense, LCA provides criteria for decision-making on issues such as product development, policymaking, strategic planning, among others. LCA has been promoted in different European directives as a robust quantitative tool, and a keystone in decision-making by producers and stakeholders.

International Standards Organization (ISO) 14040 and 14044 (ISO, 2006a, 2006b) defines LCA as a technique for assessing the environmental aspects and potential impacts associated with a product or service, by:

- compiling an inventory of relevant inputs and outputs of a product system;
- evaluating the potential environmental impacts associated with those inputs and outputs; and
- interpreting the results of the inventory analysis and impact assessment phases in relation to the objectives of the study.

The methodology is formalized by the ISO and focused on the evaluation of the environmental burden of the studied process or product according to different parameters, such as waste produced, or energy and materials consumed.

To reach these objectives, information on inputs and outputs of the whole process need to be gathered and processed. The standardized LCA framework encompasses four phases (ISO, 2006a) (Figure 14.1):

1. *Goal and scope definition:* This is the first step of the study, and probably the most important, given that the purpose, scope, and main hypothesis considered are defined in this stage.

First, the goal has to be defined, as well as the kind of decisions that will be made from the results obtained. Second, the specific scope is determined. The scope should be well defined to ensure that the breadth, depth, and detail of the study are compatible and adequate to address the stated goal. This activity implies defining the system, its boundaries (conceptual, geographical, and temporal), quality of date, the main hypothesis, and the study limitations. A key matter on the scope is to define the functional unit (ISO, 2006a). This is the unit of the product or service whose environmental impacts will be assessed or compared (it is usually

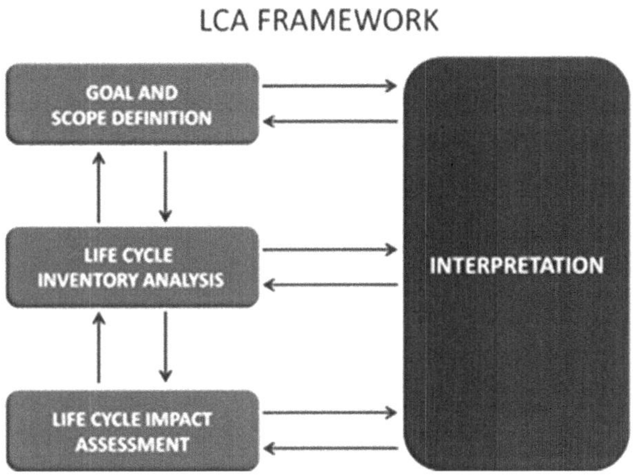

Figure 14.1 Stages of the life cycle assessment (Source: ISO, 2006a).

expressed in terms of amount of product but should be related to the amount of product needed to perform a given function).

The system boundaries delimit the unit processes, which will be included into the system. This action is partially based on choices that should be detailed and justified to provide confidence in the analysis. The system boundaries should define that which stages, process units, and flows will be included in the study.

2. *Life Cycle Inventory (LCI):* This phase is a technical process of data collection, to quantify the inputs and outputs of the system, as defined on the scope. In this stage, all emissions released to the environment (i.e. air, water, soil, and solid waste) and resource consumption (e.g energy and raw materials) along the whole production life cycle, as defined in the scope and in relation to the functional unit (ISO, 2006a) will be gathered. The main steps are: (i) data collection, (ii) relevant and nonrelevant element identification, (iii) mass and energy balances, and (iv) system burdens allocation. LCI results are the sum of elementary flows entering the system and releasing into the environment.

3. *Life Cycle Impact Assessment (LCIA):* During this stage, LCI results are translated, applying an impact assessment method, into environmental impacts at the midpoint or at the endpoint. In brief, it is the process to identify and characterize the potential effects produced in the environment by the system under analysis. A specialized software will be used for this purpose. LCI results will be assigned to the impact categories and potential environmental impacts will be calculated in each category (e.g., global warming, acidification, eutrophication, resource depletion, human health, among others). LCIA consist of four steps: classification, characterization, normalization, and weighting (ISO, 2006a).

4. *Interpretation:* In this phase, the findings obtained are presented in a synthetic way, showing the critical sources of impacts and the possible options to

reduce them. The interpretation is useful for indicating the results consistently according to all the aspects defined during the goal and scope stage. First of all, significant issues need to be identified (main process contributing most to the results). The interpretation requires consistency checks, ensuring that there is complete information. Sensitivity checks should also be run. The uncertainty and accuracy of results are also addressed at this stage.

These steps are clearly ordered; however, LCA studies are iterative, which means that LCA operations are repeated, to approximate the results paying special attention to the most relevant processes, resources, and emissions. It is a common practice carrying out one to three iterations before reaching the final results.

3 Valorization of By-Products from an LCA Perspective

In existing studies on food waste valorization, the upstream burden is often not considered. Indeed, there is an exclusion of the production stage of the waste and only the downstream management steps are considered. It is called the "zero-burden-assumption," and shows relevance for comparative studies on two disposal systems with the same type of waste (Oldfield & Holden, 2014). This assumption is also relevant considering the fact that waste has a negative economic value because of the disposal cost. However, this approach remains controversial, since it keeps being used in non-comparative studies or when there are discrepancies in the amount or type of waste being included in the systems. Although the heterogeneity through space and time of waste flows makes it hard to include the upstream stages of the food waste in the system boundaries (Häckl & Kunz, 2018; ISO, 2006a).

Nevertheless, depending on the model of recycling, assuming zero impacts from upstream stages can be relevant. In short, two processes are usually defined: open-loop and closed-loop recycling. In the case of extraction of valuable components from food by-products, the open-loop recycling model is used, since the valuable product is not reused as raw material for the same food production. In this case, recovery of compounds from by-products can be seen as an exchange between two different systems, food by-products becoming a raw material in a second system (Olofsson & Börjesson, 2018). The open-loop recycling system can be described as a multifunctionality problem and it results in separating the two system boundaries (i.e., system A for food production and system B for food waste/by-product valorization). One solution is to expand the system boundaries and include both products and the another is to make allocation of the environmental impacts between the systems. One allocation approach very close from the "zero-burden-assumption" is the cutoff method.

It can be assumed that the environmental impact associated with a product must be direct and then associated with the impact of the initial food production to the system A. It results in a cutoff of the environmental impacts associated with the primary material extraction for system B. This approach is not representative of the multifunctionality of the system. However, it might not be such a problem in the case of waste valorization, since food by-products often have no economic value and the system could be seen, in fact, as not multifunctional: the cutoff becomes a simple economic allocation. Nonetheless, another possible issue can appear with the cutoff method. For instance, if previous studies already take into account the reduction of environmental burden in system A, thanks to the extracted valuable compounds, it will be wrong to not consider the raw material impacts in the system B (Olofsson & Börjesson, 2018).

When the "zero-burden-assumption" is not well founded for a study, the allocation must be made based on physical or other relationships. The method of system expansion can also deal with the multifunctionality of the system. This method is based on substitution: it considers that the by-product replaces other similar products and improves the overall environmental performance of the system. In this case, the comparable product that could be replaced by residual biomass type is identified and the potential value of the waste is highlighted (Olofsson & Börjesson, 2018). In LCAs, about biomass valorization, allocation is the most common method. More than one basis (mass, energy, and economic criteria) are often used for allocation between streams. However, allocating more environmental impacts to by-products is controversial, since the main product drives production. In this sense, it could be not be intuitive to choose energy allocation when some residues have high-fat content in comparison with low-fat products. For the same reason, mass allocation is sometimes disregarded. The economic value of products and by-products could then be interesting to make allocation, but it can be complex to predict the future cost of by-products, especially at present with the increase of their use for biofuel production. In the end, both the mass and economic allocations are criticized in cases where they do not represent accurately the details of the targeted processes. In this sense, the allocation choice should depend on each situation and depict as precisely as possible on the market, the environmental burdens, and the dynamics of the agri-food industry. In the other hand, substitution or system expansion is recommended by ISO 14044 (ISO, 2006b) for by-products. Mass and energy values are the most commonly applied bases for substitution, but more specific properties of by-products can also be considered. In the case of pig feed substitution, it can take as an example the substitution made on digestible energy content or gelatin strength (Olofsson & Börjesson, 2018).

The functional unit is often defined according to the amount of waste to be treated, whereas defining it as the output of a valorization process would be more interesting. Considering that to reach a circular economy is desirable, qualitative interest of recycled material should overtake the quantitative value, which is less representative of the actual situation. Moreover, considering the mass of waste managed as a functional unit does not allow the comparison of value-added components extracted from by-products (Oldfield et al, 2018).

4 Environmental Impacts of Innovative Green Extraction Procedures

Green extraction procedures present several environmental benefits over traditional extraction methods, mainly due to their high selectivity for the targeted compound, their fast and easy operating conditions, and their low energy consumption and CO_2 emissions (Satari & Karimi, 2018).

Considering the relevance of these techniques, during the past few years some research was focused on the assessment of these procedures from the environmental point of view.

4.1 Ultrasounds-Assisted Extraction

Ultrasound-assisted extraction (UAE) is widely used for valorization of agri-food raw materials, by-products, or wastes that contain valuable bioactive compounds, such as aromas, pigments, antioxidants, and other organic and mineral compounds. It is considered to be a green extraction process that accelerate heat and mass transfer. Ultrasound waves interact with the plant material and alter its physical and chemical properties. The cavitational effect facilitates the release of valuable compounds and the cell walls are disrupted, so that the mass transport is eased (Chemat et al., 2017). It provides high extraction yield and rate, small extraction time, and high processing throughput (Kyriakopoulou et al., 2015) (Table 14.1). This technique is compatible with any solvent, and thus food grade solvents, such as water or ethanol can be used (Panja, 2017). The extraction has been used, for example, by Vauchel et al. (2018) for the extraction of antioxidant polyphenols from chicory grounds (by-products of chicory beverage industrial production) (Table 14.1). The valuable product from this study could find an application as natural antioxidant in food and cosmetic industries. The study highlighted the fact that heating the solvent lead to higher environmental impacts in all the analyzed indicators (e.g., climate change, ozone depletion, acidification, land use, among others), since the increase in energy consumption affects the extraction phase. Using ethanol is also more damaging to the environment than water at the materials stage. However, ethanol improves the extraction efficiency and negative effects might be counterbalanced (Vauchel et al., 2018). Ultrasound assistance shows a positive impact in comparison with temperature and ethanol, and enables reduction of environmental impacts to 25%. UAE was also reported to extract pectin from citrus processing waste (CPW) more efficiently than conventional extraction techniques (Table 14.1).

4.2 Compressed Fluids-Based Extraction

Within this category of extraction, pressurized liquid extraction (PLE), supercritical fluid extraction (SFE), and gas-expanded liquids extraction are the most common.

Table 14.1 Summary of LCA on food green extraction studies

TITLE	AUTHOR	YEAR	FUNCTIONAL UNIT	EXTRACTION TECHNIQUE	ANALYSIS
Life cycle analysis of β-carotene extraction techniques	Kyriakopoulou et al.	2015	1 kg of β-carotene extract	Microwave extraction/ ultrasound extraction/ conventional solvent extraction	Microwave extraction shows the best environmental impacts in comparison with UAE and CSE for β-carotene extraction from carrots. It minimizes the carbon footprint of the final product providing high yields in short time. The benefits also include solvent, energy, time, and cost saving
Life cycle assessment of green pilot-scale extraction processes to obtain potent antioxidants from rosemary leaves	Rodríguez-Meizoso et al.	2012	10 g of dry rosemary extract	Supercritical fluid extraction	– CO_2 and ethanol in SFE = no important impact (low volumes and green solvent) – Electricity consumption = key factor, affects all environmental impact categories, between 96.3 (human toxicity) and 99.7% (terrestrial eco-toxicity) of the total impact of each category – The impact of CO_2 on global warming is greater when coming from the electricity consumption than from the amount needed for extraction – Type of primary energy sources in the production of electricity is then a key factor in LCA for green processes
Comparative LCA of ultrasound-assisted extraction of polyphenols from chicory grounds under different operational conditions	Vauchel et al.	2018	0.55 L of extract 18 exhibiting 220 µmol Trolox equivalent antioxidant capacity	Ultrasound-assisted extraction	Heating the solvent leads to higher impacts in all impact categories since the increase in energy consumption affects extraction phase. Using ethanol is also more damaging to the environment than water at the materials stage. However, ethanol improves extraction efficiency and negative effects might be counterbalanced. Ultrasound assistance shows a positive impact in comparison with temperature and ethanol, and enables environmental impacts reduction of 25%
Life cycle assessment of supercritical CO_2 extraction of caffeine from coffee beans	De Marco et al.	2018	1 kg of decaf coffee constituted by 60/40 Arabica/Robusta blend and 11.4 g of pure caffeine	Supercritical CO_2 extraction	Importance of the electricity source. Significant reduction of all the impact categories when photovoltaic panels are used

These techniques consider the use of green solvents with a reduction in volume (Herrero & Ibañez, 2018).

SFE is used for the extraction of oils and fats, antioxidants, pigments, and aromas by percolation. High pressures are required for oil extraction (280 bars), whereas aromatic fractions can be extracted at moderate conditions (Chemat et al., 2017). This process can become very energy intensive if the solvent used requires high critical temperature. Therefore, energy is necessary to raise the solvent temperature above the critical point. That is the reason why water is not considered as a possibility in SFE, as well as flammable solvents. However, CO_2 is a perfect candidate for SFE because has a very low critical temperature (31°C), it is nontoxic, cheap, nonflammable, chemically inert, and available in high quality and quantity. In addition, CO_2 is recovered after extraction and recycled, thanks to a compressor (Panja, 2017). In an LCA by Rodríguez-Meizoso et al. (2012) on green pilot-scale extraction processes for obtaining antioxidants from rosemary leaves, the use of CO_2 and ethanol in SFE shows no important impact mainly due to the very low volumes used and the fact that they are green solvent (Table 14.1). On the contrary, electricity consumption is the key factor and affects all environmental impact categories under study, with a contribution from 96.3 (human toxicity) to 99.7% (terrestrial ecotoxicity) of the total impact of each category. Thus, the impact of CO_2 on global warming is greater when coming from the electricity consumption than from the amount needed for extraction. The type of primary energy sources used in the production of electricity is then a key factor in LCA for green processes (Rodríguez-Meizoso et al., 2012). De Marco et al. (2018) also demonstrated that electricity consumption is a relevant aspect. Indeed, the substitution of a portion (20–40%) of the Italian Energy mix with photovoltaic panels allows a significant decrease in all impact categories for the supercritical CO_2 extraction of caffeine from coffee beans (De Marco et al., 2018) (Table 14.1). On the other side, Satari and Karimi (2018) demonstrated that the most efficient techniques for the extraction of carotenoids from CPW also appeared to be environmentally friendly. In this regard, supercritical CO_2 and hydrolytic enzymes show high rates of extraction (Satari & Karimi, 2018). Finally, PLE can be considered as a green process because it uses mainly water (pressurized hot water extraction or subcritical water extraction) and aqueous alcohols, which are environmentally friendly and nontoxic solvents. PLE is also energy efficient because the liquid is heated at a lower temperature (50–200 °C) than its vaporization temperature; thanks to an increase in pressure (Panja, 2017). Nevertheless, it is important to mention that a few LCAs have been done yet on this topic.

4.3 Pulsed Electric Field (Electroporation)

PEF extraction consists in a critical acceleration of the solid–liquid extraction. It allows an extraction of compounds from plant cells, such as colorants, sucrose, polyphenols, and other secondary metabolites (Chemat et al., 2017). PEF is a nonthermal, selective, and clean process, with a short extraction time. Nevertheless, the electro-resistive heating occurring during the PEF discharge causes elevation of the temperature of the food sample. To find a solution for this problem, a cooling system must be used and it implies a dissipation of significant amount of energy (Doona, 2010). Electrotechnologies, such as PEFs, are water saving, and thus minimize the production of wastewater. They are also eco-friendly when renewable sources of energy are used for producing electricity, even if they present lower energy consumption in comparison with conventional extraction and pretreatment technologies. In soft materials, such as peals and pulps, the energy needed by PEF treatment varies between 1 and 15 kJ/kg, whereas 20–100 kJ/kg are needed by conventional extraction techniques for achieving the same results. In comparison, hard materials, such as seeds require 100–800 kJ/kg with PEFs. This high energetic efficiency enables a reduction in the overall energetic consumption. PEF also implies an increase in recovery and bioprocessing efficiency and a potential elimination of solvent use. In this sense, PEF is likely to reduce the use of nonrenewable resources and can be seen as a green extraction technique (Rocha et al., 2018). However, according to the research done so far, no environmental LCA has been done yet on PEF extraction on by-products from the food industry.

4.4 Microwave-Assisted Extraction

Microwave-assisted extraction (MAE) shows many advantages compared to the previous methods, for example, the equipment size is reduced, it is also easy to use, fast, and consume low amounts of solvent (Razzaghi et al., 2018). A simulation of an MAE plant for the production of D-limonene and pectin from orange peels at industrial-scale has been performed by Pfaltzgraff et al. (2013). They conclude that for an input of 10,000 t of waste, 152 and 1,080 t of D-limonene and pectin per year were obtained, respectively (Pfaltzgraff et al., 2013). Within MAE, it can distinguish two specific class of techniques: solvent-free microwave hydrodistillation (SFME) and microwave hydrodiffusion and gravity (MHG). SFME is used for the extraction of essential oils from aromatic plants and fruits. It was developed in 2004 by Lucchesi et al. (2004) and it rejects less CO_2 in atmosphere per gram of essential oil than traditional method. MHG allows extraction of essential oils without distillation and evaporation, which are very energy-consuming processes (Herrero & Ibañez, 2018; ISO, 2006a).

Microwave extraction shows a best environmental performance in comparison with UAE and conventional solvent extraction (CSE) for β-carotene extraction from carrots. It minimizes the carbon footprint of the final product providing high yields in sort time. The benefits also include solvent, energy, time, and cost saving (Kyriakopoulou et al., 2015). Microwave heating is more rapid and effective than traditional solvent extraction because heat is transferred directly to the product, so that the valuable compounds are released more quickly and easily from cells. Moreover, this technique uses less solvent and time than traditional solvent extraction (Dorta et al., 2013). Jacotet-Navarro et al. (2015) compared MAE and UAE with conventional processes at a lab-scale for extraction of rosmarinic, carnosic, and ursolic acids from rosemary. On energy consumption, conventional processes, such as 5 h reflux, are the most energy-consuming technique (850 kWh/kg extract) due to the long extraction duration. The carbon emissions are also high with 680 kg CO_2/kg extract. In comparison, ultrasound extraction, which lasts only 30 min, presents the lowest values with an energy consumption of 23 kWh/kg extract and carbon emissions of 19 kg CO_2/kg extract. Microwave also shows low values but higher than ultrasound extraction (Jacotet-Navarro et al., 2015). In another study, performed by Satari and Karimi (2018), MAE appeared to be a more efficient method than UAE and traditional solvent extraction for phenolic acid present in CPW. The extraction efficiency and antioxidant activity were found to be higher. In addition, pectin from CPW was extracted faster and with less energy consumption than conventional techniques (Satari & Karimi, 2018).

4.5 Enzyme-Assisted Extraction

Enzymatic extraction is nowadays widely used as an environmentally friendly technique for the extraction of green-labeled pectin from CPW (Satari & Karimi, 2018). It is based on the disruption of the cell wall integrity; thanks to specific enzymes, those degrade the major components of plant cell wall and fibers (i.e., celluloses, hemicelluloses, and pectin). With the degradation of cell walls, the bioactive components are released in the solvent. Besides the fact that enzyme-assisted extraction (EAE) implies a long processing time and a high-energy consumption, it is still considered as a greener extraction process than acid- or alkaline-assisted extraction. In addition, the product presents superior and preserved quality (Pojić et al., 2018). Cheng et al. (2018) assessed that EAE was an alternative to reduce the environmental impacts and maintain a high oil recovery from soybean seeds. Indeed, it has lower greenhouse gases and criteria pollutants emissions than a basic expelling process (2.35 kg CO2, 3.52 g CH4 and 0.04 g N2O per 1 kg of soybean oil production), which requires high energy consumption. Additionally, EAE has the potential to be a greener process than hexane extraction because of lower environmental impacts. However, EAE has still to be improved to be a cleaner and an eco-friendlier process by lowering the energy requirements (Cheng et al., 2018).

5 Regulatory Issues

The perspective of reaching a sustainable circular bio-based economy is linked with various challenges. Regulatory foresight and identification of legal barriers to the uptake of the product are essential and need anticipation. Bioeconomy is defined as the production of renewable biological resources and their transformation into food, feed, bioenergy, and bio-based compounds; thanks to efficient and innovative technologies. In this sense, bioeconomy rests on biomass use. The European Standard EN 16575:2014 defines biomass as a material of biological origin, excluding geological and fossilized formations. Bioeconomy is promoted at European and national levels. The first policy framework on bioeconomy at a European level, Innovative for Sustainable Growth: A Bioeconomy for Europe (European Commission, 2012), was defined in 2012 by the EC. The strategy of this document relies on five interconnected societal challenges: guarantee food security, control natural resource sustainability, scale down dependence on nonrenewable resources, reduce and adapt to climate change, and support European competitiveness creating new jobs. This policy framework aims at improving biological resources management in Europe, and thus facilitate market uptake of new bio-based products. Another recent European strategy was defined in 2015 as "Growth Within: a circular economy vision for a competitive Europe." This directive again highlights the importance of a sustainable approach to economic growth. In this vein, lots of European countries are already applying bioeconomy principles in their economic strategies and promoting new innovative technologies for valuable compounds recovery from biowaste. Bio-based economy implies the challenges about food security. For instance, one could be tempted to produce dedicated crop to comply with the demand of biomass and then compromise direct food production by stressing land availability. This is where organic waste and residues can intervene and provide a sustainable route to overcome these issues. Biomass availability and innovative technology foresight are widely studied: trends are established on biomass availability and new technologies are anticipated. Market foresight is also increasingly studied by assessing preferences of consumers and social acceptance, but there is a lack of research about policy and regulation foresight. Indeed, moving toward a bioeconomy would imply an exploration of the social, economic, environmental, and technological changes, as well as an anticipation of the emerging policy challenges and opportunities. Regulatory uncertainty and unsupportive regulation have been identified as the two main barriers to the development of bio-based products on new markets. It has been reported that regulatory framework governing was regularly changed and sometimes incoherent about the use of biomass for further use in valorization processes. This is the reason why policies and regulations should be better foreseen to anticipate needed changes (Morone et al., 2017).

The huge quantities of waste generated each year imply changes in upstream and downstream processing strategies, and thus demand new creative and innovative process design, effective execution, and properly planned systems (Anal, 2017). Valorization routes of biomass can be divided into three categories: i) direct use in the case of food is unaltered or barely altered; ii) material recovery, such as chemical extraction and energy recovery; and, iii) energy recovery, thanks to incineration or anaerobic digestion. According to the waste hierarchy, reuse must be the first consideration, followed by material recovery and energy recovery. Using waste food for animal feeding is the simplest valorization route and requires very few resources. Nevertheless, lots of by-products are not edible and this is where material recovery intervenes (Poltronieri & D'Urso, 2016). These valorization routes need, of course, sanitary regulation, especially in the case of valorization of food waste into new food components. However, there is a lack of clarity in the definition of by-products. By-products can be defined as products coming from a process for which the aim is the production of another main product. The following conditions also have to be met to complete the definition: the products must be further used, it can be used without any further processing step, and the substance is an integral part of the process production and further use must be legal, which means that it must fulfil all the health and environmental legal requirements for their use and will not imply further environmental and human health impacts. Even with this definition, the difference between waste and by-product remains blurred (Morone et al., 2017).

Furthermore, there are some European legal requirements that define the possibility to use food waste as components of food. The permissibility of a food product used as an input substance must be assessed by qualifying the quality of the product and defining the future use of the output product. If the output is a food or a food ingredient, the input must be a food or a food ingredient that complies with Article 2 of EC Regulation No, 178/2002 (European Commission, 2002). This regulation lays down the general principles and requirements of food law, establishes the European Food Safety Authority, and lay down procedures in matter of food safety. It states, among others, that food shall not be residues or contaminants; hence, the interest of by-product designation instead of waste.

The European Union (EU) regulations take into consideration food processed by new technologies. The Novel Food Regulation (EU) 2015/2283 (European Commission, 2015b) defines novel food as a product obtained with a process that was not used for food production before 15 May 1997 and that significantly affects food composition, structure, nutritional value, metabolism, or level of undesirable substances. Thus, not all food processed by a new production process is necessarily "novel." In this sense, "novel food" can be manufactured by production processes, such as all green extraction processes that have been very newly developed. The requirement for market uptake of novel food is that it should be safe for consumers, properly labeled, and not be nutritionally disadvantageous in the case it replaces another food (European Commission, 2018). The regulatory status must be clarified by the party who wants to market the food. Furthermore, new processing technologies are regulatory encouraged by the EU because they bring benefits to the consumer, enhance food security, and minimize environmental impact of food production. In terms of food by-products and food waste management, the Regulation (EC) 2015/2283 on novel food and food ingredients draws the legal framework for their utilization and exploitation (Pojić et al., 2018). Then, the producer of the substance has to submit a request to the Member State where the product will be sold for the first time and wait for its authorization (Anal, 2017).

If the product is likely to have an effect on human health, a risk assessment will be carried out by the European Food Safety Authority (EFSA). Finally, after receiving a favorable vote from the Standing Committee and a publication by the Commission, the food product will be able to enter the EU market (European Commission, 2018).

In terms of labeling, the novel product must fulfil the legal requirement of the Regulation (EC) No 1169/2011. The name of the product and if necessary the condition of use must be specified in the label. Green extraction procedure can provide compounds for dietary supplements or can be used in other food. Then, the nutrition and health claims of these new products should be done in accordance with the Health and Nutrition Claims Regulation (EC) No 1924/2006 (European Commission, 2018).

Nevertheless, most of the efforts made during the last years by the EU were more about reduction and prevention of food waste and waste management, rather than food waste valorization. It is in accordance with the Waste Framework Directive (2008/98/EC) (European Commission, 2008), which describes the priority of the different method of managing waste. The matter with food waste valorization in Europe is that it is classified as nonhazardous waste and managed like any other type of nonhazardous waste, even though valuable components can be extracted. However, the EU waste directive target to totally avoid landfilling for biodegradable waste by 2025 (Anal, 2017).

6 Conclusion

According to the studies discussed in this chapter, the green extraction techniques present best environmental performances than conventional extraction methods. The good performances are linked with the high yields of extraction, the low energy consumption, and the use of no or green solvents. In particular, microwave extraction shows lower environmental impacts than ultrasound extraction, being solvent, energy, time, and cost saving. Electricity consumption is a key factor for green extraction processes, which corresponds to the major source of CO_2 emissions in SFE. In this sense, heating the solvent in ultrasound extraction influences significantly the global environmental impacts of the process. The use of ethanol, depending on the volume, appears to be more damaging to the environment than water. Nevertheless, extraction efficiency is significantly improved when ethanol is used instead of water. To conclude, electricity seems to be a hotspot in LCA of green extraction techniques. To improve the energy efficiency and minimize resources depletion, renewable energy sources, such as hydroelectric power, should be used.

At a regulatory level, it can be concluded that the further use of by-products for food ingredients must comply with the definition of the regulation that lays down the general principles and requirements of food law, establishes the EFSA, and lays down procedures in matters of food safety. In addition, the components extracted with green extraction techniques are defined as *Novel Food* by another regulation established in 2015. The requirements for the market uptake of *Novel Food* are that they must be safe for consumers, properly labeled, and not nutritiously disadvantageous when aiming at replacing other food. This regulation, also, more precisely defines the possibility of using by-products and waste for novel food production. On the other hand, for selling a substance, which is produced by green extraction technique, the producer must wait for the approbation of the Member State, where the product will be sold for the first time and eventually suffer a risk assessment from the EFSA. The final approbation is given by the Standing Committee. Regarding labeling procedures, they should be the same as for other dietary supplements or food components and should respect the Health and Nutrition Claims Regulation.

Finally, this chapter also shows how recent are environmental LCA on green extraction techniques. Even though, the studies performed so far show several benefits compared with conventional extraction techniques (e.g., they are faster and use less harmful solvent), not clear conclusions can be drawn. In addition, lots of studies can be found about new techniques and processes using microwave, compressed fluids, or ultrasound extraction on food waste presented as *green*; nonetheless, very few studies really verify this claim. In this sense, there is a real need in assessing these technologies to give them more credibility and to improve them in order to slowly replace the conventional extraction techniques.

References

Anal, A.K. Food Processing By-products and Their Utilization. John Wiley & Sons: (2017).

Banaszewska, A., Cruijssen, F.C.A.M., Claassen, G.D.H. and Van der Vorst, J.G.A.J. Effect and key factors of byproducts valorization: The case of dairy industry. *Journal of Dairy Science* 97(4): (2014) 1893–1908.

Chemat, F., Rombaut, N., Meullemiestre, A., Turk, M., Perino, S., Fabiano-Tixier, A.S. and Abert-Vian, M. Review of green food processing techniques. Preservation, transformation, and extraction. *Innovative Food Science & Emerging Technologies* 41: (2017) 357–377.

Cheng, M.H., Sekhon, J.J., Rosentrater, K.A., Wang, T., Jung, S. and Johnson, L.A. Environmental impact assessment of soybean oil production: Extruding-expelling process, hexane extraction and aqueous extraction. *Food and Bioproducts Processing* 108: (2018) 58–68.

De Marco, I., Riemma, S. and Iannone, R. Life cycle assessment of supercritical CO_2 extraction of caffeine from coffee beans. *The Journal of Supercritical Fluids* 133: (2018) 393–400.

Doona, C. J., ed. Case Studies in Novel Food Processing Technologies: Innovations in Processing, Packaging, and Predictive Modelling. Woodhead Publishing: (2010).

Dorta, E., Lobo, M.G. and González, M. Improving the efficiency of antioxidant extraction from mango peel by using microwave-assisted extraction. *Plant Foods for Human Nutrition* 68(2): (2013) 190–199.

European Commission. Regulation (EC) No 178/2002 of the European Parliament and of the Council of 28 January 2002 laying down the general principles and requirements of food law, establishing the European Food Safety Authority and laying down procedures in matters of food safety. *Official Journal of the European Communities* 31: (2002) 1–24.

European Commission. Directive 2008/98/EC of the European Parliament and of the Council of 19 November 2008 on waste. *Official Journal of the European Union* L 312(13): (2008) 22–11.

European Commission. Directorate-General for Research and Innovation. Innovating for Sustainable Growth: A Bioeconomy for Europe. Publications Office of the European Union: (2012).

European Commission. Closing the loop – An EU action plan for the circular economy, COM. (2015a) 0614.

European Commission. Regulation (EU) 2015/2283 of the European Parliament and of the Council of 25 November 2015 on novel foods. *Official Journal of the European* L 327: (2015b) 1–22.

European Commission. Novel food. Policies, information and services. (2018) [Online]. Available: https://ec.europa.eu/food/safety/novel_food_en. [Accessed: 26-Jul–2018].

FAO, Food and Agriculture Organization of the United Nations. Global food losses and food waste – Extent, causes and prevention. (2011) www.fao.org/docrep/014/mb060e/mb060e02.pdf.

Häckl, K. and Kunz, W. Some aspects of green solvents. *Comptes Rendus Chimie* 21(6): (2018) 572–580.

Herrero, M. and Ibañez, E. Green extraction processes, biorefineries and sustainability: Recovery of high added-value products from natural sources. *The Journal of Supercritical Fluids* 134: (2018) 252–259.

ISO, International Organization for Standardization. Environmental Management: Life Cycle Assessment; Principles and Framework. No. 2006. ISO: (2006a).

ISO, International Organization for Standardization. Environmental Management: Life Cycle Assessment: Requirements and Regulations. ISO: (2006b).

Jacotet-Navarro, M., Rombaut, N., Fabiano-Tixier, A.S., Danguien, M., Bily, A. and Chemat, F. Ultrasound versus microwave as green processes for extraction of rosmarinic, carnosic and ursolic acids from rosemary. *Ultrasonics Sonochemistry* 27: (2015) 102–109.

Kyriakopoulou, K., Papadaki, S. and Krokida, M. Life cycle analysis of β-carotene extraction techniques. *Journal of Food Engineering* 167: (2015) 51–58.

Lucchesi, M. E., F. Chemat, and J. Smadja. Solvent-free microwave extraction of essential oil from aromatic herbs: comparison with conventional hydro-distillation. *Journal of Chromatography A* 1043(2): (2004) 323–327.

Morone, P., F. Papendiek, and V. E. Tartiu. *Food waste reduction and valorisation*. Springer, 2017.

Oldfield, T. and Holden, N.M. An evaluation of upstream assumptions in food-waste life cycle assessments. *In Proceedings of the 9th International Conference on Life Cycle Assessment in the Agri-Food Sector*: (2014) 8–10.

Oldfield, T.L., White, E. and Holden, N.M. The implications of stakeholder perspective for LCA of wasted food and green waste. *Journal of Cleaner Production* 170: (2018) 1554–1564.

Olofsson, J., and Börjesson, P. Residual biomass as resource–Life-cycle environmental impact of wastes in circular resource systems. *Journal of Cleaner Production* 196: (2018) 997–1006.

Panja, P. Green extraction methods of food polyphenols from vegetable materials. *Current Opinion in Food Science* 23: (2017) 173–182.

Pfaltzgraff, L.A., Cooper, E.C., Budarin, V. and Clark, J.H. Food waste biomass: A resource for high-value chemicals. *Green Chemistry* 15(2): (2013) 307–314.

Pojić, M., Mišan, A. and Tiwari, B. Eco-innovative technologies for extraction of proteins for human consumption from renewable protein sources of plant origin. *Trends in Food Science & Technology* 75: (2018) 93–104.

Poltronieri, P. and D'Urso, O.F. Biotransformation of Agricultural Waste and By-products: The Food, Feed, Fibre. Fuel (4F) Economy. Elsevier: (2016) 1–357.

Razzaghi, S.E., Arabhosseini, A., Turk, M., Soubrat, T., Cendres, A., Kianmehr, M.H., Perino, S. and Chemat, F. Operational efficiencies of six microwave based extraction methods for orange peel oil. *Journal of Food Engineering* 241: (2019) 26–32.

Rocha, C.M., Genisheva, Z., Ferreira-Santos, P., Rodrigues, R., Vicente, A.A., Teixeira, J.A. and Pereira, R.N. Electric field-based technologies for valorization of bioresources. *Bioresource Technology* 254: (2018) 325–339.

Rodríguez-Meizoso, I., Castro-Puyana, M., Börjesson, P., Mendiola, J.A., Turner, C. and Ibáñez, E. Life cycle assessment of green pilot-scale extraction processes to obtain potent antioxidants from rosemary leaves. *The Journal of Supercritical Fluids* 72: (2012) 205–212.

Roy, P., Nei, D., Orikasa, T., Xu, Q., Okadome, H., Nakamura, N. and Shiina, T. A review of life cycle assessment (LCA) on some food products. *Journal of Food Engineering* 90(1): (2009) 1–10.

Satari, B. and Karimi, K. Citrus processing wastes: Environmental impacts, recent advances, and future perspectives in total valorization. *Resources, Conservation and Recycling* 129: (2018) 153–167.

UN, United Nations. Sustainable development goals. 17 goals to transform our world. (2015) [Accessed: 26-Jul–2018].

Vauchel, P., Colli, C., Pradal, D., Philippot, M., Decossin, S., Dhulster, P. and Dimitrov, K. Comparative LCA of ultrasound-assisted extraction of polyphenols from chicory grounds under different operational conditions. *Journal of Cleaner Production* 196: (2018) 1116–1123.

Index

I

J

K

L